Mikko Lipasti

Mikko Lipasti has been an assistant professor at the University of Wisconsin-Madison since 1999, where he is actively pursuing various research topics in the realms of processor, system, and memory architecture. He has advised a total of 17 graduate students, including two completed Ph.D. theses and numerous M.S. projects, and has published more than 30 papers in top computer architecture conferences and journals. He is most well known for his seminal Ph.D. work in value prediction. His research program has received in excess of $2 million in support through multiple grants from the National Science Foundation as well as financial support and equipment donations from IBM, Intel, AMD, and Sun Microsystems.

The Eta Kappa Nu Electrical Engineering Honor Society selected Mikko as the country's Outstanding Young Electrical Engineer for 2002. He is also a member of the IEEE and the Tau Beta Pi engineering honor society. He received his B.S. in computer engineering from Valparaiso University in 1991, and M.S. (1992) and Ph.D. (1997) degrees in electrical and computer engineering from Carnegie Mellon University. Prior to beginning his academic career, he worked for IBM Corporation in both software and future processor and system performance analysis and design guidance, as well as operating system kernel implementation. While at IBM he contributed to system and microarchitectural definition of future IBM server computer systems. He has served on numerous conference and workshop program committees and is co-organizer of the annual Workshop on Duplicating, Deconstructing, and Debunking (WDDD). He has filed seven patent applications, six of which are issued U.S. patents; won the Best Paper Award at MICRO-29; and has received IBM Invention Achievement, Patent Issuance, and Technical Recognition Awards.

Mikko has been happily married since 1991 and has a nine-year-old daughter and a six-year old son. In his spare time, he enjoys regular exercise, family bike rides, reading, and volunteering his time at his local church and on campus as an English-language discussion group leader at the International Friendship Center.

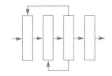

Modern Processor Design

Fundamentals of Superscalar Processors

John Paul Shen
Intel Corporation

Mikko H. Lipasti
University of Wisconsin

Boston   Burr Ridge, IL   Dubuque, IA   Madison, WI   New York   San Francisco   St. Louis
Bangkok   Bogotá   Caracas   Kuala Lumpur   Lisbon   London   Madrid   Mexico City
Milan   Montreal   New Delhi   Santiago   Seoul   Singapore   Sydney   Taipei   Toronto

Higher Education

MODERN PROCESSOR DESIGN: FUNDAMENTALS OF SUPERSCALAR PROCESSORS

Published by McGraw-Hill, a business unit of The McGraw-Hill Companies, Inc., 1221 Avenue of the Americas, New York, NY 10020. Copyright © 2005 by The McGraw-Hill Companies, Inc. All rights reserved. No part of this publication may be reproduced or distributed in any form or by any means, or stored in a database or retrieval system, without the prior written consent of The McGraw-Hill Companies, Inc., including, but not limited to, in any network or other electronic storage or transmission, or broadcast for distance learning.

Some ancillaries, including electronic and print components, may not be available to customers outside the United States.

This book is printed on acid-free paper.

1 2 3 4 5 6 7 8 9 0 DOC/DOC 0 9 8 7 6 5 4

ISBN 0–07–057064–7

Publisher: *Elizabeth A. Jones*
Senior Sponsoring Editor: *Carlise Paulson*
Developmental Editor: *Michelle L. Flomenhoft*
Marketing Manager: *Dawn R. Bercier*
Project Manager: *Jodi Rhomberg*
Senior Production Supervisor: *Laura Fuller*
Lead Media Project Manager: *Audrey A. Reiter*
Media Technology Producer: *Eric A. Weber*
Senior Coordinator of Freelance Design: *Michelle D. Whitaker*
Cover Designer: *Elise Lansdon*
Compositor: *Interactive Composition Corporation*
Typeface: *10.5/12 Times Roman*
Printer: *R. R. Donnelley Crawfordsville, IN*

Library of Congress Cataloging-in-Publication Data

Shen, John Paul.
   Modern processor design : fundamentals of superscalar processors / John Paul Shen, Mikko H. Lipasti.—1 st ed.
     p. cm.
   Includes index.
   ISBN 0–07–057064–7
   1. Microprocessors—Design and construction. I. Lipasti, Mikko H. II. Title.

TK7895.M5S52   2005
621.39'16—dc22

2004050406
CIP

www.mhhe.com

To

Our parents:
Paul and Sue Shen
Tarja and Simo Lipasti

Our spouses:
Amy C. Shen
Erica Ann Lipasti

Our children:
Priscilla S. Shen, Rachael S. Shen, and Valentia C. Shen
Emma Kristiina Lipasti and Elias Joel Lipasti

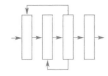

Table of Contents

	Table of Contents	iii
	Additional Resources	ix
	Preface	x

1 Processor Design — 1

- 1.1 The Evolution of Microprocessors — 2
- 1.2 Instruction Set Processor Design — 4
  - 1.2.1 Digital Systems Design — 4
  - 1.2.2 Architecture, Implementation, and Realization — 5
  - 1.2.3 Instruction Set Architecture — 6
  - 1.2.4 Dynamic-Static Interface — 8
- 1.3 Principles of Processor Performance — 10
  - 1.3.1 Processor Performance Equation — 10
  - 1.3.2 Processor Performance Optimizations — 11
  - 1.3.3 Performance Evaluation Method — 13
- 1.4 Instruction-Level Parallel Processing — 16
  - 1.4.1 From Scalar to Superscalar — 16
  - 1.4.2 Limits of Instruction-Level Parallelism — 24
  - 1.4.3 Machines for Instruction-Level Parallelism — 27
- 1.5 Summary — 32

2 Pipelined Processors — 39

- 2.1 Pipelining Fundamentals — 40
  - 2.1.1 Pipelined Design — 40
  - 2.1.2 Arithmetic Pipeline Example — 44
  - 2.1.3 Pipelining Idealism — 48
  - 2.1.4 Instruction Pipelining — 51
- 2.2 Pipelined Processor Design — 54
  - 2.2.1 Balancing Pipeline Stages — 55
  - 2.2.2 Unifying Instruction Types — 61
  - 2.2.3 Minimizing Pipeline Stalls — 71
  - 2.2.4 Commercial Pipelined Processors — 87
- 2.3 Deeply Pipelined Processors — 94
- 2.4 Summary — 97

3 Memory and I/O Systems — 105

- 3.1 Introduction — 105
- 3.2 Computer System Overview — 106
- 3.3 Key Concepts: Latency and Bandwidth — 107

	3.4	Memory Hierarchy	110
		3.4.1 Components of a Modern Memory Hierarchy	111
		3.4.2 Temporal and Spatial Locality	113
		3.4.3 Caching and Cache Memories	115
		3.4.4 Main Memory	127
	3.5	Virtual Memory Systems	136
		3.5.1 Demand Paging	138
		3.5.2 Memory Protection	141
		3.5.3 Page Table Architectures	142
	3.6	Memory Hierarchy Implementation	145
	3.7	Input/Output Systems	153
		3.7.1 Types of I/O Devices	154
		3.7.2 Computer System Busses	161
		3.7.3 Communication with I/O Devices	165
		3.7.4 Interaction of I/O Devices and Memory Hierarchy	168
	3.8	Summary	170
4	**Superscalar Organization**		**177**
	4.1	Limitations of Scalar Pipelines	178
		4.1.1 Upper Bound on Scalar Pipeline Throughput	178
		4.1.2 Inefficient Unification into a Single Pipeline	179
		4.1.3 Performance Lost Due to a Rigid Pipeline	179
	4.2	From Scalar to Superscalar Pipelines	181
		4.2.1 Parallel Pipelines	181
		4.2.2 Diversified Pipelines	184
		4.2.3 Dynamic Pipelines	186
	4.3	Superscalar Pipeline Overview	190
		4.3.1 Instruction Fetching	191
		4.3.2 Instruction Decoding	195
		4.3.3 Instruction Dispatching	199
		4.3.4 Instruction Execution	203
		4.3.5 Instruction Completion and Retiring	206
	4.4	Summary	209
5	**Superscalar Techniques**		**217**
	5.1	Instruction Flow Techniques	218
		5.1.1 Program Control Flow and Control Dependences	218
		5.1.2 Performance Degradation Due to Branches	219
		5.1.3 Branch Prediction Techniques	223
		5.1.4 Branch Misprediction Recovery	228
		5.1.5 Advanced Branch Prediction Techniques	231
		5.1.6 Other Instruction Flow Techniques	236
	5.2	Register Data Flow Techniques	237
		5.2.1 Register Reuse and False Data Dependences	237
		5.2.2 Register Renaming Techniques	239
		5.2.3 True Data Dependences and the Data Flow Limit	244

		5.2.4	The Classic Tomasulo Algorithm	246
		5.2.5	Dynamic Execution Core	254
		5.2.6	Reservation Stations and Reorder Buffer	256
		5.2.7	Dynamic Instruction Scheduler	260
		5.2.8	Other Register Data Flow Techniques	261
	5.3	Memory Data Flow Techniques		262
		5.3.1	Memory Accessing Instructions	263
		5.3.2	Ordering of Memory Accesses	266
		5.3.3	Load Bypassing and Load Forwarding	267
		5.3.4	Other Memory Data Flow Techniques	273
	5.4	Summary		279
6	**The PowerPC 620**			**301**
	6.1	Introduction		302
	6.2	Experimental Framework		305
	6.3	Instruction Fetching		307
		6.3.1	Branch Prediction	307
		6.3.2	Fetching and Speculation	309
	6.4	Instruction Dispatching		311
		6.4.1	Instruction Buffer	311
		6.4.2	Dispatch Stalls	311
		6.4.3	Dispatch Effectiveness	313
	6.5	Instruction Execution		316
		6.5.1	Issue Stalls	316
		6.5.2	Execution Parallelism	317
		6.5.3	Execution Latency	317
	6.6	Instruction Completion		318
		6.6.1	Completion Parallelism	318
		6.6.2	Cache Effects	318
	6.7	Conclusions and Observations		320
	6.8	Bridging to the IBM POWER3 and POWER4		322
	6.9	Summary		324
7	**Intel's P6 Microarchitecture**			**329**
	7.1	Introduction		330
		7.1.1	Basics of the P6 Microarchitecture	332
	7.2	Pipelining		334
		7.2.1	In-Order Front-End Pipeline	334
		7.2.2	Out-of-Order Core Pipeline	336
		7.2.3	Retirement Pipeline	337
	7.3	The In-Order Front End		338
		7.3.1	Instruction Cache and ITLB	338
		7.3.2	Branch Prediction	341
		7.3.3	Instruction Decoder	343
		7.3.4	Register Alias Table	346
		7.3.5	Allocator	353

7.4	The Out-of-Order Core		355
	7.4.1	Reservation Station	355
7.5	Retirement		357
	7.5.1	The Reorder Buffer	357
7.6	Memory Subsystem		361
	7.6.1	Memory Access Ordering	362
	7.6.2	Load Memory Operations	363
	7.6.3	Basic Store Memory Operations	363
	7.6.4	Deferring Memory Operations	363
	7.6.5	Page Faults	364
7.7	Summary		364
7.8	Acknowledgments		365

8 Survey of Superscalar Processors — 369

8.1	Development of Superscalar Processors		369
	8.1.1	Early Advances in Uniprocessor Parallelism: The IBM Stretch	369
	8.1.2	First Superscalar Design: The IBM Advanced Computer System	372
	8.1.3	Instruction-Level Parallelism Studies	377
	8.1.4	By-Products of DAE: The First Multiple-Decoding Implementations	378
	8.1.5	IBM Cheetah, Panther, and America	380
	8.1.6	Decoupled Microarchitectures	380
	8.1.7	Other Efforts in the 1980s	382
	8.1.8	Wide Acceptance of Superscalar	382
8.2	A Classification of Recent Designs		384
	8.2.1	RISC and CISC Retrofits	384
	8.2.2	Speed Demons: Emphasis on Clock Cycle Time	386
	8.2.3	Brainiacs: Emphasis on IPC	386
8.3	Processor Descriptions		387
	8.3.1	Compaq / DEC Alpha	387
	8.3.2	Hewlett-Packard PA-RISC Version 1.0	392
	8.3.3	Hewlett-Packard PA-RISC Version 2.0	395
	8.3.4	IBM POWER	397
	8.3.5	Intel i960	402
	8.3.6	Intel IA32—Native Approaches	405
	8.3.7	Intel IA32—Decoupled Approaches	409
	8.3.8	x86-64	417
	8.3.9	MIPS	417
	8.3.10	Motorola	422
	8.3.11	PowerPC—32-bit Architecture	424
	8.3.12	PowerPC—64-bit Architecture	429
	8.3.13	PowerPC-AS	431
	8.3.14	SPARC Version 8	432
	8.3.15	SPARC Version 9	435

	8.4	Verification of Superscalar Processors	439
	8.5	Acknowledgments	440

9 Advanced Instruction Flow Techniques — 453

	9.1	Introduction	453
	9.2	Static Branch Prediction Techniques	454
		9.2.1 Single-Direction Prediction	455
		9.2.2 Backwards Taken/Forwards Not-Taken	456
		9.2.3 Ball/Larus Heuristics	456
		9.2.4 Profiling	457
	9.3	Dynamic Branch Prediction Techniques	458
		9.3.1 Basic Algorithms	459
		9.3.2 Interference-Reducing Predictors	472
		9.3.3 Predicting with Alternative Contexts	482
	9.4	Hybrid Branch Predictors	491
		9.4.1 The Tournament Predictor	491
		9.4.2 Static Predictor Selection	493
		9.4.3 Branch Classification	494
		9.4.4 The Multihybrid Predictor	495
		9.4.5 Prediction Fusion	496
	9.5	Other Instruction Flow Issues and Techniques	497
		9.5.1 Target Prediction	497
		9.5.2 Branch Confidence Prediction	501
		9.5.3 High-Bandwidth Fetch Mechanisms	504
		9.5.4 High-Frequency Fetch Mechanisms	509
	9.6	Summary	512

10 Advanced Register Data Flow Techniques — 519

	10.1	Introduction	519
	10.2	Value Locality and Redundant Execution	523
		10.2.1 Causes of Value Locality	523
		10.2.2 Quantifying Value Locality	525
	10.3	Exploiting Value Locality without Speculation	527
		10.3.1 Memoization	527
		10.3.2 Instruction Reuse	529
		10.3.3 Basic Block and Trace Reuse	533
		10.3.4 Data Flow Region Reuse	534
		10.3.5 Concluding Remarks	535
	10.4	Exploiting Value Locality with Speculation	535
		10.4.1 The Weak Dependence Model	535
		10.4.2 Value Prediction	536
		10.4.3 The Value Prediction Unit	537
		10.4.4 Speculative Execution Using Predicted Values	542
		10.4.5 Performance of Value Prediction	551
		10.4.6 Concluding Remarks	553
	10.5	Summary	554

11 Executing Multiple Threads — 559

- 11.1 Introduction — 559
- 11.2 Synchronizing Shared-Memory Threads — 562
- 11.3 Introduction to Multiprocessor Systems — 565
    - 11.3.1 Fully Shared Memory, Unit Latency, and Lack of Contention — 566
    - 11.3.2 Instantaneous Propagation of Writes — 567
    - 11.3.3 Coherent Shared Memory — 567
    - 11.3.4 Implementing Cache Coherence — 571
    - 11.3.5 Multilevel Caches, Inclusion, and Virtual Memory — 574
    - 11.3.6 Memory Consistency — 576
    - 11.3.7 The Coherent Memory Interface — 581
    - 11.3.8 Concluding Remarks — 583
- 11.4 Explicitly Multithreaded Processors — 584
    - 11.4.1 Chip Multiprocessors — 584
    - 11.4.2 Fine-Grained Multithreading — 588
    - 11.4.3 Coarse-Grained Multithreading — 589
    - 11.4.4 Simultaneous Multithreading — 592
- 11.5 Implicitly Multithreaded Processors — 600
    - 11.5.1 Resolving Control Dependences — 601
    - 11.5.2 Resolving Register Data Dependences — 605
    - 11.5.3 Resolving Memory Data Dependences — 607
    - 11.5.4 Concluding Remarks — 610
- 11.6 Executing the Same Thread — 610
    - 11.6.1 Fault Detection — 611
    - 11.6.2 Prefetching — 613
    - 11.6.3 Branch Resolution — 614
    - 11.6.4 Concluding Remarks — 615
- 11.7 Summary — 616

Index — 623

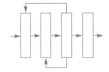

Additional Resources

In addition to the comprehensive coverage within the book, a number of additional resources are available with Shen/Lipasti's MODERN PROCESSOR DESIGN through the book's website at **www.mhhe.com/shen**.

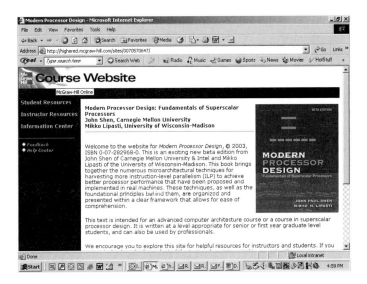

Instructor Resources

- **Solutions Manual**—A complete set of solutions for the chapter-ending homework problems are provided.
- **PowerPoint Slides**—Two sets of MS PowerPoint slides, from Carnegie Mellon University and the University of Wisconsin-Madison, can be downloaded to supplement your lecture presentations.
- **Figures**—A complete set of figures from the book are available in eps format. These figures can be used to create your own presentations.
- **Sample Homework Files**—A set of homework assignments with answers from Carnegie Mellon University are provided to supplement your own assignments.
- **Sample Exams**—A set of exams with answers from Carnegie Mellon University are also provided to supplement your own exams.
- **Links to www.simplescalar.com**—We provide several links to the SimpleScalar tool set, which are available free for non-commercial academic use.

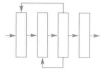

Preface

This book emerged from the course *Superscalar Processor Design,* which has been taught at Carnegie Mellon University since 1995. Superscalar Processor Design is a mezzanine course targeting seniors and first-year graduate students. Quite a few of the more aggressive juniors have taken the course in the spring semester of their junior year. The prerequisite to this course is the Introduction to Computer Architecture course. The objectives for the Superscalar Processor Design course include: (1) to teach modern processor design skills at the microarchitecture level of abstraction; (2) to cover current microarchitecture techniques for achieving high performance via the exploitation of instruction-level parallelism (ILP); and (3) to impart insights and hands-on experience for the effective design of contemporary high-performance microprocessors for mobile, desktop, and server markets. In addition to covering the contents of this book, the course contains a project component that involves the microarchitectural design of a future-generation superscalar microprocessor.

During the decade of the 1990s many microarchitectural techniques for increasing clock frequency and harvesting more ILP to achieve better processor performance have been proposed and implemented in real machines. This book is an attempt to codify this large body of knowledge in a systematic way. These techniques include deep pipelining, aggressive branch prediction, dynamic register renaming, multiple instruction dispatching and issuing, out-of-order execution, and speculative load/store processing. Hundreds of research papers have been published since the early 1990s, and many of the research ideas have become reality in commercial superscalar microprocessors. In this book, the numerous techniques are organized and presented within a clear framework that facilitates ease of comprehension. The foundational principles that underlie the plethora of techniques are highlighted.

While the contents of this book would generally be viewed as graduate-level material, the book is intentionally written in a way that would be very accessible to undergraduate students. Significant effort has been spent in making seemingly complex techniques to appear as quite straightforward through appropriate abstraction and hiding of details. The priority is to convey clearly the key concepts and fundamental principles, giving just enough details to ensure understanding of implementation issues without massive dumping of information and quantitative data. The hope is that this body of knowledge can become widely possessed by not just microarchitects and processor designers but by most B.S. and M.S. students with interests in computer systems and microprocessor design.

Here is a brief summary of the chapters.

Chapter 1: Processor Design

This chapter introduces the art of processor design, the instruction set architecture (ISA) as the specification of the processor, and the microarchitecture as the implementation of the processor. The dynamic/static interface that separates compile-time software and run-time hardware is defined and discussed. The goal of this chapter is not to revisit in depth the traditional issues regarding ISA design, but to erect the proper framework for understanding modern processor design.

Chapter 2: Pipelined Processors

This chapter focuses on the concept of pipelining, discusses instruction pipeline design, and presents the performance benefits of pipelining. Pipelining is usually introduced in the first computer architecture course. Pipelining provides the foundation for modern superscalar techniques and is presented in this chapter in a fresh and unique way. We intentionally avoid the massive dumping of bar charts and graphs; instead, we focus on distilling the foundational principles of instruction pipelining.

Chapter 3: Memory and I/O Systems

This chapter provides a larger context for the remainder of the book by including a thorough grounding in the principles and mechanisms of modern memory and I/O systems. Topics covered include memory hierarchies, caching, main memory design, virtual memory architecture, common input/output devices, processor-I/O interaction, and bus design and organization.

Chapter 4: Superscalar Organization

This chapter introduces the main concepts and the overall organization of superscalar processors. It provides a "big picture" view for the reader that leads smoothly into the detailed discussions in the next chapters on specific superscalar techniques for achieving performance. This chapter highlights only the key features of superscalar processor organizations. Chapter 7 provides a detailed survey of features found in real machines.

Chapter 5: Superscalar Techniques

This chapter is the heart of this book and presents all the major microarchitecture techniques for designing contemporary superscalar processors for achieving high performance. It classifies and presents specific techniques for enhancing instruction flow, register data flow, and memory data flow. This chapter attempts to organize a plethora of techniques into a systematic framework that facilitates ease of comprehension.

Chapter 6: The PowerPC 620

This chapter presents a detailed analysis of the PowerPC 620 microarchitecture and uses it as a case study to examine many of the issues and design tradeoffs introduced in the previous chapters. This chapter contains extensive performance data of an aggressive out-of-order design.

Chapter 7: Intel's P6 Microarchitecture

This is a case study chapter on probably the most commercially successful contemporary superscalar microarchitecture. It is written by the Intel P6 design team led by Bob Colwell and presents in depth the P6 microarchitecture that facilitated the implementation of the Pentium Pro, Pentium II, and Pentium III microprocessors. This chapter offers the readers an opportunity to peek into the mindset of a top-notch design team.

Chapter 8: Survey of Superscalar Processors

This chapter, compiled by Prof. Mark Smotherman of Clemson University, provides a historical chronicle on the development of superscalar machines and a survey of existing superscalar microprocessors. The chapter was first completed in 1998 and has been continuously revised and updated since then. It contains fascinating information that can't be found elsewhere.

Chapter 9: Advanced Instruction Flow Techniques

This chapter provides a thorough overview of issues related to high-performance instruction fetching. The topics covered include historical, currently used, and proposed advanced future techniques for branch prediction, as well as high-bandwidth and high-frequency fetch architectures like trace caches. Though not all such techniques have yet been adopted in real machines, future designs are likely to incorporate at least some form of them.

Chapter 10: Advanced Register Data Flow Techniques

This chapter highlights emerging microarchitectural techniques for increasing performance by exploiting the program characteristic of *value locality*. This program characteristic was discovered recently, and techniques ranging from software memoization, instruction reuse, and various forms of value prediction are described in this chapter. Though such techniques have not yet been adopted in real machines, future designs are likely to incorporate at least some form of them.

Chapter 11: Executing Multiple Threads

This chapter provides an introduction to thread-level parallelism (TLP), and provides a basic introduction to multiprocessing, cache coherence, and high-performance implementations that guarantee either sequential or relaxed memory ordering across multiple processors. It discusses single-chip techniques like multithreading and on-chip multiprocessing that also exploit thread-level parallelism. Finally, it visits two emerging technologies—implicit multithreading and preexecution—that attempt to extract thread-level parallelism automatically from single-threaded programs.

In summary, Chapters 1 through 5 cover fundamental concepts and foundational techniques. Chapters 6 through 8 present case studies and an extensive survey of actual commercial superscalar processors. Chapter 9 provides a thorough overview of advanced instruction flow techniques, including recent developments in advanced branch predictors. Chapters 10 and 11 should be viewed as advanced topics chapters that highlight some emerging techniques and provide an introduction to multiprocessor systems.

This is the first edition of the book. An earlier beta edition was published in 2002 with the intent of collecting feedback to help shape and hone the contents and presentation of this first edition. Through the course of the development of the book, a large set of homework and exam problems have been created. A subset of these problems are included at the end of each chapter. Several problems suggest the use of the

Simplescalar simulation suite available from the Simplescalar website at **http://www.simplescalar.com**. A companion website for the book contains additional support material for the instructor, including a complete set of lecture slides (**www.mhhe.com/shen**).

Acknowledgments

Many people have generously contributed their time, energy, and support toward the completion of this book. In particular, we are grateful to Bob Colwell, who is the lead author of Chapter 7, Intel's P6 Microarchitecture. We also acknowledge his coauthors, Dave Papworth, Glenn Hinton, Mike Fetterman, and Andy Glew, who were all key members of the historic P6 team. This chapter helps ground this textbook in practical, real-world considerations. We are also grateful to Professor Mark Smotherman of Clemson University, who meticulously compiled and authored Chapter 8, Survey of Superscalar Processors. This chapter documents the rich and varied history of superscalar processor design over the last 40 years. The guest authors of these two chapters added a certain radiance to this textbook that we could not possibly have produced on our own. The PowerPC 620 case study in Chapter 6 is based on Trung Diep's Ph.D. thesis at Carnegie Mellon University. Finally, the thorough survey of advanced instruction flow techniques in Chapter 9 was authored by Gabriel Loh, largely based on his Ph.D. thesis at Yale University.

In addition, we want to thank the following professors for their detailed, insightful, and thorough review of the original manuscript. The inputs from these reviews have significantly improved the first edition of this book.

- David Andrews, *University of Arkansas*
- Angelos Bilas, *University of Toronto*
- Fred H. Carlin, *University of California at Santa Barbara*
- Yinong Chen, *Arizona State University*
- Lynn Choi, *University of California at Irvine*
- Dan Connors, *University of Colorado*
- Karel Driesen, *McGill University*
- Alan D. George, *University of Florida*
- Arthur Glaser, *New Jersey Institute of Technology*
- Rajiv Gupta, *University of Arizona*
- Vincent Hayward, *McGill University*
- James Hoe, *Carnegie Mellon University*
- Lizy Kurian John, *University of Texas at Austin*
- Peter M. Kogge, *University of Notre Dame*
- Angkul Kongmunvattana, *University of Nevada at Reno*
- Israel Koren, *University of Massachusetts at Amherst*
- Ben Lee, *Oregon State University*
- Francis Leung, *Illinois Institute of Technology*
- Walid Najjar, *University of California Riverside*
- Vojin G. Oklabdzija, *University of California at Davis*
- Soner Onder, *Michigan Technological University*
- Parimal Patel, *University of Texas at San Antonio*
- Jih-Kwon Peir, *University of Florida*
- Gregory D. Peterson, *University of Tennessee*
- Amir Roth, *University of Pennsylvania*
- Kevin Skadron, *University of Virginia*
- Mark Smotherman, *Clemson University*
- Miroslav N. Velev, *Georgia Institute of Technology*
- Bin Wei, *Rutgers University*
- Anthony S. Wojcik, *Michigan State University*
- Ali Zaringhalam, *Stevens Institute of Technology*
- Xiaobo Zhou, *University of Colorado at Colorado Springs*

This book grew out of the course *Superscalar Processor Design* at Carnegie Mellon University. This course has been taught at CMU since 1995. Many teaching assistants of this course have left their indelible touch in the contents of this book. They include Bryan Black, Scott Cape, Yuan Chou, Alex Dean, Trung Diep, John Faistl, Andrew Huang, Deepak Limaye, Chris Nelson, Chris Newburn, Derek Noonburg, Kyle Oppenheim, Ryan Rakvic, and Bob Rychlik. Hundreds of students have taken this course at CMU; many of them provided inputs that also helped shape this book. Since 2000, Professor James Hoe at CMU has taken this course even further. We both are indebted to the nurturing we experienced while at CMU, and we hope that this book will help perpetuate CMU's historical reputation of producing some of the best computer architects and processor designers.

A draft version of this textbook has also been used at the University of Wisconsin since 2000. Some of the problems at the end of each chapter were actually contributed by students at the University of Wisconsin. We appreciate their test driving of this book.

John Paul Shen, *Director,*
Microarchitecture Research, Intel Labs, Adjunct Professor,
ECE Department, Carnegie Mellon University

Mikko H. Lipasti, *Assistant Professor,*
ECE Department, University of Wisconsin

June 2004
Soli Deo Gloria

CHAPTER 1

Processor Design

CHAPTER OUTLINE

1.1 The Evolution of Microprocessors
1.2 Instruction Set Processor Design
1.3 Principles of Processor Performance
1.4 Instruction-Level Parallel Processing
1.5 Summary

References
Homework Problems

Welcome to contemporary microprocessor design. In its relatively brief lifetime of 30+ years, the microprocessor has undergone phenomenal advances. Its performance has improved at the astounding rate of doubling every 18 months. In the past three decades, microprocessors have been responsible for inspiring and facilitating some of the major innovations in computer systems. These innovations include embedded microcontrollers, personal computers, advanced workstations, handheld and mobile devices, application and file servers, web servers for the Internet, low-cost supercomputers, and large-scale computing clusters. Currently more than 100 million microprocessors are sold each year for the mobile, desktop, and server markets. Including embedded microprocessors and microcontrollers, the total number of microprocessors shipped each year is well over one billion units.

Microprocessors are *instruction set processors* (ISPs). An ISP executes instructions from a predefined instruction set. A microprocessor's functionality is fully characterized by the instruction set that it is capable of executing. All the programs that run on a microprocessor are encoded in that instruction set. This predefined instruction set is also called the *instruction set architecture* (ISA). An ISA serves as an interface between software and hardware, or between programs and processors. In terms of processor design methodology, an ISA is the *specification* of a design while a microprocessor or ISP is the *implementation* of a design. As with all forms of engineering design, microprocessor design is inherently a creative process that involves subtle tradeoffs and requires good intuition and clever insights.

This book focuses on contemporary superscalar microprocessor design at the microarchitecture level. It presents existing and proposed microarchitecture techniques in a systematic way and imparts foundational principles and insights, with the hope of training new microarchitects who can contribute to the effective design of future-generation microprocessors.

1.1 The Evolution of Microprocessors

The first microprocessor, the Intel 4004, was introduced in 1971. The 4004 was a 4-bit processor consisting of approximately 2300 transistors with a clock frequency of just over 100 kilohertz (kHz). Its primary application was for building calculators. The year 2001 marks the thirtieth anniversary of the birth of microprocessors. High-end microprocessors, containing up to 100 million transistors with a clock frequency reaching 2 gigahertz (GHz), are now the building blocks for supercomputer systems and powerful client and server systems that populate the Internet. Within a few years microprocessors will be clocked at close to 10 GHz and each will contain several hundred million transistors.

The three decades of the history of microprocessors tell a truly remarkable story of technological advances in the computer industry; see Table 1.1. The evolution of the microprocessor has pretty much followed the famed Moore's law, observed by Gordon Moore in 1965, that the number of devices that can be integrated on a single piece of silicon will double roughly every 18 to 24 months. In a little more than 30 years, the number of transistors in a microprocessor chip has increased by more than four orders of magnitude. In that same period, microprocessor performance has increased by more than five orders of magnitude. In the past two decades, microprocessor performance has been doubling every 18 months, or an increase by a factor of 100 in each decade. Such phenomenal performance improvement is unmatched by that in any other industry.

In each of the three decades of its existence, the microprocessor has played major roles in the most critical advances in the computer industry. During the first decade, the advent of the 4-bit microprocessor quickly led to the introduction of the

Table 1.1

The amazing decades of the evolution of microprocessors

	1970–1980	1980–1990	1990–2000	2000–2010
Transistor count	2K–100K	100K–1M	1M–100M	100M–2B
Clock frequency	0.1–3 MHz	3–30 MHz	30 MHz–1 GHz	1–15 GHz
Instructions/cycle	0.1	0.1–0.9	0.9–1.9	1.9–2.9

8-bit microprocessor. These narrow bit-width microprocessors evolved into self-contained microcontrollers that were produced in huge volumes and deployed in numerous embedded applications ranging from washing machines, to elevators, to jet engines. The 8-bit microprocessor also became the heart of a new popular computing platform called the *personal computer* (PC) and ushered in the PC era of computing.

The decade of the 1980s witnessed major advances in the architecture and microarchitecture of 32-bit microprocessors. Instruction set design issues became the focus of both academic and industrial researchers. The importance of having an instruction set architecture that facilitates efficient hardware implementation and that can leverage compiler optimizations was recognized. Instruction pipelining and fast cache memories became standard microarchitecture techniques. Powerful scientific and engineering workstations based on 32-bit microprocessors were introduced. These workstations in turn became the workhorses for the design of subsequent generations of even more powerful microprocessors.

During the decade of the 1990s, microprocessors became the most powerful and most popular form of computers. The clock frequency of the fastest microprocessors exceeded that of the fastest supercomputers. Personal computers and workstations became ubiquitous and essential tools for productivity and communication. Extremely aggressive microarchitecture techniques were devised to achieve unprecedented levels of microprocessor performance. Deeply pipelined machines capable of achieving extremely high clock frequencies and sustaining multiple instructions executed per cycle became popular. Out-of-order execution of instructions and aggressive branch prediction techniques were introduced to avoid or reduce the number of pipeline stalls. By the end of the third decade of microprocessors, almost all forms of computing platforms ranging from personal handheld devices to mainstream desktop and server computers to the most powerful parallel and clustered computers are based on the building blocks of microprocessors.

We are now heading into the fourth decade of microprocessors, and the momentum shows no sign of abating. Most technologists agree that Moore's law will continue to rule for at least 10 to 15 years more. By 2010, we can expect microprocessors to contain more than 1 billion transistors with clocking frequencies greater than 10 GHz. We can also expect new innovations in a number of areas. The current focus on *instruction-level parallelism* (ILP) will expand to include *thread-level parallelism* (TLP) as well as *memory-level parallelism* (MLP). Architectural features that historically belong to large systems, for example, multiprocessors and memory hierarchies, will be implemented on a single chip. Many traditional "macroarchitecture" issues will now become microarchitecture issues. Power consumption will become a dominant performance impediment and will require new solutions at all levels of the design hierarchy, including fabrication process, circuit design, logic design, microarchitecture design, and software run-time environment, in order to sustain the same rate of performance improvements that we have witnessed in the past three decades.

The objective of this book is to introduce the fundamental principles of microprocessor design at the microarchitecture level. Major techniques that have been developed and deployed in the past three decades are presented in a comprehensive way. This book attempts to codify a large body of knowledge into a systematic framework. Concepts and techniques that may appear quite complex and difficult to decipher are distilled into a format that is intuitive and insightful. A number of innovative techniques recently proposed by researchers are also highlighted. We hope this book will play a role in producing a new generation of microprocessor designers who will help write the history for the fourth decade of microprocessors.

1.2 Instruction Set Processor Design

The focus of this book is on designing instruction set processors. Critical to an instruction set processor is the instruction set architecture, which specifies the functionality that must be implemented by the instruction set processor. The ISA plays several crucial roles in instruction set processor design.

1.2.1 Digital Systems Design

Any engineering design starts with a specification with the objective of obtaining a good design or an implementation. *Specification* is a behavioral description of what is desired and answers the question "What does it do?" while *implementation* is a structural description of the resultant design and answers the question "How is it constructed?" Typically the design process involves two fundamental tasks: synthesis and analysis. *Synthesis* attempts to find an implementation based on the specification. *Analysis* examines an implementation to determine whether and how well it meets the specification. Synthesis is the more creative task that searches for possible solutions and performs various tradeoffs and design optimizations to arrive at the best solution. The critical task of analysis is essential in determining the correctness and effectiveness of a design; it frequently employs simulation tools to perform design validation and performance evaluation. A typical design process can require the traversing of the analysis-synthesis cycle numerous times in order to arrive at the final best design; see Figure 1.1.

In digital systems design, specifications are quite rigorous and design optimizations rely on the use of powerful software tools. Specification for a combinational logic circuit takes the form of boolean functions that specify the relationship

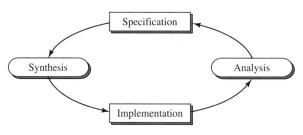

Figure 1.1
Engineering Design.

between input and output variables. The implementation is typically an optimized two-level AND-OR design or a multilevel network of logic gates. The optimization attempts to reduce the number of logic gates and the number of levels of logic used in the design. For sequential circuit design, the specification is in the form of state machine descriptions that include the specification of the state variables as well as the output and next state functions. Optimization objectives include the reduction of the number of states and the complexity of the associated combinational logic circuits. Logic minimization and state minimization software tools are essential. Logic and state machine simulation tools are used to assist the analysis task. These tools can verify the logic correctness of a design and determine the critical delay path and hence the maximum clocking rate of the state machine.

The design process for a microprocessor is more complex and less straightforward. The specification of a microprocessor design is the instruction set architecture, which specifies a set of instructions that the microprocessor must be able to execute. The implementation is the actual hardware design described using a *hardware description language* (HDL). The primitives of an HDL can range from logic gates and flip-flops, to more complex modules, such as decoders and multiplexers, to entire functional modules, such as adders and multipliers. A design is described as a schematic, or interconnected organization, of these primitives.

The process of designing a modern high-end microprocessor typically involves two major steps: microarchitecture design and logic design. Microarchitecture design involves developing and defining the key techniques for achieving the targeted performance. Usually a performance model is used as an analysis tool to assess the effectiveness of these techniques. The performance model accurately models the behavior of the machine at the clock cycle granularity and is able to quantify the number of machine cycles required to execute a benchmark program. The end result of microarchitecture design is a high-level description of the organization of the microprocessor. This description typically uses a *register transfer language* (RTL) to specify all the major modules in the machine organization and the interactions between these modules. During the logic design step, the RTL description is successively refined by the incorporation of implementation details to eventually yield the HDL description of the actual hardware design. Both the RTL and the HDL descriptions can potentially use the same description language. For example, Verilog is one such language. The primary focus of this book is on microarchitecture design.

1.2.2 Architecture, Implementation, and Realization

In a classic textbook on computer architecture by Blaauw and Brooks [1997] the authors defined three fundamental and distinct levels of abstraction: architecture, implementation, and realization. *Architecture* specifies the functional behavior of a processor. *Implementation* is the logical structure or organization that performs the architecture. *Realization* is the physical structure that embodies the implementation.

Architecture is also referred to as the instruction set architecture. It specifies an instruction set that characterizes the functional behavior of an instruction set processor. All software must be mapped to or encoded in this instruction set in order to be executed by the processor. Every program is compiled into a sequence of instructions in this instruction set. Examples of some well-known architectures are IBM 360, DEC VAX, Motorola 68K, PowerPC, and Intel IA32. Attributes associated with an architecture include the assembly language, instruction format, addressing modes, and programming model. These attributes are all part of the ISA and exposed to the software as perceived by the compiler or the programmer.

An implementation is a specific design of an architecture, and it is also referred to as the *microarchitecture*. An architecture can have many implementations in the lifetime of that ISA. All implementations of an architecture can execute any program encoded in that ISA. Examples of some well-known implementations of the above-listed architecture are IBM 360/91, VAX 11/780, Motorola 68040, PowerPC 604, and Intel P6. Attributes associated with an implementation include pipeline design, cache memories, and branch predictors. Implementation or microarchitecture features are generally implemented in hardware and hidden from the software. To develop these features is the job of the microprocessor designer or the microarchitect.

A realization of an implementation is a specific physical embodiment of a design. For a microprocessor, this physical embodiment is usually a chip or a multi-chip package. For a given implementation, there can be various realizations of that implementation. These realizations can vary and differ in terms of the clock frequency, cache memory capacity, bus interface, fabrication technology, packaging, etc. Attributes associated with a realization include die size, physical form factor, power, cooling, and reliability. These attributes are the concerns of the chip designer and the system designer who uses the chip.

The primary focus of this book is on the implementation of modern microprocessors. Issues related to architecture and realization are also important. Architecture serves as the specification for the implementation. Attributes of an architecture can significantly impact the design complexity and the design effort of an implementation. Attributes of a realization, such as die size and power, must be considered in the design process and used as part of the design objectives.

1.2.3 Instruction Set Architecture

Instruction set architecture plays a very crucial role and has been defined as a contract between the software and the hardware, or between the program and the machine. By having the ISA as a contract, programs and machines can be developed independently. Programs can be developed that target the ISA without requiring knowledge of the actual machine implementation. Similarly, machines can be designed that implement the ISA without concern for what programs will run on them. Any program written for a particular ISA should be able to run on any machine implementing that same ISA. The notion of maintaining the same ISA across multiple implementations of that ISA was first introduced with the IBM S/360 line of computers [Amdahl et al., 1964].

Having the ISA also ensures software portability. A program written for a particular ISA can run on all the implementations of that same ISA. Typically given an ISA, many implementations will be developed over the lifetime of that ISA, or multiple implementations that provide different levels of cost and performance can be simultaneously developed. A program only needs to be developed once for that ISA, and then it can run on all these implementations. Such program portability significantly reduces the cost of software development and increases the longevity of software. Unfortunately this same benefit also makes migration to a new ISA very difficult. Successful ISAs, or more specifically ISAs with a large software installed base, tend to stay around for quite a while. Two examples are the IBM 360/370 and the Intel IA32.

Besides serving as a reference targeted by software developers or compilers, ISA serves as the specification for processor designers. Microprocessor design starts with the ISA and produces a microarchitecture that meets this specification. Every new microarchitecture must be validated against the ISA to ensure that it performs the functional requirements specified by the ISA. This is extremely important to ensure that existing software can run correctly on the new microarchitecture.

Since the advent of computers, a wide variety of ISAs have been developed and used. They differ in how operations and operands are specified. Typically an ISA defines a set of instructions called *assembly instructions*. Each instruction specifies an operation and one or more operands. Each ISA uniquely defines an *assembly language*. An assembly language program constitutes a sequence of assembly instructions. ISAs have been differentiated according to the number of operands that can be explicitly specified in each instruction, for example two-address or three-address architectures. Some early ISAs use an *accumulator* as an implicit operand. In an accumulator-based architecture, the accumulator is used as an implicit source operand and the destination. Other early ISAs assume that operands are stored in a stack [last in, first out (LIFO)] structure and operations are performed on the top one or two entries of the stack. Most modern ISAs assume that operands are stored in a multientry register file, and that all arithmetic and logical operations are performed on operands stored in the registers. Special instructions, such as load and store instructions, are devised to move operands between the register file and the main memory. Some traditional ISAs allow operands to come directly from both the register file and the main memory.

ISAs tend to evolve very slowly due to the inertia against recompiling or redeveloping software. Typically a twofold performance increase is needed before software developers will be willing to pay the overhead to recompile their existing applications. While new extensions to an existing ISA can occur from time to time to accommodate new emerging applications, the introduction of a brand new ISA is a tall order. The development of effective compilers and operating systems for a new ISA can take on the order of 10+ years. The longer an ISA has been in existence and the larger the installed base of software based on that ISA, the more difficult it is to replace that ISA. One possible exception might be in certain special application domains where a specialized new ISA might be able to provide significant performance boost, such as on the order of 10-fold.

Unlike the glacial creep of ISA innovations, significantly new microarchitectures can be and have been developed every 3 to 5 years. During the 1980s, there were widespread interests in ISA design and passionate debates about what constituted the best ISA features. However, since the 1990s the focus has shifted to the implementation and to innovative microarchitecture techniques that are applicable to most, if not all, ISAs. It is quite likely that the few ISAs that have dominated the microprocessor landscape in the past decades will continue to do so for the coming decade. On the other hand, we can expect to see radically different and innovative microarchitectures for these ISAs in the coming decade.

1.2.4 Dynamic-Static Interface

So far we have discussed two critical roles played by the ISA. First, it provides a contract between the software and the hardware, which facilitates the independent development of programs and machines. Second, an ISA serves as the specification for microprocessor design. All implementations must meet the requirements and support the functionality specified in the ISA. In addition to these two critical roles, each ISA has a third role. Inherent in the definition of every ISA is an associated definition of an interface that separates what is done *statically* at compile time versus what is done *dynamically* at run time. This interface has been called the *dynamic-static interface* (DSI) by Yale Patt and is illustrated in Figure 1.2 [Melvin and Patt, 1987].

The DSI is a direct consequence of having the ISA serve as a contract between the software and the hardware. Traditionally, all the tasks and optimizations done in the static domain at compile time involve the software and the compiler, and are considered *above* the DSI. Conversely, all the tasks and optimizations done in the dynamic domain at run time involve the hardware and are considered *below* the DSI. All the architecture features are specified in the ISA and are therefore exposed to the software above the DSI in the static domain. On the other hand, all the implementation features of the microarchitecture are below the DSI and operate in the dynamic domain at run time; usually these are completely hidden from the software and the compiler in the static domain. As stated earlier, software development can take place above the DSI independent of the development of the microarchitecture features below the DSI.

A key issue in the design of an ISA is the placement of the DSI. In between the application program written in a high-level language at the top and the actual hardware of the machine at the bottom, there can be different levels of abstractions where the DSI can potentially be placed. The placement of the DSI is correlated

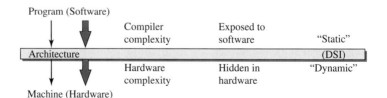

Figure 1.2
The Dynamic-Static Interface.

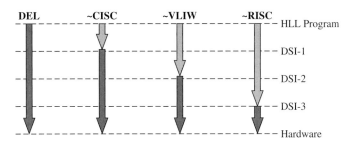

Figure 1.3
Conceptual Illustration of Possible Placements of DSI in ISA Design.

with the decision of what to place above the DSI and what to place below the DSI. For example, performance can be achieved through optimizations that are carried out above the DSI by the compiler as well as through optimizations that are performed below the DSI in the microarchitecture. Ideally the DSI should be placed at a level that achieves the best synergy between static techniques and dynamic techniques, i.e., leveraging the best combination of compiler complexity and hardware complexity to obtain the desired performance. This DSI placement becomes a real challenge because of the constantly evolving hardware technology and compiler technology.

In the history of ISA design, a number of different placements of the DSI have been proposed and some have led to commercially successful ISAs. A conceptual illustration of possible placements of the DSI is shown in Figure 1.3. This figure is intended not to be highly rigorous but simply to illustrate that the DSI can be placed at different levels. For example, Mike Flynn has proposed placing the DSI very high and doing everything below the DSI, such that a program written in a high-level language can be directly executed by a *directly executable language machine* [Flynn and Hoevel, 1983]. A complex instruction set computer (CISC) ISA places the DSI at the traditional assembly language, or macrocode, level. In contrast, a reduced instruction set computer (RISC) ISA lowers the DSI and expects to perform more of the optimizations above the DSI via the compiler. The lowering of the DSI effectively exposes and elevates what would have been considered microarchitecture features in a CISC ISA to the ISA level. The purpose of doing this is to reduce the hardware complexity and thus achieve a much faster machine [Colwell et al., 1985].

The DSI provides an important separation between architecture and implementation. Violation of this separation can become problematic. As an ISA evolves and extensions are added, the placement of the DSI is implicitly shifted. The lowering of the DSI by promoting a former implementation feature to the architecture level effectively exposes part of the original microarchitecture to the software. This can facilitate optimizations by the compiler that lead to reduced hardware complexity. However, hardware technology changes rapidly and implementations must adapt and evolve to take advantage of the technological changes. As implementation styles and techniques change, some of the older techniques or microarchitecture features may become ineffective or even undesirable. If some of these older features were promoted to the ISA level, then they become part of the ISA and there will exist installed software base or legacy code containing these features. Since all future implementations must support the entire ISA to ensure the portability of all existing code, the unfortunate consequence is that all future implementations must continue to support those ISA features that had been promoted earlier, even if they are now ineffective and even undesirable. Such mistakes have been made with real ISAs. The lesson learned from these mistakes is that a strict separation of architecture and microarchitecture must be maintained in a disciplined fashion. Ideally, the architecture or ISA should only contain features necessary to express the functionality or the semantics of the software algorithm, whereas all the features that are employed to facilitate better program performance should be relegated to the implementation or the microarchitecture domain.

The focus of this book is not on ISA design but on microarchitecture techniques, with almost exclusive emphasis on performance. ISA features can influence the design effort and the design complexity needed to achieve high levels of performance. However, our view is that in contemporary high-end microprocessor design, it is the microarchitecture, and not the ISA, that is the dominant determinant of microprocessor performance. Hence, the focus of this book is on microarchitecture techniques for achieving high performance. There are other important design objectives, such as power, cost, and reliability. However, historically performance has received the most attention, and there is a large body of knowledge on techniques for enhancing performance. It is this body of knowledge that this book is attempting to codify.

1.3 Principles of Processor Performance

The primary design objective for new leading-edge microprocessors has been performance. Each new generation of microarchitecture seeks to significantly improve on the performance of the previous generation. In recent years, reducing power consumption has emerged as another, potentially equally important design objective. However, the demand for greater performance will always be there, and processor performance will continue to be a key design objective.

1.3.1 Processor Performance Equation

During the 1980s several researchers independently discovered or formulated an equation that clearly defines processor performance and cleanly characterizes the fundamental factors that contribute to processor performance. This equation has come to be known as the *iron law* of processor performance, and it is shown in Equation (1.1). First, the processor performance equation indicates that a processor's performance is measured in terms of how long it takes to execute a particular program (*time/program*). Second, this measure of time/program or *execution time* can be formulated as a product of three terms: *instructions/program, cycles/instruction,* and *time/cycle.* The first term indicates the total number of dynamic instructions that need to be executed for a particular program; this term is also referred to as the *instruction count*. The second term indicates on average (averaging over the entire execution of the program) how many machine cycles are consumed to execute each instruction; typically this term is denoted as the *CPI* (cycles per instruction). The third term indicates the length of time of each machine cycle, namely, the *cycle time* of the machine.

$$\frac{1}{\text{Performance}} = \frac{\text{time}}{\text{program}} = \frac{\text{instructions}}{\text{program}} \times \frac{\text{cycles}}{\text{instruction}} \times \frac{\text{time}}{\text{cycle}} \quad (1.1)$$

The shorter the program's execution time, the better the performance. Looking at Equation 1.1, we can conclude that processor performance can be improved by reducing the magnitude of any one of the three terms in this equation. If the instruction count can be reduced, there will be fewer instructions to execute and the execution time will be reduced. If CPI is reduced, then on average each instruction will consume fewer machine cycles. If cycle time can be reduced, then each cycle will consume less time and the overall execution time is reduced. It might seem from this equation that improving performance is quite trivial. Unfortunately, it is not that straightforward. The three terms are not all independent, and there are complex interactions between them. The reduction of any one term can potentially increase the magnitude of the other two terms. The relationship between the three terms cannot be easily characterized. Improving performance becomes a real challenge involving subtle tradeoffs and delicate balancing acts. It is exactly this challenge that makes processor design fascinating and at times more of an art than a science. Section 1.3.2 will examine more closely different ways to improve processor performance.

1.3.2 Processor Performance Optimizations

It can be said that all performance optimization techniques boil down to reducing one or more of the three terms in the processor performance equation. Some techniques can reduce one term while leaving the other two unchanged. For example, when a compiler performs optimizations that eliminate redundant and useless instructions in the object code, the instruction count can be reduced without impacting the CPI or the cycle time. As another example, when a faster circuit technology or a more advanced fabrication process is used that reduces signal propagation delays, the machine cycle time can potentially be reduced without impacting the instruction count or the CPI. Such types of performance optimization techniques are always desirable and should be employed if the cost is not prohibitive.

Other techniques that reduce one of the terms can at the same time increase one or both of the other terms. For these techniques, there is performance gain only if the reduction in one term is not overwhelmed by the increase in the other terms. We can examine these techniques by looking at the reduction of each of the three terms in the processor performance equation.

There are a number of ways to reduce the instruction count. First, the instruction set can include more complex instructions that perform more work per instruction. The total number of instructions executed can decrease significantly. For example, a program in a RISC ISA can require twice as many instructions as one in a CISC ISA. While the instruction count may go down, the complexity of the execution unit can increase, leading to a potential increase of the cycle time. If deeper pipelining is used to avoid increasing the cycle time, then a higher branch misprediction penalty can result in higher CPI. Second, certain compiler optimizations can result in fewer instructions being executed. For example, unrolling loops can reduce the number of loop closing instructions executed. However, this can lead to an increase in the static code size, which can in turn impact the instruction cache hit rate, which can in turn increase the CPI. Another similar example is the in-lining of function calls. By eliminating calls and returns, fewer instructions are executed, but the code size can significantly expand. Third, more recently researchers have proposed the dynamic elimination of redundant computations via microarchitecture techniques. They have observed that during program execution, there are frequent repeated executions of the same computation with the same data set. Hence, the result of the earlier computation can be buffered and directly used without repeating the same computation [Sodani and Sohi, 1997]. Such computation reuse techniques can reduce the instruction count, but can potentially increase the complexity in the hardware implementation which can lead to the increase of cycle time. We see that decreasing the instruction count can potentially lead to increasing the CPI and/or cycle time.

The desire to reduce CPI has inspired many architectural and microarchitectural techniques. One of the key motivations for RISC was to reduce the complexity of each instruction in order to reduce the number of machine cycles required to process each instruction. As we have already mentioned, this comes with the overhead of an increased instruction count. Another key technique to reduce CPI is instruction pipelining. A pipelined processor can overlap the processing of multiple instructions. Compared to a nonpipelined design and assuming identical cycle times, a pipelined design can significantly reduce the CPI. A shallower pipeline, that is, a pipeline with fewer pipe stages, can yield a lower CPI than a deeper pipeline, but at the expense of increased cycle time. The use of cache memory to reduce the average memory access latency (in terms of number of clock cycles) will also reduce the CPI. When a conditional branch is taken, stalled cycles can result from having to fetch the next instruction from a nonsequential location. Branch prediction techniques can reduce the number of such stalled cycles, leading to a reduction of CPI. However, adding branch predictors can potentially increase the cycle time due to the added complexity in the fetch pipe stage, or even increase the CPI if a deeper pipeline is required to maintain the same cycle time. The emergence of superscalar processors allows the processor pipeline to simultaneously process multiple instructions in each pipe stage. By being able to sustain the execution of multiple instructions in every machine cycle, the CPI can be significantly reduced. Of course, the complexity of each pipe stage can increase, leading to a potential increase of cycle time or the pipeline depth, which can in turn increase the CPI.

The key microarchitecture technique for reducing cycle time is pipelining. Pipelining effectively partitions the task of processing an instruction into multiple stages. The latency (in terms of signal propagation delay) of each pipe stage determines the machine cycle time. By employing deeper pipelines, the latency of each pipe stage, and hence the cycle time, can be reduced. In recent years, aggressive pipelining has been the major technique used in achieving phenomenal increases of clock frequency of high-end microprocessors. As can be seen in Table 1.1, during the most recent decade, most of the performance increase has been due to the increase of the clock frequency.

There is a downside to increasing the clock frequency through deeper pipelining. As a pipeline gets deeper, CPI can go up in three ways. First, as the front end of the pipeline gets deeper, the number of pipe stages between fetch and execute increases. This increases the number of penalty cycles incurred when branches are mispredicted, resulting in the increase of CPI. Second, if the pipeline is so deep that a primitive *arithmetic-logic unit* (ALU) operation requires multiple cycles, then the necessary latency between two dependent instructions, even with result-forwarding hardware, will be multiple cycles. Third, as the clock frequency increases with deeper *central processing unit* (CPU) pipelines, the latency of memory, in terms of number of clock cycles, can significantly increase. This can increase the average latency of memory operations and thus increase the overall CPI. Finally, there is hardware and latency overhead in pipelining that can lead to diminishing returns on performance gains. This technique of getting higher frequency via deeper pipelining has served us well for more than a decade. It is not clear how much further we can push it before the requisite complexity and power consumption become prohibitive.

As can be concluded from this discussion, achieving a performance improvement is not a straightforward task. It requires interesting tradeoffs involving many and sometimes very subtle issues. The most talented microarchitects and processor designers in the industry all seem to possess the intuition and the insights that enable them to make such tradeoffs better than others. It is the goal, or perhaps the dream, of this book to impart not only the concepts and techniques of superscalar processor design but also the intuitions and insights of superb microarchitects.

1.3.3 Performance Evaluation Method

In modern microprocessor design, hardware prototyping is infeasible; most designers use simulators to do performance projection and ensure functional correctness during the design process. Typically two types of simulators are used: functional simulators and performance simulators. *Functional simulators* model a machine at the architecture (ISA) level and are used to verify the correct execution of a program. Functional simulators actually interpret or *execute* the instructions of a program. *Performance simulators* model the microarchitecture of a design and are used to measure the number of machine cycles required to execute a program. Usually performance simulators are concerned not with the semantic correctness of instruction execution, but only with the timing of instruction execution.

Performance simulators can be either trace-driven or execution-driven; as illustrated in Figure 1.4. *Trace-driven* performance simulators process pregenerated traces to determine the cycle count for executing the instructions in the traces. A trace captures the dynamic sequence of instructions executed and can be generated in three different ways; see Figure 1.4(a). One way is via *software instrumentation,*

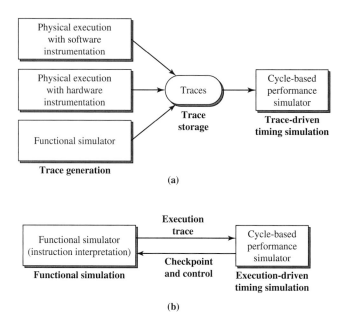

Figure 1.4
Performance Simulation Methods: (a) Trace-Driven Simulation; (b) Execution-Driven Simulation.

which inserts special instructions into a program prior to run time so that when the instrumented program is executed on a physical system, the inserted instructions will produce the dynamic execution trace. Another way is via *hardware instrumentation,* which involves putting special hardware probes to monitor the system bus and to record the actual execution trace when a program is executed on the system. Software instrumentation can significantly increase the code size and the program execution time. Hardware instrumentation requires the monitoring hardware and is seriously limited by the buffering capacity of the monitoring hardware. The third trace generation method uses a *functional simulator* to simulate the execution of a program. During simulation, hooks are embedded in the simulator to record the dynamic execution trace. For all three methods, once traces are generated, they can be stored for subsequent repeated use by trace-driven performance simulators.

Execution-driven performance simulators overcome some of the limitations of trace-driven performance simulators; see Figure 1.4(b). Instead of using pregenerated traces, an execution-driven performance simulator is interfaced to a functional simulator, and the two simulators work in tandem. During simulation, the functional simulator executes the instructions and passes information associated with the executed instructions to the performance simulator. The performance simulator then tracks the timing of these instructions and their movement through the pipeline stages. It has the ability to issue directives to the functional simulator to checkpoint the simulation state and to later resume from the checkpointed state. The checkpoint capability allows the simulation of speculative instructions, such as instructions following a branch prediction. More specifically, execution-driven simulation can simulate the mis-speculated instructions, such as the instructions following a mispredicted branch, going through the pipeline. In trace-driven simulation, the pre-generated trace contains only the actual (nonspeculative) instructions executed, and a trace-driven simulator cannot account for the instructions on a mis-speculated path and their potential contention for resources with other (nonspeculative) instructions. Execution-driven simulators also alleviate the need to store long traces. Most modern performance simulators employ the execution-driven paradigm. The most advanced execution-driven performance simulators are supported by functional simulators that are capable of performing full-system simulation, that is, the simulation of both application and operating system instructions, the memory hierarchy, and even input/output devices.

The actual implementation of the microarchitecture model in a performance simulator can vary widely in terms of the amount and details of machine resources that are explicitly modeled. Some performance models are merely cycle counters that assume unlimited resources and simply calculate the total number of cycles needed for the execution of a trace, taking into account inter-instruction dependences. Others explicitly model the organization of the machine with all its component modules. These performance models actually simulate the movement of instructions through the various pipeline stages, including the allocation of limited machine resources in each machine cycle. While many performance simulators claim to be "cycle-accurate," the methods they use to model and track the activities in each machine cycle can be quite different.

While there is heavy reliance on performance simulators during the early design stages of a microprocessor, the validation of the accuracy of performance simulators is an extremely difficult task. Typically the performance model or simulator is implemented in the early phase of the design and is used to do initial tradeoffs of various microarchitecture features. During this phase there isn't a reference that can be used to validate the performance model. As the design progresses and an RTL model of the design is developed, the RTL model can be used as a reference to validate the accuracy of the performance model. However, simulation using the RTL model is very slow, and therefore only very short traces can be used. During the entire design process, discipline is essential to concurrently evolve the performance model and the RTL model to ensure that the performance model is tracking all the changes made in the RTL model. It is also important to do post-silicon validation of the performance model so that it can be used as a good starting point for the next-generation design. Most performance simulators used in academic research are never validated. These simulators can be quite complex and, just like all large pieces of software, can contain many bugs that are difficult to eliminate. It is quite likely that a large fraction of the performance data published in many research papers using unvalidated performance models is completely erroneous. Black argues convincingly for more rigorous validation of processor simulators [Black and Shen, 1998].

Other than the difficulty of validating their accuracy, another problem associated with performance simulators is the extremely long simulation times that are often required. Most contemporary performance evaluations involve the simulation of many benchmarks and a total of tens to hundreds of billion instructions. During the early phase of the design, performance simulators are used extensively to support the exploration of numerous tradeoffs, which require many simulation runs using different sets of parameters in the simulation model. For execution-driven performance simulators that have fairly detailed models of a complex machine, a slowdown factor of four to five orders of magnitude is rather common. In other words, to simulate a single machine cycle of the target machine, that is the machine being modeled, can require the execution of 10,000 to 100,000 machine cycles on the host machine. A large set of simulation runs can sometimes take many days to complete, even using a large pool of simulation machines.

1.4 Instruction-Level Parallel Processing

Instruction-level parallel processing can be informally defined as the concurrent processing of multiple instructions. Traditional sequential processors execute one instruction at a time. A leading instruction is completed before the next instruction is processed. To a certain extent, pipelined processors achieve a form of instruction-level parallel processing by overlapping the processing of multiple instructions. As many instructions as there are pipeline stages can be concurrently in flight at any one time. Traditional sequential (CISC) processors can require an average of about 10 machine cycles for processing each instruction, that is CPI = 10. With pipelined (RISC) processors, even though each instruction may still require multiple cycles to complete, by overlapping the processing of multiple instructions in the pipeline, the effective average CPI can be reduced to close to one if a new instruction can be initiated every machine cycle.

With scalar pipelined processors, there is still the limitation of fetching and initiating at most one instruction into the pipeline every machine cycle. With this limitation, the best possible CPI that can be achieved is one; or inversely, the best possible throughput of a scalar processor is one instruction per cycle (IPC). A more aggressive form of instruction-level parallel processing is possible that involves fetching and initiating multiple instructions into a wider pipelined processor every machine cycle. While the decade of the 1980s adopted CPI = 1 as its design objective for single-chip microprocessors, the goal for the decade of the 1990s was to reduce CPI to below one, or to achieve a throughput of IPC greater than one. Processors capable of IPC greater than one are termed *superscalar* processors. This section presents the overview of instruction-level parallel processing and provides the bridge between scalar pipelined processors and their natural descendants, the superscalar processors.

1.4.1 From Scalar to Superscalar

Scalar processors are pipelined processors that are designed to fetch and issue at most one instruction every machine cycle. *Superscalar* processors are those that are designed to fetch and issue multiple instructions every machine cycle. This subsection presents the basis and motivation for evolving from scalar to superscalar processor implementations.

1.4.1.1 Processor Performance.
In Section 1.3.1 we introduced the iron law of processor performance, as shown in Equation (1.1). That equation actually represents the inverse of performance as a product of *instruction count,* average *CPI,* and the clock *cycle time.* We can rewrite that equation to directly represent performance as a product of the inverse of *instruction count,* average *IPC* (IPC = 1/CPI), and the clock *frequency,* as shown in Equation (1.2). Looking at this equation, we see that performance can be increased by increasing the IPC, increasing the frequency, or decreasing the instruction count.

$$\text{Performance} = \frac{1}{\text{instruction count}} \times \frac{\text{instructions}}{\text{cycle}} \times \frac{1}{\text{cycle time}} = \frac{\text{IPC} \times \text{frequency}}{\text{instruction count}} \quad (1.2)$$

Instruction count is determined by three contributing factors: the instruction set architecture, the compiler, and the operating system. The ISA and the amount of work encoded into each instruction can strongly influence the total number of instructions executed for a program. The effectiveness of the compiler can also strongly influence the number of instructions executed. The operating system functions that are invoked by the application program effectively increase the total number of instructions executed in carrying out the execution of the program.

Average IPC (instructions per cycle) reflects the average instruction throughput achieved by the processor and is a key measure of microarchitecture effectiveness. Historically, the inverse of IPC, that is, CPI (cycles per instruction), has been used to indicate the average number of machine cycles needed to process each instruction. The use of CPI was popular during the days of scalar pipelined processors. The performance penalties due to various forms of pipeline stalls can be cleanly stated as different CPI overheads. Back then, the ultimate performance goal for scalar pipelined processors was to reduce the average CPI to one. As we move into the superscalar domain, it becomes more convenient to use IPC. The new performance goal for superscalar processors is to achieve an average IPC greater than one. The bulk of the microarchitecture techniques presented in this book target the improvement of IPC.

Frequency is strongly affected by the fabrication technology and circuit techniques. Increasing the number of pipeline stages can also facilitate higher clocking frequencies by reducing the number of logic gate levels in each pipe stage. Traditional pipelines can have up to 20 levels of logic gates in each pipe stage; most contemporary pipelines have only 10 or fewer levels. To achieve high IPC in superscalar designs, the pipeline must be made wider to allow simultaneous processing of multiple instructions in each pipe stage. The widening of the pipeline increases the hardware complexity and the signal propagation delay of each pipe stage. Hence, with a wider pipeline, in order to maintain the same frequency an even deeper pipeline may be required. There is a complex tradeoff between making pipelines wider and making them deeper.

1.4.1.2 Parallel Processor Performance.
As we consider the parallel processing of instructions in increasing processor performance, it is insightful to revisit the classic observation on parallel processing commonly referred to as *Amdahl's law*

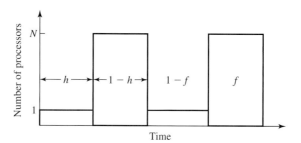

Figure 1.5
Scalar and Vector Processing in a Traditional Supercomputer.

[Amdahl, 1967]. Traditional supercomputers are parallel processors that perform both scalar and vector computations. During scalar computation only one processor is used. During vector computation all N processors are used to perform operations on array data. The computation performed by such a parallel machine can be depicted as shown in Figure 1.5, where N is the number of processors in the machine and h is the fraction of time the machine spends in scalar computation. Conversely, $1 - h$ is the fraction of the time the machine spends in vector computation.

One formulation of Amdahl's law states that the efficiency E of the parallel machine is measured by the overall utilization of the N processors or the fraction of time the N processors are busy. Efficiency E can be modeled as

$$E = \frac{h + N \times (1-h)}{N} = \frac{h + N - Nh}{N} = 1 - h \times \left(1 - \frac{1}{N}\right) \quad (1.3)$$

As the number of processors N becomes very large, the efficiency E approaches $1 - h$, which is the fraction of time the machine spends in vector computation. As N becomes large, the amount of time spent in vector computation becomes smaller and smaller and approaches zero. Hence, as N becomes very large, the efficiency E approaches zero. This means that almost all the computation time is taken up with scalar computation, and further increase of N makes very little impact on reducing the overall execution time.

Another formulation of this same principle is based on the amount of work that can be done in the vector computation mode, or the *vectorizability* of the program. As shown in Figure 1.5, f represents the fraction of the program that can be parallelized to run in vector computation mode. Therefore, $1 - f$ represents the fraction of the program that must be executed sequentially. If T is the total time required to run the program, then the relative speedup S can be represented as

$$S = \frac{1}{T} = \frac{1}{(1-f) + (f/N)} \quad (1.4)$$

where T is the sum of $(1 - f)$, the time required to execute the sequential part, and f/N, the time required to execute the parallelizable part of the program. As N becomes very large, the second term of this sum approaches zero, and the total execution time is dictated by the amount of time required to execute the sequential part. This is commonly referred to as the *sequential bottleneck;* that is, the time spent in sequential execution or scalar computation becomes a limit to how much overall performance improvement can be achieved via the exploitation of parallelism. As N increases or as the machine parallelism increases, the performance will become more and more sensitive to and dictated by the sequential part of the program.

The efficiency of a parallel processor drops off very quickly as the number of processors is increased. Furthermore, as the *vectorizability,* i.e., the fraction of the program that can be parallelized, of a program drops off slightly from 100%, the efficiency drop-off rate increases. Similarly the overall speedup drops off very quickly when f, the vectorizability of the program, drops even just very slightly from 100%. Hence, the overall performance improvement is very sensitive to the vectorizability of the program; or to state it another way, the overall speedup due to parallel processing is strongly dictated by the sequential part of the program as the machine parallelism increases.

1.4.1.3 Pipelined Processor Performance.
Harold Stone proposed that a performance model similar to that for parallel processors can be developed for pipelined processors [Stone, 1987]. A typical execution profile of a pipelined processor is shown in Figure 1.6(a). The machine parallelism parameter N is now the depth of the pipeline, that is, the number of stages in the pipeline. There are three phases

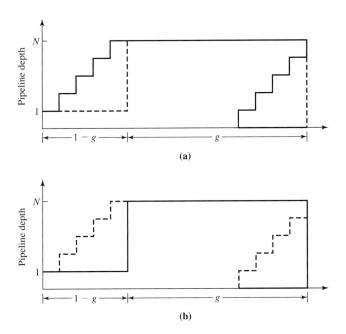

Figure 1.6

Idealized Pipelined Execution Profile: (a) Actual; (b) Modeled.

in this execution profile. The first phase is the pipeline filling phase during which the first sequence of N instructions enters the pipeline. The second phase is the pipeline full phase, during which the pipeline is full and represents the steady state of the pipeline. This is assuming that there is no pipeline disruption, and therefore represents the perfect pipeline execution profile. The third phase is the pipeline draining phase, during which no new instruction is entering the pipeline and the pipeline is finishing the instructions still present in the pipeline stages.

For modeling purposes, we can modify the execution profile of Figure 1.6(a) to the execution profile of Figure 1.6(b) by moving some of the work done during the pipeline filling phase to the pipeline draining phase. The total amount of work remains the same; that is, the areas within the two profiles are equal. The number of pipeline stages is N, the fraction of the time that all N pipeline stages are utilized is g, and $1 - g$ is the fraction of time when only one pipeline stage is utilized. Essentially $1 - g$ can be viewed as the fraction of time when only one instruction is moving through the pipeline; that is, there is no overlapping of instructions in the pipeline.

Unlike the idealized pipeline execution profile, the realistic pipeline execution profile must account for the stalling cycles. This can be done as shown in Figure 1.7(a). Instead of remaining in the pipeline full phase for the duration of the entire execution, this steady state is interrupted by pipeline stalls. Each stall effectively induces a new pipeline draining phase and a new pipeline filling phase, as shown in Figure 1.7(a), due to the break in the pipeline full phase. Similar modification can be performed on this execution profile to result in the modified profile of

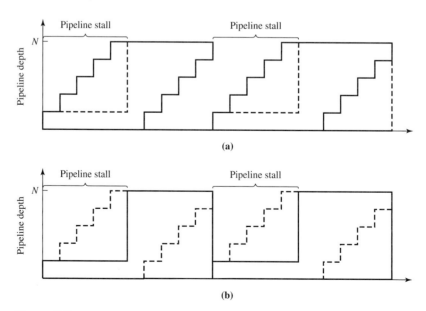

Figure 1.7
Realistic Pipeline Execution Profile: (a) Actual; (b) Modeled.

Figure 1.7(b) by moving part of the work done in the two pipeline filling phases to the two pipeline draining phases. Now the modified profile of Figure 1.7(b) resembles the execution profile of parallel processors as shown in Figure 1.5.

With the similarity of the execution profiles, we can now borrow the performance model of parallel processors and apply it to pipelined processors. Instead of being the number of processors, N is now the number of pipeline stages, or the maximum speedup possible. The parameter g now becomes the fraction of time when the pipeline is filled, and the parameter $1 - g$ now represents the fraction of time when the pipeline is stalled. The speedup S that can be obtained is now

$$S = \frac{1}{(1-g) + (g/N)} \quad (1.5)$$

Note that g, the fraction of time when the pipeline is full, is analogous to f, the vectorizability of the program in the parallel processor model. Therefore, Amdahl's law can be analogously applied to pipelined processors. As g drops off just slightly from 100%, the speedup or the performance of a pipelined processor can drop off very quickly. In other words, the actual performance gain that can be obtained through pipelining can be strongly degraded by just a small fraction of stall cycles. As the degree of pipelining N increases, the fraction of stall cycles will become increasingly devastating to the actual speedup that can be achieved by a pipeline processor. Stall cycles in pipelined processors are now the key adversary and are analogous to the sequential bottleneck for parallel processors. Essentially, stall cycles constitute the pipelined processor's sequential bottleneck.

Equation (1.5) is a simple performance model for pipelined processors based on Amdahl's law for parallel processors. It is assumed in this model that whenever the pipeline is stalled, there is only one instruction in the pipeline, or it effectively becomes a sequential nonpipelined processor. The implication is that when a pipeline is stalled no overlapping of instructions is allowed; this is effectively equivalent to stalling the pipeline for N cycles to allow the instruction causing the stall to completely traverse the pipeline. We know, however, that with clever design of the pipeline, such as with the use of forwarding paths, to resolve a hazard that causes a pipeline stall, the number of penalty cycles incurred is not necessarily N and most likely less than N. Based on this observation, a refinement to the model of Equation (1.5) is possible.

$$S = \frac{1}{\frac{g_1}{1} + \frac{g_2}{2} + \cdots + \frac{g_N}{N}} \quad (1.6)$$

Equation (1.6) is a generalization of Equation (1.5) and provides a refined model for pipelined processor performance. In this model, g_i represents the fraction of time when there are i instructions in the pipeline. In other words, g_i represents the fraction of time when the pipeline is stalled for $(N - i)$ penalty cycles. Of course, g_N is the fraction of time when the pipeline is full.

This pipelined processor performance model is illustrated by applying it to the six-stage TYP pipeline in Chapter 2. Note that the TYP pipeline has a load penalty of one cycle and a branch penalty of four cycles. Based on the statistics from the IBM study presented in Chapter 2, the typical percentages of load and branch instructions are 25% and 20%, respectively. Assuming that the TYP pipeline is designed with a bias for a branch not taken, only 66.6% of the branch instructions, those that are actually taken, will incur the branch penalty. Therefore, only 13% of the instructions (branches) will incur the four-cycle penalty and 25% of the instructions (loads) will incur the one-cycle penalty. The remaining instructions (62%) will incur no penalty cycles. The performance of the TYP pipeline can be modeled as shown in Equation (1.7).

$$S_{TYP} = \frac{1}{\frac{0.13}{(6-4)} + \frac{0.25}{(6-1)} + \frac{0.62}{6}} = \frac{1}{\frac{0.13}{2} + \frac{0.25}{5} + \frac{0.62}{6}} = \frac{1}{0.22} = 4.5 \quad (1.7)$$

The resultant performance of the six-stage TYP pipeline processor is a factor of 4.5 over that of the sequential or nonpipelined processor. Note that the TYP is a six-stage pipeline with the theoretical speedup potential of 6. The actual speedup based on our model of Equation (1.6) is 4.5, as shown in Equation (1.7), which can be viewed as the *effective degree of pipelining* of the TYP pipeline. Essentially the six-stage TYP processor behaves as a perfect pipeline with 4.5 pipeline stages. The difference between 6 and 4.5 reflects the difference between the potential (peak) pipeline parallelism and the achieved (actual) pipeline parallelism.

1.4.1.4 The Superscalar Proposal.
We now restate Amdahl's law that models the performance of a parallel processor:

$$S = \frac{1}{(1-f) + (f/N)} \quad (1.8)$$

This model gives the performance or speedup of a parallel system over that of a nonparallel system. The *machine parallelism* is measured by N, the number of processors in the machine, and reflects the maximum number of tasks that can be simultaneously performed by the system. The parameter f, however, is the vectorizability of the program which reflects the *program parallelism*. The formulation of this model is influenced by traditional supercomputers that contain a scalar unit and a vector unit. The vector unit, consisting of N processors, executes the vectorizable portion of the program by performing N tasks at a time. The nonvectorizable portion of the program is then executed in the scalar unit in a sequential fashion. We have already observed the oppressive tyranny of the nonvectorizable portion of the program on the overall performance that can be obtained through parallel processing.

The assumption that the nonvectorizable portion of the program must be executed sequentially is overly pessimistic and not necessary. If some, even low, level of parallelism can be achieved for the nonvectorizable portion of the program, the severe impact of the sequential bottleneck can be significantly moderated. Figure 1.8 illustrates this principle. This figure, taken from an IBM technical report coauthored

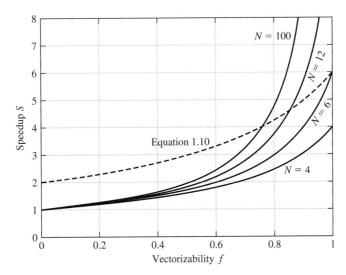

Figure 1.8
Easing of the Sequential Bottleneck with Instruction-Level Parallelism for Nonvectorizable Code.
Source: Agerwala and Cocke, 1987.

by Agerwala and Cocke [1987], plots the speedup as a function of f, the vectorizability of a program, for several values of N, the maximum parallelism of the machine. Take the example of the case when $N = 6$. The speedup is

$$S = \frac{1}{(1-f) + (f/6)} \quad (1.9)$$

Examining the curve for Equation (1.9) in Figure 1.8, we see that the speedup is equal to 6 if f is 100%, that is, perfectly vectorizable. As f drops off from 100%, the speedup drops off very quickly; as f becomes 0%, the speedup is one; that is, no speedup is obtained. With higher values of N, this speedup drop-off rate gets significantly worse, and as f approaches 0%, all the speedups approach one, regardless of the value of N. Now assume that the minimum degree of parallelism of 2 can be achieved for the nonvectorizable portion of the program. The speedup now becomes

$$S = \frac{1}{\frac{(1-f)}{2} + \frac{f}{6}} \quad (1.10)$$

Examining the curve for Equation (1.10) in Figure 1.8, we see that it also starts at a speedup of 6 when f is 100%, but drops off more slowly than the curve for Equation (1.9) when f is lowered from 100%. In fact this curve crosses over the curve for Equation (1.8) with $N = 100$ when f is approximately 75%. This means that for cases with f less than 75%, it is more beneficial to have a system with maximum parallelism of only 6, that is $N = 6$, but a minimum parallelism of two for the nonvectorizable portion, than a system with maximum parallelism of $N = 100$ with sequential execution of the nonvectorizable portion. The vectorizability of a program f is a complex function involving the application algorithm, the programming language, the compiler, and the architecture. Other than those for scientific applications involving mostly numerical computations, most programs for general-purpose computing tend not to have very high vectorizability. It is safe to say that most general-purpose programs have f less than 75%, and for many, significantly less.

One primary motivation for designing superscalar processors is to develop general-purpose processors that can achieve some (perhaps low relative to vectorizing) level of parallelism for a wide range of application programs. The goal is to ensure that some degree of instruction-level parallelism can be achieved for all portions of the program so as to moderate the severe impact of the sequential bottleneck. Of course, the highly vectorizable programs will continue to achieve good speedup via parallelism. Note that the curve for Equation (1.10) is always higher than that for Equation (1.9) even at high values of f, and is higher than other curves for large values of N at lower values of f. The goal for superscalar processors is to achieve generalized instruction-level parallelism and the consequent speedup for all types of application programs, including those that are not necessarily vectorizable.

1.4.2 Limits of Instruction-Level Parallelism

In Equation (1.10), parallelism of degree 6 can be achieved for the f fraction of the program and parallelism of degree 2 can be achieved for the remaining $1 - f$ fraction of the program. The speedup S can be viewed as the *aggregate degree of parallelism* that can be achieved for the entire program. For example, if the parameter f is 50% and the peak parallelism N is 6, then the speedup or the aggregate degree of parallelism is

$$S = \frac{1}{\frac{(1-f)}{2} + \frac{f}{6}} = \frac{1}{\frac{0.5}{2} + \frac{0.5}{6}} = 3 \qquad (1.11)$$

The implication of Equation (1.11) is that effectively an overall or aggregate degree of parallelism of 3 is achieved for the entire program. Applying this result at the instruction level, we see that Equation (1.11) indicates that an average of three instructions can be simultaneously executed at a time. For traditional vector computation, the number of operations that can be simultaneously performed is largely determined by the size of the vectors or arrays, or essentially the data set size. For general-purpose unstructured programs, the key question is, what aggregate degree of instruction-level parallelism can potentially be achieved?

Instruction-level parallelism can be informally defined as the aggregate degree of parallelism (measured by the number of instructions) that can be achieved by the concurrent execution of multiple instructions. Possible limits of ILP have been investigated for almost three decades. Numerous experimental studies have been performed that yield widely varying results on purported limits of ILP. The following table provides a sample listing of reported limits in order of increasing degrees of ILP.

Study	ILP Limit
Weiss and Smith, 1984	1.58
Sohi and Vajapeyam, 1987	1.81
Tjaden and Flynn, 1970	1.86
Tjaden and Flynn, 1973	1.96
Uht and Wedig, 1986	2.0
Smith et al., 1989	2.0
Jouppi and Wall, 1989	2.4
Johnson, 1991	2.5
Acosta et al., 1986	2.79
Wedig, 1982	3.0
Butler et al., 1991	5.8
Melvin and Patt, 1991	6
Wall, 1991	7
Kuck et al., 1972	8
Riseman and Foster, 1972	51
Nicolau and Fisher, 1984	90

This listing is certainly not exhaustive, but clearly illustrates the diversity and possible inconsistency of the research findings. Most of these are limit studies making various idealized assumptions. The real challenge is how to achieve these levels of ILP in realistic designs. The purported limits are also not monotonic with respect to chronological order. During the decade of the 1990s the debate on the limits of ILP replaced the RISC vs. CISC debate of the 1980s [Colwell et al., 1985]. This new debate on the limit of ILP is still not settled.

1.4.2.1 Flynn's Bottleneck. One of the earliest studies done at Stanford University by Tjaden and Flynn in 1970 concluded that the ILP for most programs is less than 2. This limit has been informally referred to as *Flynn's bottleneck*. This study focused on instruction-level parallelism that can be found within basic block boundaries. Since crossing basic block boundaries involves crossing control dependences, which can be dependent on run-time data, it is assumed that the basic blocks must be executed sequentially. Because of the small size of most basic blocks, typically the degree of parallelism found is less than 2. This result or Flynn's bottleneck has since been confirmed by several other studies.

One study in 1972 that confirmed this result was by Riseman and Foster [1972]. However, they extended their study to examine the degree of ILP that can be achieved if somehow control dependences can be surmounted. This study reported various degrees of parallelism that can be achieved if various numbers of control dependences can be overcome. If the number of control dependences that can be overcome is unlimited, then the limit of ILP is around 51. This study highlights the strong influence of control dependences on the limits of ILP.

1.4.2.2 Fisher's Optimism.
At the other end of the spectrum is a study performed by Nicolau and Fisher in 1984 at Yale University. This study hints at almost unlimited amounts of ILP in many programs. The benchmarks used in this study tend to be more numerical, and some of the parallelisms measured were due to data parallelism resulting from array-type data sets. An idealized machine model capable of executing many instructions simultaneously was assumed. While some idealized assumptions were made in this study, it does present a refreshing optimistic outlook on the amount of ILP that can be harvested against the pessimism due to Flynn's bottleneck. We informally refer to this purported limit on ILP as *Fisher's optimism*.

Initially this optimism was received with a great deal of skepticism. A number of subsequent events somewhat vindicated this study. First a prototype machine model called the VLIW (very long instruction word) processor was developed along with a supporting compiler [Fisher, 1983]. Subsequently, a commercial venture (Multiflow, Inc.) was formed to develop a realistic VLIW machine, which resulted in the Multiflow TRACE computer. The TRACE machines were supported by a powerful VLIW compiler that employs *trace scheduling* (developed by Josh Fisher et al.) to extract instruction-level parallelism [Fisher, 1981]. Multiflow, Inc., was reasonably successful and eventually had an installed base of more than 100 machines. More importantly, the short-lived commercial TRACE machines were the first general-purpose uniprocessors to achieve an average IPC greater than one. Although the actual levels of ILP achieved by the TRACE machines were far less than the limits published earlier by Nicolau and Fisher in 1984, they did substantiate the claims that there are significant amounts of ILP that can be harvested beyond the previously accepted limit of 2.

1.4.2.3 Contributing Factors.
Many of the studies on the limits of ILP employ different experimental approaches and make different assumptions. Three key factors contribute to the wide range of experimental results: benchmarks used, machine models assumed, and compilation techniques employed. Each study adopts its own set of benchmarks, and frequently the results are strongly influenced by the benchmarks chosen. Recently, the Standard Performance Evaluation Corporation (SPEC) benchmark suites have become widely adopted, and most manufacturers of processors and computing systems provide SPEC ratings for their systems. While strict guidelines exist for manufacturers to report the SPEC ratings on their products (see www.spec.org), there are still quite nonuniform uses of the SPEC ratings by researchers. There are also strong indications that the SPEC benchmark suites are only appropriate for workstations running scientific and engineering applications, and are not relevant for other application domains such as commercial transaction processing and embedded real-time computing.

The second key factor that contributes to the confusion and controversy on the limits of ILP is the assumptions made by the various studies about the machine model. Most of the limit studies assume idealized machine models. For example, the cache memory is usually not considered or is assumed to have a 100% hit rate with one-cycle latency. Some models assume infinite-sized machines with infinite register files. Usually one-cycle latency is assumed for all operation and functional unit types. Other studies employ more realistic machine models, and these usually resulted in more pessimistic, and possibly unnecessarily pessimistic, limits. Of course, there is also a great deal of nonuniformity in the instruction set architectures used. Some are fictitious architectures, and others use existing architectures. The architectures used also tend to have a strong influence on the experimental results.

Finally, the assumptions made about the compilation techniques used are quite diverse. Many of the studies do not include any consideration about the compiler; others assume infinitely powerful compilers. Frequently, these studies are based on dynamic traces collected on real machines. Simulation results based on such traces are not only dependent on the benchmarks and architectures chosen, but also strongly dependent on the compilers used to generate the object code. The potential contribution of the compilation techniques to the limits of ILP is an ongoing area of research. There is currently a significant gap between the assumed capabilities of all-powerful compilers and the capabilities of existing commercially available compilers. Many anticipate that many more advancements can be expected in the compilation domain.

Probably the safest conclusion drawn from the studies done so far is that the real limit of ILP is beyond that being achieved on current machines. There is room for more and better research. The assumption of any specific limit is likely to be premature. As more powerful and efficient microarchitectures are designed and more aggressive compilation techniques are developed, the previously made assumptions may have to be changed and previously purported limits may have to be adjusted upward.

1.4.3 Machines for Instruction-Level Parallelism

Instruction-level parallelism is referred to as *fine-grained parallelism* relative to other forms of coarse-grained parallelism involving the concurrent processing of multiple program fragments or computing tasks. Machines designed for exploiting general ILP are referred to as ILP machines and are typically uniprocessors with machine resource parallelisms at the functional unit level. A classification of ILP machines was presented by Norm Jouppi in 1989 [Jouppi and Wall, 1989]. ILP machines are classified according to a number of parameters.

- *Operation latency (OL)*. The number of machine cycles until the result of an instruction is available for use by a subsequent instruction. The reference instruction used is a simple instruction that typifies most of the instructions in the instruction set. The operation latency is the number of machine cycles required for the execution of such an instruction.

- *Machine parallelism (MP)*. The maximum number of simultaneously executing instructions the machine can support. Informally, this is the maximum number of instructions that can be simultaneously in flight in the pipeline at any one time.

- *Issue latency (IL)*. The number of machine cycles required between issuing two consecutive instructions. Again the reference instructions are simple instructions. In the present context, *issuing* means the initiating of a new instruction into the pipeline.

- *Issue parallelism (IP).* The maximum number of instructions that can be issued in every machine cycle.

In Jouppi's classification, the scalar pipelined processor is used as the baseline machine. The classification also uses a generic four-stage instruction pipeline for illustration. These stages are

1. IF (instruction fetch)
2. DE (instruction decode)
3. EX (execute)
4. WB (write back)

The EX stage is used as a reference for the determination of the operation latency. The scalar pipelined processor, used as the baseline machine, is defined to be a machine with OL = 1 cycle and IL = 1 cycle. This baseline machine, with its instruction processing profile illustrated in Figure 1.9, can issue one new instruction into the pipeline in every cycle, and a typical instruction requires one machine cycle for its execution. The corresponding MP is equal to k, the number of stages in the pipeline; in Figure 1.9 MP = 4. The IP is equal to one instruction per cycle. Notice all four of these parameters are static parameters of the machine and do not take into account the dynamic behavior that depends on the program being executed.

When we discuss the performance or speedup of ILP machines, this baseline machine is used as the reference. Earlier in this chapter we referred to the speedup that can be obtained by a pipelined processor over that of a sequential nonpipelined processor that does not overlap the processing of multiple instructions. This form of speedup is restricted to comparison within the domain of scalar processors and focuses on the increased throughput that can be obtained by a (scalar) pipelined processor with respect to a (scalar) nonpipelined processor. Beginning with Chapter 3,

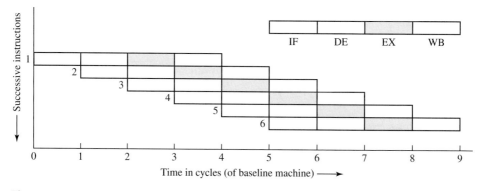

Figure 1.9
Instruction Processing Profile of the Baseline Scalar Pipelined Machine.

which deals with ILP machines, the form of speedup referred to is the performance of an ILP machine compared to the scalar pipelined processor, which is used as the new reference machine.

1.4.3.1 Superpipelined Machines. A *superpipelined* machine is defined with respect to the baseline machine and is a machine with higher degrees of pipelining than the baseline machine. In a superpipelined machine, the machine cycle time is shorter than that of the baseline machine and is referred to as the *minor cycle* time. The cycle time of a superpipelined machine is $1/m$ of the baseline cycle time, or equivalently there are m minor cycles in the baseline cycle. A superpipelined machine is characterized by OL = 1 cycle = m minor cycles and IL = 1 minor cycle. In other words, the simple instruction still requires one baseline cycle, equal to m minor cycles, for execution, but the machine can issue a new instruction in every minor cycle. Consequently, IP = 1 instruction/minor cycle = m instructions/cycle, and MP = $m \times k$. The instruction processing profile of a superpipelined machine is shown in Figure 1.10.

A superpipelined machine is a pipelined machine in which the degree of pipelining is beyond that dictated by the operation latency of the simple instructions. Essentially superpipelining involves pipelining of the execution stage into multiple stages. An "underpipelined" machine cannot issue instructions as fast as they are executed. On the other hand, a superpipelined machine issues instructions faster than they are executed. A superpipelined machine of degree m, that is, one that takes m minor cycles to execute a simple operation, can potentially achieve better performance than that of the baseline machine by a factor of m. Technically, traditional pipelined computers that require multiple cycles for executing simple operations should be classified as superpipelined. For example, the latency for performing fixed-point addition is three cycles in both the CDC 6600 [Thornton, 1964] and the CRAY-1 [Russell, 1978], and new instructions can be issued in every cycle. Hence, these are really superpipelined machines.

In a way, the classification of superpipelined machines is somewhat artificial, because it depends on the choice of the baseline cycle and the definition of a simple operation. The key characteristic of a superpipelined machine is that the

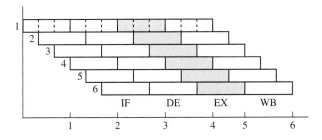

Figure 1.10

Instruction Processing Profile of a Superpipelined Machine of Degree $m = 3$.

result of an instruction is not available to the next $m - 1$ instructions. Hence, a superpipelined processor can be viewed simply as a more deeply pipelined processor with some restrictions on the placement of forwarding paths. In a standard pipelined processor, the implicit assumption is that the sources and destinations of forwarding paths can be the outputs and inputs, respectively, of any of the pipeline stages. If a superpipelined machine is viewed as a deeply pipelined machine with $m \times k$ stages, then the outputs of some of the stages cannot be accessed for forwarding and the inputs of some of the stages cannot receive forwarded data. The reason for this is that some of the operations that require multiple minor cycles and multiple pipeline stages to complete are primitive operations, in the sense of being noninterruptible for the purpose of data forwarding. This is really the key distinction between pipelined and superpipelined machines. In this book, outside of this section, there is no special treatment of superpipelined machines as a separate class of processors distinct from pipelined machines.

The 64-bit MIPS R4000 processor is one of the first processors claimed to be "superpipelined." Internally, the R4000 has eight physical stages in its pipeline, as shown in Figure 1.11, with a physical machine cycle time of 10 nanoseconds (ns) [Bashteen et al., 1991, Mirapuri et al., 1992]. However, the chip requires a 50-MHz clock input and has an on-chip clock doubler. Consequently, the R4000 uses 20 ns as its baseline cycle, and it is considered superpipelined of degree 2 with respect to a four-stage baseline machine with a 20-ns cycle time. There are two minor cycles to every baseline cycle. In the case of the R4000, the multicycle primitive operations are the cache access operations. For example, the first two physical stages (IF and IS) are required to perform the I-cache access, and similarly the DF and DS physical stages are required for D-cache access. These are noninterruptible operations; no data forwarding can involve the buffers between the IF and IS stages or the buffers between the DF and DS stages. Cache accesses, here considered "simple" operations, are pipelined and require an operation latency of two (minor) cycles. The issue latency for the entire pipeline is one (minor) cycle; that is, one new instruction can be issued every

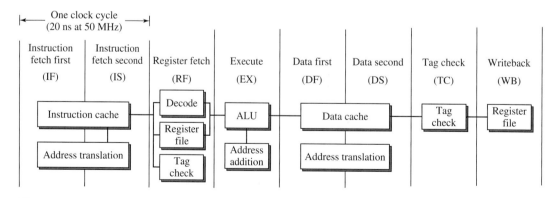

Figure 1.11
The "Superpipelined" MIPS R4000 8-Stage Pipeline.

10 ns. Potentially the R4000 can achieve a speedup over the baseline four-stage pipeline by a factor of 2.

1.4.3.2 Superscalar Machines. Superscalar machines are extensions of the baseline scalar pipelined machines and are characterized by OL = 1 cycle, IL = 1 cycle, and IP = n instructions/cycle. The machine cycle is the same as the baseline cycle; there are no minor cycles. A simple operation is executed in one cycle. In every cycle, multiple instructions can be issued. The superscalar degree is determined by the issue parallelism n, the maximum number of instructions that can be issued in every cycle. The instruction processing profile of a superscalar machine is illustrated in Figure 1.12. Compared to a scalar pipelined processor, a superscalar machine of degree n can be viewed as having n pipelines or a pipeline that is n times wider in the sense of being able to carry n instructions in each pipeline stage instead of one. A superscalar machine has MP = $n \times k$. It has been shown that a superpipelined machine and a superscalar machine of the same degree have the same machine parallelism and can achieve roughly the same level of performance.

There is no reason why a superscalar machine cannot also be superpipelined. The issue latency can be reduced to $1/m$ of the (baseline) cycle while maintaining the issue parallelism of n instructions in every (minor) cycle. The total issue parallelism or throughput will be $n \times m$ instructions per (baseline) cycle. The resultant machine parallelism will become MP = $n \times m \times k$, where n is the superscalar degree, m is the superpipelined degree, and k is the degree of pipelining of the baseline machine. Alternatively the machine parallelism can be viewed as MP = $n \times (m \times k)$, representing a superscalar machine with $m \times k$ pipeline stages. Such a machine can be equivalently viewed as a more deeply pipelined processor of $m \times k$ stages with superscalar degree n, without having to invoke the tedious term "superscalar-superpipelined" machine; and we won't.

1.4.3.3 Very-Long-Instruction-Word Machines. Quite similar to the superscalar machines is another class of ILP machines called VLIW (very long instruction word) machines by Josh Fisher [Fisher, 1983]. The intent and performance objectives are very similar for these two classes of machines; the key difference lies in the

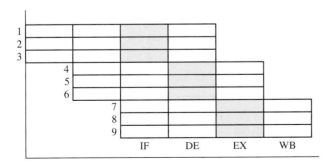

Figure 1.12

Instruction Processing Profile of a Superscalar Machine of Degree $n = 3$.

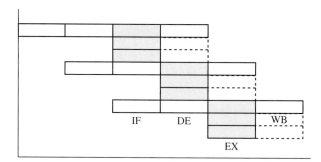

Figure 1.13
Instruction Processing Profile of a VLIW Machine of Degree $n = 3$.

placement of the dynamic-static interface (DSI) or the partitioning of what is done at run time via hardware mechanisms and what is done at compile time via software means. The instruction processing profile of a VLIW machine is illustrated in Figure 1.13.

Unlike in a superscalar machine, the IF and DE stages of a VLIW machine need not be replicated to support the simultaneous processing, that is, fetching and decoding, of n separate instructions. In a superscalar machine, the decision of which n instructions are to be issued into the execute stage is made at run time. For a VLIW machine, such an instruction-issuing decision is made at compile time, and the n instructions to be simultaneously issued into the execute stage are determined by the compiler and stored appropriately in the program memory as a very long instruction word.

Superscalar and VLIW machines represent two different approaches to the same ultimate goal, which is achieving high processor performance via instruction-level parallel processing. The two approaches have evolved through different historical paths and from different perspectives. It has been suggested that these two approaches are quite synergistic, and there is a strong motivation for pursuing potential integration of the two approaches. This book focuses on dynamic techniques implemented in the microarchitecture; hence, we will not address in depth VLIW features that rely on aggressive compile-time techniques.

1.5 Summary

Microprocessors have had an unparalleled impact on the computer industry. The changes that have taken place during the lifetime of microprocessors (30+ years) have been phenomenal. Microprocessors are now entering their fourth decade. It is fascinating to speculate on what we can expect from microprocessors in this coming decade.

Although it was the fad of past decades, instruction set architecture (ISA) design is no longer a very interesting topic. We have learned a great deal about how to design an elegant and scalable ISA. However, code compatibility and the software installed base are more crucial in determining the longevity of an ISA. It has been amply shown that any ISA deficiency can be overcome by microarchitecture techniques. Furthermore, with the emergence of portable bytecodes and dynamic just-in-time (JIT) compilation, the meaning of ISA and the consequent placement of the dynamic-static interface (DSI) will become quite blurred.

In the coming decade, microarchitecture will be where the action is. As the chip integration density approaches 1 billion transistors on a die, many of the traditional (macro)architecture features, such as the memory subsystem, multiple processors, and input/output subsystem, will become on-die issues and hence become effectively microarchitecture issues. Traditional system-level architecture will become part of chip-level design. We can expect to see the integration of multiple processors, the cache memory hierarchy, the main memory controller (and possibly even the main memory), input/output devices, and network interface devices on one chip.

We can expect to see many new innovative microarchitecture techniques. As we approach and possibly exceed the 10-GHz clocking speed, we will need to rethink many of the fundamentals of microarchitecture design. A simple ALU operation may take multiple cycles. A sophisticated branch predictor can require up to 10 cycles. Main memory latency will be 1000+ cycles long. It may take tens of clock cycles to traverse the entire die. What we currently think of as very aggressive pipelining will be viewed as rather elementary.

Future microprocessors will become single-chip computing systems that will need to exploit various forms of parallelism. These systems will need to go beyond instruction-level parallelism to harvest thread-level parallelism (TLP) in the workload. Perhaps the most important will be the pursuit of memory-level parallelism (MLP) in being able to process many simultaneous memory accesses. As main memory latency becomes three orders of magnitude slower than the CPU cycle time, we will need to find clever ways of trading the memory bandwidth to mitigate the severe negative impact of long memory latency on overall performance. The main challenge will become the movement of data, not the operations performed on the data.

REFERENCES

Acosta, R., J. Kilestrup, and H. Torng: "An instruction issuing approach to enhancing performance in multiple functional unit processors," *IEEE Trans. on Computers,* C35, 9, 1986, pp. 815–828.

Agerwala, T., and J. Cocke: "High performance reduced instruction set processors," Technical report, IBM Computer Science, 1987.

Amdahl, G.: "Validity of the single processor approach to achieving large scale computing capabilities," *AFIPS Conf. Proc.,* 1967, pp. 483–485.

Amdahl, G., G. Blaauw, and F. P. Brooks, Jr.: "Architecture of the IBM System/360," *IBM Journal of Research and Development,* 8, 1964, pp. 87–101.

Bashteen, A., I. Lui, and J. Mullan: "A superpipeline approach to the MIPS architecture," *Proc. COMPCON Spring 91,* 1991, pp. 325–333.

Blaauw, G., and F. P. Brooks, Jr.: *Computer Architecture: Concepts and Evolution.* Reading, MA: Addison-Wesley, 1997.

Black, B., and J. P. Shen: "Calibration of microprocessor performance models," *Computer,* 31, 5, 1998, pp. 59–65.

Butler, M., T.-Y. Yeh, Y. Patt, M. Alsup, H. Scales, and M. Shebanow: "Instruction level parallelism is greater than two," *Proc. 18th Int. Symposium on Computer Architecture,* 1991, pp. 276–286.

Colwell, R., C. Hitchcock, E. Jensen, H. B. Sprunt, and C. Kollar: "Instructions sets and beyond: computers, complexity, and controversy," *IEEE Computer,* 18, 9, 1985, pp. 8–19.

Fisher, J.: "Trace scheduling: A technique for global microcode compaction. *IEEE Trans. on Computers,*" C-30, 7, 1981, pp. 478–490.

Fisher, J. A.: "Very long instruction word architectures and the ELI-512," Technical Report YLU 253, Yale University, 1983.

Flynn, M., and L. Hoevel: "Execution architecture: the DELtran experiment," *IEEE Trans. on Computers,* C-32, 2, 1983, pp. 156–175.

Johnson, M.: *Superscalar Microprocessor Design.* Englewood Cliffs, NJ: Prentice Hall, 1991.

Jouppi, N. P., and D. W. Wall: "Available instruction-level parallelism for superscalar and superpipelined machines," *Proc. Third Int. Conf. on Architectural Support for Programming Languages and Operating Systems (ASPLOS-III),* 1989, pp. 272–282.

Kuck, D., Y. Muraoka, and S. Chen: "On the number of operations simultaneously executable in Fortran-like programs and their resulting speedup," *IEEE Trans. on Computers,* C-21, 1972, pp. 1293–1310.

Melvin, S., and Y. Patt: "Exploiting fine-grained parallelism through a combination of hardware and software techniques," *Proc. 18th Int. Symposium on Computer Architecture,* 1991, pp. 287–296.

Melvin, S. W., and Y. Patt: "A clarification of the dynamic/static interface," *Proc. 20th Annual Hawaii Int. Conf. on System Sciences,* 1987, pp. 218–226.

Mirapuri, S., M. Woodacre, and N. Vasseghi: "The MIPS R4000 processor," *IEEE Micro,* 12, 2, 1992, pp. 10–22.

Nicolau, A., and J. Fisher: "Measuring the parallelism available for very long instruction word architectures," *IEEE Transactions on Computers,* C-33, 1984, pp. 968–976.

Riseman, E. M., and C. C. Foster: "The inhibition of potential parallelism by conditional jumps," *IEEE Transactions on Computers,* 1972, pp. 1405–1411.

Russell, R. M.: "The Cray-1 Computer System," *Communications of the ACM,* 21, 1, 1978, pp. 63–72.

Smith, M. D., M. Johnson, and M. A. Horowitz: "Limits on multiple instruction issue," *Proc. Third Int. Conf. on Architectural Support for Programming Languages and Operating Systems (ASPLOS-III),* 1989, pp. 290–302.

Sodani, A., and G. S. Sohi: "Dynamic instruction reuse," *Proc. 24th Annual Int. Symposium on Computer Architecture,* 1997, pp. 194–205.

Sohi, G., and S. Vajapeyam: "Instruction issue logic for high-performance, interruptible pipelined processors," *Proc. 14th Annual Int. Symposium on Computer Architecture,* 1987, pp. 27–34.

Stone, H.: *High-Performance Computer Architecture.* Reading, MA: Addison-Wesley, 1987.

Thornton, J. E.: "Parallel operation in the Control Data 6600," *AFIPS Proc. FJCC, part 2,* 26, 1964, pp. 33–40.

Tjaden, G., and M. Flynn: "Representation of concurrency with ordering matrices," *IEEE Trans. on Computers,* C-22, 8, 1973, pp. 752–761.

Tjaden, G. S., and M. J. Flynn: "Detection and parallel execution of independent instructions," *IEEE Transactions on Computers,* C19, 10, 1970, pp. 889–895.

Uht, A., and R. Wedig: "Hardware extraction of low-level concurrency from a serial instruction stream," *Proc. Int. Conf. on Parallel Processing,* 1986, pp. 729–736.

Wall, D.: "Limits of instruction-level parallelism," *Proc. 4th Int. Conf. on Architectural Support for Programming Languages and Operating Systems,* 1991, pp. 176–188.

Wedig, R.: Detection of Concurrency in Directly Executed Language Instruction Streams. PhD thesis, Stanford University, 1982.

Weiss, S., and J. Smith: "Instruction issue logic in pipelined supercomputers," *Proc. 11th Annual Symposium on Computer Architecture,* 1984, pp. 110–118.

HOMEWORK PROBLEMS

P1.1 Using the resources of the World Wide Web, list the top five reported benchmark results for SPECINT2000, SPECFP2000, and TPC-C.

P1.2 Graph SPECINT2000 vs. processor frequency for two different processor families (e.g., AMD Athlon and HP PA-RISC) for as many frequencies as are posted at www.spec.org. Comment on performance scaling with frequency, pointing out any anomalies and suggesting possible explanations for them.

P1.3 Explain the differences between architecture, implementation, and realization. Explain how each of these relates to processor performance as expressed in Equation (1.1).

P1.4 As silicon technology evolves, implementation constraints and tradeoffs change, which can affect the placement and definition of the dynamic-static interface (DSI). Explain why architecting a branch delay slot [as in the millions of instructions per second (MIPS) architecture] was a reasonable thing to do when that architecture was introduced, but is less attractive today.

P1.5 Many times, implementation issues for a particular generation end up determining tradeoffs in instruction set architecture. Discuss at least one historical implementation constraint that explains why CISC instruction sets were a sensible choice in the 1970s.

P1.6 A program's run time is determined by the product of instructions per program, cycles per instruction, and clock frequency. Assume the following instruction mix for a MIPS-like RISC instruction set: 15% stores, 25% loads, 15% branches, and 35% integer arithmetic, 5% integer shift, and 5% integer multiply. Given that load instructions require two cycles, branches require four cycles, integer ALU instructions require one cycle, and integer multiplies require ten cycles, compute the overall CPI.

P1.7 Given the parameters of Problem 6, consider a strength-reducing optimization that converts multiplies by a compile-time constant into a sequence of shifts and adds. For this instruction mix, 50% of the multiplies can be converted to shift-add sequences with an average length of three instructions. Assuming a fixed frequency, compute the change in instructions per program, cycles per instruction, and overall program speedup.

P1.8 Recent processors like the Pentium 4 processors do not implement single-cycle shifts. Given the scenario of Problem 7, assume that $s = 50\%$ of the additional instructions introduced by strength reduction are shifts, and shifts now take four cycles to execute. Recompute the cycles per instruction and overall program speedup. Is strength reduction still a good optimization?

P1.9 Given the assumptions of Problem 8, solve for the break-even ratio s (percentage of additional instructions that are shifts). That is, find the value of s (if any) for which program performance is identical to the baseline case without strength reduction (Problem 6).

P1.10 Given the assumptions of Problem 8, assume you are designing the shift unit on the Pentium 4 processor. You have concluded there are two possible implementation options for the shift unit: four-cycle shift latency at a frequency of 2 GHz, or two-cycle shift latency at 1.9 GHz. Assume the rest of the pipeline could run at 2 GHz, and hence the two-cycle shifter would set the entire processor's frequency to 1.9 GHz. Which option will provide better overall performance?

P1.11 Using Amdahl's law, compute speedups for a program that is 85% vectorizable for a system with 4, 8, 16, and 32 processors. What would be a reasonable number of processors to build into a system for running such an application?

P1.12 Using Amdahl's law, compute speedups for a program that is 98% vectorizable for a system with 16, 64, 256, and 1024 processors. What would be a reasonable number of processors to build into a system for running such an application?

P1.13 Replot the graph in Figure 1.8 on page 23 for each of the ILP limits shown in the list of studies in Section 1.4.2. What conclusions can you draw from the graphs you created?

P1.14 Compare and contrast these two ILP limit studies by reading the relevant papers and explaining why the limits are so different: Jouppi and Wall [1989] vs. Wall [1991].

P1.15 In 1995, the IBM AS/400 line of computers transitioned from a CISC instruction set to a RISC instruction set. Because of the simpler instruction set, the realizable clock frequency for a given technology generation and the CPI metric improved dramatically. However, for the same reason, the number of instructions per program also increased noticeably. Given the following parameters, compute the total performance improvement that occurred with this transition. Furthermore, compute the break-even clock frequency, break-even cycles per instruction, and break-even code expansion ratios for this transition, assuming the other two factors are held constant.

Performance Factor	AS/400 CISC (IMPI) (Actual)	AS/400 RISC (PowerPC) (Actual)	Actual Ratio	Break-even Ratio
Relative frequency	50 MHz	125 MHz	2.5	?
Cycles per instruction	7	3	0.43	?
Relative instructions per program (dynamic count)	1000	3300	3.3	?

P1.16 MIPS (millions of instructions per second) was commonly used to gauge computer system performance up until the 1980s. Explain why it can be a very poor measure of a processor's performance. Are there any circumstances under which it is a valid measure of performance? If so, describe those circumstances.

P1.17 MFLOPS (millions of floating-point operations per second) was commonly used to gauge computer system performance up until the 1980s. Explain why it can be a very poor measure of a processor's performance. Are there any circumstances under which it is a valid measure of performance? If so, describe those circumstances.

Terms and Buzzwords

These problems are similar to the "Jeopardy Game" on TV. The answers are shown and you are to provide the *best* correct questions. For each answer there may be more than one *appropriate* question; you need to provide the *best* one.

P1.18 A: Instruction-level parallelism within a basic block is typically upper bounded by 2.

Q: What is _____ ?

P1.19 A: It will significantly reduce the machine cycle time, but can increase the branch penalty.

Q: What is _____ ?

P1.20 A: Describes the speedup achievable when some fraction of the program execution is not parallelizable.

Q: What is _____ ?

P1.21 A: A widely used solution to Flynn's bottleneck.

Q: What is _____ ?

P1.22 A: The best way to describe a computer system's performance.

Q: What is _____ ?

P1.23 A: This specifies the number of registers, available addressing modes, and instruction opcodes.

Q: What is _____ ?

P1.24 A: This determines a processor's configuration and number of functional units.

Q: What is _____ ?

P1.25 A: This is a type of processor that relies heavily on the compiler to statically schedule independent instructions.

Q: What is _____ ?

P1.26 A: This is a type of processor where results of instructions are not available until two or more cycles after the instruction begins execution.

Q: What is _____ ?

P1.27 A: This is a type of processor that attempts to execute more than one instruction at the same time.

Q: What is _____ ?

P1.28 A: This important study showed that instruction-level parallelism was abundant, if only control dependences could somehow be overcome.

Q: What is _____ ?

P1.29 A: This is a type of processor that executes high-level languages without the aid of a compiler.

Q: What is _____ ?

P1.30 A: This approach to processor simulation requires substantial storage space.

Q: What is _____ ?

CHAPTER 2

Pipelined Processors

CHAPTER OUTLINE

2.1 Pipelining Fundamentals
2.2 Pipelined Processor Design
2.3 Deeply Pipelined Processors
2.4 Summary

References
Homework Problems

Pipelining is a powerful implementation technique for enhancing system throughput without requiring massive replication of hardware. It was first employed in the early 1960s in the design of high-end mainframes. Instruction pipelining was first introduced in the IBM 7030, nicknamed the Stretch computer [Bloch, 1959, Bucholtz, 1962]. Later the CDC 6600 incorporated both pipelining and the use of multiple functional units [Thornton, 1964].

During the 1980s, pipelining became the cornerstone of the RISC approach to processor design. Most of the techniques that constituted the RISC approach are directly or indirectly related to the objective of efficient pipelining. Since then, pipelining has been effectively applied to CISC processors as well. The Intel i486 was the first pipelined implementation of the IA32 architecture [Crawford, 1990]. Pipelined versions of Digital's VAX and Motorola's M68K architectures were also quite successful commercially.

Pipelining is a technique that is now widely employed in the design of instruction set processors. This chapter focuses on the design of (scalar) pipelined processors. Many of the approaches and techniques related to the design of pipelined processors, such as pipeline interlock mechanisms for hazard detection and resolution, are foundational to the design of superscalar processors.

The current trend is toward very deep pipelines. Pipeline depth has increased from less than 10 to more than 20. Deep pipelines are necessary for achieving very high clock frequencies. This has been a very effective means of gaining greater processor performance. There are some indications that this trend will continue.

2.1 Pipelining Fundamentals

This section presents the motivations and the fundamental principles of pipelining. Historically there are two major types of pipelines: *arithmetic pipelines* and *instruction pipelines*. While instruction pipelines are the focus of this book, we begin by examining an arithmetic pipeline example. Arithmetic pipelines more readily illustrate a set of idealized assumptions underlying the principles of pipelined designs. We term these idealized assumptions the *pipelining idealism*. It is dealing with the discrepancy between these idealized assumptions and realistic considerations in instruction pipelining that makes pipelined processor design so interesting.

2.1.1 Pipelined Design

This subsection introduces the foundational notions of pipelined design. The motivations and limitations of pipelining are presented. A theoretical model, proposed by Peter Kogge, of optimal pipelining from the hardware design perspective is described [Kogge, 1981].

2.1.1.1 Motivations. The primary motivation for pipelining is to increase the throughput of a system with little increase in hardware. The *throughput*, or bandwidth, of a system is measured in terms of the number of tasks performed per unit time, and it characterizes the performance of the system. For a system that operates on one task at a time, the throughput P is equal to $1/D$, where D is the latency of a task or the delay associated with the performance of a task by the system. The throughput of a system can be increased by pipelining if there are many tasks that require the use of the same system. The actual latency for each task still remains the same or may even increase slightly.

Pipelining involves partitioning the system into multiple stages with added buffering between the stages. These stages and the interstage buffers constitute the *pipeline*. The computation carried out by the original system is decomposed into k subcomputations, carried out in the k stages of the pipeline. A new task can start into the pipeline as soon as the previous task has traversed the first stage. Hence, instead of initiating a new task every D units of time, a new task can be initiated every D/k units of time, where k is the number of stages in the pipeline, and the processing of k computations is now overlapped in the pipeline. It is assumed that the original latency of D has been evenly partitioned into k stages and that no additional delay is introduced by the added buffers. Given that the total number of tasks to be processed is very large, the throughput of a pipelined system can potentially approach k times that of a nonpipelined system. This potential performance increase by a factor of k by simply adding new buffers in a k-stage pipeline is the primary attraction of the pipelined design. Figure 2.1 illustrates the potential k-fold increase of throughput in a k-stage pipelined system.

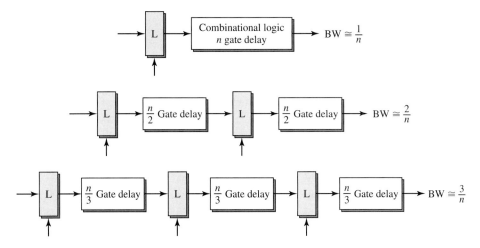

Figure 2.1
Potential k-Fold Increase of Throughput in a k-Stage Pipelined System.

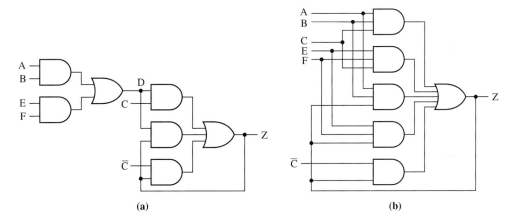

Figure 2.2
The Earle Latch and Its Incorporation into Logic Without Incurring Additional Gate Delay: (a) Earle Latch Following the Combinational Logic; (b) Earle Latch Integrated with the Combinational Logic.

So far we have assumed that the addition of interstage buffers does not introduce any additional delay. This is not unrealistic. The Earle latch shown in Figure 2.2(a) was designed and used in the IBM 360/91 for buffering between stages of carry-save adders in the pipelined multiply unit. In the Earle latch, the output Z follows the input D when clock C = 1. When the clock goes low, the value at D is latched at Z through the latching loop, and then the output Z becomes insensitive to further changes at D. Proper hold time is required on the D input to ensure proper latching. The middle AND gate ensures glitch-free operation; the product term represented by this AND gate "covers" a potential hazard. A hazard is a spurious pulse caused by a race condition involving simultaneous change of multiple signals. The top and bottom inputs to the OR gate can potentially change simultaneously in opposite directions. Under such a condition, if the OR gate does not have the middle (redundant) input, a spurious pulse (the hazard) can potentially appear at the output of the OR gate. The Earle latch has this desirable glitch-free operation feature. Furthermore, the Earle latch can be integrated into the logic function so as not to incur any additional gate delay. Figure 2.2(b) illustrates how the latching function can be merged into the last two AND-OR levels of the combinational logic circuit resulting in no additional gate delay for the addition of the latch. The circuit in Figure 2.2(b) performs the same logic function as that of Figure 2.2(a) without incurring two additional gate delays for latching. The increase of gate fan-in by one can slightly increase the delay through these gates.

2.1.1.2 Limitations. Since the performance gained in a pipelined design is proportional to the depth, that is, the number of stages, of a pipeline, it might seem that the best design is always to maximize the number of stages of a pipelined system. However, due to clocking constraints, there are physical limitations to how finely an original computation can be partitioned into pipeline stages.

Each stage of a pipeline can be viewed as a piece of combinational logic F followed by a set of latches L. Signals must propagate through F and be latched at L. Let T_M be the maximum propagation delay through F, that is, the delay through the longest signal path; let T_m be the minimum propagation delay through F, that is, the delay through the shortest signal path. Let T_L be the additional time needed for proper clocking. Delay T_L can include the necessary setup and hold times to ensure proper latching, as well as the potential clock skews, that is, the worst-case disparity between the arrival times of the clock edge at different latches. If the first set of signals X_1 is applied at the inputs to the stage at time T_1, then the outputs of F must be valid at $T_1 + T_M$. For proper latching at L, the signals at the outputs of F must continue to be valid until $T_1 + T_M + T_L$. When the second set of signals X_2 is applied at the inputs to F at time T_2, it takes at least until $T_2 + T_m$ for the effects to be felt at the latches L. To ensure that the second set of signals does not overrun the first set, it is required that

$$T_2 + T_m > T_1 + T_M + T_L \qquad (2.1)$$

which means that the earliest possible arrival of X_2 at the latches must not be sooner than the time required for the proper latching of X_1. This inequality can be rewritten as

$$T_2 - T_1 > T_M - T_m + T_L \qquad (2.2)$$

where $T_2 - T_1$ is effectively the minimum clocking period T. Therefore, the clocking period T must be greater than $T_M - T_m + T_L$, and the maximum clocking rate cannot exceed $1/T$.

Based on the foregoing analysis, two factors limit the clocking rate. One is the difference between the maximum and minimum propagation delays through the logic, namely, $T_M - T_m$. The other is the additional time required for proper clocking, namely, T_L. The first factor can be eliminated if all signal propagation paths are of the same length. This can be accomplished by padding the short paths. Hence, $T_M - T_m$ is close to zero. The second factor is dictated by the need to latch the results of the pipeline stages. Proper latching requires the propagation of a signal through a feedback loop and the stabilizing of that signal value in the loop. Another contribution to T_L is the worst-case clock skew. The clock signal may arrive at different latches at slightly different times due to the generation and distribution of the clock signals to all the latches. In a fully synchronous system, this worst-case clock skew must be accounted for in the clocking period. Ultimately, the limit of how deeply a synchronous system can be pipelined is determined by the minimum time required for latching and the uncertainty associated with the delays in the clock distribution network.

2.1.1.3 Tradeoff. Clocking constraints determine the ultimate physical limit to the depth of pipelining. Aside from this limit, maximum pipeline depth may not be the optimal design when cost, or pipelining overhead, is considered. In the hardware design of a pipelined system, the tradeoff between cost and performance must be considered. A cost/performance tradeoff model for pipelined design has been proposed by Peter Kogge and is summarized here [Kogge, 1981]. Models for both cost and performance are proposed. The cost of a nonpipelined design is denoted as G. This cost can be in terms of gate count, transistor count, or silicon real estate. The cost C for a k-stage pipelined design is equal to

$$C = G + k \times L \qquad (2.3)$$

where k is the number of stages in the pipeline, L is the cost of adding each latch, and G is the cost of the original nonpipelined hardware. Based on this cost model, the pipeline cost C is a linear function of k, the depth of the pipeline. Basically, the cost of a pipeline goes up linearly with respect to the depth of the pipeline.

Assume that the latency in the nonpipelined system is T. Then the performance of the nonpipelined design is $1/T$, the computation rate. The performance P of the pipelined design can be modeled as $1/(T/k + S)$, where T is the latency of the original nonpipelined design and S is the delay due to the addition of the latch. Assuming that the original latency T can be evenly divided into k stages, $(T/k + S)$ is the delay associated with each stage and is thus the clocking period of the pipeline. Consequently, $1/(T/k + S)$ is equal to the clocking rate and the throughput of the pipelined design. Hence, the performance of the pipelined design is

$$P = \frac{1}{(T/k + S)} \qquad (2.4)$$

Note that P is a nonlinear function of k.

Given these models for cost and performance, the expression for the cost/performance ratio is

$$\frac{C}{P} = \frac{G + k \times L}{\frac{1}{(T/k + S)}} \qquad (2.5)$$

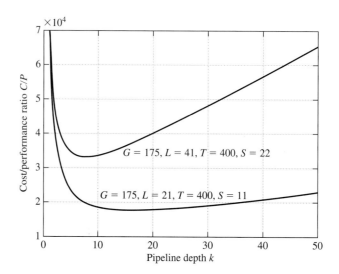

Figure 2.3
Cost/Performance Tradeoff Model for Pipelined Designs.

This expression can be rewritten as

$$\frac{C}{P} = LT + GS + LSk + \frac{GT}{k} \qquad (2.6)$$

which is plotted in Figure 2.3 for two sets of sample values of G, L, T, and S.

Equation (2.6) expresses the cost/performance ratio as a function of k. The first derivative can be taken and set equal to zero to determine the value of k that will produce the minimal cost/performance ratio. This value of k, shown in Equation (2.7), is the optimal pipelining depth in terms of the other parameters.

$$k_{opt} = \sqrt{\frac{GT}{LS}} \qquad (2.7)$$

Given this expression for the optimal value of k, a pipelined design with $k < k_{opt}$ can be considered as underpipelined in that further pipelining or increasing the pipeline depth is beneficial and the increased cost is justified by the increase of performance. On the other hand, $k > k_{opt}$ indicates an overpipelined design in which there is a diminishing return of performance for the increased cost of pipelining. The foregoing tradeoff model is based purely on hardware design considerations; there is no consideration of the dynamic behavior of the pipeline or the computations being performed. We will take up these issues later, beginning in Section 2.2.

2.1.2 Arithmetic Pipeline Example

There are two major types of pipelines: arithmetic pipelines and instruction pipelines. Although instruction pipeline design is the focus of this chapter, we will begin by looking at an arithmetic pipeline example. Arithmetic pipelines clearly illustrate the effectiveness of pipelining without having to deal with some of the complex issues involved in instruction pipeline design. These complex issues will be addressed in subsequent sections of this chapter.

2.1.2.1 Floating-Point Multiplication. The design of a pipelined floating-point multiplier is used as the example. This "vintage" board-level design is taken from a classic text by Shlomo Waser and Mike Flynn [Waser and Flynn, 1982]. (Even though this design assumes 1980 technology, nonetheless it still serves as an effective vehicle to illustrate arithmetic pipelining.) This design assumes a 64-bit floating-point format that uses the excess-128 notation for the exponent e (8 bits) and the sign-magnitude fraction notation with the hidden bit for the mantissa m (57 bits, including the hidden bit).

The floating-point multiplication algorithm implemented in this design is as follows.

1. Check to see if any operand is zero. If it is, the result is immediately set to zero.

2. Add the two characteristics (physical bit patterns of the exponents) and correct for the excess-128 bias, that is, $e_1 + (e_2 - 128)$.

3. Perform fixed-point multiplication of the two mantissas m_1 and m_2.

4. Normalize the product of the mantissas, which involves shifting left by one bit and decrementing the exponent by 1. (The normalized representation of the mantissa has no leading zeros.)

5. Round the result by adding 1 to the first guard bit (the bit immediately to the right of the least-significant bit of the mantissa). This is effectively rounding up. If the mantissa overflows, then the mantissa must be shifted right one bit and the exponent incremented by 1 to maintain the normalized representation for the mantissa.

Figure 2.4 illustrates in the functional block diagram the nonpipelined design of the floating-point multiplier. The input latches store the two operands to be multiplied. At the next clock the product of the two operands will be stored in the output latches.

The fixed-point mantissa multiplier represents the most complex module in this design and consists of three submodules for partial product generation, partial product reduction, and final reduction. The hardware complexity, in terms of the number of integrated circuit (IC) chips, and the propagation delay, in nanoseconds, of each submodule can be obtained.

- *Partial product generation.* Simultaneous generation of the partial products can be performed using 8×8 hardware multipliers. To generate all the partial products, 34 such 8×8 multipliers are needed. The delay involved is 125 ns.

- *Partial product reduction.* Once all the partial products are generated, they must be reduced or summed. A summing circuit called the (5, 5, 4) counter can be used to reduce two columns of 5 bits each into a 4-bit sum. A (5, 5, 4) counter can be implemented using a $1 \text{K} \times 4$ read-only memory (ROM) with

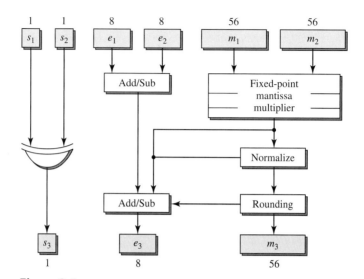

Figure 2.4
A Nonpipelined Floating-Point Multiplier. Waser and Flynn, 1982.

a delay of 50 ns. Three levels of (5, 5, 4) counters are needed to reduce all the partial products. Hence a total of 72 such 1K × 4 ROMs are needed, incurring a total delay of 150 ns.

- *Final reduction.* Once all the partial products have been reduced down to two partial products a final level of reduction can be implemented using fast carry-lookahead (CLA) adders to produce the final result. Sixteen 4-bit adder chips with CLA plus five 4-bit CLA units are needed for this final reduction step. A total of 21 IC chips and a 55-ns delay are required.

Two additional modules are needed for the mantissa section, namely, a shifter for performing normalization (2 chips, 20-ns delay) and an incrementer for performing rounding (15 chips, 50-ns delay). The Add/Sub modules in the exponent section require another 4 chips; their delays are unimportant because they are not in the critical delay path. An additional 17 and 10 chips are needed for implementing the input and output latches, respectively. The total chip counts and critical delays of the modules in the nonpipelined design are summarized in Table 2.1.

Based on the tabulation in Table 2.1, the nonpipelined design of the floating-point multiplier requires 175 chips and can be clocked at 2.5 MHz with a clock period of 400 ns. This implies that the nonpipelined design can achieve a throughput of 2.5 MFLOPS (million floating-point operations per second).

2.1.2.2 Pipelined Floating-Point Multiplier. The nonpipelined design of the floating-point multiplier can be pipelined to increase its throughput. In this example, we will assume that there is no pipelining within a submodule; that is, the finest granularity for partitioning into pipeline stages is at the submodule level. We now examine the delays associated with each of the (sub)modules in the critical delay path. These delays are shown in the third column of Table 2.1. The partial product

Table 2.1
Chip counts and critical delays of the modules in the nonpipelined floating-point multiplier design.

Module	Chip Count	Delay, ns
Partial product generation	34	125
Partial product reduction	72	150
Final reduction	21	55
Normalization	2	20
Rounding	15	50
Exponent section	4	
Input latches	17	
Output latches	10	
Total	175	400

Source: Waser and Flynn, 1982.

reduction submodule has the longest delay, 150 ns; this delay then determines the delay of a stage in the pipeline. The five (sub)modules in the critical path can be partitioned into three fairly even stages with delays of 125 ns (partial product generation), 150 ns (partial product reduction), and 125 ns (final reduction, normalization, and rounding). The resultant three-stage pipelined design is shown in Figure 2.5.

In determining the actual clocking rate of the pipelined design, we must consider clocking requirements. Assuming that edge-triggered registers are used for buffering between pipeline stages, we must add the clock-edge-to-register-output delay of 17 ns and the setup time of 5 ns to the stage delay of 150 ns. This results in the minimum clocking period of 172 ns. Therefore, instead of clocking at the rate of 2.5 MHz, the new pipelined design can be clocked at the rate of 5.8 MHz. This represents a factor of 2.3 increase in throughput. Note, however, that the latency for performing each multiplication has increased slightly, from 400 to 516 ns.

The only additional hardware required for the pipelined design is the edge-triggered register chips for buffering between pipeline stages. On top of the original 175 IC chips, an additional 82 IC chips are required. Using chip count as a measure of hardware complexity, the total of 257 IC chips represents an increase of 45% in terms of hardware complexity. This 45% increase in hardware cost resulted in a 130% increase in performance. Clearly, this three-stage pipelined design of the floating-point multiplier is a win over the original nonpipelined design.

This example assumes board-level implementations using off-the-shelf parts. Given today's chip technology, this entire design can be easily implemented as a small module on a chip. While a board-level implementation of the floating-point multiplier may be viewed as outdated, the purpose of this example is to succinctly illustrate the effectiveness of pipelining using a published specific design with actual latency and hardware cost parameters. In fact, the upper curve in Figure 2.3 reflects the parameters from this example.

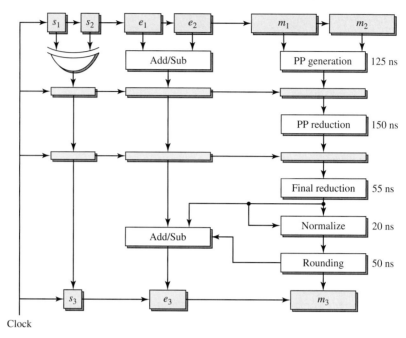

Figure 2.5
A Pipelined Floating-Point Multiplier.
Source: Waser and Flynn, 1982.

2.1.3 Pipelining Idealism

Recall that the motivation for a *k*-stage pipelined design is to achieve a *k*-fold increase in throughput, as illustrated in Figure 2.1. However, in the foregoing example, the three-stage pipelined floating-point multiplier only achieved a factor of 2.3 increase in throughput. The main reason for falling short of the three-fold increase of throughput is that the *k*-fold increase in throughput for a *k*-stage pipelined design represents the ideal case and is based on three idealized assumptions, which we referred to as the *pipelining idealism*. The understanding of pipelining idealism is crucial to the appreciation of pipelined designs. The unavoidable deviations from this idealism in real pipelines make pipelined designs challenging. The solutions for dealing with this idealism-realism gap comprise the interesting techniques for pipelined designs. The three points of pipelining idealism are

1. *Uniform subcomputations.* The computation to be performed can be evenly partitioned into uniform-latency subcomputations.

2. *Identical computations.* The same computation is to be performed repeatedly on a large number of input data sets.

3. *Independent computations.* All the repetitions of the same computation are mutually independent.

2.1.3.1 Uniform Subcomputations. The first point of pipelining idealism states that the computation to be pipelined can be evenly partitioned into k uniform-latency subcomputations. This means that the original design can be evenly partitioned into k balanced (i.e., having the same latency) pipeline stages. If the latency of the original computation, and hence the clocking period of the nonpipelined design, is T, then the clocking period of a k-stage pipelined design is exactly T/k, which is the latency of each of the k stages. Given this idealized assumption, the k-fold increase in throughput is achieved due to the k-fold increase of the clocking rate.

This idealized assumption may not be true in an actual pipelined design. It may not be possible to partition the computation into perfectly balanced stages. We see in our floating-point multiplier example that the latency of 400 ns of the original computation is partitioned into three stages with latencies of 125, 150, and 125 ns, respectively. Clearly the original latency has not been evenly partitioned into three balanced stages. Since the clocking period of a pipelined design is dictated by the stage with the longest latency, the stages with shorter latencies in effect will incur some inefficiency or penalty. In our example, the first and third stages have an inefficiency of 25 ns each; we called such inefficiency within pipeline stages, the *internal fragmentation* of pipeline stages. Because of such internal fragmentation, the total latency required for performing the same computation will increase from T to T_f, and the clocking period of the pipelined design will be no longer T/k but T_f/k. In our example the performance of the three subcomputations will require 450 ns instead of the original 400 ns, and the clocking period will be not 133 ns (400/3 ns) but 150 ns.

There is a secondary implicit assumption, namely, that no additional delay is introduced by the introduction of buffers between pipeline stages and that no additional delay is required for ensuring proper clocking of the pipeline stages. Again, this assumption may not be true in actual designs. In our example, an additional 22 ns is required to ensure proper clocking of the pipeline stages, which resulted in the cycle time of 172 ns for the three-stage pipelined design. The ideal cycle time for a three-stage pipelined design would have been 133 ns. The difference between 172 and 133 ns for the clocking period accounts for the shortfall from the idealized three-fold increase of throughput.

The first point of pipelining idealism basically assumes two things: (1) There is no inefficiency introduced due to the partitioning of the original computation into multiple subcomputations; and (2) there is no additional delay caused by the introduction of the interstage buffers and the clocking requirements. In chip-level design the additional delay incurred for proper pipeline clocking can be minimized by employing latches similar to the Earle latch. The partitioning of a computation into balanced pipeline stages constitutes the first challenge of pipelined design. The goal is to achieve stages as balanced as possible to minimize internal fragmentation. Internal fragmentation due to imperfectly balanced pipeline stages is the primary cause of deviation from the first point of pipelining idealism. This deviation becomes a form of pipelining overhead and leads to the shortfall from the idealized k-fold increase of throughput in a k-stage pipelined design.

2.1.3.2 Identical Computations.
The second point of pipelining idealism states that many repetitions of the same computation are to be performed by the pipeline. The same computation is repeated on multiple sets of input data; each repetition requires the same sequence of subcomputations provided by the pipeline stages. For our floating-point multiplier example, this means that many pairs of floating-point numbers are to be multiplied and that each pair of operands is sent through the same three pipeline stages. Basically this assumption implies that all the pipeline stages are used by every repetition of the computation. This is certainly true for our example.

This assumption holds for the floating-point multiplier example because this pipeline performs only one function, that is, floating-point multiplication. If a pipeline is designed to perform multiple functions, this assumption may not hold. For example, an arithmetic pipeline can be designed to perform both addition and multiplication. In a multiple-function pipeline, not all the pipeline stages may be required by each of the functions supported by the pipeline. It is possible that a different subset of pipeline stages is required for performing each of the functions and that each computation may not require all the pipeline stages. Since the sequence of data sets traverses the pipeline in a synchronous manner, some data sets will not require some pipeline stages and effectively will be idling during those stages. These unused or idling pipeline stages introduce another form of pipeline inefficiency that can be called *external fragmentation* of pipeline stages. Similar to internal fragmentation, external fragmentation is a form of pipelining overhead and should be minimized in multifunction pipelines. For the pipelined floating-point multiplier example, there is no external fragmentation.

The second point of pipelining idealism effectively assumes that all pipeline stages are always utilized. Aside from the implication of having no external fragmentation, this idealized assumption also implies that there are many sets of data to be processed. It takes k cycles for the first data set to reach the last stage of the pipeline; these cycles are referred to as the pipeline *fill time*. After the last data set has entered the first pipeline stage, an additional k cycles are needed to drain the pipeline. During pipeline fill and drain times, not all the stages will be busy. The main reason for assuming the processing of many sets of input data is that the pipeline fill and drain times constitute a very small fraction of the total time. Hence, the pipeline stages can be considered, for all practical purposes, to be always busy. In fact, the throughput of 5.8 MFLOPS for the pipelined floating-point multiplier is based on this assumption.

2.1.3.3 Independent Computations.
The third point of pipelining idealism states that the repetitions of computation, or simply computations, to be processed by the pipeline are independent. This means that all the computations that are concurrently resident in the pipeline stages are independent, that is, have no data or control dependences between any pair of the computations. This assumption permits the pipeline to operate in "streaming" mode, in that a later computation

need not wait for the completion of an earlier computation due to a dependence between them. For our pipelined floating-point multiplier this assumption holds. If there are multiple pairs of operands to be multiplied, the multiplication of a pair of operands does not depend on the result from another multiplication. These pairs can be processed by the pipeline in streaming mode.

For some pipelines this point may not hold. A later computation may require the result of an earlier computation. Both of these computations can be concurrently resident in the pipeline stages. If the later computation has entered the pipeline stage that needs the result while the earlier computation has not reached the pipeline stage that produces the needed result, the later computation must wait in that pipeline stage. This waiting is referred to as a *pipeline stall*. If a computation is stalled in a pipeline stage, all subsequent computations may have to be stalled as well. Pipeline stalls effectively introduce idling pipeline stages, and this is essentially a dynamic form of external fragmentation and results in the reduction of pipeline throughput. In designing pipelines that need to process computations that are not necessarily independent, the goal is to produce a pipeline design that minimizes the amount of pipeline stalls.

2.1.4 Instruction Pipelining

The three points of pipelining idealism are three idealized assumptions about pipelined designs. For the most part, in arithmetic pipelines the reality is not far from these idealized assumptions. However, for instruction pipelining the gap between realism and idealism is greater. It is the bridging of this gap that makes instruction pipelining interesting and challenging. In designing pipelined processors, these three points become the three major challenges. These three challenges are now briefly introduced and will be addressed in depth in Section 2.2 on pipelined processor design. These three challenges also provide a nice road map for keeping track of all the pipelined processor design techniques.

2.1.4.1 Instruction Pipeline Design.

The three points of pipelining idealism become the objectives, or desired goals, for designing instruction pipelines. The processing of an instruction becomes the computation to be pipelined. This computation must be partitioned into a sequence of fairly uniform subcomputations that will result in fairly balanced pipeline stages. The latency for processing an instruction is referred to as the *instruction cycle;* the latency of each pipeline stage determines the *machine cycle*. The instruction cycle can be viewed as a logical concept that specifies the processing of an instruction. The execution of a program with many instructions involves the repeated execution of this computation. The machine cycle is a physical concept that involves the clocking of storage elements in digital logic circuits, and it is essentially the clocking period of the pipeline stages.

We can view the earlier floating-point multiplier as an example of a very simple processor with only one instruction, namely, floating-point multiply. The instruction

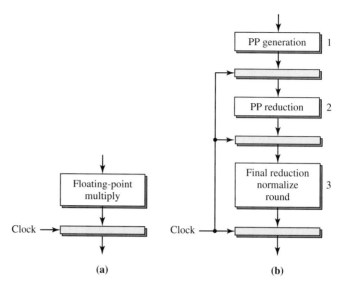

Figure 2.6
A Simple Illustration of Instruction Cycle vs. Machine Cycle.

cycle involves the performance of a floating-point multiply; see Figure 2.6(a). This computation can be naturally partitioned, based on obvious functional unit boundaries, into the following five subcomputations.

1. Partial product generation (125 ns).
2. Partial product reduction (150 ns).
3. Final reduction (55 ns).
4. Normalization (20 ns).
5. Rounding (50 ns).

For the purpose of pipelining, we had grouped the last three subcomputations into one subcomputation. This resulted in the three pipeline stages shown in Figure 2.6(b). The instruction cycle of Figure 2.6(a) has been mapped into the three machine cycles of Figure 2.6(b), resulting in a three-stage pipelined design. We can refer to the instruction cycle as an *architected* (logical) primitive which is specified in the instruction set architecture, whereas the machine cycle is a *machine* (physical) primitive and is specified in the microarchitecture. The pipelined design of Figure 2.6(b) is an *implementation* of the *architecture* specified in Figure 2.6(a).

A main task of instruction pipelining can be stated as the mapping of the logical instruction cycle to the physical machine cycles. In other words, the computation represented by the instruction cycle must be partitioned into a sequence of subcomputations to be carried out by the pipeline stages. To perform this mapping or partitioning effectively, the three points of pipelining idealism must be considered.

Uniform Subcomputations. The partitioning of the instruction cycle to multiple machine cycles can be called *stage quantization,* and it should be performed to minimize internal fragmentation of the pipeline stages. If care is not taken in stage quantization, the internal fragmentation introduced can quickly undermine the efficiency of the pipeline. This first point of pipelining idealism leads to the first challenge of instruction pipelining, namely, the need to balance the pipeline stages. The more balanced the pipeline stages are, the less will be the internal fragmentation.

Identical Computations. Unlike a single-function arithmetic pipeline, an instruction pipeline is inherently a multifunction pipeline, in that it must be able to process different instruction types. Different instruction types will require slightly different sequences of subcomputations and consequently different hardware resources. The second challenge of instruction pipelining involves the efficient coalescing or unifying of the different resource requirements of different instruction types. The pipeline must be able to support the processing of all instruction types, while minimizing unused or idling pipeline stages for each instruction type. This essentially is equivalent to minimizing the external fragmentation.

Independent Computations. Again, unlike an arithmetic pipeline that processes array data, an instruction pipeline processes instructions that are not necessarily independent of one another. Hence, the instruction pipeline must have built-in mechanisms to detect the occurrences of dependences between instructions and to ensure that such dependences are not violated. The enforcing of interinstruction dependences may incur penalties in the form of pipeline stalls. Recall that pipeline stalls are a dynamic form of external fragmentation which reduces the throughput of the pipeline. Therefore, the third challenge of instruction pipelining is the minimizing of pipeline stalls.

2.1.4.2 Instruction Set Architecture Impacts. Before we address the three major challenges of instruction pipelining in earnest, it might be enlightening to briefly consider the impacts that instruction set architectures (ISAs) can have on instruction pipelining. Again, the three points of pipelining idealism are considered in turn.

Uniform Subcomputations. The first challenge of balancing the pipeline stages implies that a set of uniform subcomputations must be identified. Looking at all the subcomputations that are involved in the processing of an instruction, one must identify the one critical subcomputation that requires the longest latency and cannot be easily further partitioned into multiple finer subcomputations. In pipelined processor design, one such critical subcomputation is the accessing of main memory. Because of the disparity of speed between the processor and main memory, memory accessing can be the critical subcomputation. To support more efficient instruction pipelining, addressing modes that involve memory access should be minimized, and fast cache memories that can keep up with the processor speed should be employed.

Identical Computations. The second challenge of unifying the resource requirements of different instruction types is one of the primary motivations for the RISC architectures. By reducing the complexity and diversity of the different instruction types, the task of unifying different instruction types is made easier. Complex addressing modes not only require additional accesses to memory, but also increase the diversity of resource requirements. To unify all these resource requirements into one instruction pipeline is extremely difficult, and the resultant pipeline can become very inefficient for many of the instructions with less complex resource requirements. These instructions would have to pay the external fragmentation overhead in that they underutilize the stages in the pipeline. The unifying of instruction types for a pipelined implementation of a RISC architecture is clean and results in an efficient instruction pipeline with little external fragmentation.

Independent Computations. The third challenge of minimizing pipeline stalls due to interinstruction dependences is probably the most fascinating area of pipelined processor design. For proper operation, an instruction pipeline must detect and enforce interinstruction dependences. Complex addressing modes, especially those that involve memory accessing, can make dependence detection very difficult due to the memory reference specifiers. In general, register dependences are easier to check because registers are explicitly specified in the instruction. Clean and symmetric instruction formats can facilitate the decoding of the instructions and the detection of dependences. Both the detection and the enforcement of dependences can be done either statically at compile time or dynamically at run time. The decision of what to do at compile time vs. run time involves the definition of the dynamic-static interface (DSI). The placement of the DSI induces interesting and subtle tradeoffs. These tradeoffs highlight the intimate relationship between compilers and (micro)architectures and the importance of considering both in the design of processors.

2.2 Pipelined Processor Design

In designing instruction pipelines or pipelined processors, the three points of pipelining idealism manifest as the three primary design challenges. Dealing with these deviations from the idealized assumptions becomes the primary task in designing pipelined processors. The three points of pipelining idealism and the corresponding three primary challenges for pipelined processor design are as follows:

1. Uniform subcomputations $\Rightarrow$ balancing pipeline stages
2. Identical computations $\Rightarrow$ unifying instruction types
3. Independent computations $\Rightarrow$ minimizing pipeline stalls

These three challenges are addressed in turn in Subsections 2.2.1 to 2.2.3. These three challenges provide a nice framework for presenting instruction pipelining techniques. All pipelined processor design techniques can be viewed as efforts in addressing these three challenges.

2.2.1 Balancing Pipeline Stages

In pipelined processor design, the computation to be pipelined is the work to be done in each instruction cycle. A typical instruction cycle can be functionally partitioned into the following five generic subcomputations.

1. Instruction fetch (IF)
2. Instruction decode (ID)
3. Operand(s) fetch (OF)
4. Instruction execution (EX)
5. Operand store (OS)

A typical instruction cycle begins with the fetching of the next instruction to be executed, which is followed by the decoding of the instruction to determine the work to be performed by this instruction. Usually one or more operands are specified and need to be fetched. These operands can reside in the registers or in memory locations depending on the addressing modes used. Once the necessary operands are available, the actual operation specified by the instruction is performed. The instruction cycle ends with the storing of the result produced by the specified operation. The result can be stored in a register or in a memory location, again depending on the addressing mode specified. In a sequential processor, this entire sequence of subcomputations is then repeated for the next instruction. During these five generic subcomputations some side effects can also occur as part of the execution of this instruction. Usually these side effects take the form of certain modifications to the machine state. These changes to the machine state are referred to as *side effects* because these effects are not necessarily explicitly specified in the instruction. The implementation complexity and resultant latency for each of the five generic subcomputations can vary significantly depending on the actual ISA specified.

2.2.1.1 Stage Quantization. One natural partitioning of the instruction cycle for pipelining is based on the five generic subcomputations. Each of the five generic subcomputations is mapped to a pipeline stage, resulting in a five-stage instruction pipeline; see Figure 2.7. We called this example pipeline the GENERIC (GNR) instruction pipeline. In the GNR pipeline, the logical instruction cycle has been mapped into five physical machine cycles. The machine cycles/instruction cycle ratio of 5 reflects the degree of pipelining and gives some indication of the granularity of the pipeline stages.

The objective of stage quantization is to partition the instruction cycle into balanced pipeline stages so as to minimize internal fragmentation in the pipeline stages. Stage quantization can begin with the natural functional partition of the instruction cycle, for example, the five generic subcomputations. Multiple subcomputations with short latencies can be grouped into one new subcomputation to achieve more balanced stages. For example, the three subcomputations—final reduction, normalization, and rounding—of the floating-point multiplication computation are grouped

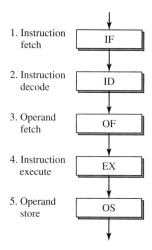

Figure 2.7
The Five-Stage GENERIC (GNR) Instruction Pipeline.

into one subcomputation in the pipelined design of Figure 2.6(b). Similarly, some of the five generic subcomputations of a typical instruction cycle can be grouped to achieve more balanced stages. For example, if an instruction set architecture employs fixed instruction length, simple addressing modes, and orthogonal fields in the instruction format, then both the IF and ID subcomputations should be quite straightforward and relatively simple compared to the other three subcomputations. These two subcomputations can potentially be combined into one new subcomputation, resulting in four subcomputations that are more balanced in terms of their required latencies. Based on these four subcomputations a four-stage instruction pipeline can be implemented; see Figure 2.8(a). In fact, the combining of the IF and ID subcomputations is employed in the MIPS R2000/R3000 pipelined processors [Moussouris et al., 1986, Kane, 1987]. This approach essentially uses the subcomputation with the longest latency as a reference and attempts to group other subcomputations with shorter latencies into a new subcomputation with comparable latency as the reference. This will result in a coarser-grained machine cycle and a lower degree of pipelining.

Instead of combining subcomputations with short latencies, an opposite approach can be taken to balance the pipeline stages. A given subcomputation with extra-long latency can be further partitioned into multiple subcomputations of shorter latencies. This approach uses the subcomputation with the shortest latency as the reference and attempts to subdivide long-latency subcomputations into many finer-grained subcomputations with latencies comparable to the reference. This will result in a finer-grained machine cycle and a higher degree of pipelining. For example, if an ISA employs complex addressing modes that may involve accessing the memory for both the OF and OS subcomputations, these two subcomputations can incur long latencies and can therefore be further subdivided into multiple subcomputations.

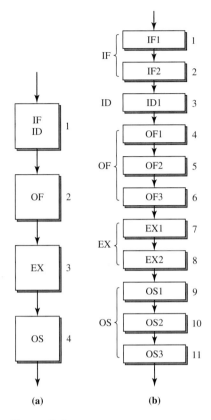

Figure 2.8
(a) A Four-Stage Instruction Pipeline Example.
(b) An 11-Stage Instruction Pipeline Example.

Additionally, some operations to be performed in the EX subcomputation may be quite complex and can be further subdivided into multiple subcomputations as well. Figure 2.8(b) illustrates such an instruction pipeline with an 11-stage design. Both the OF and OS subcomputations are mapped into three pipeline stages, while the IF and EX subcomputations are mapped into two pipeline stages. Essentially, the ID subcomputation is used as the reference to achieve balanced stages.

The two methods presented for stage quantization are (1) *merge* multiple subcomputations into one and (2) *subdivide* a subcomputation into multiple subcomputations. A combination of both methods can also be used in the design of an instruction pipeline. As shown in the previous discussion, the instruction set architecture can have a significant impact on stage quantization. In all cases, the goal of stage quantization is to minimize the overall internal fragmentation. For example, assume that the total latency for the five generic subcomputations is 280 ns and that the resultant machine cycle times for the 4-stage design of Figure 2.8(a) and the 11-stage design of Figure 2.8(b) are 80 and 30 ns, respectively. Consequently, the total latency for the 4-stage pipeline is 320 ns (80 ns × 4) and the total latency for the 11-stage pipeline is 330 ns (30 ns × 11). The difference between the new total latency and the original total latency of 280 ns represents the internal fragmentation. Hence, the internal fragmentation for the 4-stage design is 40 ns (320 ns − 280 ns), and the internal fragmentation for the 11-stage design is 50 ns (330 ns − 280 ns). It can be concluded that the 4-stage design is more efficient than the 11-stage design in terms of incurring less overhead due to internal fragmentation. Of course, the 11-stage design yields a throughput that is 9.3 (280 ns / 30 ns) times that of a nonpipelined design, while the 4-stage design's throughput is only 3.5 (280 ns / 80 ns) times that of a nonpipelined design. As can be seen in both designs, the internal fragmentation has hindered the attainment of the idealized throughput increase by factors of 11 and 4 for the 11-stage and the 4-stage pipelines, respectively.

2.2.1.2 Hardware Requirements.
In most realistic engineering designs, the goal is not simply to achieve the best possible performance, but to achieve the best performance/cost ratio. Hence, in addition to simply maximizing the throughput (performance) of an instruction pipeline, hardware requirements (cost) must be considered. In general, higher degrees of pipelining will incur greater costs in terms of hardware requirements. Clearly there is the added cost due to the additional buffering between pipeline stages. We have already seen in the model presented in Section 2.1.1.3 that there is a point beyond which further pipelining yields diminishing returns due to the overhead of buffering between pipeline stages. Besides this buffering overhead, there are other, and more significant, hardware requirements for highly pipelined designs.

In assessing the hardware requirements for an instruction pipeline, the first thing to keep in mind is that for a k-stage instruction pipeline, in the worst case, or actually best case in terms of performance, there are k instructions concurrently present in the pipeline. There will be an instruction resident in each pipeline stage, with a total of k instructions all in different phases of the instruction cycle. Hence, the entire pipeline must have enough hardware to support the concurrent processing of k instructions in the k pipeline stages. The hardware requirements fall into three categories: (1) the logic required for control and data manipulation in each stage, (2) register-file ports to support concurrent register accessing by multiple stages, and (3) memory ports to support concurrent memory accessing by multiple stages.

We first examine the four-stage instruction pipeline of Figure 2.8(a). Assuming a load/store architecture, a typical register-register instruction will need to read the two register operands in the first stage and store the result back to a register in the fourth stage. A load instruction will need to read from memory in the second stage, while a store instruction will need to write to memory in the fourth stage. Combining the requirements for all four stages, a register file with two read ports and one write port will be required, and a data memory interface capable of performing one memory read and one memory write in every machine cycle will be required. In addition, the first stage needs to read from the instruction memory in every cycle for instruction fetch. If a unified (instruction and data) memory is used, then this memory must be able to support two read accesses and one write access in every machine cycle.

Similar analysis of hardware requirements can be performed for the 11-stage instruction pipeline of Figure 2.8(b). To accommodate slow instruction memory, the IF generic subcomputation is subdivided and mapped to two pipeline stages, namely, the IF1 and IF2 stages. Instruction fetch is initiated in IF1 and completes in IF2. Even though instruction fetch takes two machine cycles, it is pipelined; that is, while the first instruction is completing the fetching in IF2, the second instruction can begin fetching in IF1. This means that the instruction memory must be able to support two concurrent accesses, by IF1 and IF2 pipeline stages, in every machine cycle. Similarly, the mapping of both the OF and OS generic subcomputations to three pipeline stages each implies that at any one time there could be up to six instructions in the pipeline, all in the process of accessing the data memory. Hence, the data memory must be able to support six independent concurrent accesses without conflict in every machine cycle. This can potentially require a six-ported data memory. Furthermore, if the instruction memory and the data memory are unified into one memory unit, an eight-ported memory unit can potentially be required. Such multiported memory units are extremely expensive to implement. Less expensive solutions, such as using interleaved memory with multiple banks, that attempt to simulate true multiported functionality usually cannot guarantee conflict-free concurrent accesses at all times.

As the degree of pipelining, or the pipeline depth, increases, the amount of hardware resources needed to support such a pipeline increases significantly. The most significant increases of hardware resources are the additional ports to the register file(s) and the memory unit(s) needed to support the increased degree of concurrent accesses to these data storage units. Furthermore, to accommodate long memory access latency, the memory access subcomputation must be pipelined. However, the physical pipelining of memory accessing beyond two machine cycles can become quite complex, and frequently conflict-free concurrent accesses must be compromised.

2.2.1.3 Example Instruction Pipelines. The stage quantization of two commercial pipelined processors is presented here to provide illustrations of real instruction pipelines. The MIPS R2000/R3000 RISC processors employ a five-stage instruction pipeline, as shown in Figure 2.9(a). The MIPS architecture is a load/store architecture. The IF and ID generic subcomputations are merged into the IF stage, which will require one memory (I-cache) read in every machine cycle. The OF generic subcomputation is carried out in both the RD and MEM stages. For ALU instructions that access only register operands, operand fetch is done in the RD stage and requires the reading of two registers. For load instructions, the operand fetch also requires accessing the memory (D-cache) and is carried out in the MEM stage, which is the only stage in the pipeline that can access the D-cache. The OS generic subcomputation is carried out in the MEM and WB stages. Store instructions must access the D-cache and are done in the MEM stage. ALU and load instructions write their results back to the register file in the WB stage.

MIPS processors normally employ separate instruction and data caches. In every machine cycle the R2000/R3000 pipeline must support the concurrent accesses of

60 MODERN PROCESSOR DESIGN

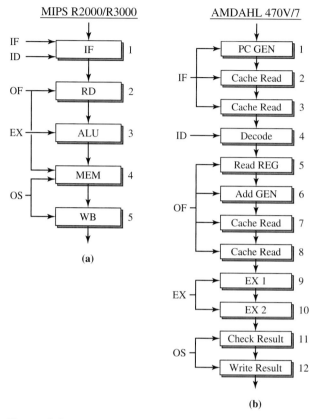

Figure 2.9
Two Commercial Instruction Pipelines: (a) MIPS R2000/R3000 Five-Stage Pipeline; (b) AMDAHL 470V/7 12-Stage Pipeline.

one I-cache read by the IF stage and one D-cache read (for a load instruction) or write (for a store instruction) by the MEM stage. Note that with the split cache configuration, both the I-cache and the D-cache need not be multiported. On the other hand, if both instructions and data are stored in the same cache, the unified cache will need to be dual-ported to support this pipeline. The register file must provide adequate ports to support two register reads by the RD stage and one register write by the WB stage in every machine cycle.

Figure 2.9(b) illustrates the 12-stage instruction pipeline of the AMDAHL 470V/7. The IF generic subcomputation is implemented in the first three stages. Because of the complex addressing modes that must be supported, the OF generic subcomputation is mapped into four stages. Both the EX and OS generic subcomputations are partitioned into two pipeline stages. In stage 1 of this 12-stage pipeline, the address of the next sequential instruction is computed. Stage 2 initiates cache access to read the instruction; stage 3 loads the instruction from the cache into the

I-unit (instruction unit). Stage 4 decodes the instruction. Two general-purpose registers are read during stage 5; these registers are used as address registers. Stage 6 computes the address of an operand in memory. Stage 7 initiates cache access to read the memory operand; stage 8 loads the operand from the cache into the I-unit and also reads register operands. Stages 9 and 10 are the two execute stages in the E-unit (execute unit). In Stage 11 error checking is performed on the computed result. The final result is stored into the destination register in stage 12.

This 12-stage pipeline must support the concurrent accesses of two register reads by stage 5 and one register write by stage 12 in every machine cycle, along with four cache memory reads by stages 2, 3, 7, and 8 in every machine cycle. The memory subsystem of this pipelined processor is clearly much more complicated than that of the MIPS R2000/R3000 pipeline.

The current trend in pipelined processor design is toward higher degrees of pipelining with deeper pipeline depth. This produces finer-grained pipelined stages that can be clocked at higher rates. While four or five stages are common in first-generation pipelined RISC processors, instruction pipelines with more than ten stages are becoming commonplace. There is also the trend toward implementing multiple pipelines with different numbers of stages. This is the subject of superscalar processor design, which will be addressed in Chapter 4.

2.2.2 Unifying Instruction Types

The second point of pipelining idealism assumes that the same computation is to be performed repeatedly by the pipeline. For most instruction pipelines, this idealized assumption of repetition of identical computations does not hold. While the instruction pipeline repeatedly processes instructions, there are different types of instructions involved. Although the instruction cycle is repeated over and over, repetitions of the instruction cycle may involve the processing of different instruction types. Different instruction types have different resource requirements and may not require the exact same sequence of subcomputations. The instruction pipeline must be able to support the different requirements and must provide a superset of all the subcomputations needed by all the instruction types. Each instruction type may not require all the pipeline stages in the instruction pipeline. For each instruction type, the unnecessary pipeline stages become a form of inefficiency or overhead for that instruction type; such inefficiency or overhead has been referred to as *external fragmentation* of the pipeline in Section 2.1.3.2. The goal for unifying instruction types, the key challenge resulting from the second point of pipelining idealism, is to minimize the external fragmentations for all the instruction types.

2.2.2.1 Classification of Instruction Types.
To perform a computation, a computer must do three generic tasks:

1. Arithmetic operation
2. Data movement
3. Instruction sequencing

These three generic tasks are carried out by the processing of instructions in the processor. The arithmetic operation task involves the performing of arithmetic and logical operations on specified operands. This is the most obvious part of performing a computation and has often been equated to computation. A processor can support a large variety of arithmetic operations. The data movement task is responsible for moving the operands and the results between storage locations. Typically there is a hierarchy of storage locations, and explicit instructions are used to move the data among these locations. The instruction sequencing task is responsible for the sequencing of instructions. Typically a computation is specified in a program consisting of many instructions. The performance of the computation involves the processing of a sequence of instructions. This sequencing of instructions, or the program flow, can be explicitly specified by the instructions themselves.

How these three generic tasks are assigned to the various instructions of an ISA is a key component of instruction set design. A very complex instruction can be specified that actually performs all three of these generic tasks. In a typical horizontally microcoded machine, every microinstruction has fields that are used to specify all three of these generic tasks. In more traditional instruction set architectures known as complex instruction set computer (CISC) architectures, many of the instructions carry out more than one of these three generic tasks.

Influenced by the RISC research of the 1980s, most recent instruction set architectures all share some common attributes. These recent architectures include Hewlett-Packard's Precision architecture, IBM's Power architecture, IBM/Motorola's PowerPC architecture, and Digital's Alpha architecture. These modern ISAs tend to have fixed-length instructions, symmetric instruction formats, load/store architectures, and simple addressing modes. Most of these attributes are quite compatible with instruction pipelining. For the most part, this book adopts and assumes such a typical RISC architecture in its examples and illustrations.

In a typical modern RISC architecture, the instruction set employs a dedicated instruction type for each of the three generic tasks; each instruction only carries out one of the three generic tasks. Based on the three generic tasks, instructions can be classified into three types:

1. *ALU instructions.* For performing arithmetic and logical operations.

2. *Load/store instructions.* For moving data between registers and memory locations.

3. *Branch instructions.* For controlling instruction sequencing.

ALU instructions perform arithmetic and logical operations strictly on register operands. Only load and store instructions can access the data memory. Both the load/store and branch instructions employ fairly simple addressing modes. Typically only register-indirect with an offset addressing mode is supported. Often PC-relative addressing mode is also supported for branch instructions. In the following detailed specification of the three instruction types, the use of an instruction cache (I-cache) and a data cache (D-cache) is also assumed.

Table 2.2
Specification of ALU instruction type

Generic Subcomputations	ALU Instruction Type	
	Integer Instruction	**Floating-Point Instruction**
IF	Fetch instruction (access I-cache).	Fetch instruction (access I-cache).
ID	Decode instruction.	Decode instruction.
OF	Access register file.	Access FP register file.
EX	Perform ALU operation.	Perform FP operation.
OS	Write back to register file.	Write back to FP register file.

The semantics of each of the three instruction types can be specified based on the sequence of subcomputations performed by that instruction type. This specification can begin with the five generic subcomputations (Section 2.2.1) with subsequent further refinements. Eventually, these subcomputations specify the sequence of register transfers used for hardware implementation. For convenience, ALU instructions are further divided into integer and floating-point instructions. The semantics of the ALU instructions are specified in Table 2.2.

In a load/store architecture, load and store instructions are the only instructions that access the data memory. A load instruction moves data from a memory location into a register; a store instruction moves data from a register to a memory location. In the specification of the semantics of load and store instructions in Table 2.3, it is assumed that the only addressing mode is register-indirect with an offset. This addressing mode computes an effective address by adding the content of a register with the offset specified in the immediate field of the instruction.

Comparing the specifications in Tables 2.2 and 2.3, we can observe that the sequences of subcomputations required for ALU and load/store instruction types are similar but not exactly the same. ALU instructions need not generate memory addresses. On the other hand, load/store instructions, other than having to generate the effective address, do not have to perform explicit arithmetic or logical operations. They simply move data between registers and memory locations. Even between load and store instructions there are subtle differences. For the load instruction, the OF generic subcomputation expands into three subcomputations involving accessing the register file for the base address, generating the effective address, and accessing the memory location. Similarly for the store instruction, the OS generic subcomputation consists of two subcomputations involving generating the effective address and storing a register operand into a memory location. This assumes that the base address and the register operand are both accessed from the register file during the OF generic subcomputation.

Finally the sequences of subcomputations that specify the unconditional jump and the conditional branch instructions are presented in Table 2.4. A similar

Table 2.3
Specification of load/store instruction type

Generic Subcomputations	Load/Store Instruction Type	
	Load Instruction	**Store Instruction**
IF	Fetch instruction (access I-cache).	Fetch instruction (access I-cache).
ID	Decode instruction.	Decode instruction.
OF	Access register file (base address). Generate effective address (base + offset). Access (read) memory location (access D-cache).	Access register file (register operand, and base address).
EX		
OS	Write back to register file.	Generate effective address (base + offset). Access (write) memory location (access D-cache).

Table 2.4
Specification of branch instruction type

Generic Subcomputations	Branch Instruction Type	
	Jump (unconditional) Instruction	**Conditional Branch Instruction**
IF	Fetch instruction (access I-cache).	Fetch instruction (access I-cache).
ID	Decode instruction.	Decode instruction.
OF	Access register file (base address). Generate effective address (base + offset).	Access register file (base address). Generate effective address (base + offset).
EX		Evaluate branch condition.
OS	Update program counter with target address.	If condition is true, update program counter with target address.

addressing mode as that for the load/store instructions is employed for the branch instructions. A PC-relative addressing mode can also be supported. In this addressing mode, the address of the target of the branch (or jump) instruction is generated by adding a displacement to the current content of the program counter. Typically this displacement can be either a positive or a negative value, to facilitate both forward and backward branches.

Examining the specifications of the three major instruction types in Tables 2.2 to 2.4, we see that the initial subcomputations for all three types are quite similar. However, there are differences in the later subcomputations. For example, ALU instructions do not access data memory, and hence for them no memory address generation is needed. On the other hand, load/store and branch instruction types share the same required subcomputation of effective address generation. Load/store instructions must access the data memory, while branch instructions must provide the address of the target instruction. We also see that for a conditional branch instruction, in addition to generating the effective address, evaluation of the branch condition must be performed. This can involve simply the checking of a status bit generated by an earlier instruction, or it can require the performance of an arithmetic operation on a register operand as part of the processing of the branch instruction, or it can involve checking the value of a specified register.

Based on the foregoing specifications of the instruction semantics, resource requirements for the three major instruction types can be determined. While the three instruction types share some commonality in terms of the fetching and decoding of the instructions, there are differences between the instruction types. These differences in the instruction semantics will lead to differences in the resource requirements.

2.2.2.2 Coalescing of Resource Requirements. The challenge of unifying the different instruction types involves the efficient coalescing of the different resource requirements into one instruction pipeline that can accommodate all the instruction types. The objective is to minimize the total resources required by the pipeline and at the same time maximize the utilization of all the resources in the pipeline. The procedure for unifying different instruction types can be informally stated as consisting of the following three steps.

1. Analyze the sequence of subcomputations of each instruction type, and determine the corresponding resource requirements.

2. Find commonality between instruction types, and merge common subcomputations to share the same pipeline stage.

3. If there exists flexibility, without violating the instruction semantics, shift or reorder the subcomputations to facilitate further merging.

This procedure of unifying instruction types can be illustrated by applying it to the instruction types specified in Tables 2.2 to 2.4. For simplicity and clarity, floating-point instructions and unconditional jumps are not considered. Summary specifications of the ALU, load, store, and branch instruction types are repeated in Figure 2.10. The four sequences of subcomputations required by these four instruction types are taken from Tables 2.2 to 2.4 and are summarized in the four columns on the left-hand side of Figure 2.10. We now apply the unifying procedure from the top down, by examining the four sequences of subcomputations and the associated

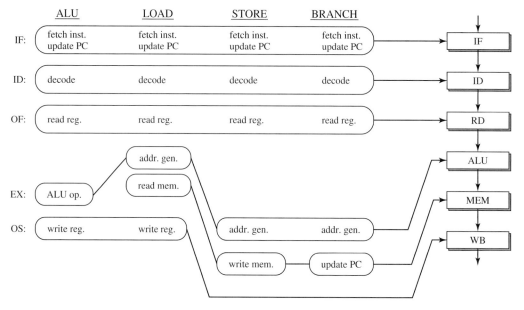

Figure 2.10
Unifying of ALU, Load, Store, and Branch Instruction Types into a Six-Stage Instruction Pipeline, Henceforth Identified as the TYPICAL (TYP) Instruction Pipeline.

hardware resources required to support them. This procedure results in the definition of the stages of an instruction pipeline.

All four instruction types share the same common subcomputations for IF and ID. Hence, the first two subcomputations for all four instruction types can be easily merged and used to define the first two pipeline stages, labeled IF and ID, for instruction fetching and instruction decoding.

All four instruction types also read from the register file for the OF generic subcomputation. ALU instructions access the two register operands. Load and branch instructions access a register to obtain the base address. Store instructions access a register to obtain the register operand and another register for the base address. In all four cases either one or two registers are read. These similar subcomputations can be merged into the third stage of the pipeline, called RD, for reading up to two registers from the register file. The register file must be capable of supporting two independent and concurrent reads in every machine cycle.

ALU instructions require an ALU functional unit for performing the necessary arithmetic and logical operations. While load, store, and branch instructions do not need to perform such operations, they do need to generate an effective address for accessing memory. It can be observed that the address generation task can be performed by the ALU functional unit. Hence, these subcomputations can be merged into the fourth stage of the pipeline, called ALU, which consists primarily of the ALU functional unit for performing arithmetic/logical operations or effective address generation.

Both the load and store instruction types need to access the data memory. Hence a pipeline stage must be devoted to this subcomputation. The fifth stage of the pipeline, labeled MEM, is included for this purpose.

Both the ALU and load instruction types must write a result back to the register file as their last subcomputation. An ALU instruction writes the result of the operation performed on the register operands into a destination register. A load instruction loads into the destination register the data fetched from memory. No memory access is required by an ALU instruction; hence, the writing back to the destination register can theoretically take place immediately after the ALU stage. However, for the purpose of unifying with the register write-back subcomputation of the load instruction type, the register write-back subcomputation for ALU instructions is delayed by one pipeline stage and takes place in the sixth pipeline stage, named WB. This incurs one idle machine cycle for ALU instructions in the MEM pipeline stage. This is a form of external fragmentation and introduces some inefficiency in the pipeline.

For conditional branch instructions, the branch condition must be determined prior to updating the program counter. Since the ALU functional unit is used to perform effective address generation, it cannot be used to perform the branch condition evaluation. If the branch condition evaluation involves only the checking of a register to determine if it is equal to zero, or if it is positive or negative, then only a simple comparator is needed. This comparator can be added, and the earliest pipeline stage in which it can be added is the ALU stage, that is, after the reference register is read in the RD stage. Hence, the earliest pipeline stage in which the program counter can be updated with the branch target address, assuming the conditional branch is taken, is during the MEM stage, that is, after the target address is computed and the branch condition is determined in the ALU stage.

The foregoing coalescing of resource requirements for the different instruction types resulted in the six-stage instruction pipeline shown in the right-hand side of Figure 2.10. This instruction pipeline is identified as the TYPICAL (TYP) instruction pipeline and is used in the remainder of this chapter as an illustration vehicle. Other than the one idling pipeline stage (MEM) for ALU instructions, store and branch instruction types also incur some external fragmentation. Both store and branch instructions do not need to write back to a register and are idling during the WB stage. Overall this six-stage instruction pipeline is quite efficient. Load instructions use all six stages of the pipeline; the other three instruction types use five of the six stages.

In unifying different instruction types into one instruction pipeline, there are three optimization objectives. The first is to minimize the total resources required to support all the instruction types. In a way, the objective is to determine the pipeline that is analogous to the *least common multiple* of all the different resource requirements. The second objective is to maximize the utilization of all the pipeline stages by the different instruction types, in other words, to minimize the idling stages incurred by each instruction type. Idling stages lead to external fragmentation and result in inefficiency and throughput penalty. The third objective is to minimize the overall latency for each of the instruction types. Hence, if an idling stage is unavoidable for a particular instruction type and there is flexibility in terms of the placement of that idling stage, then it is always better to place it at the end of the pipeline. This will allow the instruction to effectively complete earlier and reduce the overall latency for that instruction type.

2.2.2.3 Instruction Pipeline Implementation.
In the six-stage TYP instruction pipeline (Figure 2.10), there are potentially six different instructions simultaneously present or "in flight" in the pipeline at any one time. Each of the six instructions is going through one of the pipeline stages. The register file must support two reads (by the instruction in the RD stage) and one write (by the instruction in the WB stage) in every machine cycle. The I-cache must support one read in every machine cycle. Unless interrupted by a branch instruction, the IF stage continually increments the program counter and fetches the next sequential instruction from the I-cache. The D-cache must support one memory read or memory write in every machine cycle. Only the MEM stage accesses the D-cache; hence, at any time only one instruction in the pipeline can be accessing the data memory.

The pipeline diagram in Figure 2.10 is only a logical representation of the six-stage TYP instruction pipeline and illustrates only the ordering of the six pipeline stages. The actual physical organization of the TYP instruction pipeline is shown in Figure 2.11, which is the functional block diagram of the TYP pipelined processor

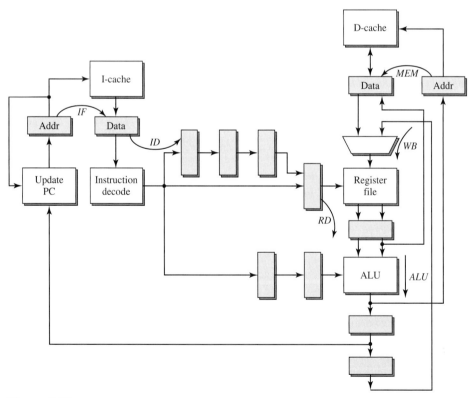

Figure 2.11
The Physical Organization of the Six-Stage TYP Instruction Pipeline.

implementation. In this diagram the buffers between the pipeline stages are explicitly identified. The logical buffer between two particular pipeline stages can actually involve multiple physical buffers distributed in this diagram. The single logical path that traverses the six pipeline stages in sequence actually involves multiple physical paths in this diagram. The progression of each instruction through the pipeline must be traced along these physical paths.

The physical organization of the six-stage TYP instruction pipeline in Figure 2.11 looks more complex than it really is. To help digest it, we can first examine the pipeline's interfaces to the register file and the memory subsystem. Assuming a split cache organization, that is, separate caches for storing instructions and data, two single-ported caches, one I-cache and one D-cache, are needed. The memory subsystem interface of the TYP pipeline is quite simple and efficient, and resembles most scalar pipelined processors. The IF stage accesses the I-cache, and the MEM stage accesses the D-cache, as shown in Figure 2.12. The I-cache can support the fetch of one instruction in every machine cycle; a miss in the I-cache will stall the pipeline. In the MEM stage of the pipeline, a load (store) instruction performs a read (write) from (to) the D-cache. Note that it is assumed here that the latency for accessing the D-cache, and the I-cache, is within one machine cycle. As caches become larger and processor logic becomes more deeply pipelined, maintaining this one machine cycle latency for the caches will become more difficult.

The interface to the multiported register file is shown in Figure 2.13. Only the RD and the WB stages access the register file. In every machine cycle, the register

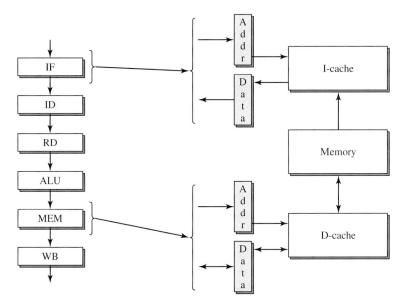

Figure 2.12
The Six-Stage TYP Instruction Pipeline's Interface to the Memory Subsystem.

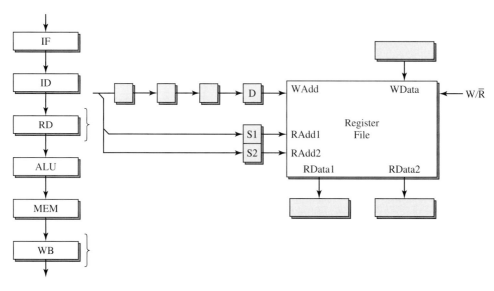

Figure 2.13
The Six-Stage TYP Instruction Pipeline's Interface to the Multiported Register File.

file must support potentially two register reads by the RD stage and one register write by the WB stage. Hence, a multiported register file with two read ports and one write port is required. Such a register file is illustrated in Figure 2.13. It has three address ports, two data output ports, and one data input port for supporting two reads and one write in every machine cycle. The instruction that is performing the register write is in the WB stage and precedes the instruction that is performing the register reads in the RD stage by three machine cycles or intervening instructions. Consequently, there are three additional pipeline stage buffers at the register write address port to ensure that the register write address specifying the destination register to be written arrives at the register file write address port at exactly the same time as the data to be written are arriving at the input data port of the register file. Three-ported register files are not very complex. However, as the number of ports increases beyond three, the hardware complexity increases very rapidly. This is especially true for increasing the number of write ports due to circuit design limitations. Multiported register files with up to 20 some ports are feasible and can be found in some high-end microprocessors.

If we look at the logical diagram of the six-stage TYP instruction pipeline of Figure 2.10, it appears that every instruction flows through the single linear path through the six pipeline stages. However, different sets of physical paths in the physical organization of the TYP instruction pipeline of Figure 2.11 are traversed by different instruction types. Some of the flow path segments are labeled in Figure 2.11 to show which pipeline stages they are associated with. Essentially some of the pipeline stages are physically distributed in the physical organization diagram of the pipeline.

The six-stage TYP instruction pipeline is quite similar to two other instruction pipelines, namely, the MIPS R2000/R3000 and the instructional DLX processor used in the popular textbook by John Hennessy and David Patterson [2003]. Both are five-stage pipelines. The MIPS pipeline combines the IF and ID stages of the TYP pipeline into one pipeline stage. The DLX pipeline combines the ID and RD stages of the TYP pipeline into one pipeline stage. The other four stages are essentially the same for all three pipelines. The TYP pipeline is used in the remainder of this chapter as a running example.

2.2.3 Minimizing Pipeline Stalls

The third point of pipelining idealism assumes that the computations that are performed by the pipeline are mutually independent. In a k-stage pipeline, there can be k different computations going on at any one time. For an instruction pipeline, there can be up to k different instructions present or in flight in the pipeline at any one time. These instructions may not be independent of one another; in fact, usually there are dependences between the instructions in flight. Having independent instructions in the pipeline facilitates the streaming of the pipeline; that is, instructions move through the pipeline without encountering any pipeline stalls. When there are inter-instruction dependences, they must be detected and resolved. The resolution of these dependences can require the stalling of the pipeline. The challenge and design objective is to minimize such pipeline stalls and the resultant throughput degradation.

2.2.3.1 Program Dependences and Pipeline Hazards. At the ISA abstraction level, a program is specified as a sequence of assembly language instructions. A typical instruction can be specified as a function $i: T \leftarrow S1\ op\ S2$, where the domain of instruction i is $D(i) = \{S1, S2\}$, the range is $R(i) = \{T\}$, and the mapping from the domain to the range is defined by op, the operation. Given two instructions i and j, with j following i in the lexical ordering of the two instructions, a data dependence can exist between i and j, or j can be data-dependent on i, denoted $i\delta j$, if one of the following three conditions exists.

$$R(i) \cap D(j) \neq \emptyset \qquad (2.8)$$
$$R(j) \cap D(i) \neq \emptyset \qquad (2.9)$$
$$R(i) \cap R(j) \neq \emptyset \qquad (2.10)$$

The first condition implies that instruction j requires an operand that is in the range of instruction i. This is referred to as the read-after-write (RAW) or *true* data dependence and is denoted $i\delta_d j$. The implication of a true data dependence is that instruction j cannot begin execution until instruction i completes. The second condition indicates that an operand required by i is in the range of j, or that instruction j will modify the variable which is an operand of i. This is referred to as the write-after-read (WAR) or *anti* data dependence and is denoted $i\delta_a j$. The existence of an anti-dependence requires that instruction j not complete prior to the execution of instruction i; otherwise, instruction i will get the wrong operand. The third condition indicates that both instructions i and j share a common variable in their range, meaning that both will modify that same variable. This is referred to as the write-after-write (WAW) or *output* data dependence and is denoted $i\delta_o j$. The existence of an output dependence requires that instruction j not complete before the completion of instruction i; otherwise, instructions subsequent to j that have the same variable in their domains will receive the wrong operand. Clearly, the read-after-read case involves both instructions i and j accessing the same operand and is harmless regardless of the relative order of the two accesses.

These three possible ways for the domains and ranges of two instructions to overlap induce the three types of possible data dependences between two instructions, namely, true (RAW), anti (WAR), and output (WAW) data dependences. Since, in assembly code, the domains and the ranges of instructions can be variables residing in either the registers or memory locations, the common variable in a dependence can involve either a register or a memory location. We refer to them as *register* dependences and *memory* dependences. In this chapter we focus primarily on register dependences. Figure 2.14 illustrates the RAW, WAR, and WAW register data dependences.

Other than data dependences, a *control* dependence can exist between two instructions. Given instructions i and j, with j following i, j is control-dependent on i, denoted $i\delta_c j$, if whether instruction j is executed or not depends on the outcome of the execution of instruction i. Control dependences are consequences of the control flow structure of the program. A conditional branch instruction causes uncertainty on instruction sequencing. Instructions following a conditional branch can have control dependences on the branch instruction.

An assembly language program consists of a sequence of instructions. The semantics of this program assume and depend on the sequential execution of the instructions. The sequential listing of the instructions implies a sequential precedence between adjacent instructions. If instruction i is followed by instruction $i + 1$ in the program listing, then it is assumed that first instruction i is executed, and then instruction $i + 1$ is executed. If such sequential execution is followed, the semantic correctness of the program is guaranteed. To be more precise, since an instruction cycle can involve multiple subcomputations, the implicit assumption is that all the subcomputations of instruction i are carried out before any of the

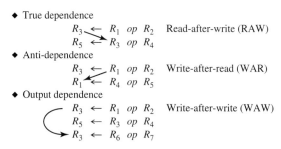

Figure 2.14
Illustration of RAW, WAR, and WAW Data Dependences.

subcomputations of instruction $i + 1$ can begin. We called this the *total sequential execution* of the program; that is, all the subcomputations of the sequence of instructions are carried out sequentially.

Given a pipelined processor with k pipeline stages, the processing of k instructions is overlapped in the pipeline. As soon as instruction i finishes its first subcomputation and begins its second subcomputation, instruction $i + 1$ begins its first subcomputation. The k subcomputations, corresponding to the k pipeline stages, of a particular instruction are overlapped with subcomputations of other instructions. Hence, the total sequential execution does not hold. While total sequential execution is sufficient to ensure semantic correctness, it is not a necessary requirement for semantic correctness. The total sequential execution implied by the sequential listing of instructions is an overspecification of the semantics of a program. The essential requirement in ensuring that the program semantics are not violated is that all the inter-instruction dependences not be violated. In other words, if there exists a dependence between two instructions i and j, with j following i in the program listing, then the reading/writing of the common variable by instructions i and j must occur in original sequential order. In pipelined processors, if care is not taken, there is the potential that program dependences can be violated. Such potential violations of program dependences are called *pipeline hazards*. All pipeline hazards must be detected and resolved for correct program execution.

2.2.3.2 Identification of Pipeline Hazards.
Once all the instruction types are unified into an instruction pipeline and the functionality for all the pipeline stages is defined, analysis of the instruction pipeline can be performed to identify all the pipeline hazards that can occur in that pipeline. Pipeline hazards are consequences of both the organization of the pipeline and inter-instruction dependences. The focus of this chapter is on *scalar* instruction pipelines. By definition, a scalar instruction pipeline is a single pipeline with multiple pipeline stages organized in a linear sequential order. Instructions enter the pipeline according to the sequential order specified by the program listing. Except when pipeline stalls occur, instructions flow through a scalar instruction pipeline in the lockstep fashion; that is, each instruction advances to the next pipeline stage with every machine cycle. For scalar instruction pipelines, necessary conditions on the pipeline organization for the occurrence of pipeline hazards due to data dependences can be determined.

A pipeline hazard is a potential violation of a program dependence. Pipeline hazards can be classified according to the type of program dependence involved. A WAW hazard is a potential violation of an output dependence. A WAR hazard is a potential violation of an anti-dependence. A RAW hazard is a potential violation of a true data dependence. A data dependence involves the reading and/or writing of a common variable by two instructions. For a hazard to occur, there must exist at least two pipeline stages in the pipeline which can contain two instructions that can simultaneously access the common variable.

Figure 2.15 illustrates the necessary conditions on the pipeline organization for the occurrence of WAW, WAR, and RAW hazards. These necessary conditions apply to hazards caused by both memory and register data dependences (only register

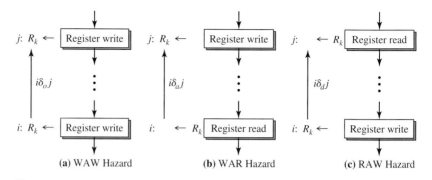

Figure 2.15
Necessary Conditions on the Pipeline Organization for the Occurrence of (a) WAW Hazards, (b) WAR Hazards, and (c) RAW Hazards.

dependences are illustrated in the figure). In order for a WAW hazard to occur due to an output dependence $i\delta_o j$, there must exist at least two pipeline stages that can perform two simultaneous writes to the common variable; see Figure 2.15(a). If only one stage in the pipeline can write to that variable, then no hazard can occur because both writes—in fact all writes—to that variable will be performed by that pipeline stage according to the original sequential order specified by the program listing. Figure 2.15(b) specifies that in order for a WAR hazard to occur, there must exist at least two stages in the pipeline, with an earlier stage x and a later stage y, such that stage x can write to that variable and stage y can read that variable. In order for the anti-dependence $i\delta_a j$ to be violated, instruction j must perform the write, that is, reach stage x, prior to instruction i performing the read or reaching stage y. If this necessary condition does not hold, it is impossible for instruction j, a trailing instruction, to perform a write prior to instruction i completing its read. For example, if there exists only one pipeline stage that can perform both the read and write to that variable, then all accesses to that variable are done in the original sequential order, and hence no WAR hazard can occur. In the case where the stage performing the read is earlier in the pipeline than the stage performing the write, the leading instruction i must complete its read before the trailing instruction j can possibly perform the write in a later stage in the pipeline. Again, no WAR hazard can occur in such a pipeline. In actuality, the necessary conditions presented in Figure 2.15 are also sufficient conditions and can be considered as characterizing conditions for the occurrence of WAW, WAR, and RAW pipeline hazards.

Figure 2.15(c) specifies that in order for a RAW hazard to occur due to a true data dependence $i\delta_d j$, there must exist two pipeline stages x and y, with x occurring earlier in the pipeline than y, such that stage x can perform a read and stage y can perform a write to the common variable. With this pipeline organization, the dependence $i\delta_d j$ can be violated if the trailing instruction j reaches stage x prior to the leading instruction i reaching stage y. Arguments similar to that used for WAR hazards can be applied to show that if this necessary condition does not hold, then no RAW hazard can occur. For example, if only one pipeline stage performs all the reads and writes, then effectively total sequential execution is carried out and no hazard can occur. If the stage performing the read is positioned later in the pipeline than the stage performing the write, then RAW hazards can never occur; the reason is that all the writes of leading instructions will be completed before the trailing instructions perform their reads.

Since pipeline hazards are caused by potential violations of program dependences, a systematic procedure for identifying all the pipeline hazards that can occur in an instruction pipeline can be formulated by considering each dependence type in turn. The specific procedure employed in this chapter examines program dependences in the following order.

1. Memory data dependence
   a. Output dependence
   b. Anti-dependence
   c. True data dependence

2. Register data dependence
   a. Output dependence
   b. Anti-dependence
   c. True data dependence

3. Control dependence

We illustrate this procedure by applying it to the six-stage TYP instruction pipeline. First, memory data dependences are considered. A memory data dependence involves a common variable stored in memory that is accessed (either read or write) by two instructions. Given a load/store architecture, memory data dependences can only occur between load/store instructions. To determine whether pipeline hazards can occur due to memory data dependences, the processing of load/store instructions by the pipeline must be examined. Assuming a split cache design, in the TYP pipeline, only the MEM stage can access the D-cache. Hence, all accessing of memory locations by load/store instructions must and can only occur in the MEM stage; there is only one stage in the pipeline that performs reads and writes to the data memory. Based on the necessary conditions presented in Figure 2.15 no pipeline hazards due to memory data dependences can occur in the TYP pipeline. Essentially, all accesses to the data memory are performed sequentially, and the processing of all load/store instructions is done in the total sequential execution mode. Therefore, for the TYP pipeline, there are no pipeline hazards due to memory data dependences.

Register data dependences are considered next. To determine pipeline hazards that can occur due to register data dependences, all pipeline stages that can access the register file must be identified. In the TYP pipeline, all register reads occur in the RD stage and all register writes occur in the WB stage. An output (WAW) dependence, denoted $i\delta_o j$, indicates that an instruction i and a subsequent instruction j both share the same destination register. To enforce the output dependence, instruction i must write to that register first; then instruction j can write to that same register. In the TYP pipeline, only the WB stage can perform writes to the register file. Consequently, all register writes are performed in sequential order by the WB stage; and according to the necessary condition of Figure 2.15(a), no pipeline hazards due to output dependences can occur in the TYP pipeline.

An anti (WAR) dependence, denoted $i\delta_a j$, indicates that instruction i is reading from a register that is the destination register of a subsequent instruction j. It must be ensured that instruction i reads that register before instruction j writes into that register. The only way that an anti-dependence can cause a pipeline hazard is if the trailing instruction j can perform a register write earlier than instruction i can perform its register read. This is an impossibility in the TYP pipeline because all register reads occur in the RD stage, which is earlier in the pipeline than the WB stage, the only stage in which register writes can occur. Hence, the necessary condition of Figure 2.15(b) does not exist in the TYP pipeline. Consequently, no pipeline hazards due to anti-dependences can occur in the TYP pipeline.

The only type of register data dependences that can cause pipeline hazards in the TYP pipeline are the true data dependences. The necessary condition of Figure 2.15(c) exists in the TYP pipeline because the pipeline stage RD that performs register reads is positioned earlier in the pipeline than the WB stage that performs register writes. A true data dependence, denoted $i\delta_d j$, involves instruction i writing into a register and a trailing instruction j reading from that same register. If instruction j immediately follows instruction i, then when j reaches the RD stage, instruction i will still be in the ALU stage. Hence, j cannot read the register operand that is the result of instruction i until i reaches the WB stage. To enforce this data dependence, instruction j must be prevented from entering the RD stage until instruction i has completed the WB stage. RAW pipeline hazards can occur for true data dependences because a trailing instruction can reach the register read stage in the pipeline prior to the leading instruction completing the register write stage in the pipeline.

Finally, control dependences are considered. Control dependences involve control flow changing instructions, namely, conditional branch instructions. The outcome of a conditional branch instruction determines whether the next instruction to be fetched is the next sequential instruction or the target of the conditional branch instruction. Essentially there are two candidate instructions that can follow a conditional branch. In an instruction pipeline, under normal operation, the instruction fetch stage uses the content of the program counter to fetch the next instruction, and then increments the content of the program counter to point to the next sequential instruction. This task is repeated in every machine cycle by the instruction fetch stage to keep the pipeline filled. When a conditional branch instruction is fetched, potential disruption of this sequential flow can occur. If the conditional branch is not taken, then the continued fetching by the instruction fetch stage of the next sequential instruction is correct. However, if the conditional branch is actually taken, then the fetching of the next sequential instruction by the instruction fetch stage will be incorrect. The problem is that this ambiguity cannot be resolved until the condition for branching is known.

A control dependence can be viewed as a form of register data (RAW) dependence involving the program counter (PC). A conditional branch instruction writes into the PC, whereas the fetching of the next instruction involves reading of the PC. The conditional branch instruction updates the PC with the address of the target instruction if the branch is taken; otherwise, the PC is updated with the address of the next sequential instruction. In the TYP pipeline, the updating of the PC with the target instruction address is performed in the MEM stage, whereas the IF stage uses the content of the PC to fetch the next instruction. Hence, the IF stage performs reads on the PC register, and the MEM stage which occurs later in the pipeline performs writes to the PC register. This ordering of the IF and MEM stages, according to Figure 2.15(c), satisfies the necessary condition for the occurrence of RAW hazards involving the PC register. Therefore, a control hazard exists in the TYP pipeline, and it can be viewed as a form of RAW hazard involving the PC.

2.2.3.3 Resolution of Pipeline Hazards.
Given the organization of the TYP pipeline, the only type of pipeline hazards due to data dependences that can occur are the RAW hazards. In addition, pipeline hazards due to control dependences can occur. All these hazards involve a leading instruction i that writes to a register (or PC) and a trailing instruction j that reads that register. With the presence of pipeline hazards, mechanisms must be provided to resolve these hazards, that is, ensure that the corresponding data dependences are not violated. With regard to each RAW hazard in the TYP pipeline, it must be ensured that the read occurs after the write to the common register, or the *hazard register*.

To resolve a RAW hazard, the trailing instruction j must be prevented from entering the pipeline stage in which the hazard register is read by j, until the leading instruction i has traversed the pipeline stage in which the hazard register is written by i. This is accomplished by stalling the earlier stages of the pipeline, namely all the stages prior to the stage performing a register read, thus preventing instruction j from entering the critical register read stage. The number of machine cycles by which instruction j must be held back is, in the worst case, equal to the distance between the two critical stages of the pipeline, that is, the stages performing read and write to the hazard register. In the case of the TYP pipeline, if the leading instruction i is either an ALU or a load instruction, the critical register write stage is the WB stage and the critical register read stage for all trailing instruction types is the RD stage. The distance between these two critical stages is three cycles; hence, the worst-case penalty is three cycles, as shown in Table 2.5. The worst-case penalty is incurred

Table 2.5
Worst-case penalties due to RAW hazards in the TYP pipeline

	Leading Instruction Type (i)		
	ALU	Load	Branch
Trailing instruction types (j)	ALU, Load/Store, Br.	ALU, Load/Store, Br.	ALU, Load/Store, Br.
Hazard register	Int. register (Ri)	Int. register (Ri)	PC
Register write stage (i)	WB (stage 6)	WB (stage 6)	MEM (stage 5)
Register read stage (j)	RD (stage 3)	RD (stage 3)	IF (stage 1)
RAW distance or penalty	3 cycles	3 cycles	4 cycles

when instruction j immediately follows instruction i in the original program listing; that is, j is equal to $i + 1$. In this case, instruction j must be stalled for three cycles in the ID stage and is allowed to enter the RD stage three cycles later as instruction i exits the WB stage. If the trailing instruction j does not immediately follow instruction i, that is, if there are intervening instructions between i and j, then the penalty will be less than three cycles. It is assumed that the intervening instructions do not depend on instruction i. The actual number of penalty cycles incurred is thus equal to $3 - s$, where s is the number of intervening instructions. For example, if there are three instructions between i and j, then no penalty cycle is incurred. In this case, instruction j will be entering the RD stage just as instruction i is exiting the WB stage, and no stalling is required to satisfy the RAW dependence.

For control hazards, the leading instruction i is a branch instruction, which updates the PC in the MEM stage. The fetching of the trailing instruction j requires the reading of the PC in the IF stage. The distance between these two stages is four cycles; hence, the worst-case penalty is four cycles. When a conditional branch instruction is encountered, all further fetching of instructions is stopped by stalling the IF stage until the conditional branch instruction completes the MEM stage in which the PC is updated with the branch target address. This requires stalling the IF stage for four cycles. Further analysis reveals that this stalling is only necessary if the conditional branch is actually taken. If it turns out that the conditional branch is not taken, then the IF stage could have continued its fetching of the next sequential instructions. This feature can be included in the pipeline design, so that following a conditional branch instruction, the instruction fetching is not stalled. Effectively, the pipeline assumes that the branch will not be taken. In the event that the branch is taken, the PC is updated with the branch target in the MEM stage and all the instructions residing in earlier pipeline stages are deleted, or flushed, and the next instruction fetched is the branch target. With such a design, the four-cycle penalty is incurred only when the conditional branch is actually taken, and there is no penalty cycle otherwise.

Similar to RAW hazards due to register data dependence, the four-cycle penalty incurred by a control hazard can be viewed as the worst-case penalty. If instructions that are not control-dependent on instruction i can be inserted between instruction i and instruction j, the control-dependent instruction, then the actual number of penalty cycles incurred can be reduced by the number of instructions inserted. This is the concept of *delayed branches*. Essentially these penalty cycles are filled by useful instructions that must be executed regardless of whether the conditional branch is taken. The actual number of penalty cycles is $4 - s$, where s is the number of control-independent instructions that can be inserted between instructions i and j. Delayed branches or the filling of penalty cycles due to branches makes it difficult to implement the earlier technique of assuming that the branch is not taken and allowing the IF stage to fetch down the sequential path. The reason is that mechanisms must be provided to distinguish the filled instructions from the actual normal sequential instructions. In the event that the branch is actually taken, the filled instructions need not be deleted, but the normal sequential instructions must be deleted because they should not have been executed.

2.2.3.4 Penalty Reduction via Forwarding Paths.

So far we have implicitly assumed that the only mechanism available for dealing with hazard resolution is to stall the dependent trailing instruction and ensure that the writing and reading of the hazard register are done in their normal sequential order. More aggressive techniques are available in the actual implementation of the pipeline that can help reduce the penalty cycles incurred by pipeline hazards. One such technique involves the incorporation of *forwarding paths* in the pipeline.

With respect to pipeline hazards, the leading instruction i is the instruction on which the trailing instruction j depends. For RAW hazards, instruction j needs the result of instruction i for its operand. Figure 2.16 illustrates the processing of the leading instruction i in the case when i is an ALU instruction or a load instruction. If the leading instruction i is an ALU instruction, the result needed by instruction j is actually produced by the ALU stage and is available when instruction i completes the ALU stage. In other words, the operand needed by instruction j is actually available at the output of the ALU stage when instruction i exits the ALU stage, and j need not wait two more cycles for i to exit the WB stage. If the output of the ALU stage can be made available to the input side of the ALU stage via a physical forwarding path, then the trailing instruction j can be allowed to enter the ALU stage as soon as the leading instruction i leaves the ALU stage. In this case, instruction j need not access the dependent operand by reading the register file in the RD stage; instead, it can obtain the dependent operand by accessing the output of the ALU stage. With the addition of this forwarding path and the associated control logic, the worst-case penalty incurred is now zero cycles when the leading instruction is an ALU instruction. Even if the trailing instruction is instruction $i + 1$, no stalling is needed because instruction $i + 1$ can enter the ALU stage as instruction i leaves the ALU stage just as a normal pipeline operation.

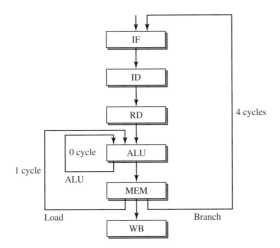

Figure 2.16

Incorporation of Forwarding Paths in the TYP Pipeline to Reduce ALU and Load Penalties.

In the case that the leading instruction is a load instruction rather than an ALU instruction, a similar forwarding path can be incorporated to reduce the penalty cycles incurred due to a leading load instruction and a dependent trailing instruction. Examining Figure 2.16 reveals that if the leading instruction is a load instruction, the result of this load instruction, that is, the content of the memory location being loaded into the register, is available at the output of the MEM stage when the load instruction completes the MEM stage. Again, a forwarding path can be added from the output of the MEM stage to the input of the ALU stage to support the requirement of the trailing instruction. The trailing instruction can enter the ALU stage as soon as the leading load instruction completes the MEM stage. This effectively reduces the worst-case penalty due to a leading load instruction from three cycles down to just one cycle. In the worst case, the dependent instruction is instruction $i + 1$, i.e., $j = i + 1$. In normal pipeline processing when instruction i is in the ALU stage, instruction $i + 1$ will be in the RD stage. When instruction i advances to the MEM stage, instruction $i + 1$ must be held back at the RD stage via stalling the earlier stages of the pipeline. However, in the next cycle when instruction i exits the MEM stage, with the forwarding path from the output of the MEM stage to the input of the ALU stage, instruction $i + 1$ can be allowed to enter the ALU stage. In effect, instruction $i + 1$ is only stalled for one cycle in the RD stage; hence the worst-case penalty is one cycle. With the incorporation of the forwarding paths the worst-case penalties for RAW hazards can be reduced as shown in Table 2.6.

The penalty due to a RAW hazard with an ALU instruction as the leading instruction is referred to as the *ALU penalty*. Similarly, the penalty due to a leading load instruction is referred to as the *load penalty*. For the TYP pipeline, with forwarding paths added, the ALU penalty is zero cycles. In effect, when the leading instruction is an ALU instruction, no penalty is incurred. Note that the source of the forwarding path is the output of the ALU stage, this being the earliest point where the result of instruction i is available. The destination of the forwarding path is the input to the ALU stage, this being the latest point where the result from instruction i is

Table 2.6

Worst-case penalties due to RAW hazards in the TYP pipeline when forwarding paths are used

	Leading Instruction Type (i)		
	ALU	**Load**	**Branch**
Trailing instruction types (j)	ALU, Load/Store, Br.	ALU, Load/Store, Br.	ALU, Load/Store, Br.
Hazard register	Int. register (Ri)	Int. register (Ri)	PC
Register write stage (i)	WB (stage 6)	WB (stage 6)	MEM (stage 5)
Register read stage (j)	RD (stage 3)	RD (stage 3)	IF (stage 1)
Forward from outputs of:	ALU, MEM, WB	MEM, WB	MEM
Forward to input of:	ALU	ALU	IF
Penalty w/ forwarding paths	0 cycles	1 cycle	4 cycles

needed by instruction j. A forwarding path from the earliest point a result is available to the latest point that result is needed by a dependent instruction is termed the *critical forwarding path,* and it represents the best that can be done in terms of reducing the hazard penalty for that type of leading instruction.

In addition to the critical forwarding path, additional forwarding paths are needed. For example, forwarding paths are needed that start from the outputs of the MEM and WB stages and end at the input to the ALU stage. These two additional forwarding paths are needed because the dependent instruction j could potentially be instruction $i + 2$ or instruction $i + 3$. If $j = i + 2$, then when instruction j is ready to enter the ALU stage, instruction i will be exiting the MEM stage. Hence, the result of instruction i, which still has not been written back to the destination register and is needed by instruction j, is now available at the output of the MEM stage and must be forwarded to the input of the ALU stage to allow instruction j to enter that stage in the next cycle. Similarly, if $j = i + 3$, the result of instruction i must be forwarded from the output of the WB stage to the input of the ALU stage. In this case, although instruction i has completed the write back to the destination register, instruction j has already traversed the RD stage and is ready to enter the ALU stage. Of course, in the case that $j = i + 4$, the RAW dependence is easily satisfied via the normal reading of the register file by j without requiring the use of any forwarding path. By the time j reaches the RD stage, i will have completed the WB stage.

If the leading instruction is a load instruction, the earliest point at which the result of instruction i is available is at the output of the MEM stage, and the latest point where this result is needed is at the input to the ALU stage. Hence the critical forwarding path for a leading load instruction is from the output of the MEM stage to the input of the ALU stage. This represents the best that can be done, and in this case the incurring of the one cycle penalty is unavoidable. Again, another forwarding path from the output of the WB stage to the input of the ALU stage is needed in case the dependent trailing instruction is ready to enter the ALU stage when instruction i is exiting the WB stage.

Table 2.6 indicates that no forwarding path is used to reduce the penalty due to a branch instruction. If the leading instruction i is a branch instruction and given the addressing mode assumed for the TYP pipeline, the earliest point where the result is available is at the output of the MEM stage. For branch instructions, the branch target address and the branch condition are generated in the ALU stage. It is not until the MEM stage that the branch condition is checked and that the target address of the branch is loaded into the PC. Consequently, only after the MEM stage can the PC be used to fetch the branch target. On the other hand, the PC must be available at the beginning of the IF stage to allow the fetching of the next instruction. Hence the latest point where the result is needed is at the beginning of the IF stage. As a result the critical forwarding path, or the best that can be done, is the current penalty path of updating the PC with the branch target in the MEM stage and starting the fetching of the branch target in the next cycle if the branch is taken. If, however, the branch condition can be generated early enough in the ALU stage to allow updating the PC with the branch target address toward the end of the ALU stage, then in that case the branch penalty can be reduced from four cycles to three cycles.

2.2.3.5 Implementation of Pipeline Interlock.
The resolving of pipeline hazards via hardware mechanisms is referred to as *pipeline interlock*. Pipeline interlock hardware must detect all pipeline hazards and ensure that all the dependences are satisfied. Pipeline interlock can involve stalling certain stages of the pipeline as well as controlling the forwarding of data via the forwarding paths.

With the addition of forwarding paths, the scalar pipeline is no longer a simple sequence of pipeline stages with data flowing from the first stage to the last stage. The forwarding paths now provide potential feedback paths from outputs of later stages to inputs of earlier stages. For example, the three forwarding paths needed to support a leading ALU instruction involved in a pipeline hazard are illustrated in Figure 2.17. These are referred to as *ALU forwarding paths*. As the leading ALU instruction i traverses down the pipeline stages, there could be multiple trailing instructions that are data (RAW) dependent on instruction i. The right side of Figure 2.17 illustrates how multiple dependent trailing instructions are satisfied during three consecutive machine cycles. During cycle t1, instruction i forwards its result to dependent instruction $i + 1$ via forwarding path a. During the next cycle, t2, instruction i forwards its result to dependent instruction $i + 2$ via forwarding path b. If instruction $i + 2$ also requires the result of instruction $i + 1$, this result can also be forwarded to $i + 2$ by $i + 1$ via forwarding path a during this cycle. During cycle t3, instruction i can forward its result to instruction $i + 3$ via forwarding path c. Again, path a or path b can also be activated during this cycle if instruction $i + 3$ also requires the result of $i + 2$ or $i + 1$, respectively.

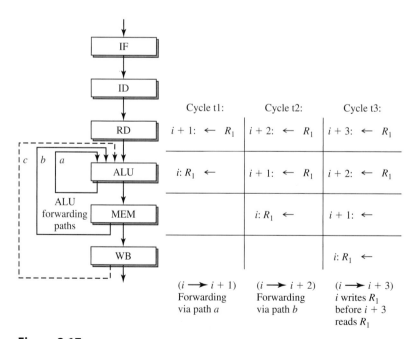

Figure 2.17
Forwarding Paths for Supporting Pipeline Hazards Due to an ALU Leading Instruction.

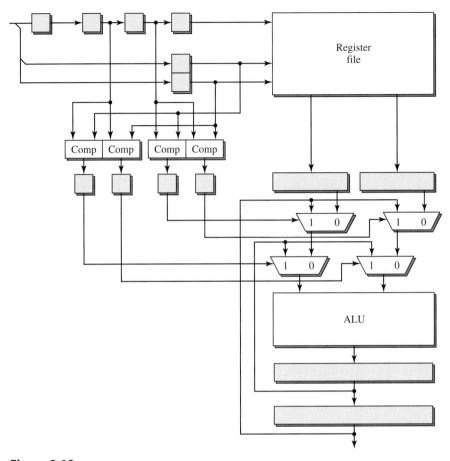

Figure 2.18
Implementation of Pipeline Interlock for RAW Hazards Involving a Leading ALU Instruction.

The physical implementation of the logical diagram of Figure 2.17 is shown in Figure 2.18. Note that RAW hazards are detected using comparators that compare the register specifiers of consecutive instructions. Four 5-bit (assuming 32 registers) comparators are shown in Figure 2.18. If the trailing instruction j is currently in the RD stage, that is, attempting to read its two register operands, then the first two comparators (to the left) are checking for possible RAW dependences between instruction j and instruction $j - 1$, which is now in the ALU stage. These two comparators are comparing the two source register specifiers of j with the destination register specifier of $j - 1$. At the same time the other two comparators (to the right) are checking for possible RAW dependences between j and $j - 2$, which is now in the MEM stage. These two comparators are comparing the two source register specifiers of j with the destination register specifier of $j - 2$. The outputs of these four comparators are used as control signals in the next cycle for activating the appropriate forwarding paths if dependences are detected.

Forwarding path *a* is activated by the first pair of comparators if any RAW dependences are detected between instructions j and $j - 1$. Similarly, forwarding path *b* is activated by the outputs of the second pair of comparators for satisfying any dependences between instructions j and $j - 2$. Both paths can be simultaneously activated if j depends on both $j - 1$ and $j - 2$.

Forwarding path *c* of Figure 2.17 is not shown in Figure 2.18; the reason is that this forwarding path may not be necessary if appropriate care is taken in the design of the multiported register file. If the physical design of the three-ported (two reads and one write) register file performs first the write and then the two reads in each cycle, then the third forwarding path is not necessary. Essentially instruction j will read the new, and correct, value of the dependent register when it traverses the RD stage. In other words, the forwarding is performed internally in the register file. There is no need to wait for one more cycle to read the dependent register or to forward it from the output of the WB stage to the input of the ALU stage. This is a reasonable design choice, which can reduce either the penalty cycle by one or the number of forwarding paths by one, and it is actually implemented in the MIPS R2000/R3000 pipeline.

To reduce the penalty due to pipeline hazards that involve leading load instructions, another set of forwarding paths is needed. Figure 2.19 illustrates the two forwarding paths needed when the leading instruction involved in a pipeline hazard is a load instruction. These are referred to as *load forwarding paths*. Forwarding path *d* forwards the output of the MEM stage to the input of the ALU stage, whereas path *e* forwards the output of the WB stage to the input of the ALU stage. When the leading

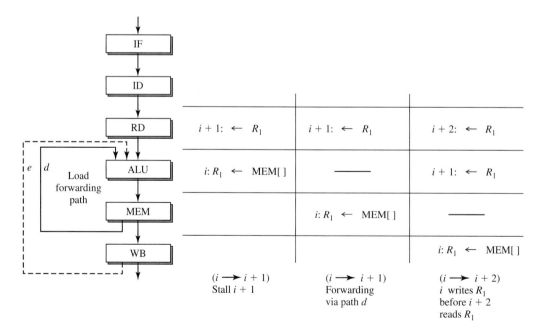

Figure 2.19
Forwarding Paths for Supporting Pipeline Hazards Due to a Leading Load Instruction.

instruction i reaches the ALU stage, if instruction $i + 1$ is dependent on instruction i, it must be stalled in the RD stage for one cycle. In the next cycle, when instruction i is exiting the MEM stage, its result can be forwarded to the ALU stage via path d to allow instruction $i + 1$ to enter the ALU stage. In case instruction $i + 2$ also depends on instruction i, the same result is forwarded in the next cycle via path e from the WB stage to the ALU stage to allow instruction $i + 2$ to proceed into the ALU stage without incurring another stall cycle. Again, if the multiported register file performs first the write and then the read, then forwarding path e will not be necessary. For example, instruction $i + 2$ will read the result of instruction i in the RD stage while instruction i is simultaneously performing a register write in the WB stage.

The physical implementation of all the forwarding paths for supporting pipeline hazards due to both ALU and load leading instructions is shown in Figure 2.20. Forwarding path e is not shown, assuming that the register file is designed to perform first the write and then the read in each cycle. Note that while both ALU forwarding path b

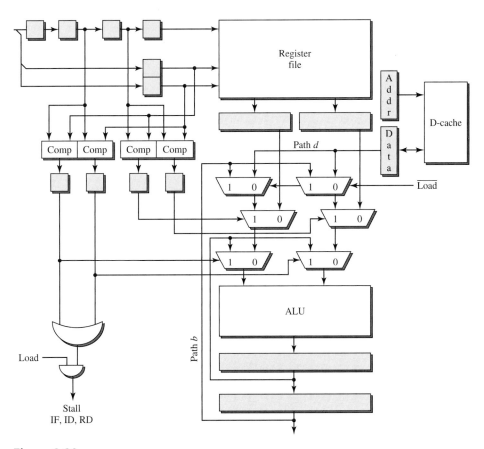

Figure 2.20
Implementation of Pipeline Interlock for RAW Hazards Involving ALU and Load Instructions.

and load forwarding path d are shown in Figure 2.17 and Figure 2.19, respectively, as going from the output of the MEM stage to the input of the ALU stage, these are actually two different physical paths, as shown in Figure 2.20. These two paths feed into the first pair of multiplexers, and only one of the two can be selected depending on whether the leading instruction in the MEM stage is an ALU or a load instruction. Forwarding path b originates from the buffer in the MEM stage that contains the output of the ALU from the previous machine cycle. Forwarding path d originates from the buffer in the MEM stage that contains the data accessed from the D-cache.

The same two pairs of comparators are used to detect register dependences regardless of whether the leading instruction is an ALU or a load instruction. Two pairs of comparators are required because the interlock hardware must detect possible dependences between instructions i and $i + 1$ as well as between instructions i and $i + 2$. When the register file is designed to perform first a write and then a read in each cycle, a dependence between instructions i and $i + 3$ is automatically satisfied when they traverse the WB and RD stages, respectively. The output of the first pair of comparators is used along with a signal from the ID stage indicating that the leading instruction is a load to produce a control signal for stalling the first three stages of the pipeline for one cycle if a dependence is detected between instructions i and $i + 1$, and that instruction i is a load.

Pipeline interlock hardware for the TYP pipeline must also deal with pipeline hazards due to control dependences. The implementation of the interlock mechanism for supporting control hazards involving a leading branch instruction is shown in Figure 2.21. Normally, in every cycle the IF stage accesses the I-cache to

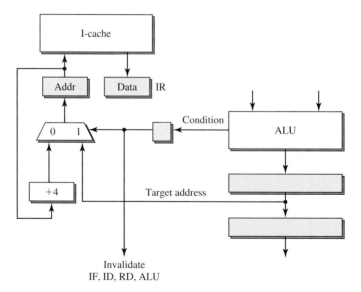

Figure 2.21

Implementation of Pipeline Interlock for Hazards Involving a Branch Instruction.

fetch the next instruction and at the same time increments the PC in preparation for fetching the next sequential instruction. When a branch instruction is fetched in the IF stage and then decoded to be such an instruction in the ID stage, the IF stage can be stalled until the branch instruction traverses the ALU stage, in which both the branch condition and the target address are generated. In the next cycle, corresponding to the MEM stage of the branch instruction, the branch condition is used to load the branch target address into the PC via the right side of the PC multiplexer if the branch is taken. This results in a four-cycle penalty whenever a branch instruction is encountered. Alternatively, the branch instruction can be assumed to be not taken, and the IF stage continues to fetch subsequent instructions along the sequential path. In the case when the branch instruction is determined to be taken, the PC is updated with the branch target during the MEM stage of the branch instruction, and the sequential instructions in the IF, ID, RD, and ALU stages are invalidated and flushed from the pipeline. In this case, the four-cycle penalty is incurred only when the branch is actually taken. If the branch turns out to be not taken, then no penalty cycle is incurred.

2.2.4 Commercial Pipelined Processors

Pipelined processor design has become a mature and widely adopted technology. The compatibility of the RISC philosophy with instruction pipelining is well known and well exploited. Pipelining has also been successfully applied to CISC architectures. This subsection highlights two representative pipelined processors. The MIPS R2000/R3000 pipeline is presented as representative of RISC pipeline processors [Moussouris et al., 1986; Kane, 1987]. The Intel i486 is presented as representative of CISC pipelined processors [Crawford, 1990]. Experimental data from an IBM study on RISC pipelined processors done by Tilak Agerwala and John Cocke in 1987 are presented as representative of the characteristics and the performance capabilities of scalar pipelined processors [Agerwala and Cocke, 1987].

2.2.4.1 RISC Pipelined Processor Example. MIPS is a RISC architecture with 32-bit instructions. There are three different instruction formats as shown in Figure 2.22.

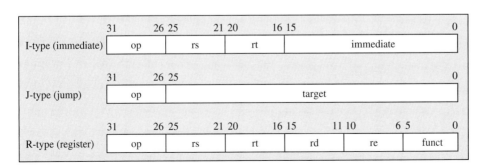

Figure 2.22
Instruction Formats Used in the MIPS Instruction Set Architecture.

Instructions can be divided into four types.

- *Computational* instructions perform arithmetic, logical, and shift operations on register operands. They can employ the R-type format if all the operands and the result are registers, or the I-type format if one of the operands is specified in the immediate field of the instruction.
- *Load/store* instructions move data between the memory and registers. They employ the I-type format. The only addressing mode is the base register plus the signed offset stored in the immediate field.
- *Jump and branch* instructions steer the control flow of the program. Jumps are unconditional and use the J-type format to jump to an absolute address composed of the 26-bit target and the high-order 4 bits of the PC. Branches are conditional and use the I-type format to specify the target address as the PC plus the 16-bit offset in the immediate field.
- *Other* instructions in the instruction set are used to perform operations in the coprocessors and other special system functions. Coprocessor 0 (CP0) is the system control coprocessor. CP0 instructions manipulate the memory management and exception handling facilities. Floating-point instructions are implemented as coprocessor instructions and are performed by a separate floating-point processor.

The MIPS R2000/R3000 pipeline is a five-stage instruction pipeline quite similar to the TYP pipeline. However, each pipeline stage is further divided into two separate phases, identified as phase one ($\phi 1$) and phase two ($\phi 2$). The functions performed by each of the five stages and their phases are described in Table 2.7.

There are a number of interesting features in this five-stage pipeline. The I-cache access, which requires an entire cycle, actually takes place during $\phi 2$ of

Table 2.7
Functionality of the MIPS R2000/R3000 five-stage pipeline

Stage Name	Phase	Function Performed
1. IF	$\phi 1$	Translate virtual instruction address using TLB.
	$\phi 2$	Access I-cache using physical address.
2. RD	$\phi 1$	Return instructions from I-cache; check tags and parity.
	$\phi 2$	Read register file; if a branch, generate target address.
3. ALU	$\phi 1$	Start ALU operation; if a branch, check branch condition.
	$\phi 2$	Finish ALU operation; if a load/store, translate virtual address.
4. MEM	$\phi 1$	Access D-cache.
	$\phi 2$	Return data from D-cache; check tags and parity.
5. WB	$\phi 1$	Write register file.
	$\phi 2$	—

the IF stage and φ1 of the RD stage. One translation lookaside buffer (TLB) is used to do address translation for both the I-cache and the D-cache. The TLB is accessed during φ1 of the IF stage, for supporting I-cache access and is accessed during φ2 of the ALU stage, for supporting D-cache access, which takes place during the MEM cycle. The register file performs first a write (φ1 of WB stage), and then a read (φ2 of RD stage) in every machine cycle. This pipeline requires a three-ported (two reads and one write) register file and a single-ported I-cache and a single-ported D-cache to support the IF and MEM stages, respectively.

With forwarding paths from the outputs of the ALU and the MEM stages back to the input of the ALU stage, no ALU leading hazards will incur a penalty cycle. The load penalty, that is, the worst-case penalty incurred by a load leading hazard, is only one cycle with the forwarding path from the output of the MEM stage to the input of the ALU stage. The branch penalty is also only one cycle. This is made possible due to several features of the R2000/R3000 pipeline. First, branch instructions use only PC-relative addressing mode. Unlike a register which must be accessed during the RD stage, the PC is available after the IF stage. Hence, the branch target address can be calculated, albeit using a separate adder, during the RD stage. The second feature is that no explicit condition code bit is generated and stored. The branch condition is generated during φ1 of the ALU stage by comparing the contents of the referenced register(s). Normally with the branch condition being generated in the ALU stage (stage 3) and the instruction fetch being done in the IF stage (stage 1), the expected penalty would be two cycles. However, in this particular pipeline design the I-cache access actually does not start until φ2 of the IF stage. With the branch condition being available at the end of φ1 of the ALU stage and since the I-cache access does not begin until φ2 of the IF stage, the branch target address produced at the end of the RD stage can be steered by the branch condition into the PC prior to the start of I-cache access in the middle of the IF stage. Consequently only a one-cycle penalty is incurred by branch instructions.

Compared to the six-stage TYP pipeline, the five-stage MIPS R2000/R3000 pipeline is a better design in terms of the penalties incurred due to pipeline hazards. Both pipelines have the same ALU and load penalties of zero cycles and one cycle, respectively. However, due to the above stated features in its design, the MIPS R2000/R3000 pipeline incurs only one cycle, instead of four cycles, for its branch penalty. Influenced by and having benefited from the RISC research done at Stanford University, the MIPS R2000/R3000 has a very clean design and is a highly efficient pipelined processor.

2.2.4.2 CISC Pipelined Processor Example. In 1978 Intel introduced one of the first 16-bit microprocessors, the Intel 8086. Although preceded by earlier 8-bit microprocessors from Intel (8080 and 8085), the 8086 began an evolution that would eventually result in the Intel IA32 family of object code compatible microprocessors. The Intel IA32 is a CISC architecture with variable-length instructions and complex addressing modes, and it is by far the most dominant architecture today in terms of sales volume and the accompanying application software base. In 1985, the Intel 386, the 32-bit version of the IA32 family, was introduced [Crawford, 1986]. The first pipelined version of the IA32 family, the Intel 486, was introduced in 1989.

Table 2.8
Functionality of the Intel 486 five-stage pipeline

Stage Name	Function Performed
1. Instruction fetch	Fetch instruction from the 32-byte prefetch queue (prefetch unit fills and flushes prefetch queue).
2. Instruction decode-1	Translate instruction into control signals or microcode address. Initiate address generation and memory access.
3. Instruction decode-2	Access microcode memory. Output microinstruction to execute unit.
4. Execute	Execute ALU and memory accessing operations.
5. Register write-back	Write back result to register.

While the original 8086 chip had less than 30K transistors, the 486 chip has more than 1M transistors. The 486 is object code compatible with all previous members of the IA32 family, and it became the most popular microprocessor used for personal computers in the early 1990s [Crawford, 1990].

The 486 implemented a five-stage instruction pipeline. The functionality of the pipeline stages is described in Table 2.8. An instruction prefetch unit, via the bus interface unit, prefetches 16-byte blocks of instructions into the prefetch queue. During the instruction fetch stage, each instruction is fetched from the 32-byte prefetch queue. Instruction decoding is performed in two stages. Hardwired control signals as well as microinstructions are produced during instruction decoding. The execute stage performs both ALU operations as well as cache accesses. Address translation and effective address generation are carried out during instruction decoding; memory accessing is completed in the execute stage. Hence, a memory load followed immediately by a use does not incur any penalty cycle; output of the execute stage is forwarded to its input. However, if an instruction that produces a register result is followed immediately by another instruction that uses the same register for address generation, then a penalty cycle is necessary because address generation is done during instruction decoding. The fifth stage in the pipeline performs a register write-back. Floating-point operations are carried out by an on-chip floating-point unit and can incur multiple cycles for their execution.

With the five-stage instruction pipeline, the 486 can execute many IA32 instructions in one cycle without using microcode. Some instructions require the accessing of micro-instructions and multiple cycles. The 486 clearly demonstrates the performance improvement that can be obtained via instruction pipelining. Based on typical instruction mix and the execution times for the frequently used IA32 instructions, the Intel 386 is able to achieve an average cycles per instruction (CPI) of 4.9 [Crawford, 1986]. The pipelined Intel 486 can achieve an average CPI of about 1.95. This represents a speedup by a factor of about 2.5. In our terminology, the five-stage i486 achieved an effective degree of pipelining of 2.5. Clearly, significant pipelining overhead is involved, primarily due to the complexity of the IA32 instruction set architecture and the burden of ensuring object code compatibility. Nonetheless, for a CISC architecture, the speedup obtained is quite respectable. The 486 clearly demonstrated the feasibility of pipelining a CISC architecture.

2.2.4.3 Scalar Pipelined Processor Performance. A report documenting the IBM experience with pipelined RISC machines by Tilak Agerwala and John Cocke in 1987 provided an assessment of the performance capability of scalar pipelined RISC processors [Agerwala and Cocke, 1987]. Some of the key observations from that report are presented here. In this study, it is assumed that the I-cache and D-cache are separate. The I-cache can supply one instruction per cycle to the processor. Only load/store instructions access the D-cache. In this study, the hit rates for both caches are assumed to be 100%. The default latency for both caches is one cycle. The following characteristics and statistics are used in the study.

1. Dynamic instruction mix
   a. ALU: 40% (register-register)
   b. Loads: 25%
   c. Stores: 15%
   d. Branches: 20%

2. Dynamic branch instruction mix
   a. Unconditional: 33.3% (always taken)
   b. Conditional—taken: 33.3%
   c. Conditional—not taken: 33.3%

3. Load scheduling
   a. Cannot be scheduled: 25% (no delay slot filled)
   b. Can be moved back one or two instructions: 65% (fill two delay slots)
   c. Can be moved back one instruction: 10% (fill one delay slot)

4. Branch scheduling
   a. Unconditional: 100% schedulable (fill one delay slot)
   b. Conditional: 50% schedulable (fill one delay slot)

The performance of a processor can be estimated using the average cycles per instruction. The idealized goal of a scalar pipeline processor is to achieve a CPI = 1. This implies that the pipeline is processing or completing, on the average, one instruction in every cycle. The IBM study attempted to quantify how closely this idealized goal can be reached. Initially, it is assumed that there is no ALU penalty and that the load and branch penalties are both two cycles. Given the dynamic instruction mix, the CPI overheads due to these two penalties can be computed.

- Load penalty overhead: $0.25 \times 2 = 0.5$ CPI
- Branch penalty overhead: $0.20 \times 0.66 \times 2 = 0.27$ CPI
- **Resultant CPI:** $1.0 + 0.5 + 0.27 = $ **1.77** CPI

Since 25% of the dynamic instructions are loads, if we assume each load incurs the two-cycle penalty, the CPI overhead is 0.5. If the pipeline assumes that branch instructions are not taken, or biased for not taken, then only the 66.6% of the branch instructions that are taken will incur the two-cycle branch penalty. Taking into account both the load and branch penalties, the expected CPI is 1.77. This is far from the idealized goal of CPI = 1.

Assuming that a forwarding path can be added to bypass the register file for load instructions, the load penalty can be reduced from two cycles down to just one cycle. With the addition of this forwarding path, the CPI can be reduced to $1.0 + 0.25 + 0.27 = 1.52$.

In addition, the compiler can be employed to schedule instructions into the load and branch penalty slots. Assuming the statistics presented in the preceding text, since 65% of the loads can be moved back by one or two instructions and 10% of the loads can be moved back by one instruction, a total of 75% of the load instructions can be scheduled, or moved back, so as to eliminate the load penalty of one cycle. For 33.3% of the branch instructions that are unconditional, they can all be scheduled to reduce the branch penalty for them from two cycles to one cycle. Since the pipeline is biased for not taken branches, the 33.3% of the branches that are conditional and not taken incur no branch penalty. For the remaining 33.3% of the branches that are conditional and taken, the assumption is that 50% of them are schedulable, that is, can be moved back one instruction. Hence 50% of the conditional branches that are taken will incur only a one-cycle penalty, and the other 50% will incur the normal two-cycle penalty. The new CPI overheads and the resultant CPI are shown here.

- Load penalty overhead: $0.25 \times 0.25 \times 1 = 0.0625$ CPI
- Branch penalty overhead: $0.20 \times [0.33 \times 1 + 0.33 \times 0.5 \times 1 + 0.33 \times 0.5 \times 2] = 0.167$ CPI
- **Resultant CPI:** $1.0 + 0.063 + 0.167 =$ **1.23** CPI

By scheduling the load and branch penalty slots, the CPI overheads due to load and branch penalties are significantly reduced. The resultant CPI of 1.23 is approaching the idealized goal of CPI = 1. The CPI overhead due to the branch penalty is still significant. One way to reduce this overhead further is to consider ways to reduce the branch penalty of two cycles. From the IBM study, instead of using the register-indirect mode of addressing, 90% of the branches can be coded as PC-relative. Using the PC-relative addressing mode, the branch target address generation can be done without having to access the register file. A separate adder can be included to generate the target address in parallel with the register read stage. Hence, for the branch instructions that employ PC-relative addressing, the branch penalty can be reduced by one cycle. For the 33.3% of the branches that are unconditional, they are 100% schedulable. Hence, the branch penalty is only one cycle. If 90% of them can be made PC-relative and consequently eliminate the branch penalty, then only the remaining 10% of the unconditional branches will incur the branch penalty of one cycle. The corresponding

Table 2.9
Conditional branch penalties considering PC-relative addressing and scheduling of penalty slot

PC-relative Addressing	Schedulable	Branch Penalty
Yes (90%)	Yes (50%)	0 cycles
Yes (90%)	No (50%)	1 cycle
No (10%)	Yes (50%)	1 cycle
No (10%)	No (50%)	2 cycles

CPI overhead for unconditional branches is then $0.20 \times 0.33 \times 0.10 \times 1 = 0.0066$ CPI.

With the employment of the PC-relative addressing mode, the fetch stage is no longer biased for the not taken branches. Hence all conditional branches can be treated in the same way, regardless of whether they are taken. Depending on whether a conditional branch can be made PC-relative and whether it can be scheduled, there are four possible cases. The penalties for these four possible cases for conditional branches are shown in Table 2.9.

Including both taken and not taken ones, 66.6% of the branches are conditional. The CPI overhead due to conditional branches is derived by considering the cases in Table 2.9 and is equal to

$$0.20 \times 0.66 \times \{[0.9 \times 0.5 \times 1] + [0.1 \times 0.5 \times 1] + [0.1 \times 0.5 \times 2]\} = 0.079 \text{ CPI}$$

Combining the CPI overheads due to unconditional and conditional branches results in the total CPI overhead due to branch penalty of $0.0066 + 0.079 = 0.0856$ CPI. Along with the original load penalty, the new overheads and the resultant overall CPI are shown here.

- Load penalty overhead: 0.0625 CPI
- Branch penalty overhead: 0.0856 CPI
- **Resultant CPI:** $1.0 + 0.0625 + 0.0856 =$ **1.149** CPI

Therefore, with a series of refinements, the original CPI of 1.77 is reduced to 1.15. This is quite close to the idealized goal of CPI = 1. One way to view this is that CPI = 1 represents the ideal instruction pipeline, in which a new instruction is entered into the pipeline in every cycle. This is achievable only if the third point of pipelining idealism is true, that is, all the instructions are independent. In real programs there are inter-instruction dependences. The CPI = 1.15 indicates that only a 15% overhead or inefficiency is incurred in the design of a realistic instruction pipeline that can deal with inter-instruction dependences. This is quite impressive and reflects the effectiveness of instruction pipelining.

2.3 Deeply Pipelined Processors

Pipelining is a very effective means of improving processor performance, and there are strong motivations for employing deep pipelines. A deeper pipeline increases the number of pipeline stages and reduces the number of logic gate levels in each pipeline stage. The primary benefit of deeper pipelines is the ability to reduce the machine cycle time and hence increase the clocking frequency. During the 1980s most pipelined microprocessors had four to six pipeline stages. Contemporary high-end microprocessors have clocking frequencies in the multiple-gigahertz range, and pipeline depths have increased to more than 20 pipeline stages. Pipelines have gotten not only deeper, but also wider, such as superscalar processors. As pipelines get wider, there is increased complexity in each pipeline stage, which can increase the delay of each pipeline stage. To maintain the same clocking frequency, a wider pipeline will need to be made even deeper.

There is a downside to deeper pipelines. With a deeper pipeline the penalties incurred for pipeline hazard resolution can become larger. Figure 2.23 illustrates what can happen to the ALU, load, and branch penalties when a pipeline becomes wider and much deeper. Comparing the shallow and the deep pipelines, we see that the ALU penalty increases from zero cycles to one cycle, the load penalty increases from one cycle to four cycles, and most importantly, the branch penalty goes from three cycles to eleven cycles. With increased pipeline penalties, the average CPI increases. The potential performance gain due to the higher clocking frequency of a deeper pipeline can be ameliorated by the increase of CPI. To ensure overall performance improvement with a deeper pipeline, the increase in clocking frequency must exceed the increase in CPI.

There are two approaches that can be used to mitigate the negative impact of the increased branch penalty in deep pipelines; see Figure 2.24. Among the three pipeline penalties, the branch penalty is the most severe because it spans all the front-end pipeline stages. With a mispredicted branch, all the instructions in the front-end pipeline stages must be flushed. The first approach to reduce the branch

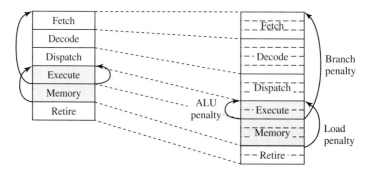

Figure 2.23

Impact on ALU, Load, and Branch Penalties with Increasing Pipeline Depth.

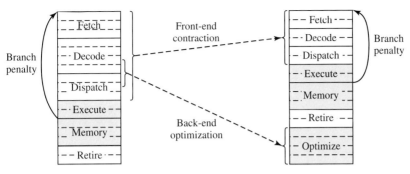

Figure 2.24
Mitigating the Branch Penalty Impact of Deep Pipelines.

penalty is to reduce the number of pipeline stages in the front end. For example, a CISC architecture with variable instruction length can require very complex instruction decoding logic that can require multiple pipeline stages. By using a RISC architecture, the decoding complexity is reduced, resulting in fewer front-end pipeline stages. Another example is the use of pre-decoding logic prior to loading instructions into the I-cache. Pre-decoded instructions fetched from the I-cache require less decoding logic and hence fewer decode stages.

The second approach is to move some of the front-end complexity to the back end of the pipeline, resulting in a shallower front end and hence a smaller branch penalty. This has been an active area of research. When a sequence of instructions is repeatedly executed by a pipeline, the front-end pipeline stages repeatedly perform the same work of fetching, decoding, and dispatching on the same instructions. Some have suggested that the result of the work done can be cached and reused without having to repeat the same work. For example, a block of decoded instructions can be stored in a special cache. Subsequent fetching of these same instructions can be done by accessing this cache, and the decoding pipeline stage(s) can be bypassed. Other than just caching these decoded instructions, additional optimization can be performed on these instructions, leading to further elimination of the need for some of the front-end pipeline stages. Both the caching and the optimization can be implemented in the back end of the pipeline without impacting the front-end depth and the associated branch penalty. In order for deep pipelines to harvest the performance benefit of a higher clocking frequency, the pipeline penalties must be kept under control.

There are different forms of tradeoffs involved in designing deep pipelines. As indicated in Section 2.1, a k-stage pipeline can potentially achieve an increase of throughput by a factor of k relative to a nonpipelined design. When cost is taken into account, there is a tradeoff involving cost and performance. This tradeoff dictates that the optimal value of k not be arbitrarily large. This is illustrated in Figure 2.3. This form of tradeoff deals with the hardware cost of implementing the pipeline, and it indicates that there is a pipeline depth beyond which the additional cost of pipelining cannot be justified by the diminishing return on the performance gain.

There is another form of tradeoff based on the foregoing analysis of CPI impact induced by deep pipelines. This tradeoff involves the increase of clocking frequency versus the increase of CPI. According to the iron law of processor performance (Sec. 1.3.1, Eq. 1.1), performance is determined by the product of clocking frequency and the average IPC, or the *frequency/CPI* ratio. As pipelines get deeper, frequency increases but so does CPI. Increasing the pipeline depth is profitable as long as the added pipeline depth brings about a net increase in performance. There is a point beyond which pipelining any deeper will lead to little or no performance improvement. The interesting question is, How deep can a pipeline go before we reach this point of diminishing returns?

A number of recent studies have focused on determining the optimum pipeline depth [Hartstein and Puzak, 2002, 2003; Sprangle and Carmean, 2002; Srinivasan et al., 2002] for a microprocessor. As pipeline depth increases, frequency can be increased. However the frequency does not increase linearly with respect to the increase of pipeline depth. The sublinear increase of frequency is due to the overhead of adding latches. As pipeline depth increases, CPI also increases due to the increase of branch and load penalties. Combining frequency and CPI behaviors yields the overall performance. As pipeline depth is increased, the overall performance tends to increase due to the benefit of the increased frequency. However when pipeline depth is further increased, there reaches a point where the CPI overhead overcomes the benefit of the increased frequency; any further increase of pipeline depth beyond this point can actually bring about the gradual decrease of overall performance. In a recent study, Hartstein and Puzak [2003] showed, based on their performance model, this point of diminishing return, and hence the optimum pipeline depth, occurs around pipeline depth of ~25 stages. Using more aggressive assumptions, Sprangle and Carmean [2002] showed that the optimum pipeline depth is actually around 50 stages.

If power consumption is taken into account, the optimum pipeline depth is significantly less than 25 or 50 pipe stages. The higher frequency of a deeper pipeline leads to a significant increase of power consumption. Power consumption can become prohibitive so as to render a deep pipeline infeasible, even if there is more performance to be harvested. In the same study, Hartstein and Puzak [2003] developed a new model for optimum pipeline depth by taking into account power consumption in addition to performance. They use a model based on the BIPS^3/W metric, where BIPS^3 is billions of instructions per second to the third power, and W is watt. This model essentially favors performance (BIPS) to power (W) by a ratio of 3 to 1. Given their model, the optimum pipeline depth is now more in the range of 6–9 pipe stages. Assuming lower latching overhead and with increasing leakage power, they showed the optimum pipeline depth could potentially be in the range of 10–15 pipe stages. While in recent years we have witnessed the relentless push towards ever higher clocking frequencies and ever deeper pipelines, the constraints due to power consumption and heat dissipation can become serious impediments to this relentless push.

2.4 Summary

Pipelining is a microarchitecture technique that can be applied to any ISA. It is true that the features of RISC architectures make pipelining easier and produce more efficient pipeline designs. However, pipelining is equally effective on CISC architectures. Pipelining has proved to be a very powerful technique in increasing processor performance, and in terms of pipeline depth there is still plenty of headroom. We can expect much deeper pipelines.

The key impediment to pipelined processor performance is the stalling of the pipeline due to inter-instruction dependences. A branch penalty due to control dependences is the biggest culprit. Dynamic branch prediction can alleviate this problem so as to incur the branch penalty only when a branch misprediction occurs. When a branch is correctly predicted, there is no stalling of the pipeline; however, when a branch misprediction is detected, the pipeline must be flushed.

As pipelines get deeper, the branch penalty increases and becomes the key challenge. One strategy is to reduce the branch penalty by reducing the depth of the front end of the pipeline, that is, the distance between the instruction fetch stage and the stage in which branch instructions are resolved. An alternative is to increase the accuracy of the dynamic branch prediction algorithm so that the frequency of branch misprediction is reduced; hence, the frequency of incurring the branch penalty is also reduced. We did not cover dynamic branch prediction in this chapter. This is a very important topic, and we have chosen to present branch prediction in the context of superscalar processors. We will get to it in Chapter 5.

Pipelined processor design alters the relevance of the classic view of CPU design. The classic view partitions the design of a processor into data path design and control path design. Data path design focuses on the design of the ALU and other functional units as well as the accessing of registers. Control path design focuses on the design of the state machines to decode instructions and generate the sequence of control signals necessary to appropriately manipulate the data path. This view is no longer relevant. In a pipelined processor this partition is no longer obvious. Instructions are decoded in the decode stage, and the decoded instructions, including the associated control signals, are propagated down the pipeline and used by various subsequent pipeline stages. Each pipeline stage simply uses the appropriate fields of the decoded instruction and associated control signals. Essentially there is no longer the centralized control performed by the control path. Instead, a form of distributed control via the propagation of the control signals through the pipeline stages is used. The traditional sequencing through multiple control path states to process an instruction is now replaced by the traversal through the various pipeline stages. Essentially, not only is the data path pipelined, but also the control path. Furthermore, the traditional data path and the control path are now integrated into the same pipeline.

REFERENCES

Agerwala, T., and J. Cocke: "High performance reduced instruction set processors," Technical report, IBM Computer Science, 1987.

Bloch, E.: "The engineering design of the STRETCH computer," *Proc. Fall Joint Computer Conf.,* 1959, pp. 48–59.

Bucholtz, W.: *Planning a Computer System: Project Stretch.* New York: McGraw-Hill, 1962.

Crawford, J.: "Architecture of the Intel 80386," *Proc. IEEE Int. Conf. on Computer Design: VLSI in Computers,* 1986, pp. 155–160.

Crawford, J.: "The execution pipeline of the Intel i486 CPU," *Proc. COMPCON Spring '90,* 1990, pp. 254–258.

Hartstein, A., and T. R. Puzak: "Optimum power/performance pipeline depth," *Proc. of the 36th Annual International Symposium on Microarchitecture (MICRO),* Dec. 2003.

Hartstein, A., and T. R. Puzak: "The optimum pipeline depth for a microprocessor," *Proc. of the 29th Annual International Symposium on Computer Architecture (ISCA),* June 2002.

Hennessy, J., and D. Patterson: *Computer Architecture: A Quantitative Approach,* 3rd ed., San Mateo, CA: Morgan Kaufmann Publishers, 2003.

Kane, G.: *MIPS R2000/R3000 RISC Architecture.* Englewood Cliffs, NJ: Prentice Hall, 1987.

Kogge, P.: *The Architecture of Pipelined Computers.* New York: McGraw-Hill, 1981.

Moussouris, J., L. Crudele, D. Frietas, C. Hansen, E. Hudson, R. March, S. Przybylski, and T. Riordan: "A CMOS RISC processor with integrated system functions," *Proc. COMPCON,* 1986, pp. 126–131.

Sprangle, E., and D. Carmean: "Increasing processor performance by implementing deeper pipelines," *Proc. of the 29th Annual International Symposium on Computer Architecture (ISCA),* June 2002.

Srinivasan, V., D. Brooks, M. Gschwind, P. Bose, V. Zyuban, P. N. Strenski, and P. G. Emma: "Optimizing pipelines for power and performance," *Proc. of the 35th Annual International Symposium on Microarchitecture (MICRO),* Dec. 2002.

Thornton, J. E.: "Parallel operation in the Control Data 6600," *AFIPS Proc. FJCC part 2,* vol. 26, 1964, pp. 33–40.

Waser, S., and M. Flynn: *Introduction to Arithmetic for Digital Systems Designers.* New York: Holt, Rinehart, and Winston, 1982.

HOMEWORK PROBLEMS

P2.1 Equation (2.4), which relates the performance of an ideal pipeline to pipeline depth, looks very similar to Amdahl's law. Describe the relationship between the terms in these two equations, and develop an intuitive explanation for why the two equations are so similar.

P2.2 Using Equation (2.7), the cost/performance optimal pipeline depth k_{opt} can be computed using parameters G, T, L, and S. Compute k_{opt} for the pipelined floating-point multiplier example in Section 2.1 by using the chip count as the cost terms ($G = 175$ chips and $L = 82/2 = 41$ chips per interstage latch) and the delays shown for T and S (T = 400 ns, S = 22 ns). How different is k_{opt} from the proposed pipelined design?

P2.3 Identify and discuss two reasons why Equation (2.4) is only useful for naive approximations of potential speedup from pipelining.

P2.4 Consider that you would like to add a *load-immediate* instruction to the TYP instruction set and pipeline. This instruction extracts a 16-bit immediate value from the instruction word, sign-extends the immediate value to 32 bits, and stores the result in the destination register specified in the instruction word. Since the extraction and sign-extension can be accomplished without the ALU, your colleague suggests that such instructions be able to write their results into the register in the decode (ID) stage. Using the hazard detection algorithm described in Figure 2.15, identify what additional hazards such a change might introduce.

P2.5 Ignoring pipeline interlock hardware (discussed in Problem 6), what additional pipeline resources does the change outlined in Problem 4 require? Discuss these resources and their cost.

P2.6 Considering the change outlined in Problem 4, redraw the pipeline interlock hardware shown in Figure 2.18 to correctly handle the load-immediate instructions.

P2.7 Consider that you would like to add byte-wide ALU instructions to the TYP instruction set and pipeline. These instructions have semantics that are otherwise identical to the existing word-width ALU instructions, except that the source operands are only 1 byte wide and the destination operand is only 1 byte wide. The byte-wide operands are stored in the same registers as the word-wide instructions, in the low-order byte, and the register writes must only affect the low-order byte (i.e., the high-order bytes must remain unchanged). Redraw the RAW pipeline interlock detection hardware shown in Figure 2.18 to correctly handle these additional ALU instructions.

P2.8 Consider adding a store instruction with an indexed addressing mode to the TYP pipeline. This store differs from the existing store with the *register + immediate* addressing mode by computing its effective address as the sum of two source registers, that is, stx r3,r4,r5 performs r3←MEM[r4+r5]. Describe the additional pipeline resources needed to support such an instruction in the TYP pipeline. Discuss the advantages and disadvantages of such an instruction.

P2.9 Consider adding a *load-update* instruction with *register + immediate and postupdate* addressing mode. In this addressing mode, the effective address for the load is computed as register + immediate, and the resulting address is written back into the base register. That is, lwu r3,8(r4) performs r3←MEM[r4+8]; r4←r4+8. Describe the additional pipeline resources needed to support such an instruction in the TYP pipeline.

P2.10 Given the change outlined in Problem 9, redraw the pipeline interlock hardware shown in Figure 2.20 to correctly handle the load-update instruction.

P2.11 Bypass network design: given the following ID, EX, MEM, and WB pipeline configuration, draw all necessary Mux0 and Mux1 bypass paths to resolve RAW data hazards. Assume that load instructions are always separated by at least one independent instruction [possibly a no-operation instruction (NOP)] from any instruction that reads the loaded register (hence you never stall due to a RAW hazard).

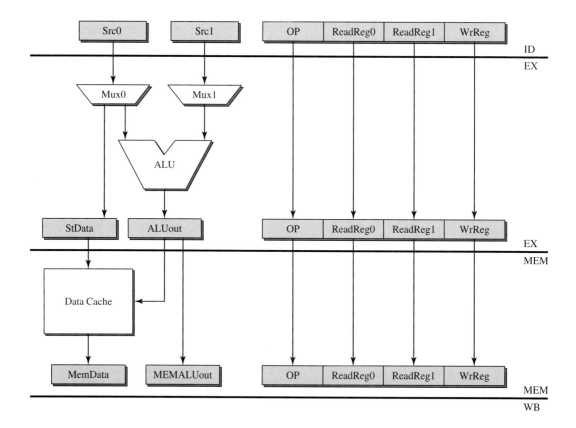

P2.12 Given the forwarding paths in Problem 11, draw a detailed design for Mux0 and Mux1 that clearly identifies which bypass paths are selected under which control conditions. Identify each input to each mux by the name of the pipeline latch that it is bypassing from. Specify precisely the boolean equations that are used to control Mux0 and Mux1. Possible inputs to the boolean equations are:

- ID.OP, EX.OP, MEM.OP = {'load', 'store', 'alu', 'other'}
- ID.ReadReg0, ID.ReadReg1 = [0..31,32] where 32 means a register is not read by this instruction
- EX.ReadReg0, etc., as in ID stage
- MEM.ReadReg0, etc., as in ID stage
- ID.WriteReg, EX.WriteReg, MEM.WriteReg = [0..31,33] where 33 means a register is not written by this instruction
- Draw Mux0 and Mux1 with labeled inputs; you do not need to show the controls using gates. Simply write out the control equations using symbolic OP comparisons, etc. [e.g., Ctrl1 = (ID.op == 'load') & (ID.WriteReg==MEM.ReadReg0)].

P2.13 Given the IBM experience outlined in Section 2.2.4.3, compute the CPI impact of the addition of a level-zero data cache that is able to supply the data operand in a single cycle, but only 75% of the time. The level-zero and level-one caches are accessed in parallel, so that when the level-zero cache misses, the level-one cache returns the result in the next cycle, resulting in one load-delay slot. Assume uniform distribution of level-zero hits across load-delay slots that can and cannot be filled. Show your work.

P2.14 Given the assumptions of Problem 13, compute the CPI impact if the level-one cache is accessed sequentially, only after the level-zero cache misses, resulting in two load-delay slots instead of one. Show your work.

P2.15 The IBM study of pipelined processor performance assumed an instruction mix based on popular C programs in use in the 1980s. Since then, object-oriented languages like C++ and Java have become much more common. One of the effects of these languages is that object inheritance and polymorphism can be used to replace conditional branches with virtual function calls. Given the IBM instruction mix and CPI shown in the following table, perform the following transformations to reflect the use of C++ and Java, and recompute the overall CPI and speedup or slowdown due to this change:

- Replace 50% of taken conditional branches with a load instruction followed by a jump register instruction (the load and jump register implement a virtual function call).
- Replace 25% of not-taken branches with a load instruction followed by a jump register instruction.

Instruction Type	Old Mix, %	Latency	Old CPI	Cycles	New Mix, %	Instructions	Cycles	New CPI
Load	25.0	2	0.50	500				
Store	15.0	1	0.15	150				
Arithmetic	30.0	1	0.30	300				
Logical	10.0	1	0.10	100				
Branch-T	8.0	3	0.24	240				
Branch-NT	6.0	2	0.12	120				
Jump	5.0	2	0.10	100				
Jump register	1.0	3	0.03	30				
Total	100.0		1.54	1540				

P2.16 In a TYP-based pipeline design with a data cache, load instructions check the tag array for a cache hit in parallel with accessing the data array to read the corresponding memory location. Pipelining stores to such a cache is more difficult, since the processor must check the tag first, before it overwrites the data array. Otherwise, in the case of a cache miss, the wrong memory location may be overwritten by the store. Design a solution to this problem that does not require sending the store down the pipe twice, or stalling the pipe for every store instruction, or dual-porting the data cache. Referring to Figure 2.15, are there any new RAW, WAR, and/or WAW memory hazards?

P2.17 The MIPS pipeline shown in Table 2.7 employs a two-phase clocking scheme that makes efficient use of a shared TLB, since instruction fetch accesses the TLB in phase one and data fetch accesses in phase two. However, when resolving a conditional branch, both the branch target address and the branch fall-through address need to be translated during phase one—in parallel with the branch condition check in phase one of the ALU stage—to enable instruction fetch from either the target or the fall-through during phase two. This seems to imply a dual-ported TLB. Suggest an architected solution to this problem that avoids dual-porting the TLB.

Problems 18 through 24: Instruction Pipeline Design

This problem explores pipeline design. As discussed earlier, pipelining involves balancing the pipe stages. Good pipeline implementations minimize both internal and external fragmentation to create simple balanced designs. Below is a nonpipelined implementation of a simple microprocessor that executes only ALU instructions, with no data hazards:

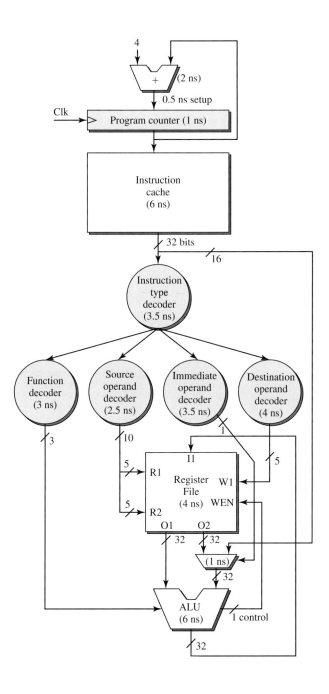

P2.18 Generate a pipelined implementation of the simple processor outlined in the figure that minimizes internal fragmentation. Each subblock in the diagram is a primitive unit that cannot be further partitioned into smaller ones. The original functionality must be maintained in the pipelined implementation. Show the diagram of your pipelined implementation. Pipeline registers have the following timing requirements:

- 0.5-ns setup time
- 1-ns delay time (from clock to output)

P2.19 Compute the latencies (in nanoseconds) of the *instruction cycle* of the nonpipelined and the pipelined implementations.

P2.20 Compute the *machine cycle* times (in nanoseconds) of the nonpipelined and the pipelined implementations.

P2.21 Compute the (potential) speedup of the pipelined implementation in Problems 18–20 over the original nonpipelined implementation.

P2.22 What microarchitectural techniques could be used to further reduce the machine cycle time of pipelined designs? Explain how the machine cycle time is reduced.

P2.23 Draw a simplified diagram of the pipeline stages in Problem 18; you should include all the necessary data forwarding paths. This diagram should be similar to Figure 2.16.

P2.24 Discuss the impact of the data forwarding paths from Problem 23 on the pipeline implementation in Problem 18. How will the timing be affected? Will the pipeline remain balanced once these forwarding paths are added? What changes to the original pipeline organization of Problem 18 might be needed?

CHAPTER 3

Memory and I/O Systems

CHAPTER OUTLINE

3.1 Introduction
3.2 Computer System Overview
3.3 Key Concepts: Latency and Bandwidth
3.4 Memory Hierarchy
3.5 Virtual Memory Systems
3.6 Memory Hierarchy Implementation
3.7 Input/Output Systems
3.8 Summary

References
Homework Problems

3.1 Introduction

The primary focus of this book is the design of advanced, high-performance processors; this chapter examines the larger context of computer systems that incorporate such processors. Basic components, such as memory systems, input and output, and virtual memory, and the ways in which they are interconnected are described in relative detail to enable a better understanding of the interactions between high-performance processors and the peripheral devices they are connected to.

Clearly, processors do not exist in a vacuum. Depending on their intended application, processors will interact with other components internal to a computer system, devices that are external to the system, as well as humans or other external entities. The speed with which these interactions occur varies with the type of communication that is necessary, as do the protocols used to communicate with them. Typically, interacting with performance-critical entities such as the memory subsystem is accomplished via proprietary, high-speed interfaces, while communication with peripheral or external devices is accomplished across industry-standard interfaces that sacrifice some performance for the sake of compatibility across multiple vendors. Usually such interfaces are balanced, providing symmetric bandwidth to and from the device. However, interacting with physical beings (such as humans) often leads to unbalanced bandwidth requirements. Even the fastest human typist can generate input rates of only a few kilobytes per second. In contrast, human visual perception can absorb more than 30 frames per second of image data, where each image contains several megabytes of pixel data, resulting in an output data rate of over 100 megabytes per second (Mbytes/s).

Just as the bandwidth requirements of various components can vary dramatically, the latency characteristics are diverse as well. For example, studies have shown that while subsecond response times (a *response time* is defined as the interval between a user issuing a command via the keyboard and observing the response on the display) are critical for the productivity of human computer users, response times much less than a second provide rapidly diminishing returns. Hence, low latency in responding to user input through the keyboard or mouse is not that critical. In contrast, modern processors operate at frequencies that are much higher than main memory subsystems. For example, a state-of-the-art personal computer has a processor that is clocked at 3 GHz today, while the synchronous main memory is clocked at only 133 MHz. This mismatch in frequency can cause the processor to starve for instructions and data as it waits for memory to supply them, hence motivating high-speed processor-to-memory interfaces that are optimized for low latency.

Section 3.2 presents an overview of modern computer systems. There are numerous interesting architectural tradeoffs in the design of hardware subsystems, interfaces, and protocols to satisfy input/output requirements that vary so dramatically. In Section 3.3, we define the fundamental metrics of bandwidth and latency and discuss some of the tradeoffs involved in designing interfaces that meet requirements for both metrics. In Section 3.4, we introduce the concept of a memory hierarchy and discuss the components used to build a modern memory hierarchy as well as the key tradeoffs and metrics used in the design process. Section 3.5 introduces the notion of virtual memory, which is critically important in modern systems that time-share physical execution resources. Finally, Section 3.7 discusses various input/output devices, their key characteristics, and the interconnects used in modern systems to allow them to communicate with each other and with the processor.

3.2 Computer System Overview

As illustrated in Figure 3.1, a typical computer system consists of a processor or CPU, main memory, and an input/output (I/O) bridge connected to a processor bus, and peripheral devices such as a network interface, a disk controller driving one or more disk drives, a display adapter driving a display, and input devices such as a keyboard or mouse, all connected to the I/O bus. The main memory provides volatile storage for programs and data while the computer is powered up. The design of efficient, high-performance memory systems using a hierarchical

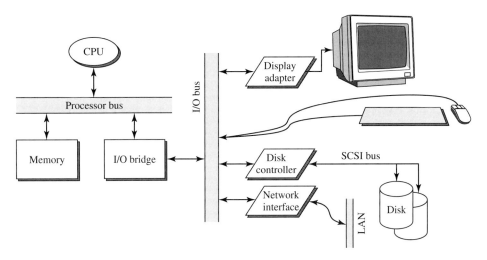

Figure 3.1
A Typical Computer System.

approach that exploits *temporal and spatial locality* is discussed in detail in Section 3.4. In contrast to volatile main memory, a disk drive provides persistent storage that survives even when the system is powered down. Disks can also be used to transparently increase effective memory capacity through the use of *virtual memory*, as described in Section 3.5. The network interface provides a physical connection for communicating across local area or wide area networks (LANs or WANs) with other computer systems; systems without local disks can also use the network interface to access remote persistent storage on file servers. The display subsystem is used to render a textual or graphical user interface on a display device such as a cathode-ray tube (CRT) or liquid-crystal display (LCD). Input devices enable a user or operator to enter data or issue commands to the computer system. We will discuss each of these types of peripheral devices (disks, network interfaces, display subsystems, and input devices) in Section 3.7.1.

Finally, a computer system must provide a means for interconnecting all these devices, as well as an interface for communicating with them. We will discuss various types of busses used to interconnect peripheral devices in Section 3.7.2 and will describe polling, interrupt-driven, and programmed means of communication with I/O devices in Section 3.7.3.

3.3 Key Concepts: Latency and Bandwidth

There are two fundamental metrics that are commonly used to characterize various subsystems, peripheral devices, and interconnections in computer systems. These two metrics are *latency*, measured in unit time, and *bandwidth*, measured in quantity per unit time. Both metrics are important for understanding the behavior of a system, so we provide definitions and a brief introduction to both in this section.

Latency is defined as the elapsed time between issuing a request or command to a particular subsystem and receiving a response or reply. It is measured either in units of time (seconds, microseconds, milliseconds, etc.) or cycles, which can be trivially translated to time given cycle time or frequency. Latency provides a measurement of the responsiveness of a particular system and is a critical metric for any subsystem that satisfies time-critical requests. An example of such a system is the memory subsystem, which must provide the processor with instructions and data; latency is critical because processors will usually stall if the memory subsystem does not respond rapidly. Latency is also sometimes called *response time* and can be decomposed into the inherent delay of a device or subsystem, called the *service time,* which forms the lower bound for the time required to satisfy a request, and the *queueing time,* which results from waiting for a particular resource to become available. Queueing time is greater than zero only when there are multiple concurrent requests competing for access to the same resource, and one or more of those requests must delay while waiting for another to complete.

Bandwidth is defined as the throughput of a subsystem; that is, the rate at which it can satisfy requests. Bandwidth is measured in quantity per unit time, where the quantity measured varies based on the type of request. At its simplest, bandwidth is expressed as the number of requests per unit time. If each request corresponds to a fixed number of bytes of data, for example, bandwidth can also be expressed as the number of bytes per unit time.

Naively, bandwidth can be defined as the inverse of latency. That is, a device that responds to a single request with latency l will have bandwidth equal to or less than $1/l$, since it can accept and respond to one request every l units of time. However, this naive definition precludes any concurrency in the handling of requests. A high-performance subsystem will frequently overlap multiple requests to increase bandwidth without affecting the latency of a particular request. Hence, bandwidth is more generally defined as the rate at which a subsystem is able to satisfy requests. If bandwidth is greater than $1/l$, we can infer that the subsystem supports multiple concurrent requests and is able to overlap their latencies with each other. Most high-performance interfaces, including processor-to-memory interconnects, standard input/output busses like peripheral component interfaces (PCIs), and device interfaces like small computer systems interface (SCSI), support multiple concurrent requests and have bandwidth significantly higher than $1/l$.

Quite often, manufacturers will also report raw or peak bandwidth numbers, which are usually derived directly from the hardware parameters of a particular interface. For example, a synchronous dynamic random-access memory (DRAM) interface that is 8 bytes wide and is clocked at 133 MHz may have a reported peak bandwidth of 1 Gbyte/s. These peak numbers will usually be substantially higher than sustainable bandwidth, since they do not account for request and response transaction overheads or other bottlenecks that might limit achievable bandwidth. Sustainable bandwidth is a more realistic measure that represents bandwidth that the subsystem can actually deliver. Nevertheless, even sustainable bandwidth might be unrealistically optimistic, since it may not account for real-life access patterns and other system components that may cause additional queueing delays, increase overhead, and reduce delivered bandwidth.

In general, bandwidth is largely driven by product-cost constraints rather than fundamental limitations of a given technology. For example, a bus can always be made wider to increase the number of bytes transmitted per cycle, hence increasing the bandwidth of the interface. This will increase cost, since the chip pin count and backplane trace count for the bus may double, and while the peak bandwidth may double, the effective or sustained bandwidth may increase by a much smaller factor. However, it is generally true that a system that is performance-limited due to insufficient bandwidth is either poorly engineered or constrained by cost factors; if cost were no object, it would usually be possible to provide adequate bandwidth.

Latency is fundamentally more difficult to improve, since it is often dominated by limitations of a particular technology, or possibly even the laws of physics. For example, the electrical characteristics of a given signaling technology used in a multidrop backplane bus will determine the maximum frequency at which that bus can operate. It follows that the minimum latency of a transaction across that bus is bounded by the cycle time corresponding to that maximum frequency. A common strategy for improving latency, short of transitioning to a newer, faster, technology, is to decompose the latency into the portions that are due to various subcomponents and attempt to maximize the concurrency of those components. For example, a modern multiprocessor system like the IBM pSeries 690 exposes concurrency in handling processor cache misses by fetching the missing block from DRAM main memory in parallel with checking other processors' caches to try and find a newer, modified copy of the block. A less aggressive approach would first check the other processors' caches and then fetch the block from DRAM only if no other processor has a modified copy. The latter approach serializes the two events, leading to increased latency whenever a block needs to be fetched from DRAM.

However, there is often a price to be paid for such attempts to maximize concurrency, since they typically require speculative actions that may ultimately prove to be unnecessary. In the preceding multiprocessor example, if a newer, modified copy is found in another processor's cache, the block must be supplied by that cache. In this case, the concurrent DRAM fetch proves to be unnecessary and consumes excess memory bandwidth and wastes energy. However, despite such cost, various forms of speculation are commonly employed in an attempt to reduce the observed latency of a request. As another example, modern processors incorporate prefetch engines that look for patterns in the reference stream and issue speculative memory fetches to bring blocks into their caches in anticipation of demand references to those blocks. In many cases, these additional speculative requests or prefetches prove to be unnecessary, and end up consuming additional bandwidth. However, when they are useful, and a subsequent demand reference occurs to a speculatively prefetched block, the latency of that reference corresponds to hitting in the cache and is much lower than if the prefetch had not occurred. Hence, average latency for all memory references can be lowered at the expense of consuming additional bandwidth to issue some number of useless prefetches.

In summary, bandwidth and latency are two fundamental attributes of computer system components, peripheral devices, and interconnection networks. Bandwidth can usually be improved by adding cost to the system, but in a well-engineered system that maximizes concurrency, latency is usually much more difficult to improve without changing the implementation technology or using various forms of speculation. Speculation can be used to improve the observed latency for a request, but this usually happens at the expense of additional bandwidth consumption. Hence, in a well-designed computer system, latency and bandwidth need to be carefully balanced against cost, since all three factors are interrelated.

3.4 Memory Hierarchy

One of the fundamental needs that a computer system must meet is the need for storage of data and program code, both while the computer is running, to support storage of temporary results, as well as while the computer is powered off, to enable the results of computation as well as the programs used to perform that computation to survive across power-down cycles. Fundamentally, such storage is nothing more than a sea of bits that is addressable by the processor. A perfect storage technology for retaining this sea of bits in a computer system would satisfy the following memory idealisms:

- *Infinite capacity.* For storing large data sets and large programs.
- *Infinite bandwidth.* For rapidly streaming these large data sets and programs to and from the processor.
- *Instantaneous or zero latency.* To prevent the processor from stalling while waiting for data or program code.
- *Persistence or nonvolatility.* To allow data and programs to survive even when the power supply is cut off.
- *Zero or very low implementation cost.*

Naturally, the system and processor designers must strive to approximate these idealisms as closely as possible so as to satisfy the performance and correctness expectations of the user. Obviously, the final factor—cost—plays a large role in how easy it is to reach these goals, but a well-designed memory system can in fact maintain the illusion of these idealisms quite successfully. This is true despite the fact that the perceived requirements for the first three—capacity, bandwidth, and latency—have been increasing rapidly over the past few decades. Capacity requirements grow because the programs and operating systems that users demand are increasing in size and complexity, as are the data sets that they operate over. Bandwidth requirements are increasing for the same reason. Meanwhile, the latency requirement is becoming increasingly important as processors continue to become faster and faster and are more easily starved for data or program code if the perceived memory latency is too long.

3.4.1 Components of a Modern Memory Hierarchy

A modern memory system, often referred to as a *memory hierarchy,* incorporates various storage technologies to create a whole that approximates each of the five memory idealisms. Figure 3.2 illustrates five typical components in a modern memory hierarchy and plots each on approximate axes that indicate their relative latency and capacity (increasing on the y axis) and bandwidth and cost per bit (increasing on the x axis). Some important attributes of each of these components are summarized in Table 3.1.

Magnetic Disks. Magnetic disks provide the most cost-efficient storage and the largest capacities of any memory technology today, costing less than one-ten-millionth of a cent per bit (i.e., roughly $1 per gigabyte of storage), while providing hundreds of gigabytes of storage in a 3.5-inch (in.) standard form factor. However, this tremendous capacity and low cost comes at the expense of limited effective bandwidth (in the tens of megabytes per second for a single disk) and extremely long latency (roughly 10 ms per random access). On the other hand, magnetic storage technologies are nonvolatile and maintain their state even when power is turned off.

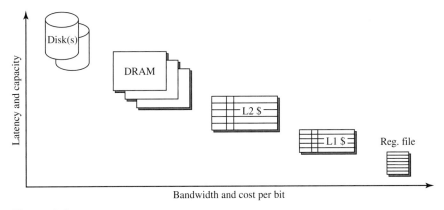

Figure 3.2
Memory Hierarchy Components.

Table 3.1
Attributes of memory hierarchy components

Component	Technology	Bandwidth	Latency	Cost per Bit ($)	Cost per Gigabyte ($)
Disk drive	Magnetic field	10+ Mbytes/s	10 ms	$< 1 \times 10^{-9}$	< 1
Main memory	DRAM	2+ Gbytes/s	50+ ns	$< 2 \times 10^{-7}$	< 200
On-chip L2 cache	SRAM	10+ Gbytes/s	2+ ns	$< 1 \times 10^{-4}$	< 100K
On-chip L1 cache	SRAM	50+ Gbytes/s	300+ ps	$> 1 \times 10^{-4}$	> 100K
Register file	Multiported SRAM	200+ Gbytes/s	300+ ps	$> 1 \times 10^{-2}$ (?)	> 10 Mbytes (?)

Main Memory. Main memory based on standard DRAM technology is much more expensive at approximately two hundred-thousandths of a cent per bit (i.e., roughly $200 per gigabyte of storage) but provides much higher bandwidth (several gigabytes per second even in a low-cost commodity personal computer) and much lower latency (averaging less than 100 ns in a modern design). We study various aspects of main memory design at length in Section 3.4.4.

Cache Memory. On-chip and off-chip cache memories, both secondary (L2) and primary (L1), utilize static random-access memory (SRAM) technology that pays a much higher area cost per storage cell than DRAM technology, resulting in much lower storage density per unit of chip area and driving the cost much higher. Of course, the latency of SRAM-based storage is much lower—as low as a few hundred picoseconds for small L1 caches or several nanoseconds for larger L2 caches. The bandwidth provided by such caches is tremendous, in some cases exceeding 100 Gbytes/s. The cost is much harder to estimate, since high-speed custom cache SRAM is available at commodity prices only when integrated with high-performance processors. However, ignoring nonrecurring expenses and considering only the $50 estimated manufacturing cost of a modern x86 processor chip like the Pentium 4 that incorporates 512K bytes of cache and ignoring the cost of the processor core itself, we can arrive at an estimated cost per bit of one hundredth of a cent per bit (i.e., roughly $100,000 per gigabyte).

Register File. Finally, the fastest, smallest, and most expensive element in a modern memory hierarchy is the register file. The register file is responsible for supplying operands to the execution units of a processor at very low latency—usually a few hundred picoseconds, corresponding to a single processor cycle—and at very high bandwidth, to satisfy multiple execution units in parallel. Register file bandwidth can approach 200 Gbytes/s in a modern eight-issue processor like the IBM PowerPC 970, that operates at 2 GHz and needs to read two and write one 8-byte operand for each of the eight issue slots in each cycle. Estimating the cost per bit in the register file is virtually impossible without detailed knowledge of a particular design and its yield characteristics; suffice it to say that it is likely several orders of magnitude higher than our estimate of $100,000 per gigabyte for on-chip cache memory.

These memory hierarchy components are attached to the processor in a hierarchical fashion to provide an overall storage system that approximates the five idealisms—infinite capacity, infinite bandwidth, zero latency, persistence, and zero cost—as closely as possible. Proper design of an effective memory hierarchy requires careful analysis of the characteristics of the processor, the programs and operating system running on that processor, and a thorough understanding of the capabilities and costs of each component in the memory hierarchy. Table 3.1 summarizes some of the key attributes of these memory hierarchy components and illustrates that bandwidth can vary by four orders of magnitude, latency can vary by eight orders of magnitude, while cost per bit can vary by seven orders of magnitude. These drastic variations, which continue to change at nonuniform rates as each technology evolves, lend themselves to a vast and incredibly dynamic design space for the system architect.

3.4.2 Temporal and Spatial Locality

How is it possible to design a memory hierarchy that reasonably approximates the infinite capacity and bandwidth, low latency, persistence, and low cost specified by the five memory idealisms? If one were to assume a truly random pattern of accesses to a vast storage space, the task would appear hopeless: the excessive cost of fast storage technologies prohibits large memory capacity, while the long latency and low bandwidth of affordable technologies violates the performance requirements for such a system. Fortunately, an empirically observed attribute of program execution called *locality of reference* provides an opportunity for designing the memory hierarchy in a manner that satisfies these seemingly contradictory requirements.

Locality of reference describes the propensity of computer programs to access the same or nearby memory locations frequently and repeatedly. Specifically, we can break locality of reference down into two dimensions: *temporal locality* and *spatial locality*. Both types of locality are common in both the instruction and data reference streams and have been empirically observed in both user-level application programs, shared library code, as well as operating system kernel code.

Temporal locality refers to accesses to the same memory location that occur close together in time; many real application programs exhibit this tendency for both program text or instruction references, as well as data references. Figure 3.3(a) annotates an example sequence of memory references with arrows representing temporal locality; each arrow connects an earlier and later memory reference to the same address. Temporal locality in the instruction reference stream can be easily explained, since it is caused by loops in program execution. As each iteration of a loop is executed, the instructions forming the body of the loop are fetched again

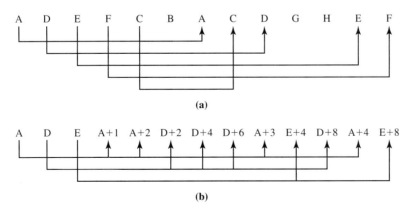

Figure 3.3

Illustration of (a) Temporal and (b) Spatial Locality.

and again. Similarly, nested or outer loops cause this repetition to occur on a coarser scale. Furthermore, even programs that contain very few discernible loop structures can still share key subroutines that are called from various locations; each time the subroutine is called, temporally local instruction references occur.

Within the data reference stream, accesses to widely used program variables lead to temporal locality, as do accesses to the current stack frame in call-intensive programs. As call-stack frames are deallocated on procedure returns and reallocated on a subsequent call, the memory locations corresponding to the top of the stack are accessed repeatedly to pass parameters, spill registers, and return function results. All this activity leads to abundant temporal locality in the data access stream.

Spatial locality refers to accesses to nearby memory locations that occur close together in time. Figure 3.3(b) annotates an example sequence of memory references with arrows representing temporal locality; an earlier reference to some address (for example, A) is followed by references to adjacent or nearby addresses (A+1, A+2, A+3, and so on). Again, most real application programs exhibit this tendency for both instruction and data references. In the instruction stream, the instructions that make up a sequential execution path through the program are laid out sequentially in program memory. Hence, in the absence of branches or jumps, instruction fetches sequence through program memory in a linear fashion, where subsequent accesses in time are also adjacent in the address space, leading to abundant spatial locality. Even when branches or jumps cause discontinuities in fetching, the targets of branches and jumps are often nearby, maintaining spatial locality, though at a slightly coarser level.

Spatial locality within the data reference stream often occurs for algorithmic reasons. For example, numerical applications that traverse large matrices of data often access the matrix elements in serial fashion. As long as the matrix elements are laid out in memory in the same order they are traversed, abundant spatial locality occurs. Applications that stream through large data files, like audio MP3 decoder or encoders, also access data in a sequential, linear fashion, leading to many spatially local references. Furthermore, accesses to automatic variables in call-intensive environments also exhibit spatial locality, since the automatic variables for a given function are laid out adjacent to each other in the stack frame corresponding to the current function.

Of course, it is possible to write programs that exhibit very little temporal or spatial locality. Such programs do exist, and it is very difficult to design a cost-efficient memory hierarchy that behaves well for such programs. If these programs or classes of applications are deemed important enough, special-purpose high-cost systems can be built to execute them. In the past, many supercomputer designs optimized for applications with limited locality of reference avoided using many of the techniques introduced in this chapter (cache memories, virtual memory, and DRAM main memory), since these techniques require locality of reference in order to be effective. Fortunately, most important applications do exhibit locality and can benefit from these techniques. Hence, the vast majority of computer systems designed today incorporate most or all of these techniques.

3.4.3 Caching and Cache Memories

The principle of *caching* instructions and data is paramount in exploiting both temporal and spatial locality to create the illusion of a fast yet capacious memory. Caches were first proposed by Wilkes [1965] and first implemented in the IBM System 360/85 in 1968 [Liptay, 1968]. Caching is accomplished by placing a small, fast, and expensive memory between the processor and a slow, large, and inexpensive main memory, and by placing instructions and data that exhibit temporal and spatial reference locality into this cache memory. References to memory locations that are cached can be satisfied very quickly, reducing average memory reference latency, while the low latency of a small cache also naturally provides high bandwidth. Hence, a cache can effectively approximate the second and third memory idealisms—infinite bandwidth and zero latency—for those references that can be satisfied from the cache. Since temporal and spatial locality are so prevalent in most programs, even small first-level caches can satisfy in excess of 90% of all references in most cases; such references are said to *hit* in the cache. Those references that cannot be satisfied from the cache are called *misses* and must be satisfied from the slower, larger, memory that is behind the cache.

3.4.3.1 Average Reference Latency.
Caching can be extended to multiple levels by adding caches of increasing capacity and latency in a hierarchical fashion, using the technologies enumerated in Table 3.1. As long as each level of the cache is able to capture a reasonable fraction of the references sent to it, the reference latency perceived by the processor is substantially lower than if all references were sent directly to the lowest level in the hierarchy. The average memory reference latency can be computed using Equation (3.1), which computes the weighted average based on the distribution of references satisfied at each level in the cache. The latency to satisfy a reference from each level in the cache hierarchy is defined as l_i, while the fraction of all references satisfied by that level is h_i.

$$\text{Latency} = \sum_{i=0}^{n} h_i \times l_i \tag{3.1}$$

This equation makes clear that as long as the hit rates h_i for the upper levels in the cache (those with low latency l_i) are relatively high, the average latency observed by the processor will be very low. For example, a two-level cache hierarchy with $h_1 = 0.95$, $l_1 = 1$ ns, $h_2 = 0.04$, $l_2 = 10$ ns, $h_3 = 0.01$, and $l_3 = 100$ ns will deliver an average latency of 0.95×1 ns $+ 0.04 \times 10$ ns $+ 0.01 \times 100$ ns $= 2.35$ ns, which is nearly two orders of magnitude faster than simply sending each reference directly to the lowest level.

3.4.3.2 Miss Rates and Cycles per Instruction Estimates.
Equation (3.1) assumes that h_i hit rates are specified as *global* hit rates, which specify the fraction of all memory references that hit in that level of the memory hierarchy. It is often useful to also understand *local* hit rates for caches, which specify the fraction of all memory references serviced by a particular cache that hit in that cache. For a first-level cache, the global and local hit rates are the same, since the first-level cache

services all references from a program. A second-level cache, however, only services those references that result in a miss in the first-level cache. Similarly, a third-level cache only services references that miss in the second-level cache, and so on. Hence, the local hit rate lh_i for cache level i is defined in Equation (3.2).

$$lh_i = \frac{h_i}{m_{i-1}} = \frac{h_i}{\left[1 - \sum_{1}^{i-1} h_i\right]} \quad (3.2)$$

Returning to our earlier example, we see that the local hit rate of the second-level cache $lh_i = 0.04/(1 - 0.95) = 0.8$. This tells us that 0.8 or 80% of the references serviced by the second-level cache were also satisfied from that cache, while $1 - 0.8 = 0.2$ or 20% were sent to the next level. This latter rate is often called a *local miss rate*, as it indicates the fraction of references serviced by a particular level in the cache that missed at that level. Note that for the first-level cache, the local and global *hit* rates are equivalent, since the first-level cache services all references. The same is true for the local and global *miss* rates of the first-level cache.

Finally, it is often useful to report cache miss rates as *per-instruction miss rates*. This metric reports misses normalized to the number of instructions executed, rather than the number of memory references performed and provides an intuitive basis for reasoning about or estimating the performance effects of various cache organizations. Given the per-instruction miss rate m_i and a specific execution-time penalty p_i for a miss in each cache in a system, one can quickly estimate the performance effect of the cache hierarchy using the memory-time-per-instruction (MTPI) metric, as defined in Equation (3.3).

$$\text{MTPI} = \sum_{i=0}^{n} m_i \times p_i \quad (3.3)$$

In this equation the p_i term is not equivalent to the latency term l_i used in Equation (3.1). Instead, it must reflect the penalty associated with a miss in level i of the hierarchy, assuming the reference can be satisfied at the next level. The miss penalties are computed as the difference between the latencies to adjacent levels in the hierarchy, as shown in Equation (3.4).

$$p_i = l_{i+1} - l_i \quad (3.4)$$

Returning to our earlier example, if $h_1 = 0.95$, $l_1 = 1$ ns, $h_2 = 0.04$, $l_2 = 10$ ns, $h_3 = 0.01$, and $l_3 = 100$ ns, then $p_1 = (l_2 - l_1) = (10 \text{ ns} - 1 \text{ ns}) = 9$ ns, which is the difference between the l_1 and l_2 latencies and reflects the additional penalty of missing the first level and having to fetch from the second level. Similarly, $p_2 = (l_3 - l_2) = (100 \text{ ns} - 10 \text{ ns}) = 90$ ns, which is the difference between the l_2 and l_3 latencies.

The m_i miss rates are also expressed as per-instruction miss rates and need to be converted from the global miss rates used earlier. To perform this conversion, we need to know the number of references performed per instruction. If we assume that each instruction is fetched individually and that 40% of instructions are either loads or stores, we have a total of $n = (1 + 0.4) = 1.4$ references per instruction. Hence, we can compute the per-instruction miss rates using Equation (3.5).

$$m_i = \frac{\left(1 - \sum_{j=1}^{i} h_j\right) \text{misses}}{\text{ref}} \times \frac{n \text{ ref}}{\text{inst}} \quad (3.5)$$

Returning to our example, we would find that $m_1 = (1 - 0.95) \times 1.4 = 0.07$ misses per instruction, while $m_2 = [1 - (0.95 + 0.04)] \times 1.4 = 0.014$ misses per instruction. Finally, substituting into Equation (3.3), we can compute the memory-time-per-instruction metric MTPI = $(0.07 \times 9 \text{ ns}) + (0.014 \times 90 \text{ ns}) = 0.63 + 1.26 = 1.89$ ns per instruction. This can also be conveniently expressed in terms of cycles per instruction by normalizing to the cycle time of the processor. For example, assuming a cycle time of 1 ns, the memory-cycles-per-instruction (MCPI) would be 1.89 cycles per instruction.

Note that our definition of MTPI in Equation (3.3) does not account for the latency spent servicing hits from the first level of cache, but only time spent for misses. Such a definition is useful in performance modeling, since it cleanly separates the time spent in the processor core from the time spent outside the core servicing misses. For example, an ideal scalar processor pipeline would execute instructions at a rate of one per cycle, resulting in a core cycles per instruction (CPI) equal to one. This CPI assumes that all memory references hit in the cache; a core CPI is also often called a *perfect* cache CPI, since the cache is perfectly able to satisfy all references with a fixed hit latency. As shown in Equation (3.6), the core CPI can be added to the MCPI computed previously to reach the actual CPI of the processor: CPI = 1.0 + 1.89 = 2.89 cycles per instruction for our recurring example.

$$\text{CPI} = \text{CoreCPI} + \text{MCPI} \quad (3.6)$$

However, one has to be careful using such equations to reason about absolute performance effects, since they do not account for any overlap or concurrency between cache misses. In Chapter 5, we will investigate numerous techniques that exist for the express purpose of maximizing overlap and concurrency, and we will see that performance approximations like Equation (3.3) are less effective at predicting the performance of cache hierarchies that incorporate such techniques.

3.4.3.3 Effective Bandwidth.
Cache hierarchies are also useful for satisfying the second memory idealism of infinite bandwidth. Each higher level in the cache hierarchy is also inherently able to provide higher bandwidth than lower levels, due to its lower access latency, so the hierarchy as a whole manages to maintain the illusion of infinite bandwidth. In our recurring example, the latency of the first-level cache is 1 ns, so a single-ported nonpipelined implementation can provide a bandwidth of 1 billion references per second. In contrast, the second level, if also not pipelined, can only satisfy one reference every 10 ns, resulting in a bandwidth of

100 million references per second. Of course, it is possible to increase concurrency in the lower levels to provide greater effective bandwidth by either multiporting or banking (see Section 3.4.4.2 for an explanation of banking or interleaving) the cache or memory, or pipelining it so that it can initiate new requests at a rate greater than the inverse of the access latency. Goodman [1983] conducted a classic study of the bandwidth benefits of caches.

3.4.3.4 Cache Organization and Design.
Each level in a cache hierarchy must be designed in a way that matches the requirements for bandwidth and latency at that level. Since the upper levels of the hierarchy must operate at speeds comparable to the processor core, they must be implemented using fast hardware techniques, necessarily limiting their complexity. Lower in the cache hierarchy, where latency is not as critical, more sophisticated schemes are attractive, and even software techniques are widely deployed. However, at all levels, there must be efficient policies and mechanisms in place for locating a particular piece or block of data, for evicting existing blocks to make room for newer ones, and for reliably handling updates to any block that the processor writes. This section presents a brief overview of some common approaches; additional implementation details are provided in Section 3.6.

Locating a Block. Each level must implement a mechanism that enables low-latency lookups to check whether or not a particular block is cache-resident. There are two attributes that determine the process for locating a block; the first is the size of the block, and the second is the organization of the blocks within the cache.

Block size (sometimes referred to as *line size*) describes the granularity at which the cache operates. Each block is a contiguous series of bytes in memory and begins on a naturally aligned boundary. For example, in a cache with 16-byte blocks, each block would contain 16 bytes, and the first byte in each block would be aligned to 16-byte boundaries in the address space, implying that the low-order 4 bits of the address of the first byte would always be zero (i.e., 0b ⋯ 0000). The smallest usable block size is the natural word size of the processor (i.e., 4 bytes for a 32-bit machine, or 8 bytes for a 64-bit machine), since each access will require the cache to supply at least that many bytes, and splitting a single access over multiple blocks would introduce unacceptable overhead into the access path. In practice, applications with abundant spatial locality will benefit from larger blocks, as a reference to any word within a block will place the entire block into the cache. Hence, spatially local references that fall within the boundaries of that block can now be satisfied as hits in the block that was installed in the cache in response to the first reference to that block.

Whenever the block size is greater than 1 byte, the low-order bits of an address must be used to find the byte or word being accessed within the block. As stated above, the low-order bits for the first byte in the block must always be zero, corresponding to a naturally aligned block in memory. However, if a byte other than the first byte needs to be accessed, the low-order bits must be used as a *block offset* to index into the block to find the right byte. The number of bits needed for the block offset is the $\log_2$ of the block size, so that enough bits are available to

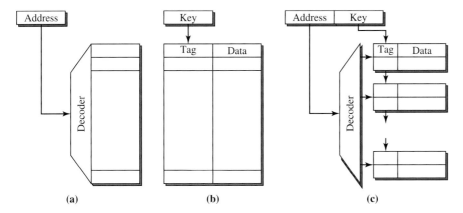

Figure 3.4
Block Placement Schemes: (a) Direct-Mapped, (b) Fully Associative, (c) Set-Associative.

span all the bytes in the block. For example, if the block size is 64 bytes, $\log_2(64) = 6$ low-order bits are used as the block offset. The remaining higher-order bits are then used to locate the appropriate block in the cache memory.

The second attribute that determines how blocks are located, *cache organization*, determines how blocks are arranged in a cache that contains multiple blocks. Figure 3.4 illustrates three fundamental approaches for organizing a cache that directly affect the complexity of the lookup process: *direct-mapped*, *fully associative*, and *set-associative*.

The simplest approach, *direct-mapped*, forces a many-to-one mapping between addresses and the available storage locations in the cache. In other words, a particular address can reside only in a single location in the cache; that location is usually determined by extracting n bits from the address and using those n bits as a direct index into one of 2^n possible locations in the cache.

Of course, since there is a many-to-one mapping, each location must also store a *tag* that contains the remaining address bits corresponding to the block of data stored at that location. On each lookup, the hardware must read the tag and compare it with the address bits of the reference being performed to determine whether a hit or miss has occurred. We describe this process in greater detail in Section 3.6.

In the degenerate case where a direct-mapped memory contains enough storage locations for every address block (i.e., the n index bits include all bits of the address), no tag is needed, as the mapping between addresses and storage locations is now one-to-one instead of many-to-one. The *register file* inside the processor is an example of such a memory; it need not be tagged since all the address bits (all bits of the register identifier) are used as the index into the register file.

The second approach, *fully associative*, allows an any-to-any mapping between addresses and the available storage locations in the cache. In this organization, any memory address can reside anywhere in the cache, and all locations must be searched to find the right one; hence, no index bits are extracted from the address to determine the storage location. Again, each entry must be tagged with the address it is currently holding, and all these tags are compared with the address of the current reference. Whichever entry matches is then used to supply the data; if no entry matches, a miss has occurred.

The final approach, *set-associative,* is a compromise between the other two. Here a many-to-few mapping exists between addresses and storage locations. On each lookup, a subset of address bits is used to generate an index, just as in the direct-mapped case. However, this index now corresponds to a *set* of entries, usually two to eight, that are searched in parallel for a matching tag. In practice, this approach is much more efficient from a hardware implementation perspective, since it requires fewer address comparators than a fully associative cache, but due to its flexible mapping policy behaves similarly to a fully associative cache. Hill and Smith [1989] present a classic evaluation of associativity in caches.

Evicting Blocks. Since each level in the cache has finite capacity, there must be a policy and mechanism for removing or *evicting* current occupants to make room for blocks corresponding to more recent references. The *replacement policy* of the cache determines the algorithm used to identify a candidate for eviction. In a direct-mapped cache, this is a trivial problem, since there is only a single potential candidate, as only a single entry in the cache can be used to store the new block, and the current occupant of that entry must be evicted to free up the entry.

In fully associative and set-associative caches, however, there is a choice to be made, since the new block can be placed in any one of several entries, and the current occupants of all those entries are candidates for eviction. There are three common policies that are implemented in modern cache designs: *first in, first out* (FIFO), *least recently used* (LRU), and *random*.

The *FIFO policy* simply keeps track of the insertion order of the candidates and evicts the entry that has resided in the cache for the longest amount of time. The mechanism that implements this policy is straightforward, since the candidate eviction set (all blocks in a fully associative cache, or all blocks in a single set in a set-associative cache) can be managed as a circular queue. The circular queue has a single pointer to the oldest entry which is used to identify the eviction candidate, and the pointer is incremented whenever a new entry is placed in the queue. This results in a single update for every miss in the cache.

However, the FIFO policy does not always match the temporal locality characteristics inherent in a program's reference stream, since some memory locations are accessed continually throughout the execution (e.g., commonly referenced global variables). Such references would experience frequent misses under a FIFO policy, since the blocks used to satisfy them would be evicted at regular intervals, as soon as every other block in the candidate eviction set had been evicted.

The *LRU policy* attempts to mitigate this problem by keeping an ordered list that tracks the recent references to each of the blocks that form an eviction set. Every time a block is referenced as a hit or a miss, it is placed on the head of this ordered list, while the other blocks in the set are pushed down the list. Whenever a block needs to be evicted, the one on the tail of the list is chosen, since it has been referenced least recently (hence the name least recently used). Empirically, this policy has been found to work quite well, but is challenging to implement, as it requires storing an ordered list in hardware and updating that list, not just on every cache miss, but on every hit as well. Quite often, a practical hardware mechanism will only implement an approximate LRU policy, rather than an exact LRU policy, due to such implementation challenges. An instance of an approximate algorithm is the not-most-recently-used (NMRU) policy, where the history mechanism must remember which block was referenced most recently and victimize one of the other blocks, choosing randomly if there is more than one other block to choose from. In the case of a two-way associative cache, LRU and NMRU are equivalent, but for higher degrees of associativity, NMRU is less exact but simpler to implement, since the history list needs only a single element (the most recently referenced block).

The final policy we consider is *random* replacement. As the name implies, under this policy a block from the candidate eviction set is chosen at random. While this may sound risky, empirical studies have shown that *random* replacement is only slightly worse than true LRU and still significantly better than FIFO. Clearly, implementing a true random policy would be very difficult, so practical mechanisms usually employ some reasonable pseudo-random approximation for choosing a block for eviction from the candidate set.

Handling Updates to a Block. The presence of a cache in the memory subsystem implies the existence of more than one copy of a block of memory in the system. Even with a single level of cache, a block that is currently cached also has a copy still stored in the main memory. As long as blocks are only read, and never written, this is not a problem, since all copies of the block have exactly the same contents. However, when the processor writes to a block, some mechanism must exist for updating all copies of the block, in order to guarantee that the effects of the write persist beyond the time that the block resides in the cache. There are two approaches for handling this problem: write-through caches and writeback caches.

A *write-through cache,* as the name implies, simply propagates each write through the cache and on to the next level. This approach is attractive due to its simplicity, since correctness is easily maintained and there is never any ambiguity about which copy of a particular block is the current one. However, its main drawback is the amount of bandwidth required to support it. Typical programs contain about 15% writes, meaning that about one in six instructions updates a block in memory. Providing adequate bandwidth to the lowest level of the memory hierarchy to write through at this rate is practically impossible, given the current and continually increasing disparity in frequency between processors and main memories. Hence, write-through policies are rarely if ever used throughout all levels of a cache hierarchy.

A write-through cache must also decide whether or not to fetch and allocate space for a block that has experienced a miss due to a write. A *write-allocate* policy implies fetching such a block and installing it in the cache, while a *write-no-allocate* policy would avoid the fetch and would fetch and install blocks only on read misses. The main advantage of a write-no-allocate policy occurs when streaming writes overwrite most or all of an entire block before any unwritten part of the block is read. In this scenario, a useless fetch of data from the next level is avoided (the fetched data is useless since it is overwritten before it is read).

A *writeback cache,* in contrast, delays updating the other copies of the block until it has to in order to maintain correctness. In a writeback cache hierarchy, an implicit priority order is used to find the most up-to-date copy of a block, and only that copy is updated. This priority order corresponds to the levels of the cache hierarchy and the order in which they are searched by the processor when attempting to satisfy a reference. In other words, if a block is found in the highest level of cache, that copy is updated, while copies in lower levels are allowed to become stale, since the update is not propagated to them. If a block is only found in a lower level, it is promoted to the top level of cache and is updated there, once again leaving behind stale copies in lower levels of the hierarchy.

The updated copy in a writeback cache is also marked with a *dirty* bit or flag to indicate that it has been updated and that stale copies exist at lower levels of the hierarchy. Ultimately, when a dirty block is evicted to make room for other blocks, it is *written back* to the next level in the hierarchy, to make sure that the update to the block persists. The copy in the next level now becomes the most up-to-date copy and must also have its dirty bit set, in order to ensure that the block will get written back to the next level when it gets evicted.

Writeback caches are almost universally deployed, since they require much less write bandwidth. Care must be taken to design these caches correctly, so that no updates are ever dropped due to losing track of a dirty cache line. We revisit writeback hierarchies in greater depth in Chapter 11 in the context of systems with multiple processors and multiple cache hierarchies.

However, despite the apparent drawbacks of write-through caches, several modern processors, including the IBM Power4 [Tendler et al., 2001] and Sun UltraSPARC III [Lauterbach and Horel, 1999], do use a write-through policy for the first level of cache. In such schemes, the hierarchy propagates all writes to the second-level cache, which is also on the processor chip. Since the next level of cache is on the chip, it is relatively easy to provide adequate bandwidth for the write-through traffic, while the design of the first-level cache is simplified, since it no longer needs to serve as the sole repository for the most up-to-date copy of a cache block and never needs to initiate writebacks when dirty blocks are evicted from it. However, to avoid excessive off-chip bandwidth consumption due to write-throughs, the second-level cache maintains dirty bits to implement a writeback policy.

Figure 3.5 summarizes the main parameters—block size, block organization, replacement policy, write policy, and write-allocation policy—that can be used to describe a typical cache design.

3.4.3.5 Cache Miss Classification.
As discussed in Section 3.4.3.1, the average reference latency delivered by a multilevel cache hierarchy can be computed as the average of the latencies of each level in the hierarchy, weighted by the global hit rate of each level. The latencies of each level are determined by the technology used and the aggressiveness of the physical design, while the miss rates are a function of the

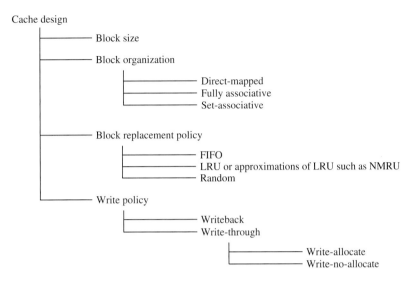

Figure 3.5
Cache Design Parameters.

organization of the cache and the access characteristics of the program that is running on the processor. Attaining a deeper understanding of the causes of cache misses in a particular cache hierarchy enables the designer to realize the shortcomings of the design and discover creative and cost-effective solutions for improving the hierarchy. The *3 C's model* proposed by Mark Hill [Hill, 1987] is a powerful and intuitive tool for classifying cache misses based on their underlying root cause. This model introduces the following mutually exclusive categories for cache misses:

- *Cold or compulsory misses.* These are due to the program's first reference to a block of memory. Such misses are considered *fundamental* since they cannot be prevented by any caching technique.
- *Capacity misses.* These are due to insufficient capacity in a particular cache. Increasing the capacity of that cache can eliminate some or all capacity misses that occur in that cache. Hence, such misses are not *fundamental,* but rather a by-product of a finite cache organization.
- *Conflict misses.* These are due to imperfect allocation of entries in a particular cache. Changing the associativity or indexing function used by a cache can increase or decrease the number of conflict misses. Hence, again, such misses are not *fundamental,* but rather a by-product of an imperfect cache organization. A fully associative cache organization can eliminate all conflict misses, since it removes the effects of limited associativity or indexing functions.

Cold, capacity, and conflict misses can be measured in a simulated cache hierarchy by simulating three different cache organizations for each cache of interest.

The first organization is the actual cache being studied; for notational convenience let's assume it experiences m_a cache misses. The second organization is a fully associative cache with the same capacity and block size as the actual cache; it experiences m_f cache misses. The third and final organization is a fully associative cache with the same block size but infinite capacity; it experiences m_c misses. The number of cold, capacity, and conflict misses can now be computed as

- Cold misses = m_c, number of misses in fully associative, infinite cache.
- Capacity misses = $m_f - m_c$, number of additional misses in finite but fully associative cache over infinite cache.
- Conflict misses = $m_a - m_f$, number of additional misses in actual cache over number of misses in fully associative, finite cache.

Cold misses are fundamental and are determined by the working set of the program in question, rather than by the cache organization. However, varying the block size directly affects the number of cold misses experienced by a cache. Intuitively, this becomes obvious by considering two extreme block sizes: a cache with a block size of one word will experience a cold miss for every unique word referenced by the program (this forms the upper bound for the number of cold misses in any cache organization), while a cache with an infinite block size will experience only a single cold miss. The latter is true because the very first reference will install all addressable memory into the cache, resulting in no additional misses of any type. Of course, practical cache organizations have a finite block size somewhere between these two endpoints, usually in the range of 16 to 512 bytes.

Capacity misses are not fundamental but are determined by the block size and capacity of the cache. Clearly, as capacity increases, the number of capacity misses is reduced, since a larger cache is able to capture a larger share of the program's working set. In contrast, as block size increases, the number of unique blocks that can reside simultaneously in a cache of fixed capacity decreases. Larger blocks tend to be utilized more poorly, since the probability that the program will access all the words in a particular block decreases as the block gets bigger, leading to a lower *effective capacity*. As a result, with fewer unique blocks and a decreased probability that all words in each block are useful, a larger block size usually results in an increased number of capacity misses. However, programs that efficiently utilize all the contents of large blocks would not experience such an increase.

Conflict misses are also not fundamental and are determined by the block size, the capacity, and the associativity of the cache. Increased capacity invariably reduces the number of conflict misses, since the probability of a conflict between two accessed blocks is reduced as the total number of blocks that can reside in the cache simultaneously increases. As with capacity misses, a larger number of smaller blocks reduces the probability of a conflict and improves the effective capacity, resulting in likely fewer conflict misses. Similarly, increased associativity will almost invariably reduce the number of conflict misses. (Problem 25 in the homework will ask you to construct a counterexample to this case.)

Table 3.2
Interaction of cache organization and cache misses

Cache Parameter	Effect on Cache Miss Rate			
	Cold Misses	Capacity Misses	Conflict Misses	Overall Misses
Reduced capacity	No effect	Increase	Likely increase	Likely increase
Increased capacity	No effect	Decrease	Likely decrease	Likely decrease
Reduced block size	Increase	Likely decrease	Likely decrease	Varies
Increased block size	Decrease	Likely increase	Likely increase	Varies
Reduced associativity	No effect	No effect	Likely increase	Likely increase
Increased associativity	No effect	No effect	Likely decrease	Likely decrease
Writeback vs. write-through	No effect	No effect	No effect	No effect
Write-no-allocate	Possible decrease	Possible decrease	Possible decrease	Possible decrease

Table 3.2 summarizes the effects of cache organizational parameters on each category of cache misses, as well as overall misses. Note that some parameters can have unexpected effects, but empirical evidence tells us that for most programs, the effects are as summarized in Table 3.2. The possible decrease noted for write-no-allocate caches is due to blocks that are only written to and never read; these blocks are never fetched into the cache, and hence never incur any type of misses. This directly reduces the number of cold misses and can indirectly reduce capacity misses, conflict misses, and overall misses.

3.4.3.6 Example Cache Hierarchy.

Figure 3.6 illustrates a typical two-level cache hierarchy, where the CPU or processor contains a register file and is directly connected to a small, fast level-one instruction cache (L1 I-$) and a small, fast level-one data cache (L1 D-$). Since these first-level or primary caches are relatively small, typically ranging from 8 up to 64K bytes, they can be accessed quickly, usually in only a single processor cycle, and they can provide enough bandwidth to keep the processor core busy with enough instructions and data. Of course, only in rare cases do they provide enough capacity to contain all the working set of a program. Inevitably, the program will issue a reference that is not found in the first-level cache. Such a reference results in a *cache miss,* or a reference that needs to be forwarded to the next level in the memory hierarchy. In the case of the example in Figure 3.6, this is the level-two cache, which contains both program text and data and is substantially larger. Modern second-level caches range from 256K bytes to 16 Mbytes, with access latencies of a few nanoseconds up to 10 or 15 ns for large, off-chip level-two caches.

Modern processors usually incorporate a second level of cache on chip, while recent processor designs like the Intel Xeon and Itanium 2 actually add a third level of cache onboard the processor chip. High-end system designs like IBM xSeries 445

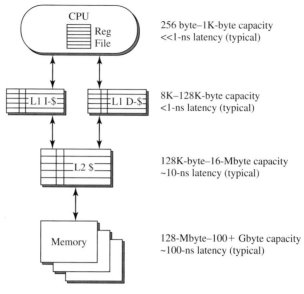

Figure 3.6
A Typical Memory Hierarchy.

multiprocessors that employ Itanium 2 processors and are intended for extremely memory-intensive server applications even include a fourth level of cache memory on the system board.

Finally, the physical memory hierarchy is backed up by DRAM that ranges in size from 128 Mbytes in entry-level desktop PCs to 100 Gbytes or more in high-end server systems. The latency for a reference that must be satisfied from DRAM is typically at least 100 ns, though it can be somewhat less in a single-processor system. Systems with multiple processors that share memory typically pay an overhead for maintaining cache coherence that increases the latency for main memory accesses, in some cases up to 1000 ns. Chapter 11 discusses many of the issues related to efficient support for coherent shared memory.

In light of the example shown in Figure 3.6, let's revisit the five *memory idealisms* introduced earlier in the chapter:

- *Infinite capacity.* For storing large data sets and large programs.
- *Infinite bandwidth.* For rapidly streaming these large data sets and programs to and from the processor.
- *Instantaneous or zero latency.* To prevent the processor from stalling while waiting for data or program code.
- *Persistence or nonvolatility.* To allow data and programs to survive even when the power supply is cut off.
- *Zero or very low implementation cost.*

We see that the highest levels of the memory hierarchy—register files and primary caches—are able to supply near-infinite bandwidth and very low average latency to the processor core, satisfying the second and third idealisms. The first idealism—infinite capacity—is satisfied by the lowest level of the memory hierarchy, since the capacities of DRAM-based memories are large enough to contain the working sets of most modern applications; for applications where this is not the case, Section 3.5 describes a technique called *virtual memory* that extends the memory hierarchy beyond random-access memory devices to magnetic disks, which provide capacities that exceed the demands of all but the most demanding applications. The fourth idealism—persistence or nonvolatility—can also be supplied by magnetic disks, which are designed to retain their state even when they are powered down. The final idealism—low implementation cost—is also satisfied, since the high per-bit cost of the upper levels of the cache hierarchy is only multiplied by a relatively small number of bits, while the lower levels of the hierarchy provide tremendous capacity at a very low cost per bit. Hence, the average cost per bit is kept near the low cost of commodity DRAM and magnetic disks, rather than the high cost of the custom SRAM in the cache memories.

3.4.4 Main Memory

In a typical modern computer system, the main memory is built from standardized commodity DRAM chips organized in a flexible, expandable manner to provide substantial capacity and expandability, high bandwidth, and a reasonably low access latency that should be only slightly higher than the access latency of the DRAM chips themselves. Since current-generation DRAM chips have a capacity of 256 megabits, a computer system with 1 Gbyte of memory would require approximately 32 DRAM chips for storage; including overhead for parity or error-correction codes to detect and tolerate soft errors would typically increase the count to 36 chips. Next-generation DRAM chips, which are just around the corner, will provide 1 gigabit of capacity each, reducing the chip count by a factor of 4. However, demand for increased memory capacity in future systems will likely keep the total number of DRAM chips required to satisfy that capacity relatively constant.

Clearly, there are many possible ways to configure a large number of DRAM chips to optimize for cost, latency, bandwidth, or expandability. Figure 3.7 illustrates one possible approach for arranging and interconnecting memory chips. In this configuration, multiple DRAM chips are mounted on a dual inline memory module (DIMM); multiple DIMMs are connected to a shared port or bank, and one or more banks are connected to a memory controller. In turn, the memory controller connects to the system's processor bus and responds to the processor's memory requests by issuing appropriate commands to one or both memory banks. Section 3.4.4.1 introduces the basic principles of DRAM chip organization, and Section 3.4.4.2 discusses several key issues in memory controller design.

3.4.4.1 DRAM Chip Organization.
DRAM chips are a commodity product that are manufactured by several competing vendors worldwide. DRAM manufacturers collaborate on standardizing the specification of the capacities and interfaces of

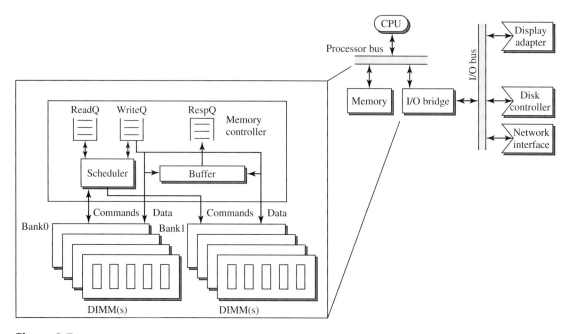

Figure 3.7
Typical Main Memory Organization.

each generation of DRAM chips in order to guarantee compatibility and consistent performance. Conceptually, the function and organization of DRAM chips is quite straightforward, since they are designed to store as many bits as possible in as compact an area as possible, while minimizing area and product cost and maximizing bandwidth. While all these factors are considered in DRAM design, historically the primary design constraints have been capacity and cost. DRAM manufacturing is an extremely competitive business, where even minor increases in product cost, potentially caused by complex designs that reduce process yields, can drive a vendor out of business. Hence, DRAM vendors are typically very conservative about adopting dramatically new or different approaches for building DRAM chips.

As semiconductor process geometries have shrunk, DRAM capacity per chip has increased at a rate more or less directly proportional to Moore's law, which predicts a doubling of devices per chip every two years or so. This has resulted in exponential growth in the capacity of DRAM chips with a fixed die size, which in turn has tended to hold product cost roughly constant. Despite the fact that device switching times improve with reduced process geometries, DRAM chip latency has not improved dramatically. This is due to the fact that DRAM access latency is dominated by wire delay, and not device switching times. Since wire delay has not improved nearly as dramatically as device switching delay, and the overall dimension of the memory array has remained largely fixed (to accommodate increased capacity), the end-to-end latency to retrieve a word from a DRAM chip has only improved at a compound rate of about 10% per year. This provides a stark contrast with the 60% compound rate of frequency growth observed for general-purpose microprocessors. This divergence in device frequency has led many computer system designers and researchers to search for new techniques that will surmount what is known as the *memory wall* [Wulf and McKee, 1995].

The other main contributor to DRAM product cost—packaging, as driven by per-chip pin count—has also remained relatively stable over the years, resulting in a dearth of dramatic improvements in bandwidth per chip. The improvements that have been made for bandwidth have been largely in the realm of enhanced signaling technology, synchronous interfaces [synchronous DRAM (SDRAM)], higher interface frequencies (e.g., PC100 which runs at 100 MHz, while PC133 runs at 133 MHz), and aggressive use of both rising and falling clock edges to transmit twice the amount of data per clock period [known as double-data rate (DDR)].

Figure 3.8 shows the internal organization of a typical DRAM chip. At its heart, there is an array of binary storage elements organized in rows and columns. The storage elements are tiny capacitors, which store a charge to represent a 1, or store no charge to represent a 0. Each capacitor-based cell is connected by a transistor to a vertical bit line that stretches from the top of the array to the bottom. The transistor is controlled by a horizontal *word line* which selects a particular row in the array for either reading or writing. The bit line is used to read the state of the cell: a charged capacitor will drive the bit line to a higher voltage; this higher voltage will be sensed by a high-gain amplifier at one end of the bit line that converts the signal to a standard logic level 1, while a discharged capacitor will drain

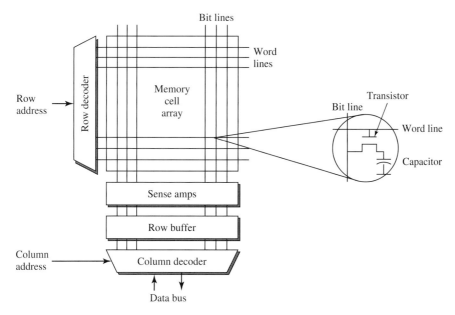

Figure 3.8
DRAM Chip Organization.

charge from the bitline, reducing its voltage to a level that is amplified to logic level 0. The high-gain analog amplifier is called a *sense amp* and relies on *bit-line precharging,* a process that presets the bit-line voltage prior to a read to an intermediate value that can swing high or low very quickly depending on the state of the accessed cell's capacitor. The bit line is also used to store a value in the cell by driving the bit line high to store a charge in the capacitor, or driving it low to drain the charge from the capacitor. Since the charge stored by a capacitor decays over time, cells in a DRAM chip must be refreshed periodically to maintain their state. This dynamic behavior lends itself to the naming of DRAM chips; the acronym stands for *dynamic random-access memory*. In contrast, SRAM or *static random-access memory,* employed in higher levels of the cache hierarchy, does not need to be refreshed since the storage cells are static complementary metal-on-semiconductor (CMOS) circuits (a pair of cross-coupled inverters) that can hold their state indefinitely, as long as power is supplied to them.

DRAM Addressing. The word lines in Figure 3.8 are used to select a row within the array to either read from that row (i.e., let the capacitors from that row drive the bit lines) or write to it (i.e., let the bit lines drive or drain the capacitors in that row). A row address of n bits must be supplied to the DRAM chip, and a decoder circuit activates one of the 2^n word lines that corresponds to the supplied row address. At each storage cell, the word line controls a pass transistor that either isolates the capacitor from the bit line or connects it to the bit line, to enable selective reading or writing of a single row. During a read, the bits in the selected row are sensed by the sense amps and are stored in a row buffer. In a subsequent command cycle, a column address of m bits must also be supplied; this is used to select one of 2^m words from within the row.

DRAM Access Latency. The latency to access a random storage location within a DRAM chip is determined by the inherent latency of the storage array augmented by the overhead required to communicate with the chip. Since DRAM chips are very cost-sensitive, and an increased pin count drives cost higher, DRAM interfaces typically share the same physical pin for several purposes. For example, a DRAM chip does not provide enough address pins to directly address any word within the array; instead, the address pins are time-multiplexed to first provide the row address when the row address strobe (RAS) control line is asserted, followed by a column address strobe (CAS) while the column address is provided. As long as there are an equal number of rows and columns, only half the number of address pins are needed, reducing cost significantly. On the other hand, two transactions are required across the interface to provide a complete address, increasing the latency of an access.

The decoupling of the row and column addresses creates an opportunity for optimizing the access latency for memory references that exhibit spatial locality. Since the DRAM chip reads the entire row into a row buffer, and then selects a word out of that row based on the column address, it is possible to provide multiple column addresses in back-to-back cycles and read bursts of words from the same row at a very high rate, usually limited only by the interface frequency and data rate.

Historically, back-to-back accesses to the same row have been called *page-mode* accesses. These same-row accesses complete much faster (up to 3 times faster in current-generation DRAM) since they do not incur the additional latency of providing a row address, decoding the row address, precharging the bit lines, and reading the row out of the memory cell array into the row buffer. This creates a performance-enhancing scheduling opportunity for the memory controller, which we will revisit in Section 3.4.4.2.

DRAM chips also share the pins of the data bus for both reads and writes. While it is easy to pipeline a stream of reads or a stream of writes across such an interface, alternating reads with writes requires a *bus turnaround*. Since the DRAM chip is driving the data bus during reads, while the memory controller is driving it during writes, care must be taken to ensure that there is never a time during which both the memory controller and DRAM chip are attempting to drive the bus, as this could result in a short-circuit condition between the respective drivers. Hence, the interface must be carefully designed to allow enough timing margin for the current bus master (e.g., the DRAM chip on a read) to stop driving the bus before the new bus master (e.g., the memory controller on a write) starts to drive the bus. This results in additional delay and reduces the efficiency of the DRAM interface.

Rambus DRAM. In the late 1990s, an alternative standard called Rambus DRAM (RDRAM) emerged from research in high-speed signaling at Stanford University. RDRAM employed many of the techniques that more recent standards for synchronous DRAM (e.g., DDR2) have since adopted, including advanced signaling, higher interface frequencies, and multiple data words per clock period. RDRAM chips also provide a *row buffer cache* which contains several of the most recently accessed DRAM rows; this increases the probability that a random access can be satisfied with much lower latency from one of the row buffer entries, avoiding the row access latency. To improve interface bandwidth, RDRAM carefully specifies the physical design for board-level traces used to connect RDRAM chips to the controller, and uses source-synchronous clocking (i.e., the clock signal travels with the data) to drive clock frequencies several times higher than more conventional SDRAM approaches. As a result of these optimizations, RDRAM is able to provide substantially higher bandwidth per pin, albeit at a higher product cost and increased design time due to the stringent physical design requirements.

High bandwidth per pin is very useful in systems that require lots of bandwidth but relatively little capacity. Examples of such systems that use RDRAM are the Sony Playstation 2 and Microsoft X-Box game controllers, which provide only 32 Mbytes of memory in their base configuration while requiring lots of memory bandwidth to support intensive three-dimensional gaming. A modest capacity requirement of only 32 Mbytes could be satisfied by a single current-generation 256-Mbit DRAM chip. However, since that DRAM chip only has a small number of data pins (between 2 and 16, typically 4 or 8), each pin must provide very high bandwidth to satisfy the system's overall bandwidth demand.

In contrast, general-purpose computer systems such as personal computers typically contain at least an order of magnitude more memory, ranging from at least 256 Mbit to several gigabytes. In such a system, per-pin bandwidth is less important, since many DRAM chips are required to provide the requisite capacity anyway, and these chips can be arranged in parallel to provide a wide interface that supplies the required bandwidth. Of course, this increases the package cost of the memory controller chip, since it has to have enough pins to access multiple DRAM chips in parallel. On the other hand, product cost can be lowered by using less aggressive and less expensive circuit board technology, since each pin signals at a lower rate.

Detailed performance evaluation and comparison of various modern DRAM technologies can be found in two recent studies [Cuppu et al., 1999; Cuppu and Jacob, 2001].

3.4.4.2 Memory Controller Organization. Memory controllers serve as the interface between a system's processor bus, which communicates reads and writes issued by the processor, and the standard DRAM command interface, which expects a tightly specified sequence of commands for supplying row addresses, column addresses, and read or write commands to each DRAM chip. There are many alternatives for arranging the mapping between the physical addresses provided by the processor and the DRAM addresses needed to access the actual storage locations. Furthermore, there are various optimizations that can be applied within the memory controller to improve read performance and increase spatial locality in the DRAM reference stream. The net effect is that the design of the memory controller can substantially impact sustainable memory bandwidth and the observed latency of memory references. The following sections discuss some of these issues in detail.

Memory Module Organization. The desired memory capacity for a system determines the total number of DRAM chips needed to provide that capacity. In practice, most systems are designed to support a range of capacities to enable use of several generations of memory chips and to allow for future expansion from an initial configuration. However, for simplicity we will assume a fixed capacity only, resulting in a fixed number of DRAM chips in the memory subsystem. Figure 3.9 illustrates four possible organizations for four DRAM chips in a system, as determined by two fundamental attributes: serial vs. parallel and interleaved vs. non-interleaved. In the top left case, the DRAM chips all share the same address, command, and data lines, but only a single chip is active at a time when it is selected via its chip select (CS) control line. In this case, the number of pins required in the controller is minimized, since all pins except for the CS control lines are shared. However, data bandwidth is limited by the data bus width of each DRAM chip. DRAM chips have 2, 4, 8, or 16 data lines—typically 4 or 8 in current-generation chips—with a price premium charged for wider interfaces. In this organization, transaction bandwidth is restricted to one concurrent command, since the address and command lines are shared across all chips.

The top right case in Figure 3.9 shows a parallel organization, in which all chips are active for all commands (hence, no chip selects), and the n-bit data busses from each chip are concatenated to form a $4n$-bit interface. This configuration provides much better data bandwidth, but at a higher memory controller cost due to

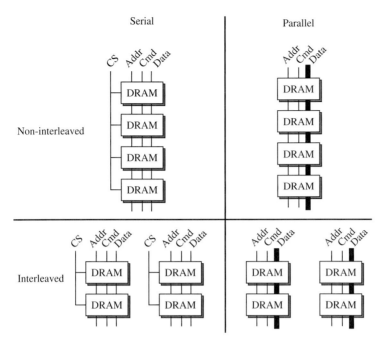

Figure 3.9
Memory Module Organization.

the private data bus pins for each DRAM chip. It provides no increase in transaction bandwidth, since the address and command lines are still shared.

The bottom left shows an *interleaved* (or *banked*) organization where half the DRAM chips are connected to one set of chip select, address, command, and data busses, while the other half is connected to a second set. This organization provides twice the data bandwidth, and also twice the transaction bandwidth, since each interleaved *bank* can now operate independently, at the cost of twice as many address and command pins. The final configuration, on the bottom right, combines the parallel and interleaved organizations, providing twice the transaction bandwidth through interleaving, and four times the data bandwidth through two $2n$-bit wide data busses. Of course, this final configuration requires the highest cost and largest number of pins in the memory controller.

There are other possible combinations of these techniques (for example, a semi-parallel scheme where only half—instead of all—DRAM chips are in parallel, and chip selects are still used to select which half drives the $2n$-bit wide data bus). In all these combinations, however, the linear physical address presented by the processor needs to be translated to an n-tuple that describes the actual physical location of the memory location being addressed. Table 3.3 summarizes the contents of the DRAM address n-tuple for several approaches for organizing memory modules. Linear physical addresses are mapped to these n-tuples by selecting appropriately sized subsets of bits from the physical address to form each element

Table 3.3
Translating linear physical address to DRAM address*

Organization (Figure 3.9)	DRAM Address Breakdown	Example Physical Address	Corresponding DRAM Address
Serial non-interleaved, 8-bit data bus	16 12 11 2 1 0 | RAS | CAS | CS |	0x4321	RAS: 0x4, CAS: 0xC8, CS: 0x1
Parallel non-interleaved, 32-bit data bus	16 12 11 2 1 0 | RAS | CAS | n/a |	0x4320	RAS: 0x4, CAS: 0xC8
Serial interleaved, 8-bit data bus	16 12 11 2 1 0 | RAS | CAS | Ba | CS |	0x8251	RAS: 0x4, CAS: 0x94, Bank: 0x0, CS: 0x0
Parallel interleaved, 16-bit data bus	16 12 11 2 1 0 | RAS | CAS | Ba | n/a |	0x8254	RAS: 0x4, CAS: 0x95, Bank: 0x0

*Examples assume 4 × 256-kbit DRAM with 8-bit data path and 8-kbit row, for a total of 128 kB of addressable memory.

of the n-tuple. In general, this selection problem has $\binom{x}{y}$ possible solutions, where x is the number of physical address bits and y is the total number of bits needed to specify all the n-tuple elements. Table 3.3 shows only one of these many possible ways of choosing the bits for each element in the DRAM address n-tuple.

Regardless of which organization is chosen, the RAS bits should be selected to maximize the number of row hits; careful study of the access patterns of important applications can reveal which address bits are the best candidates for RAS. Furthermore, in an interleaved design, the bits used for bank selection need to be selected carefully to ensure even distribution of references across the memory banks, since a poor choice of bank bits can direct all references to a single bank, negating any bandwidth benefit expected from the presence of multiple banks.

Components of a Memory Controller. As shown in Figure 3.7, a memory controller contains more than just an interface to the processor bus and an interface to the DRAM chips. It also contains logic for buffering read and write commands from the processor bus (the *ReadQ* and *WriteQ*), a response queue (*RespQ*) for buffering responses heading back to the processor, scheduling logic for issuing DRAM commands to satisfy the processor's read and write requests, and buffer space to assemble wide data responses from multiple narrow DRAM reads. This reassembly is needed whenever the processor bus issues read commands that are wider than the DRAM data interface; in such cases multiple DRAM reads have to be performed to assemble a block that matches the width of the processor's read request (which is usually the width of a cache block in the processor's cache). In a similar fashion, wide writes from the processor bus need to be decomposed into multiple narrower writes to the DRAM chips.

Although Figure 3.7 shows the memory controller as a physically separate entity, recent designs, exemplified by the AMD Opteron [Keltcher et al., 2003], integrate the memory controller directly on chip to minimize memory latency, simplify the chipset design, and reduce overall system cost. One of the drawbacks of an on-chip memory controller is that processor designs must now be synchronized with evolving memory standards. As an example, the Opteron processors must be redesigned to take advantage of the new DDR2 DRAM standard, since the onboard controller will only work the older DDR standard. In contrast, an off-chip memory controller (*North Bridge* in the Intel/PC terminology) can be more quickly redesigned and replaced to match new memory standards.

The ReadQ is used to buffer multiple outstanding reads; this decouples completion of a read from accepting the next one. Quite often the processor will issue reads in bursts, as cache misses tend to occur in clusters and accepting multiple reads into the ReadQ prevents the bus from stalling. Queueing up multiple requests may also expose more locality that the memory controller can exploit when it schedules DRAM commands. Similarly, the WriteQ prevents the bus and processor from stalling by allowing multiple writes to be outstanding at the same time. Furthermore, the WriteQ enables a latency-enhancing optimization for reads: since writes are usually not latency-critical, the WriteQ can delay them in favor of outstanding reads, allowing the reads to be satisfied first from the DRAM. The delayed writes can be retired whenever there are no pending reads, utilizing idle memory channel cycles.

Memory Reference Scheduling. Of course, reference reordering in the memory controller is subject to the same correctness requirements as a pipelined processor for maintaining read-after-write (RAW), write-after-read (WAR), and write-after-write (WAW) dependences. In effect, this means that reads cannot be reordered past pending writes to the same address (RAW), writes cannot be reordered past pending reads from the same address (WAR), and writes cannot bypass pending writes to the same address (WAW). If we are only reordering reads with respect to outstanding writes, only the RAW condition needs to be checked. If a RAW condition exists between a pending write and a newer read, the read must either stall and wait for the write to be performed against the DRAM, or the read can be satisfied directly from the write queue. Either solution will maintain correctness, while the latter should improve performance, since the latency of the read from the on-chip WriteQ will be lower than a read from an external DRAM chip.

However, in Section 3.4.4.1 we showed how DRAM chips can exploit spatial locality by fetching multiple words from the same row by issuing different column addresses to the DRAM in back-to-back cycles. These references can be satisfied much more quickly than references to different rows, which incur the latency for a row address transfer and row read in the internal DRAM array. In current-generation DRAMs, rows can be as large as 8 kilobits; an eight-wide parallel organization (extrapolating from the two-wide parallel scheme shown in Figure 3.9) would result in an 8-kilobits row in physical memory. Accesses to the same row can be satisfied much more quickly than references to other rows. Hence, the scheduling logic in advanced memory controllers will attempt to find references to the same row in the ReadQ and WriteQ and attempt to schedule them together to increase the number of *row hits*. This type of scheduling optimization can substantially reduce average DRAM read latency and improve sustained memory bandwidth, but can dramatically complicate the scheduling logic as well as the ReadQ and WriteQ bypass logic.

Current-generation DRAM chips are also internally interleaved or banked. A typical chip will contain four independent banks that replicate most of the structures shown in Figure 3.8, while sharing the external address, data, and control lines. Internal DRAM banking allows the memory controller to overlap different types of commands to each bank; for example, bank 0 can begin a bit-line precharge cycle while bank 1 is performing a row access and bank 2 is performing a column access. Furthermore, each bank has a separate row buffer, which allows the memory controller to leave multiple rows open concurrently, increasing the probability that a future request will hit an open row, reducing the access latency for that request.

Finally, banking or interleaving the DRAM interface increases the transaction bandwidth for the memory controller, since multiple banks can operate independently and concurrently, as long as the ReadQ and WriteQ contain references to different banks. High-end memory controllers in multiprocessor server systems have many independent memory banks; commodity PC systems typically have one or two.

As a final note, the parallel and interleaved organizations described here for DRAM systems can also be applied to SRAM caches in higher levels of the memory hierarchy. In particular, multibanked caches are commonly used to increase transaction bandwidth to a cache. For example, the Intel Pentium processor incorporates an eight-way interleaved primary data cache to support concurrent memory accesses from its dual pipelines [Intel Corp., 1993]. Similarly, the IBM Power four-chip multiprocessor that is described in Chapter 6 has a three-way interleaved on-chip level-2 cache to support concurrent requests from the two processor cores that are on the chip [Tendler et al., 2001].

3.5 Virtual Memory Systems

So far, we have only considered levels of the memory hierarchy that employ random-access storage technology. However, in modern high-performance computer systems, the lowest level of the memory hierarchy is actually implemented using magnetic disks as a paging device or backing store for the physical memory, comprising a virtual memory system. The backing store contains blocks of memory that have been displaced from main memory due to capacity reasons, just the same as blocks are displaced from caches and placed either in the next level of the cache hierarchy or in main memory.

Historically, virtual memory predates caches and was first introduced 40 years ago in time-shared mainframe computers to enable sharing of a precious commodity—the main memory—among multiple active programs [Kilburn et al., 1962]. Virtual memory, as the name implies, *virtualizes* main memory by separating the programmer's view of memory from the actual physical placement of blocks in memory. It does so by adding a layer of cooperating hardware and software that manages the mappings between a program's *virtual address* and the *physical address* that actually stores the data or program text being referenced. This process of address translation is illustrated in Figure 3.10. The layer of cooperating hardware and software that enables address translation is called the *virtual memory system* and is

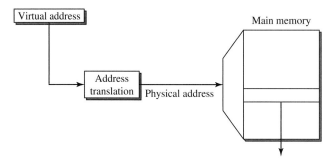

Figure 3.10
Virtual to Physical Address Translation.

responsible for maintaining the illusion that all virtually addressable memory is resident in physical memory and can be transparently accessed by the program, while also efficiently sharing the limited physical resources among competing demands from multiple active programs.

In contrast, time-sharing systems that predated or failed to provide virtual memory handicapped users and programmers by requiring them to explicitly manage physical memory as a shared resource. Portions of physical memory had to be statically allocated to concurrent programs; these portions had to be manually replaced and evicted to allocate new space; and cumbersome techniques such as data and program overlays were employed to reduce or minimize the amount of space consumed by each program. For example, a program would have to explicitly load and unload overlays that corresponded to explicit phases of program execution, since loading the entire program and data set could either overwhelm all the physical memory or starve other concurrent programs.

Instead, a virtual memory system allows each concurrent program to allocate and occupy as much memory as the system's backing store and its virtual address space allows: up to 4 Gbytes for a machine with 32-bit virtual addresses, assuming adequate backing store is available. Meanwhile, a separate *demand paging* mechanism manages the placement of memory in either the limited physical memory or in the system's capacious backing store, based on the policies of the virtual memory system. Such a system is responsible for providing the illusion that all virtually addressable memory is resident in physical memory and can be transparently accessed by the program.

The illusion of practically infinite capacity and a requirement for transparent access sound quite similar to the principles for caching described in Section 3.4.3; in fact, the underlying principles of temporal and spatial locality, as well as policies for locating, evicting, and handling updates to blocks, are all conceptually very similar in virtual memory subsystems and cache memories. However, since the relative latencies for accessing the backing store are much higher than the latencies for satisfying a cache miss from the next level of the physical memory hierarchy, the policies and particularly the mechanisms can and do differ substantially. A reference to a block that resides only in the backing store inflicts 10 ms or more of latency to read the block from disk. A pure hardware replacement scheme that stalls the processor while waiting for this amount of time would result in very poor utilization, since 10 ms corresponds to approximately 10 million instruction execution opportunities in a processor that executes one instruction per nanosecond. Hence, virtual memory subsystems are implemented as a hybrid of hardware and software, where references to blocks that reside in physical memory are satisfied quickly and efficiently by the hardware, while references that miss invoke the operating system through a *page fault* exception, which initiates the disk transfer but is also able to schedule some other, ready task to execute in the window of time between initiating and completing the disk request. Furthermore, the operating system now becomes responsible for implementing a policy for evicting blocks to allocate space for the new block being fetched from disk. We will study these issues in further detail in Section 3.5.1.

However, there is an additional complication that arises from the fact that multiple programs are sharing the same physical memory: they should somehow be protected from accessing each others' memory, either accidentally or due to a malicious program attempting to spy on or subvert another concurrent program. In a typical modern system, each program runs in its own virtual address space, which is disjoint from the address space of any other concurrent program. As long as there is no overlap in address spaces, the operating system need only ensure that no two concurrent address mappings from different programs ever point to the same physical location, and protection is ensured. However, this can limit functionality, since two programs cannot communicate via a shared memory location, and can also reduce performance, since duplicates of the same objects may need to exist in memory to satisfy the needs of multiple programs. For these two reasons, virtual memory systems typically provide mechanisms for protecting the regions of memory that they map into each program's address space; these protection mechanisms allow efficient sharing and communication to occur. We describe them further in Section 3.5.2.

Finally, a virtual memory system must provide an architected means for translating a virtual address to a physical address and a structure for storing these mappings. We outline several schemes for doing so in Section 3.5.3.

3.5.1 Demand Paging

Figure 3.11 shows an example of a single process that consumes virtual address space in three regions: for program text (to load program binaries and shared libraries); for the process stack (for activation records and automatic storage); and for the process heap (for dynamically allocated memory). Not only are these three regions noncontiguous, leaving unused holes in the virtual address space, but each of these regions can be accessed relatively sparsely. Practically speaking, only the regions that are currently being accessed need to reside in physical memory (shown as shaded in the figure), while the unaccessed or rarely accessed regions can be stored on the paging device or backing store, enabling the use of a system with a limited amount of physical memory for programs that consume large fractions of their address space, or, alternatively, freeing up main memory for other applications in a time-shared system.

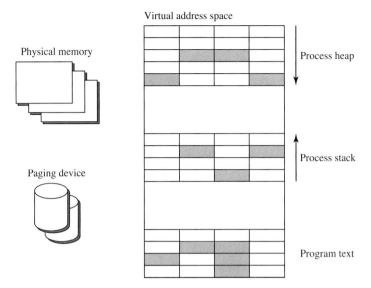

Figure 3.11
Virtual Memory System.

A virtual memory demand paging system must track regions of memory at some reasonable granularity. Just as caches track memory in blocks, a demand paging system must choose some page size as the minimum granularity for locating and evicting blocks in main memory. Typical page sizes in current-generation systems are 4K or 8K bytes. Some modern systems also support variable-sized pages or multiple page sizes to more efficiently manage larger regions of memory. However, we will restrict our discussion to fixed-size pages.

Providing a virtual memory subsystem relieves the programmer from having to manually and explicitly manage the program's use of physical memory. Furthermore, it enables efficient execution of classes of algorithms that use the virtual address space greedily but sparsely, since it avoids allocating physical memory for untouched regions of virtual memory. Virtual memory relies on *lazy allocation* to achieve this very purpose: instead of eagerly allocating space for a program's needs, it defers allocation until the program actually references the memory.

This requires a means for the program to communicate to the virtual memory subsystem that it needs to reference memory that has previously not been accessed. In a demand-paged system, this communication occurs through a page-fault exception. Initially, when a new program starts up, none of its virtual address space may be allocated in physical memory. However, as soon as the program attempts to fetch an instruction or perform a load or store from a virtual memory location that is not currently in virtual memory, a page fault occurs. The hardware registers a page fault whenever it cannot find a valid translation for the current virtual address. This is conceptually very similar to a cache memory experiencing a miss whenever it cannot find a matching tag when it performs a cache lookup.

However, a page fault is not handled implicitly by a hardware mechanism; rather, it transfers control to the operating system, which then allocates a page for the virtual address, creates a mapping between the virtual and physical addresses, installs the contents of the page into physical memory (usually by accessing the backing store on a magnetic disk), and returns control to the faulting program. The program is now able to continue execution, since the hardware can satisfy its virtual address reference from the corresponding physical memory location.

Detecting a Page Fault. To detect a page fault, the hardware must fail to find a valid mapping for the current virtual address. This requires an architected structure that the hardware searches for valid mappings before it raises a page fault exception to the operating system. The operating system's exception handler code is then invoked to handle the exception and create a valid mapping. Section 3.5.3 discusses several schemes for storing such mappings.

Page Allocation. Allocating space for a new virtual memory page is similar to allocating space for a new block in the cache, and depends on the page organization. Current virtual memory systems all use a fully-associative policy for placing virtual pages in physical memory, since it leads to efficient use of main memory, and the overhead of performing an associative search is not significant compared to the overall latency for handling a page fault. However, there must be a policy for evicting an active page whenever memory is completely full. Since a least-recently-used (LRU) policy would be too expensive to implement for the thousands of pages in a reasonably sized physical memory, some current operating systems use an approximation of LRU called the *clock algorithm*. In this scheme, each page in physical memory maintains a reference bit that is set by the hardware whenever a reference occurs to that page. The operating system intermittently clears all the reference bits. Subsequent references will set the page reference bits, effectively marking those pages that have been referenced recently. When the virtual memory system needs to find a page to evict, it randomly chooses a page from the set of pages with cleared reference bits. This scheme avoids evicting pages that have been referenced since the last time the reference bits were cleared, providing a very coarse approximation of the LRU policy.

Alternatively, the operating system can easily implement a FIFO policy for evicting pages by maintaining an ordered list of pages that have been fetched into main memory from the backing store. While not optimal, this scheme can perform reasonably well and is easy to implement since it avoids the overhead of the clock algorithm.

Once a page has been chosen for eviction, the operating system must place it in the backing store, usually by performing a write of the contents of the page to a magnetic disk. This write can be avoided if the hardware maintains a change bit or dirty bit for the page, and the dirty bit is not set. This is similar in principle to the dirty bits in a writeback cache, where only the blocks that have their dirty bit set need to be written back to the next level of the cache hierarchy when they are evicted.

Accessing the Backing Store. The backing store needs to be accessed to supply the paged contents of the virtual page that is about to be installed in physical memory.

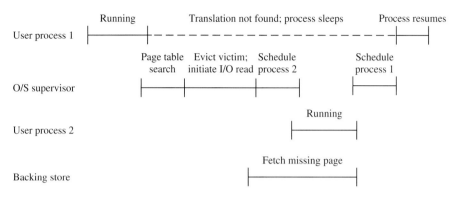

Figure 3.12
Handling a Page Fault.

Typically, this involves issuing a read to a magnetic disk, which can have a latency exceeding 10 ms. Multitasking operating systems will put a page-faulting task to sleep for the duration of the disk read and will schedule some other active task to run on the processor instead.

Figure 3.12 illustrates the steps that occur to satisfy a page fault: first, the current process 1 fails to find a valid translation for a memory location it is attempting to access; the operating system supervisor is invoked to search the page table for a valid translation via the page fault handler routine; failing to find a translation, the supervisor evicts a physical page to make room for the faulting page and initiates an I/O read to the backing store to fetch the page; the supervisor scheduler then runs to find a ready task to occupy the CPU while process 1 waits for the page fault to be satisfied; process 2 runs while the backing store completes the read; the supervisor is notified when the read completes, and runs its scheduler to find the waiting process 1; finally, process 1 resumes execution on the CPU.

3.5.2 Memory Protection

A system that time-shares the physical memory system through the use of virtual memory allows the physical memory to concurrently contain pages from multiple processes. In some scenarios, it is desirable to allow multiple processes to access the same physical page, in order to enable communication between those processes or to avoid keeping duplicate copies of identical program binaries or shared libraries in memory. Furthermore, the operating system kernel, which also has resident physical pages, must be able to protect its internal data structures from user-level programs.

The virtual memory subsystem must provide some means for protecting shared pages from defective or malicious programs that might corrupt their state. Furthermore, even when no sharing is occurring, protecting various address ranges from certain types of accesses can be useful for ensuring correct execution or for debugging new programs, since erroneous references can be flagged by the protection mechanism.

Typical virtual memory systems allow each page to be granted separate read, write, and execute permissions. The hardware is then responsible for checking that instruction fetches occur only to pages that grant execute permission, loads occur only to pages that grant read permission, and writes occur only to pages that grant write permission. These permissions are maintained in parallel with the virtual to physical translations and can only be manipulated by supervisor-state code running in the operating system kernel. Any references that violate the permissions specified for that page will be blocked, and the operating system exception handler will be invoked to deal with the problem, usually resulting in termination of the offending process.

Permission bits enable efficient sharing of read-only objects like program binaries and shared libraries. If there are multiple concurrent processes executing the same program binary, only a single copy of the program needs to reside in physical memory, since the kernel can map the same physical copy into the address space of each process. This will result in multiple virtual-physical address mappings where the physical address is the same. This is referred to as *virtual address aliasing*.

Similarly, any other read-only objects can be shared. Furthermore, programs that need to communicate with each other can request shared space from the operating system and can communicate directly with each other by writing to and reading from the shared physical address. Again, the sharing is achieved via multiple virtual mappings (one per process) to the same physical address, with appropriate read and/or write permissions set for each process sharing the memory.

3.5.3 Page Table Architectures

The virtual address to physical address mappings have to be stored in a translation memory. The operating system is responsible for updating these mappings whenever they need to change, while the processor must access the translation memory to determine the physical address for each virtual address reference that it performs. Each translation entry contains the fields shown in Figure 3.13: the virtual address, the corresponding physical address, permission bits for reading (Rp), writing (Wp), and executing (Ep), as well as reference (Ref) and change (Ch) bits, and possibly a caching-inhibited bit (Ca). The reference bit is used by the demand paging systems eviction algorithm to find pages to replace, while the change bit plays the part of a

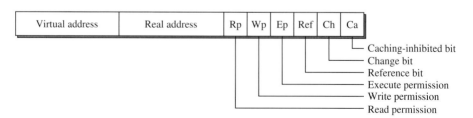

Figure 3.13
Typical Page Table Entry.

dirty bit, indicating that an eviction candidate needs to be written back to the backing store. The caching-inhibited bit is used to flag pages in memory that should not, for either performance or correctness reasons, be stored in the processor's cache hierarchy. Instead, all references to such addresses must be communicated directly through the processor bus. We will learn in Section 3.7.3 how this caching-inhibited bit is vitally important for communicating with I/O devices with memory-mapped control registers.

The translation memories are usually called page tables and can be organized either as *forward page tables* or *inverted page tables* (the latter are often called *hashed page tables* as well). At its simplest, a forward page table contains a page table entry for every possible page-sized block in the virtual address space of the process using the page table. However, this would result in a very large structure with many unused entries, since most processes do not consume all their virtual address space. Hence, forward page tables are usually structured in multiple levels, as shown in Figure 3.14. In this approach, the virtual address is decomposed into multiple sections. The highest-order bits of the address are added to the page table base register (PTBR), which points to the base of the first level of the page table. This first lookup provides a pointer to the next table; the next set of bits from the virtual address are added to this pointer to find a pointer to the next level. Finally, this pointer is added to the next set of virtual address bits to find the final leaf-level page table entry, which provides the actual physical address and permission bits corresponding to the virtual address. Of course, the multilevel page table can be extended to more than the three levels shown in Figure 3.14.

A multilevel forward page table can efficiently store translations for a sparsely populated virtual address space, since leaf nodes are only needed for those portions

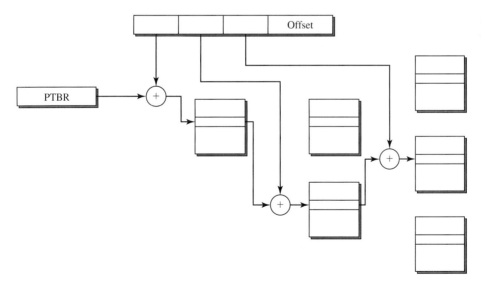

Figure 3.14
Multilevel Forward Page Table.

of the address space that are actually in use; unused tables in the middle and leaf levels are lazily allocated only when the operating system actually allocates storage in those portions of the address space. Furthermore, the page table entries themselves can be stored in virtual memory, allowing them to be paged out to the backing store. This can lead to nested page faults, when the initial page fault experiences a second page fault as it is trying to find the translation information for its virtual address. If paging of the page table is allowed, the root level of the page table needs to remain resident in physical memory to avoid an unserviceable page fault.

An alternative page table organization derives from the observation that there is little motivation to provide translation entries for more pages than can actually fit in physical memory. In an *inverted page table,* there are only enough entries to map all the physical memory, rather than enough entries to map all the virtual memory. Since an inverted page table has far fewer entries and fits comfortably into main memory, there is no need to make it pageable. Rather, the operating system can access it directly with physical addressing.

Figure 3.15 illustrates how translation entries are found in an inverted or hashed page table. The virtual address is hashed, usually by applying an exclusive-OR function to nonoverlapping portions of the virtual address, and is added to the page table base register. The resulting address is used directly as a physical address to find a set of page table entries (PTE0 through PTE3 in Figure 3.15). These page table entries are then checked sequentially to find a matching entry. Multiple entries need to be searched and provided, since it is possible for multiple virtual addresses to hash to the same location. In fact, it is possible for the number of virtual page numbers that map to the same page table entry group to exceed the capacity of the group; this results in an overflow condition that induces additional page faults. In effect, space in physical memory is now allocated in a set-associative manner, rather than a

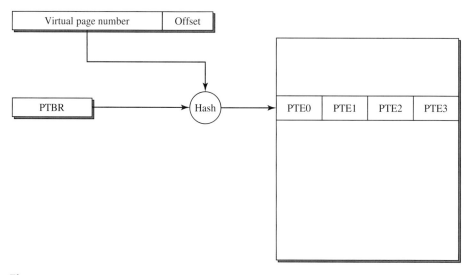

Figure 3.15
Hashed Page Table.

fully-associative manner. Fortunately, these types of conflicts are relatively rare. In the PowerPC virtual memory architecture, which uses a hashed page table, they are further mitigated by providing a secondary hashing scheme that differs substantially from the primary hash. Whenever the primary page table entry group fails to provide a valid translation, the secondary hash is used to find a second group that is also searched. The probability of failing to find a valid translation in either of the two groups is further minimized, though still not completely avoided.

One drawback of an inverted page table is that it only contains mappings for resident physical pages. Hence, pages that have been evicted from physical memory to the backing store need their mappings stored elsewhere. This is handled by the operating system, which maintains a separate software page table for tracking pages that reside in the backing store. Of course, this software page table maintains mapping from virtual addresses to the corresponding disk blocks, rather than to physical memory addresses.

As a final alternative, page tables need not be architected to reside in physical memory in a particular organization. Instead, a structure called the *translation lookaside buffer* (TLB, further described in Section 3.6 and illustrated in Figures 3.21 and 3.22) can be defined as part of the supervisor state of the processor. The TLB contains a small number (typically 64) of entries that look just like the entry illustrated in Figure 3.13, but arranged in a fully-associative fashion. The processor must provide fast associative lookup hardware for searching this structure to translate references for every instruction fetch, load, or store. Misses in an architected TLB result in page faults, which invoke the operating system. The operating system uses its own page table or other mapping structure to find a valid translation or create a new one and then updates the TLB using supervisor-mode instructions that can directly replace and update entries in the TLB. In such a scheme, the operating system can structure the page table in whatever way it deems best, since the page table is searched only by the page fault handler software, which can be modified to adapt to a variety of page table structures. This approach to handling translation misses is called a *software TLB miss handler* and is specified by the MIPS, Alpha, and SPARC instruction set architectures.

In contrast, a processor that implements an architecture that specifies the page table architecture provides a hardware state machine for accessing memory to search the page table and provide translations for all memory references. In such an architecture, the page table structure is fixed, since not just the operating system page fault handler has to access it, but a hardware state machine must also be able to search it. Such a system provides a *hardware TLB miss handler*. The PowerPC and Intel IA-32 instruction set architectures specify hardware TLB miss handlers.

3.6 Memory Hierarchy Implementation

To conclude our discussion of memory hierarchies, we address several interesting issues that arise when they are realized in hardware and interfaced to a high-performance processor. Four topics are covered in this section: memory accessing mechanisms, cache memory implementations, TLB implementations, and interaction between cache memory and the TLB.

As discussed in Section 3.4.3.4, there are three fundamental ways to access a multientry memory: indexing via an address, associative search via a tag, or a combination of the two. An *indexed memory* uses an address to index into the memory to select a particular entry; see Figure 3.4(a). A decoder is used to decode the n-bit address in order to enable one of the 2^n entries for reading or writing. There is a rigid or *direct mapping* of an address to the data which requires the data to be stored in a fixed entry in the memory. Indexed or direct-mapped memory is rigid in this mapping but less complex to implement. In contrast, an *associative memory* uses a key to search through the memory to select a particular entry; see Figure 3.4(b). Each entry of the memory has a tag field and a comparator that compares the content of its tag field to the key. When a match occurs, that entry is selected. Using this form of associative search allows the data to be flexibly stored in any location of the memory. This flexibility comes at the cost of implementation complexity. A compromise between the indexed memory and the associative memory is the *set-associative memory* which uses both indexing and associative search; see Figure 3.4(c). An address is used to index into one of the sets, while the multiple entries within a set are searched with a key to identify one particular entry. This compromise provides some flexibility in the placement of data without incurring the complexity of a fully associative memory.

Main memory is normally implemented as a large indexed memory. However, a cache memory can be implemented using any one of the three memory accessing schemes shown in Figure 3.4. When a cache memory is implemented as an indexed memory, it is referred to as a *direct-mapped cache* (illustrated in Figure 3.16). Since the direct-mapped cache is smaller and has fewer entries than the main memory, it requires fewer address bits and its smaller decoder can only decode a subset of the main memory address bits. Consequently, many main memory addresses can be mapped to the same entry in the direct-mapped cache. To ensure the selected entry contains the correct data, the remaining, i.e., not decoded,

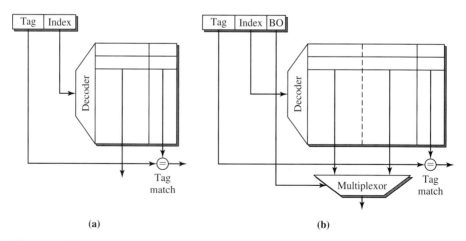

Figure 3.16
Direct-Mapped Caches: (a) Single Word Per Block; (b) Multiword Per Block.

address bits must be used to identify the selected entry. Hence in addition to the data field, each entry has an additional tag field for storing these undecoded bits. When an entry of the cache is selected, its tag field is accessed and compared with the undecoded bits of the original address to ensure that the entry contains the data being addressed.

Figure 3.16(a) illustrates a direct-mapped cache with each entry, or block, containing one word. In order to take advantage of spatial locality, the block size of a cache usually contains multiple words as shown in Figure 3.16(b). With a multi-word block, some of the bits from the original address are used to select the particular word being referenced. Hence, the original address is now partitioned into three portions: the index bits are used to select a block; the block offset bits are used to select a word within a selected block, and the tag bits are used to do a tag match against the tag stored in the tag field of the selected entry.

Cache memory can also be implemented as a fully associative or a set-associative memory, as shown in Figures 3.17 and 3.18, respectively. Fully associative caches have the greatest flexibility in terms of the placement of data in the entries of the cache. Other than the block offset bits, all other address bits are used as a key for associatively searching all the entries of the cache. This full associativity facilitates the most efficient use of all the entries of the cache, but incurs the greatest implementation complexity. Set-associative caches permit the flexible placement of data among all the entries of a set. The index bits select a particular set, the tag bits select an entry within the set, and the block offset bits select the word within the selected entry.

As discussed in Section 3.5, virtual memory requires mapping the virtual address space to the physical address space. This requires the translation of the virtual address into the physical address. Instead of directly accessing the main memory with the address generated by the processor, the virtual address generated by the processor must first be translated into a physical address. The physical address is then used to access the physical main memory, as shown in Figure 3.10.

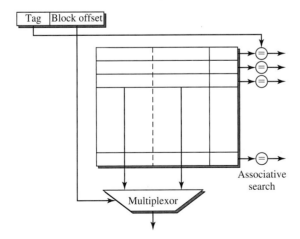

Figure 3.17
Fully Associative Cache.

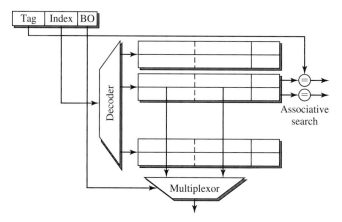

Figure 3.18
Set-Associative Cache.

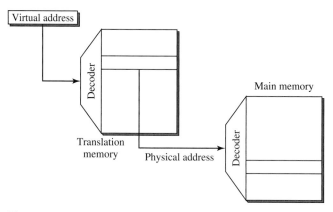

Figure 3.19
Translation of Virtual Word Address to Physical Word Address Using a Translation Memory.

As discussed in Section 3.5.3, address translation can be done using a translation memory that stores the virtual-to-real mappings; this structure is usually called a *page table*. The virtual address is used to index into or search the translation memory. The data retrieved from the selected entry in the translation memory are then used as the physical address to index the main memory. Hence, physical addresses that correspond to the virtual addresses are stored in the corresponding entries of the translation memory. Figure 3.19 illustrates the use of a translation memory to translate word addresses; i.e., it maps a virtual address of a word in the virtual address space into a physical address of a word in the physical main memory.

There are two weaknesses to the naive translation scheme shown in Figure 3.19. First, translation of word addresses will require a translation memory with the

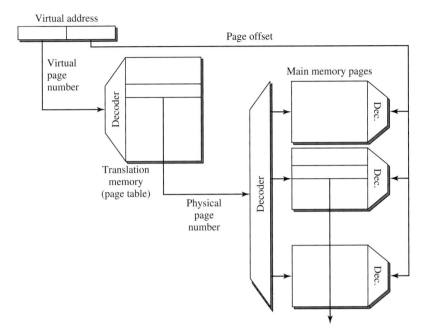

Figure 3.20
Translation of Virtual Page Address to Physical Page Address Using a Translation Memory.

same number of entries as the main memory. This can result in doubling the size of the physical main memory. Translation is usually done at a coarser granularity. Multiple (usually in powers of 2) words in the main memory can be grouped together into a *page*, and only addresses to each page need to be translated. Within the page, words can be selected using the lower-order bits of the virtual address, which form the page offset. This is illustrated in Figure 3.20. Within a virtual memory paging system, the translation memory is called the *page table*.

The second weakness of the translation memory scheme is the fact that two memory accesses are required for every main memory reference by an instruction. First the page table must be accessed to obtain the physical page number, and then the physical main memory can be accessed using the translated physical page number along with the page offset. In actual implementations the page table is typically stored in the main memory (usually in the portion of main memory allocated to the operating system); hence, every reference to memory by an instruction requires two sequential accesses to the physical main memory. This can become a serious bottleneck to performance. The solution is to cache portions of the page table in a small, fast memory called a translation lookaside buffer (TLB).

A TLB is essentially a cache memory for the page table. Just like any other cache memory, the TLB can be implemented using any one of the three memory accessing schemes of Figure 3.4. A direct-mapped TLB is simply a smaller (and faster) version of the page table. The virtual page number is partitioned into an index for the TLB and a tag; see Figure 3.21. The virtual page number is translated into the physical page number, which is concatenated with the page offset to form the physical address.

To ensure more flexible and efficient use of the TLB entries, associativity is usually added to the TLB implementation. Figure 3.22 illustrates the set-associative and fully associative TLBs. For the set-associative TLB, the virtual address bits are partitioned into three fields: index, tag, and page offset. The size of the page offset field is dictated by the page size which is specified by the architecture and the operating system. The remaining fields, i.e., index and tag, constitute the virtual

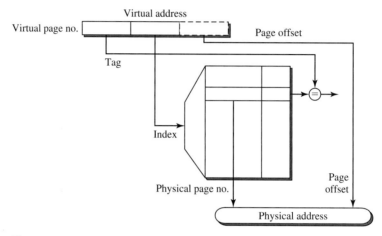

Figure 3.21
Direct-Mapped TLB.

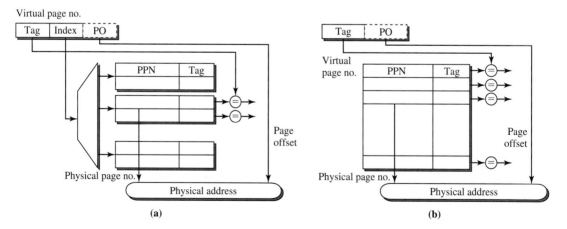

Figure 3.22
Associative TLBs: (a) Set-Associative TLB; (b) Fully Associative TLB.

page number. For the fully associative TLB, the index field is missing, and the tag field contains the virtual page number.

Caching a portion of the page table into the TLB allows fast address translation; however, *TLB misses* can occur. Not all the virtual page to physical page mappings in the page table can be simultaneously present in the TLB. When accessing the TLB, a cache miss can occur, in which case the TLB must be filled from the page table sitting in the main memory. This can incur a number of stall cycles in the pipeline. It is also possible that a TLB miss can lead to a *page fault*. A page fault occurs when the virtual page to physical page mapping does not even exist in the page table. This means that the particular page being referenced is not resident in the main memory and must be fetched from secondary storage. To service a page fault requires invoking the operating system to access the disk storage and can require potentially tens of thousands of machine cycles. Hence, when a page fault is triggered by a program, that program is suspended from execution until the page fault is serviced by the operating system. This process is illustrated in Figure 3.12.

A data cache is used to cache a portion of the main memory; a TLB is used to cache a portion of the page table. The interaction between the TLB and the data cache is illustrated in Figure 3.23. The n-bit virtual address shown in Figure 3.23 is the effective address generated by the first pipe stage. This virtual address consists of a virtual page number (v bits) and a page offset (g bits). If the TLB is a set-associative cache, the v bits of the virtual page number is further split into a k-bit index and a $(v-k)$-bit tag. The second pipe stage of the load/store unit corresponds to the accessing of the TLB using the virtual page number. Assuming there is no TLB miss, the TLB will output the physical page number (p bits), which is then concatenated with the g-bit page offset to produce the m-bit physical address where $m = p + g$ and m is not necessarily equal to n. During the third pipe stage the m-bit physical address is used to access the data cache. The exact interpretation of the

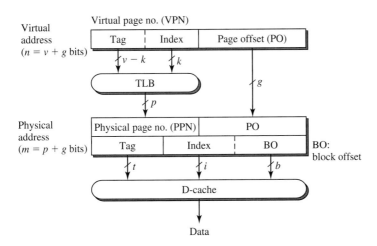

Figure 3.23
Interaction Between the TLB and the Data Cache.

m-bit physical address depends on the design of the data cache. If the data cache block contains multiple words, then the lower-order b bits are used as a block offset to select the referenced word from the selected block. The selected block is determined by the remaining $(m - b)$ bits. If the data cache is a set-associative cache, then the remaining $(m - b)$ bits are split into a t-bit tag and an i-bit index. The value of i is determined by the total size of the cache and the set associativity; i.e., there should be i sets in the set-associative data cache. If there is no cache miss, then at the end of the third pipe stage (assuming the data cache can be accessed in a single cycle) the data will be available from the data cache (assuming a load instruction is being executed).

The organization shown in Figure 3.23 has a disadvantage because the TLB must be accessed before the data cache can be accessed. Serialization of the TLB and data cache accesses introduces an overall latency that is the sum of the two latencies. Hence, one might assume that address translation and memory access are done in two separate pipe stages. The solution to this problem is to use a *virtually indexed* data cache that allows the accessing of the TLB and the data cache to be performed in parallel. Figure 3.24 illustrates such a scheme.

A straightforward way to implement a virtually indexed data cache is to use only the page offset bits to access the data cache. Since the page offset bits do not require translation, they can be used without translation. The g bits of the page offset can be used as the block offset (b bits) and the index (i bits) fields in accessing the data cache. For simplicity, let's assume that the data cache is a direct-mapped cache of 2^i entries with each entry, or block, containing 2^b words. Instead of storing the remaining bits of the virtual address, i.e., the virtual page number, as its tag field, the data cache can store the translated physical page number in its tag field. This is done at the time when a data cache line is filled. At the same time as the page offset bits are being used to access the data cache, the remaining bits of the virtual address, i.e., the virtual page number, are used to access the TLB. Assuming the TLB and data cache access latencies are comparable, at the time when the

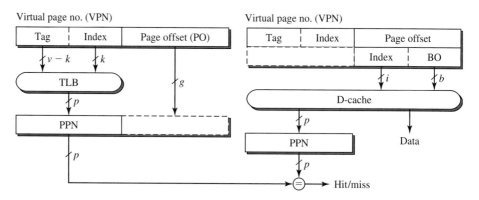

Figure 3.24
Virtually Indexed Data Cache.

physical page number from the TLB becomes available, the tag field (also containing the physical page number) of the data cache will also be available. The two *p*-bit physical page numbers can then be compared to determine whether there is a hit (matching physical page numbers) in the data cache or not. With a virtually indexed data cache, address translation and data cache access can be overlapped to reduce the overall latency. A classic paper by Wang, Baer, and Levy discusses many of the tradeoffs involved in designing a multilevel virtually addressed cache hierarchy [Wang et al., 1989].

3.7 Input/Output Systems

Obviously, a processor in isolation is largely useless to an end user and serves no practical purpose. Of course, virtually everyone has interacted with computers of various types, either directly, through a keyboard and display device, or indirectly, through the phone system or some other interface to an embedded computing system. The purpose of such interaction is to either log information into a computer system and possibly request it to perform certain computations (input) and then either observe the result or allow the computer to directly interact with external devices (output). Thus, the computer system as a whole can be thought of as a black box device with some set of inputs, provided through various interfaces, and some set of outputs, also provided through a set of interfaces. These interfaces can interact directly with a human (by capturing keystrokes on a keyboard, movements of a mouse, or even spoken commands, and by displaying text or graphics or playing audio that are comprehensible to humans) or can instead interact with other digital devices at various speeds. This section discusses some of these devices and their attributes.

Table 3.4 summarizes some attributes of common input/output devices. For each device type, the table specifies how the device is connected to the system; whether it is used for input, output, both input and output, or storage; whether it communicates with a human or some other machine; and approximate data rates for these devices. The table makes clear that I/O devices are quite diverse in their characteristics, with data rates varying by seven orders of magnitude.

Table 3.4
Types of input/output devices

Device	How/Where Connected	Input/Output/ Storage	Partner	Data Rate (kB/s)
Mouse, keyboard	Serial port	Input	Human	0.01
Graphical display	I/O bus and memory bus	Output	Human	100,000
Modem	Serial port	Input and output	Machine	2–8
LAN	I/O bus	Input and output	Machine	500–120,000
Disk	Storage bus	Storage	Machine	10,000+

3.7.1 Types of I/O Devices

This section briefly discusses the I/O devices enumerated in Table 3.4 (mouse, keyboard, graphical displays, modems, LANs, and disk drives), and also provides an overview of high-performance and fault-tolerant disk arrays.

Mouse and Keyboard. A mouse and keyboard are used to provide direct user input to the system. The keyboard and mouse devices are usually connected to the system via a low-speed serial port. The universal serial bus (USB) is an example of a standardized serial port available on many systems today. The data rates for keyboards and mice are very low, as they are limited by the speed at which humans can type on a keyboard or operate a mouse. Since the data rates are so low, keyboard and mouse input are typically communicated to the CPU via external interrupts. Every key press or movement of the mouse ultimately invokes the operating system's interrupt handler, which then samples the current state of the mouse or the keyboard to determine which key was pressed or which direction the mouse moved so it can respond appropriately. Though this may appear to create an excessive rate of interrupts that might seriously perturb the processor, the low data rates of these devices generally avoid that problem on a single-user system. However, in a large-scale time-shared system that services keyboard input from hundreds or thousands of users, the interrupt rates quickly become prohibitive. In such environments, it is not unusual to provide terminal I/O controllers that handle keyboard interrupts from users and only communicate with the main processor once a cluster of keyboard activity has been aggregated at the controller. The modern-day equivalent of this type of aggregation of interactive I/O activity occurs when users enter data into a form on their Web browser: all the data entry is captured by the user's Web browser client, and the Web server does not get involved until the user clicks on a submit button that transmits all the Web form data in a single transaction to the server. In this fashion, load on the server as well as the communication links between the client and the server is minimized, since only the aggregated information is communicated, rather than every keystroke.

Graphical Display. A graphical display conveys video or image data, illustrations, and formatted documents to the user, and also presents a user interface that simplifies the user's interaction with the system. Graphical displays must render a million or more pixels on the screen using a 24-bit color representation per pixel and usually update the screen at a rate of 60 or more frames per second. The contents of the screen are rendered in a *frame buffer* which contains a pixel-by-pixel representation of the contents of the screen. A random access memory digital-to-analog converter (*RAMDAC*) uses a high-speed interface to the frame buffer's memory and converts the digitally represented image into an analog image that is displayed on a CRT (cathode-ray tube) or LCD (liquid-crystal display) monitor. The frame buffer contents are updated by a graphics processor that typically supports various schemes for accelerating two-dimensional and three-dimensional graphics transformations. For example, dedicated hardware in the graphics processor pipeline can perform visibility checks to see if certain objects are hidden behind others, can correct for perspective in a three-dimensional environment, and can perform lighting, shading, and texture transforms to add realism to the generated image. All these transforms require extremely high bandwidth to the frame buffer memory, as well as to main memory to access the image database, where objects are represented as collections of polygons in three-dimensional space. Hence, while graphical display adapters are connected to the main I/O bus of the system to interact with the main CPU, they also often utilize a special-purpose memory port [the accelerated graphics port (AGP) is an example of such a port] to enable high memory bandwidth for performing these transforms.

Modem. Modems are used to interconnect digital systems over an analog communication line, usually a standard telephone connection. Because of the nature of standard phone lines, they are only able to provide limited bandwidth, with a maximum of 56 kbits/s with the latest standard. Hence, because of the low overall data rates, modems are usually connected to the system via a low-speed serial port, like a USB or even older RS-232 serial port.

LAN. Local area network adapters are used to connect computer systems to each other. A LAN adapter must provide a physical layer interface that converts the computer's internal signal level digital data to the signaling technology employed by the LAN interface. Fast Ethernet, running at 100 Mbits/s, dominates the industry today, while Gbit Ethernet is rapidly being adopted. LAN adapters, due to their reasonably high data rates, are usually connected directly to the I/O backplane bus of the system to provide high bandwidth access to the system's main memory and to the processor. Originally, Ethernet was conceived as a shared bus-based interconnect scheme, but over time it has evolved into a switched, point-to-point organization where each computer system has a dedicated link to a centralized switch that is responsible for routing data packets to and from each of its ports based on the destination addresses of the packets. Ethernet switches can be connected hierarchically to allow larger number of systems to communicate.

Disk Drives. Magnetic disk drives store information on a platter by changing the orientation of a magnetic field at each individually addressable location on the platter. As shown in Figure 3.25, a disk drive may contain multiple platters per spindle.

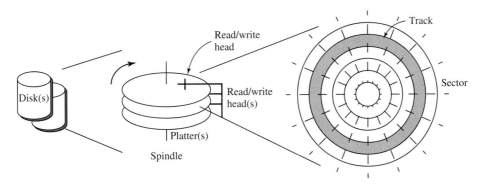

Figure 3.25
Disk Drive Structure.

Usually, data are stored on each side of the platter, with a separate read/write head for each side of each platter. Each platter has multiple concentric tracks that are divided into sectors. Read/write heads rest on a cushion of air on top of each spinning platter and seek to the desired track via a mechanical actuator. The desired sector is found by waiting for the disk to rotate to the desired position. Typical disks today rotate from 3000 to 15,000 revolutions per minute (rpm), contain anywhere from 500 to 2500 tracks with 32 or more sectors per track, and have platters with diameters ranging in size from 1 to 3.5 in.

Recent drives have moved to placing a variable number of sectors per track, where outer tracks with greater circumference have more sectors, and inner tracks with lesser circumference contain fewer tracks. This approach maintains constant areal bit density on the platter substrate, but complicates the read/write head control logic, since the linear velocity of the disk under the head varies with the track (with a higher velocity for outer tracks). Hence, the rate at which bits pass underneath the head also varies with the track, with a higher bit rate for outer tracks. In contrast, in older disks with a constant number of sectors and variable bit density, the bit rate that the head observed remained constant, independent of which track was being accessed. Some older disk drives, most notably the floppy drives in the original Apple Macintosh computers, held the bit rate and linear velocity constant by varying the rotational speed of the disk based on the position of the head, leading to an audible variation in the sound the drive generates. This approach substantially complicates the motor and its control electronics, making it infeasible for high-speed hard drives spinning at thousands of rpm, and has been abandoned in recent disk designs.

$$\text{Latency} = \text{rotational} + \text{seek} + \text{transfer} + \text{queueing} \quad (3.7)$$

As shown in Equation (3.7), the access latency for a disk drive consists of the sum of four terms: the *rotational* latency, the *seek* latency, the *transfer* latency, and *queueing* delays. Rotational latency is determined by the speed at which the disk rotates. For example, a 5400-rpm drive completes a single revolution in (60 s)/(5400 rpm) = 11.1 ms. On average, a random access will have to wait half a revolution, leading to an average rotational latency of 11.1 ms/2 = 5.5 ms for our 5400-rpm drive. The seek latency is determined by the number of tracks, the size of the platter, and the design of the seek actuator, and varies depending on the distance from the head's current position to the target track. Typical average seek latencies range from 5 to 15 ms. The transfer latency is determined by the read/write head's data transfer rate divided by the block size. Data transfer rates vary from 1 to 4 Mbytes/s or more, while typical blocks are 512 bytes; assuming a 4-Mbyte transfer rate for a 512-b block, a drive would incur 0.12 ms of transfer latency. Finally, queueing delays in the controller due to multiple outstanding requests can consume 1 ms or more of latency. The final average latency for our example drive would add up to 5.5 ms (rotational latency) + 5 ms (seek latency) + 0.1 ms (transfer latency) + 1 ms (queueing latency) = 11.6 ms.

Modern disk drives also provide cache buffers ranging in size from 2 to 8 Mbytes that are used to capture temporal and spatial locality in the disk reference stream. These operate very similarly to processor caches and are often able to satisfy a substantial fraction of all disk requests with very low latency, hence reducing the average disk latency by a considerable amount. Of course, worst-case access patterns that exhibit little spatial or temporal locality will still incur access latencies determined by the physical design of the disk, since they cannot be satisfied from the disk buffer.

Subsequent references to the same or nearby tracks or sectors can be satisfied much more quickly than the average case, since the rotational and seek latencies are minimized in those cases. Hence, modern operating systems attempt to reorder references in order to create a schedule that maximizes this type of spatial locality, hence minimizing average reference latency. As long as the operating system is aware of the physical disk layout of the blocks it is referencing, such scheduling is possible and desirable. Disk drive performance and modeling issues are discussed at length in a classic paper by Ruemmler and Wilkes [1994].

Disk Arrays. High-performance computer systems typically contain more than one disk to provide both increased capacity as well as higher bandwidth to and from the file system and the demand-paged backing store. Quite often, these disks are arranged in arrays that can be configured to provide both high performance as well as some degree of fault tolerance. In such arrays, data can be striped across multiple disks at varying levels of granularity to enable either higher data bandwidth or higher transaction bandwidth by accessing multiple disks in parallel. Figure 3.26 illustrates several approaches for striping or interleaving data across

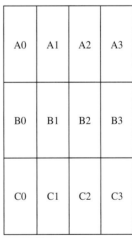

Each disk is represented by a column, each block is represented by a name (A0, A1, A2, etc.), and blocks from the same file are named with the same letter (e.g., A0, A1, and A2 are all from the same file). Independent disk arrays place related blocks on the same drive. Fine-grained interleaving subdivides each block and stripes it across multiple drives. Coarse-grained interleaving stripes related blocks across multiple drives.

Figure 3.26

Striping Data in Disk Arrays.

multiple disks at varying levels of granularity. Without interleaving, shown at left in the figure as *Independent,* blocks from a single file (e.g. A0, A1, A2) are all placed consecutively on a single disk (represented by a column in Figure 3.26. With *fine-grained* interleaving, shown in the middle of the figure, blocks are subdivided and striped across multiple disks. Finally, in *coarse-grained* interleaving, shown at right in the figure, each block is placed on a single disk, but related blocks are striped across multiple disks.

The redundant arrays of inexpensive disks (RAID) nomenclature, introduced by Patterson, Gibson, and Katz [1988] and summarized in Table 3.5, provides a useful framework for describing different approaches for combining disks into arrays. RAID level 0 provides no degree of fault tolerance, but does provide high performance by striping data across multiple disks. Because of the striping, greater aggregate read and write bandwidth is available to the objects stored on the disks.

Table 3.5
Redundant arrays of inexpensive disks (RAID) levels

RAID Level	Explanation	Overhead	Fault Tolerance	Usage and Comments
0	Data striped across disks	None	None	Widely used; fragile
1	Data mirrored	Each disk duplicated	1 of 2	Widely used; very high overhead
2	Hamming code ECC protection; data + ECC bits striped across many disks	Very high for few disks; reasonable only for a large disk array	Single disk failure	Not practical; requires too many disks to amortize cost of ECC bits
3	Data striped; single parity disk per word	Parity disk per striped block	Single disk failure	Available; high data bandwidth, poor transaction bandwidth
4	Data not striped (interleaved at block granularity); single parity disk	Parity disk per block set	Single disk failure	Available; poor write performance due to parity disk bottleneck
5	Data not striped (interleaved at block granularity); parity blocks interleaved on all disks	1 of n disk blocks used for parity (e.g., 5 disks provide data capacity of 4)	Single disk failure	Widespread; writes require updates to two disks—one for data, one for parity
6	Data not striped (interleaved at block granularity); two-dimensional parity blocks interleaved on disks	2 of n disk blocks used for parity (e.g., 6 disks provide data capacity of 4)	Multiple disk failure	Available; writes updates to three disks—one for data, one for row parity, one for column parity

Source: Patterson et al., 1988.

Furthermore, since each disk can operate independently, transaction bandwidth is also increased dramatically. However, a single disk failure will cause the entire array to fail. RAID level 1, also known as *disk mirroring,* addresses this by providing fault tolerance through mirroring of all data. This approach is simple to implement, but has very high overhead and provides no improvement in write bandwidth (since both copies must be updated), and only a doubling of read and read transaction bandwidth.

Higher levels of RAID protection use parity or error-correction codes (ECCs)[1] to reduce the overhead of fault tolerance to much less than the 100% overhead required by RAID level 1. In RAID level 2, word-level ECCs based on Hamming codes are used to identify and correct single errors. Conceptually, an ECC contains both a parity bit (used to check for a bit error in the coded word), as well as an offset that points to the data bit that is in error. Both the parity bit and offset are encoded using a Hamming code to minimize storage overhead and are used together to correct a bit error by flipping the bit at the specified offset whenever the parity bit indicates an error.

Unfortunately, the inherent overhead of word-level ECCs is high enough that RAID level 2 is impractical for all but the largest disk arrays, where large words can be spread across dozens of drives to reduce the ECC overhead. For example, the ECC *SECDED*[2] overhead for a 64-bit word size is a minimum of 7 bits, requiring a disk array with 71 drives (64 data drives and 7 ECC drives) to achieve a reasonable 11% overhead. Since the ECC SECDED overhead is much higher for smaller word sizes, RAID level 2 is rarely employed in arrays with few drives. RAID level 3 replaces the ECCs with just parity, since failing drives can typically be detected by the disk array controller without the explicit error-correction-coded offset that identifies the failing bit (modern disks include diagnostic firmware that is able to report disk failure and even predict imminent failure to the disk controller). Using only parity reduces overhead and simplifies RAID implementation. However, since data are striped at a very fine grain in RAID level 3 (at the bit level), each transaction requires the coordinated participation of all the drives in the parity set; hence, transaction bandwidth does not scale well. Instead, RAID level 4 maintains parity at a coarser block level, reducing the transaction overhead and supplying much better transaction bandwidth scaling. As illustrated in Figure 3.27, RAID level 4 places all parity blocks on the same drive. This leads to the parity drive bottleneck, since all writes must access this single drive to update their block set's parity. RAID level 5 solves the parity block bottleneck by rotating the parity blocks across all the drives, as shown at right in Figure 3.27.

RAID level 5 is widely used to provide both high performance and fault tolerance to protect against single disk failure. In RAID level 5, data blocks are independently stored across the disks in the array, while parity blocks covering a group of

---

[1] For background information on error-correcting codes, which are not covered in detail in this book, the interested reader is referred to Blahut [1983] and Rao and Fujiwara [1989].

[2] Single-error correct and dual-error detect (SECDED) codes are powerful enough to correct a single bit error and detect the occurrence of—but not correct—a dual-bit error within a protected word.

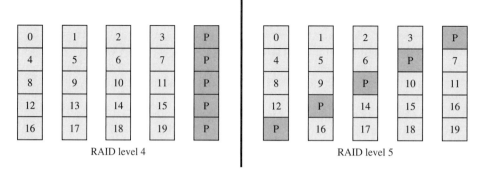

Figure 3.27
Placement of Parity Blocks in RAID Level 4 (Left) vs. RAID Level 5 (Right).

data blocks are interleaved in the drives as well. In terms of capacity overhead, this means that 1 of n disks is consumed for storing parity, in exchange for tolerance of single drive failures. Read performance is still very good, since each logical read requires only a single physical read of the actual data block. Write performance suffers slightly compared to RAID level 0, since the parity block must be updated in addition to writing the new data block. However, RAID level 5 provides a powerful combination of fault tolerance, reasonable overhead, and high performance, and is widely deployed in real systems.

Finally, RAID level 6 extends level 5 by maintaining two dimensions of parity for each block, requiring double the storage and write overhead of level 5 but providing tolerance of multiple disk failures. RAID level 6 is typically employed only in environments where data integrity is extremely important. The interested reader is referred to the original work by Patterson et al. [1988] for a more in-depth treatment of the advantages and disadvantages of the various RAID levels.

RAID controllers can be implemented either completely in hardware, with limited or no operating system involvement, or in the operating system's device driver (also known as software RAID). For example, the open-source Linux kernel supports software RAID levels 0, 1, and 5 over arrays of inexpensive, commodity integrated drive electronics (IDE) drives, and even across drives on separate LAN-connected machines that are configured as network block devices (nbd). This makes it possible to implement fault-tolerant RAID arrays using very inexpensive, commodity PC hardware.

Hardware RAID controllers, typically used in higher-end server systems, implement RAID functionality in the controller's firmware. High-end RAID controllers often also support hot-swappable drives, where a failed or failing drive can be replaced on the fly, while the RAID array remains on-line. Alternatively, a RAID array can be configured to contain hot spare drives, and the controller can automatically switch in a hot spare drive for a drive that is about to fail or has already failed (this is called automated failover). During the period of time that a failed disk is still part of the array, all accesses to blocks stored on the failed disk are satisfied by reading the parity block and the other blocks in the parity set and reconstructing the contents of the missing block using the parity function. For example, in an array employing even parity across four disks, where the failing disk is the third disk, the controller might read a parity bit of <1> and <0,1,?,1> from the remaining good drives. Since even parity implies an even number of "1" bits across the parity set, the missing "?" bit is inferred to be a "1." Since each access to the failed drive requires coordinated reads to all the remaining drives in the parity set, this on-line forward error correction process can result in very poor disk performance until the failed disk has been replaced.

In a similar fashion, a RAID array with hot spares can automatically reconstruct the contents of the failed drive and write them to the spare disk, while alerting the system operator to replace the failed drive. In a RAID array that does not support hot spares, this reconstruction process has to be conducted either off line, after the array has been powered down and the failed disk replaced, or on line, as soon as the operator has hot-swapped the failed drive with a functioning one.

3.7.2 Computer System Busses

A typical computer system provides various busses for interconnecting the components we have discussed in the preceding sections. In an ideal world, a single communication technology would satisfy the needs and requirements of all system components and I/O devices. However, for numerous practical reasons—including cost, backward compatibility, and suitability for each application—numerous interconnection schemes are employed in a single system. Figure 3.28 shows three types of busses: the processor bus, the I/O bus, and a storage bus.

Processor busses are used to connect the CPU to main memory as well as to an I/O bridge. Since CPU performance depends heavily on a high-bandwidth,

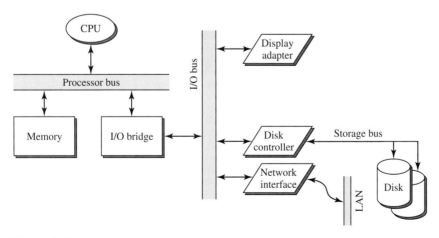

Figure 3.28
Different Types of Computer System Busses.

low-latency connection to main memory, processor busses employ leading-edge signaling technology running at very high frequencies. Processor busses are also frequently updated and improved, usually with every processor generation. Hence, all the devices that connect to the processor bus (typically the CPU, the memory controller, and the I/O bridge; often referred to as the *chip set*) need to be updated at regular intervals. Because of this de facto update requirement, there is little or no pressure on processor busses to maintain backward compatibility beyond more than one processor generation. Hence, not only does the signaling technology evolve quickly, but also the protocols used to communicate across the bus adapt quickly to take advantage of new opportunities for improving performance. Section 11.3.7 provides some additional discussion on the design of processor busses and the coherent memory interface that a modern CPU needs to provide to communicate with a processor bus. Processor busses are also designed with electrical characteristics that match very short physical distances, since the components attached to this bus are usually in very close proximity inside the physical computer package. This enables very high speed signaling technologies that would be impossible or very expensive for longer physical distances.

In contrast to the processor bus, a typical I/O bus evolves much more slowly, since backward compatibility with legacy I/O adapters is a primary design constraint. In fact, systems will frequently support multiple generations of I/O busses to enable use of legacy adapters as well as modern ones. For example, many PC systems support both the peripheral component interface (PCI) I/O bus and the industry standard architecture (ISA) bus, where the ISA bus standard stretches back 15 years into the past. Also, for cost and physical design reasons, I/O busses usually employ less aggressive signaling technology, run at much lower clock frequencies, and employ less complex communication protocols than processor busses. For example, a modern PC system might have a 533-MHz processor bus with an 8-byte datapath, while the PCI I/O bus would run at 33 MHz with a 4-byte datapath. Since most peripheral I/O devices cannot support higher data rates anyway, for cost reasons the I/O busses are less aggressive in their design. The only standard peripheral that requires much higher bandwidth is a modern graphics processing unit (or display adapter); modern PCs provide a dedicated accelerated graphics port (AGP) to supply this bandwidth to main memory, while control and other communication with the display adapter still occurs through the PCI I/O bus.

I/O busses typically need to span physical distances that are limited by the computer system enclosure; these distances are substantially longer than what the processor bus needs to span, but are still limited to less than 12 in. (30 cm) in most cases.

Finally, storage busses, used primarily to connect magnetic disk drives to the system, suffer even more from legacy issues and backward compatibility. As a result, they are often hobbled in their ability to adopt new signaling technology in a clean, straightforward fashion that does not imply less-than-elegant solutions. For example, most storage busses are limited in their use of newer technology or signaling by the oldest peer sharing that particular bus. The presence of one old device will hence limit all newer devices to the lowest common denominator of performance.

Storage busses must also be able to span much greater physical distances, since the storage devices they are connecting may reside in an external case or adjacent rack. Hence, the signaling technology and communication protocol must tolerate long transmission latencies. In the case of Fiber Channel, optical fiber links are used and can span several hundred meters, enabling storage devices to reside in separate buildings.

Simple busses support only a single concurrent transaction. Following an arbitration cycle, the device that wins the arbitration is allowed to place a command on the bus. The requester then proceeds to hold or occupy the bus until the command completes, which usually involves waiting for a response from some other entity that is connected to the same bus. Of course, if providing a response entails some long-latency event like performing a read from a disk drive, the bus is occupied for a very long time for each transaction. While such a bus is relatively easy to design, it suffers from very poor utilization due to these long wait times, during which the bus is effectively idle. In fact, virtually all modern bus designs support *split transactions,* which enable multiple concurrent requests on a single bus. On a split transaction bus, a requester first arbitrates for the bus, but then occupies the bus only long enough to issue the request, and surrenders the bus to the next user without waiting for the response. Some period of time later, the responder now arbitrates for the bus and then transmits the response as a separate bus event. In this fashion, transactions on the bus are split into two—and sometimes more than two—separate events. This interleaving of multiple concurrent requests leads to much better efficiency, since the bus can now be utilized to transfer independent requests and responses while a long-latency request is pending. Naturally, the design complexity of such a bus is much higher, since all devices connected to the bus must now be able to track outstanding requests and identify which bus transactions correspond to those requests. However, the far higher effective bandwidth that results justifies the additional complexity.

Figure 3.29 summarizes the key design parameters that describe computer system busses. First of all, the bus topology must be set as either point-to-point, which enables much higher frequency, or multidrop, which limits frequency due to the added capacitance of each electrical connection on the shared bus, but provides more flexible connectivity. Second, a particular signaling technology must be chosen to determine voltage levels, frequency, receiver/transmitter design, use of differential signals, etc. Then, several parameters related to actual data transfer must be set: the width of the data bus; whether or not the data bus lines are shared or multiplexed with the control lines; and either a single-word data transfer granularity, or support for multiword transfers, possibly including support for burst-mode operation that can saturate the data bus with back-to-back transfers. Also, a bidirectional bus that supports multiple signal drivers per data wire must provide a mechanism for turning the bus around to switch from one driver to another; this usually leads to dead cycles on the bus and reduces sustainable bandwidth (a unidirectional bus avoids this problem). Next, a clocking

164 MODERN PROCESSOR DESIGN

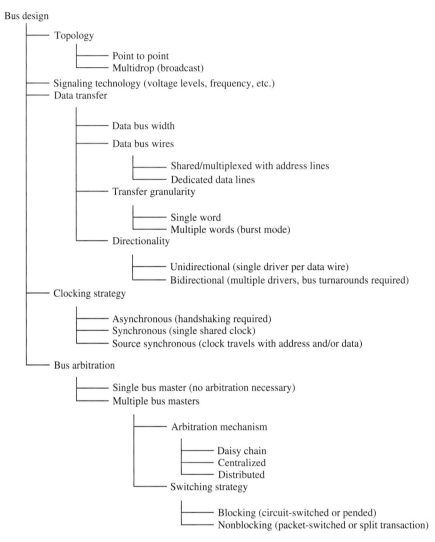

Figure 3.29
Bus Design Parameters.

strategy must also be set: the simplest option is to avoid bus clocks, instead employing handshaking sequences using request and valid lines to signal the presence of valid commands or data on the bus. As an alternative, a single shared clock can be used on a synchronous bus to avoid handshaking and improve bus utilization. Finally, an aggressive source-synchronous clocking approach can be used, where the clock travels with the data and commands, enabling the highest operating frequency and wave pipelining with multiple packets in flight at the same time. Finally, for bus designs that allow multiple bus masters to control the bus, an arbitration and switching policy must be specified. Possible arbitration mechanisms include daisy-chained arbiters, centralized arbiters, or distributed arbiters; while switching policies are either circuit-switched (also known as blocking), where a single transaction holds the bus until it completes, or packet-switched (also known as nonblocking, pipelined, or split transaction buses), where bus transactions are split into two or more packets and each packet occupies a separate slot on the bus, allowing for interleaving of packets from multiple distinct requests.

Modern high-performance bus designs are trending toward the following characteristics to maximize signaling frequency, bandwidth, and utilization: point-to-point connections with relatively few data lines to minimize crosstalk, source-synchronous clocking with support for burst mode transfers, distributed arbitration schemes, and support for split transactions. One interesting alternative bus design that has emerged recently is the simultaneous bidirectional bus: in this scheme, the bus wires have multiple pipelined source-synchronous transfers in flight at the same time, with the additional twist of signaling simultaneously in both directions across the same set of wires. Such advanced bus designs conceptually treat the digital signal as an analog waveform traveling over a well-behaved waveguide (i.e., a copper wire), and require very careful driver and receiver design that borrows concepts and techniques from the signal processing and advanced communications transceiver design communities.

3.7.3 Communication with I/O Devices

Clearly, the processor needs to communicate with I/O devices in the system using some mechanism. In practice, there are two types of communication that need to occur: control flow, which communicates commands and responses to and from the I/O device; and data flow, which actually transfers data to and from the I/O device. Control flow can further be broken down into commands which flow from the processor to the I/O device (outbound control flow), and responses signaling completion of the commands or other status information back to the processor (inbound control flow). Figure 3.30 summarizes the main attributes of I/O device communication that will be discussed in this section.

Outbound Control Flow. There are two basic approaches for communicating commands (outbound control flow) from the processor to the I/O device. The first of these is through programmed I/O: certain instruction set architectures provide specific instructions for communicating with I/O devices; for example, the Intel IA-32 instruction set provides such primitives. These programmed I/O instructions are directly connected to control registers in the I/O devices, and the I/O devices react accordingly to changes written to these control registers. The main shortcoming of this approach is that the ISA provides only a finite set of I/O port interfaces to the processor, and in the presence of multiple I/O devices, they need to be shared or virtualized in a manner that complicates operating system device driver software. Furthermore, these special-purpose instructions do not map cleanly to

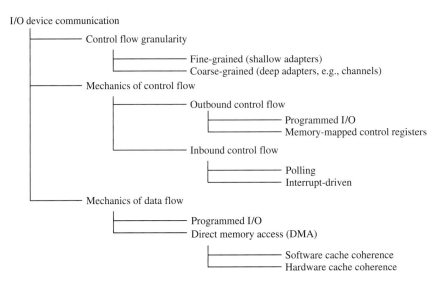

Figure 3.30
Communication with I/O Devices.

the pipelined and out-of-order designs described in this book, but require complex specialized handling within the processor core.

A more general approach for outbound control flow is to use *memory-mapped I/O*. In this approach, the device-specific control registers are mapped into the memory address space of the system. Hence, they can be accessed with conventional load and store instructions, with no special support in the ISA. However, care must be taken in cache-based systems to ensure that the effects of loads and stores to and from these memory-mapped I/O registers are actually visible to the I/O device, rather than being masked by the cache as references that hit in the cache hierarchy. Hence, virtual memory pages corresponding to memory-mapped control registers are usually marked as caching inhibited or uncacheable in the virtual memory page table (refer to Section 3.5.3 for more information on page table design). References to uncacheable pages must be routed off the processor chip and to the I/O bridge interface, which then satisfies them from the control registers of the I/O device that is mapped to the address in question.

Inbound Control Flow. For inbound control flow, i.e., responses or status information returned from the I/O devices back to the processor, there are two fundamental approaches: polling or interrupts. In a polling system, the operating system will intermittently check either a programmed I/O or memory-mapped control register to determine the status of the I/O device. While straightforward to implement, both in hardware and software, polling systems suffer from inefficiency, since the processor can spend a significant amount of time polling for completion of a long-latency event. Furthermore, since the polling activity requires communication across the processor and I/O busses of the system, these busses can become overwhelmed and begin to suffer from excessive queueing delays. Hence, a much cleaner and scalable approach involves utilizing the processor's support for external interrupts. Here, the processor is not responsible for polling the I/O device for completion. Rather, the I/O device instead is responsible for asserting the external interrupt line of the processor when it completes its activity, which then initiates the operating system's interrupt handler and conveys to the processor that the I/O is complete. The interrupt signal is routed from the I/O device, through the I/O bridge, to the processor's external interrupt controller.

Control Flow Granularity. Command and response control flow can also vary in granularity. In typical PC-based systems, most I/O devices expose a very simple interface through their control registers. They can perform fairly simple activities to support straightforward requests like reading or writing a simple block of memory to or from a peripheral device. Such devices have very fine-grained control flow, since the processor (or the operating system device driver running on the processor) has to control such devices in a very fine-grained manner, issuing many simple commands to complete a more complex transaction with the peripheral I/O device. In contrast, in the mainframe and minicomputer world, I/O devices will often expose a much more powerful and complex interface to the processor, allowing the processor to control those devices in a very coarse-grained fashion. For example, I/O channel controllers in IBM S/390-compatible mainframe systems can actually execute separate programs that contain internal control flow structures like loops and conditional branches. This richer functionality can be used to off-load such fine-grained control from the main CPU, freeing it to focus on other tasks. Modern three-dimensional graphics adapters in today's desktop PC systems are another example of I/O devices with coarse-grained control flow. The command set available to the operating system device drivers for these adapters is semantically rich and very powerful, and most of the graphics-related processing is effectively off-loaded from the main processor to the graphics adapter.

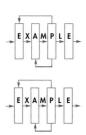

Data Flow. Data flow between the processor and I/O devices can occur in two fundamental ways. The first of these relies on instruction set support for programmed I/O. Again, the ISA must specify primitives for communicating with I/O devices, and these primitives are used not just to initiate requests and poll for completion, but also for data transfer. Hence, the processor actually needs to individually read or write each word that is transferred to or from the I/O device through an internal processor register, and move it from there to the operating system's in-memory buffer. Of course, this is extremely inefficient, as it can occupy the processor for thousands of cycles whenever large blocks of data are being transferred. These data transfers will also unnecessarily pollute and cause contention in the processor's cache hierarchy.

Instead, modern systems enable *direct memory access* (DMA) for peripheral I/O devices. In effect, these devices can issue reads and writes directly to the main memory controller, just as the processor can. In this fashion, an I/O device can update an operating system's in-memory receive buffer directly, with no intervention from the processor, and then signal the processor with an interrupt once the transfer is complete. Conversely, transmit buffers can be read directly from main memory, and a transmission completion interrupt is sent to the processor once the transmission completes.

Of course, just as with memory-mapped control registers, DMA can cause problems in cache-based systems. The operating system must guarantee that any cache blocks that the I/O device wants to read from are not currently in the processor's caches, because otherwise the I/O device may read from a stale copy of the cache block in main memory. Similarly, if the I/O device is writing to a memory location, the processor must ensure that it does not satisfy its next read from the same location from a cached copy that has now become stale, since the I/O device only updated the copy in memory. In effect, the caches must be kept *coherent* with the latest updates to and from their corresponding memory blocks. This can be done manually, in the operating system software, by using primitives in the ISA that enable flushing blocks out of cache. This approach is called *software cache coherence*.

Alternatively, the system can provide hardware that maintains coherence; such a scheme is called *hardware cache coherence*. To maintain hardware cache coherence, the I/O devices' DMA accesses must be made visible to the processor's cache hierarchy. In other words, DMA writes must either directly update matching copies in the processor's cache, or those matching copies must be marked invalid to prevent the processor from reading them in the future. Similarly, DMA reads must be satisfied from the processor's caches whenever a matching and more up-to-date copy of a block is found there, rather than being satisfied from the stale copy in main memory. Hardware cache coherence is often achieved by requiring the processor's caches to *snoop* all read and write transactions that occur across the processor-memory bus and to respond appropriately to snooped commands that match cached lines by either invalidating them (when a bus write is snooped) or supplying the most up-to-date data (when a bus read to a dirty line is snooped). Chapter 11 provides more details on hardware mechanisms for enforcing cache coherence.

3.7.4 Interaction of I/O Devices and Memory Hierarchy

As discussed in Section 3.7.3, direct memory access (DMA) by I/O devices causes a cache coherence problem in cache-based systems. Cache coherence is a more general and pervasive problem in systems that contain multiple processors, since each processor can now also update copies of blocks of memory locally in their cache, whereas the effects of those updates should be made visible to other processors in the system as well (i.e., their caches should remain coherent). We revisit this problem at length in Section 11.3 and describe cache coherence protocols that can be used to elegantly and efficiently solve this problem in systems with a few or even hundreds of processors.

MEMORY AND I/O SYSTEMS

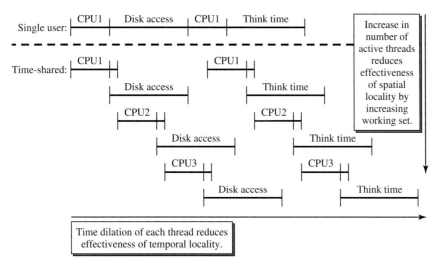

In this example, three users time-share the CPU, overlapping their CPU usage with the disk latency and think time of the other interactive users. This increases overall throughput, since the CPU is always busy, but can increase the latency observed by each user. Latency increases due to context switch overhead and queuing delay (waiting for the CPU while another user is occupying it). Temporal and spatial locality are adversely affected by time-sharing.

Figure 3.31
Time-Sharing the CPU.

However, there is another interesting interaction that occurs with the memory hierarchy due to long-latency I/O events. In our discussion of demand-paged virtual memory subsystems in Section 3.5.1, we noted that the operating system will put a faulting process to sleep while it fetches the missing page from the backing store and will schedule an alternative process to run on the processor. This process is called *time-sharing* the CPU and is illustrated in Figure 3.31. The top half of the figure shows a single process first consuming CPU, then performing a long-latency disk access, then consuming CPU time again, and finally shows *think time* while waiting for the user to respond to program output. In the bottom half of the figure, other processes with similar behavior are interleaved onto the processor while the first process is waiting for disk "access" or for user response. Clearly, much better CPU utilization results, since the CPU is no longer idle for long periods of time.

However, this increased utilization comes at a price: since each process's execution is now dilated in time due to the intervening execution of other processes, the temporal locality of each process suffers, resulting in high cache miss rates. Furthermore, the fact that the processor's memory hierarchy must now contain the working sets of all the active processes, rather than just a single active process, places great strain on the caches and reduces the beneficial effects of spatial locality. As a result, there can be substantial increases in cache miss rates and substantially worse average memory reference latency in such heavily time-shared systems.

As processors continue to increase their speed, while the latency of I/O devices improves at a glacial rate, the ratio of active processes that need to be scheduled to cover the latency of a single process's I/O event is increasing rapidly. As a result, the effects pointed out in Figure 3.31 are more and more pronounced, and the effectiveness of cache-based memory hierarchies is deteriorating. We revisit this problem in Chapter 11 as we discuss systems that execute multiple threads simultaneously.

3.8 Summary

This chapter introduces the basic concept of a memory hierarchy, discusses various technologies used to build a memory hierarchy, and covers many of the effects that a memory hierarchy has on processor performance. In addition, we have studied some of the key input and output devices that exist in systems, the technology used to implement them, and the means with which they are connected to and interact with the processor and the rest of the computer system.

We also discussed the following *memory idealisms* and showed how a well-designed memory hierarchy can provide the illusion that all of them are satisfied, at least to some extent:

- *Infinite capacity.* For storing large data sets and large programs.
- *Infinite bandwidth.* For rapidly streaming these large data sets and programs to and from the processor.
- *Instantaneous or zero latency.* To prevent the processor from stalling while waiting for data or program code.
- *Persistence or nonvolatility.* To allow data and programs to survive even when the power supply is cut off.
- *Zero or very low implementation cost.*

We have learned that the highest levels of the memory hierarchy—register files and primary caches—are able to supply near-infinite bandwidth and very low average latency to the processor core, satisfying the second and third idealisms. The first idealism—infinite capacity—is satisfied by the lowest level of the memory hierarchy, since the capacities of DRAM-based memories are large enough to contain the working sets of most modern applications; for applications where this is not the case, we learned about a technique called *virtual memory* that extends the memory hierarchy beyond random-access memory devices to magnetic disks, which provide capacities that exceed the demands of all but the most demanding applications. The fourth idealism—persistence or nonvolatility—can also be supplied by magnetic disks, which are designed to retain their state even when they are powered down. The final idealism—low implementation cost—is also satisfied, since the high per-bit cost of the upper levels of the cache hierarchy is only multiplied by a relatively small number of bits, while the lower levels of the hierarchy provide tremendous capacity at a very low cost per bit. Hence, the average cost per bit is kept near the low cost of commodity DRAM and magnetic disks, rather than the high cost of the custom SRAM in the cache memories and register files.

REFERENCES

Blahut, R. E.: *Theory and Practice of Error Control Codes*. Reading, MA: Addison-Wesley Publishing Company, 1983.

Cuppu, V., and B. L. Jacob: "Concurrency, latency, or system overhead: Which has the largest impact on uniprocesor DRAM-system performance?," *Proc. 28th Int. Symposium on Computer Architecture*, 2001, pp. 62–71.

Cuppu, V., B. L. Jacob, B. Davis, and T. N. Mudge: "A performance comparison of contemporary DRAM architectures," *Proc. 26th Int. Symposium on Computer Architecture*, 1999, pp. 222–233.

Goodman, J.: "Using cache memory to reduce processor-memory traffic," *Proc. 10th Int. Symposium on Computer Architecture*, 1983, pp. 124–131.

Hill, M., and A. Smith: "Evaluating associativity in CPU caches," *IEEE Trans. on Computers*, 38, 12, 1989, pp. 1612–1630.

Hill, M. D.: *Aspects of Cache Memory and Instruction Buffer Performance*. PhD thesis, University of California at Berkeley, Computer Science Division, 1987.

Intel Corp.: *Pentium Processor User's Manual*, Vol. 3: *Architecture and Programming Manual*. Santa Clara, CA: Intel Corp., 1993.

Keltcher, C., K. McGrath, A. Ahmed, and P. Conway: "The AMD Opteron processor for multiprocessor servers," *IEEE Micro*, 23, 2, 2003, pp. 66–76.

Kilburn, T., D. Edwards, M. Lanigan, and F. Sumner: "One-level storage systems," *IRE Transactions*, EC-11, 2, 1962, pp. 223–235.

Lauterbach, G., and T. Horel: "UltraSPARC-III: Designing third generation 64-bit performance," *IEEE Micro*, 19, 3, 1999, pp. 56–66.

Liptay, J.: "Structural aspects of the system/360 model 85, part ii," *IBM Systems Journal*, 7, 1, 1968, pp. 15–21.

Patterson, D., G. Gibson, and R. Katz: "A case for redundant arrays of inexpensive disks (RAID)," *Proc. ACM SIGMOD Conference*, 1988, pp. 109–116.

Rao, T. R. N., and E. Fujiwara: *Error-Control Coding for Computer Systems*. Englewood Cliffs, NJ: Prentice Hall, 1989.

Ruemmler, C., and J. Wilkes: "An introduction to disk drive modeling," *IEEE Computer*, 27, 3, 1994, pp. 5–15.

Tendler, J. M., S. Dodson, S. Fields, and B. Sinharoy: "IBM eServer POWER4 system microarchitecture," IBM Whitepaper, 2001.

Wang, W.-H., J.-L. Baer, and H. Levy: "Organization and performance of a two-level virtual-real cache hierarchy," *Proc. 16th Annual Int. Symposium on Computer Architecture*, 1989, pp. 140–148.

Wilkes, M.: "Slave memories and dynamic storage allocation," *IEEE Trans. on Electronic Computers*, EC-14, 2, 1965, pp. 270–271.

Wulf, W. A., and S. A. McKee: "Hitting the memory wall: Implications of the obvious," *Computer Architecture News*, 23, 1, 1995, pp. 20–24.

HOMEWORK PROBLEMS

P3.1 Given the following benchmark code and assuming a virtually-addressed fully-associative cache with infinite capacity and 64-byte blocks, compute the overall miss rate (number of misses divided by number of references). Assume that all variables except array locations reside in registers and that arrays A, B, and C are placed consecutively in memory.
```
double A[1024], B[1024], C[1024];
for(int i=0;i<1000;i += 2) {
    A[i] = 35.0 * B[i] + C[i+1];
}
```

P3.2 Given the example code in Problem 3.1 and assuming a virtually-addressed direct-mapped cache of capacity 8K-byte and 64-byte blocks, compute the overall miss rate (number of misses divided by number of references). Assume that all variables except array locations reside in registers and that arrays A, B, and C are placed consecutively in memory.

P3.3 Given the example code in Problem 3.1 and assuming a virtually-addressed two-way set-associative cache of capacity 8K-byte and 64-byte blocks, compute the overall miss rate (number of misses divided by number of references). Assume that all variables except array locations reside in registers and that arrays A, B, and C are placed consecutively in memory.

P3.4 Consider a cache with 256 bytes. The word size is 4 bytes, and the block size is 16 bytes. Show the values in the cache and tag bits after each of the following memory access operations for the following two cache organizations: direct mapped and two-way associative. Also indicate whether the access was a hit or a miss. Justify. The addresses are in hexadecimal representation. Use LRU (least recently used) replacement algorithm wherever needed.

1. Read 0010
2. Read 001C
3. Read 0018
4. Write 0010
5. Read 0484
6. Read 051C
7. Read 001C
8. Read 0210
9. Read 051C

P3.5 Describe a program that has very high temporal locality. Write pseudocode for such a program, and show that it will have a high cache hit rate.

P3.6 Describe a program that has very low temporal locality. Write pseudocode for such a program, and show that it will have a high cache miss rate.

P3.7 Write the programs of Problems 3.5 and 3.6 and compile them on a platform that supports performance counters (for example, Microsoft Windows and the Intel VTune performance counter software). Collect and report performance counter data that verifies that the program with high temporal locality experiences fewer cache misses.

P3.8 Write the programs of Problems 3.5 and 3.6 in C, and compile them using the Simplescalar compilation tools available from **http://www.simplescalar.com.** Download and compile the Simplescalar 3.0 simulation suite and use the sim-cache tool to run both programs. Verify that the program with high temporal locality experiences fewer cache misses by reporting cache miss rates from both programs.

P3.9 Describe a program that has very high spatial locality. Write pseudocode for such a program, and show that it will have a high cache hit rate.

P3.10 Describe a program that has very low spatial locality. Write pseudocode for such a program, and show that it will have a high cache miss rate.

P3.11 Write the programs of Problems 3.9 and 3.10 and compile them on a platform that supports performance counters (for example, Linux and the Intel VTune performance counter software). Collect and report performance counter data that verifies that the program with high temporal locality experiences fewer cache misses.

P3.12 Write the programs of Problems 3.9 and 3.10 in C, and compile them using the Simplescalar compilation tools available from **http://www.simplescalar.com.** Download and compile the Simple-scalar 3.0 simulation suite and use the sim-cache tool to run both programs. Verify that the program with high temporal locality experiences fewer cache misses by reporting cache miss rates from both programs.

P3.13 Consider a processor with 32-bit virtual addresses, 4K-byte pages, and 36-bit physical addresses. Assume memory is byte-addressable (i.e., the 32-bit virtual address specifies a byte in memory).

- L1 instruction cache: 64K bytes, 128-byte blocks, four-way set-associative, indexed and tagged with virtual address.
- L1 data cache: 32K bytes, 64-byte blocks, two-way set-associative, indexed and tagged with physical address, writeback.
- Four-way set-associative TLB with 128 entries in all. Assume the TLB keeps a dirty bit, a reference bit, and three permission bits (read, write, execute) for each entry.

Specify the number of offset, index, and tag bits for each of these structures in the following table. Also, compute the total size in number of bit cells for each of the tag and data arrays.

Structure	Offset Bits	Index Bits	Tag Bits	Size of Tag Array	Size of Data Array
I-cache					
D-cache					
TLB					

P3.14 Given the cache organization in Problem 3.13, explain why accesses to the data cache would take longer than accesses to the instruction cache. Suggest a lower-latency data cache design with the same capacity and describe how the organization of the cache would have to change to achieve the lower latency.

P3.15 Given the cache organization in Problem 3.13, assume the architecture requires writes that modify the instruction text (i.e., self-modifying code) to be reflected immediately if the modified instructions are fetched and executed. Explain why it may be difficult to support this requirement with this instruction cache organization.

P3.16 Assume a two-level cache hierarchy with a private level-1 instruction cache (L1I), a private level-1 data cache (L1D), and a shared level-two data cache (L2). Given local miss rates for the 4% for L1I, 7.5% for L1D, and 35% for L2, compute the global miss rate for the L2 cache.

P3.17 Assuming 1 L1I access per instruction and 0.4 data accesses per instruction, compute the misses per instruction for the L1I, L1D, and L2 caches of Problem 3.16.

P3.18 Given the miss rates of Problem 3.16 and assuming that accesses to the L1I and L1D caches take 1 cycle, accesses to the L2 take 12 cycles, accesses to main memory take 75 cycles, and a clock rate of 1 GHz, compute the average memory reference latency for this cache hierarchy.

P3.19 Assuming a perfect cache CPI (cycles per instruction) for a pipelined processor equal to 1.15 CPI, compute the MCPI and overall CPI for a pipelined processor with the memory hierarchy described in Problem 3.18 and the miss rates and access rates specified in Problems 3.16 and 3.17.

P3.20 Repeat Problem 3.16 assuming an L1I local miss rate of 7%, an L1D local miss rate of 3.5%, and an L2 local miss rate of 75%.

P3.21 Repeat Problem 3.17 given the miss rates of Problem 3.20.

P3.22 Repeat Problem 3.18 given the miss rates of Problem 3.20.

P3.23 Repeat Problem 3.19 given the miss rates of Problem 3.20.

P3.24 CPI equations can be used to model the performance of in-order superscalar processors with multilevel cache hierarchies. Compute the CPI for such a processor, given the following parameters:

- Infinite cache CPI of 1.15
- L1 cache miss penalty of 12 cycles
- L2 cache miss penalty of 50 cycles
- L1 instruction cache per-instruction miss rate of 3% (0.03 misses/instruction)
- L1 data cache per-instruction miss rate of 2% (0.02 misses/instruction).
- L2 local cache miss rate of 25% (0.25 misses/L2 reference).

P3.25 It is usually the case that a set-associative or fully associative cache has a higher hit rate than a direct-mapped cache. However, this is not always true. To illustrate this, show a memory reference trace for a program that has a higher hit rate with a two-block direct-mapped cache than a fully associative cache with two blocks.

P3.26 Download and install the Simplescalar 3.0 simulation suite and instructional benchmarks from **www.simplescalar.com.** Using the sim-cache cache simulator, plot the cache miss rates for each benchmark for the following cache hierarchy: 16K-byte two-way set-associative L1 instruction cache with 64-byte lines; 32K-byte four-way set-associative L1 data cache with 32-byte lines; 12K-byte eight-way set-associative L2 cache with 64-byte lines.

P3.27 Using the benchmarks and tools from Problem26, plot several miss-rate sensitivity curves for each of the three caches (L1I, L1D, L2) by varying each of the following parameters: cache size 0.5×, 1×, 2×, 4×; associativity 0.5×, 1×, 2×, 4×; block size 0.25×, 0.5×, 1×, 2×, 4×. Hold the other parameters fixed at the values in Problem 3.26 while varying each of the three parameters for each sensitivity curve. Based on your sensitivity curves, identify an appropriate value for each parameter near the knee of the curve (if any) for each benchmark.

P3.28 Assume a synchronous front-side processor-memory bus that operates at 100 MHz and has an 8-byte data bus. Arbitration for the bus takes one bus cycle (10 ns), issuing a cache line read command for 64 bytes of data takes one cycle, memory controller latency (including DRAM access) is 60 ns, after which data doublewords are returned in back-to-back cycles. Further assume the bus is blocking or circuit-switched. Compute the latency to fill a single 64-byte cache line. Then compute the peak read bandwidth for this processor-memory bus, assuming the processor arbitrates for the bus for a new read in the bus cycle following completion of the last read.

P3.29 Given the assumptions of Problem 3.28, assume a nonblocking (split-transaction) bus that overlaps arbitration with commands and data transfers, but multiplexes data and address lines. Assume that a read command requires a single bus cycle, and further assume that the memory controller has infinite DRAM bandwidth. Compute the peak data bandwidth for this front side bus.

P3.30 Building on the assumptions of Problem 3.29, assume the bus now has dedicated data lines and a separate arbitration mechanism for addresses/commands and data. Compute the peak data bandwidth for this front side bus.

P3.31 Consider finite DRAM bandwidth at a memory controller, as follows. Assume double-data-rate DRAM operating at 100 MHz in a parallel non-interleaved organization, with an 8-byte interface to the DRAM chips. Further assume that each cache line read results in a DRAM row miss, requiring a precharge and RAS cycle, followed by row-hit CAS cycles for each of the doublewords in the cache line. Assuming memory controller overhead of one cycle (10 ns) to initiate a read operation, and one cycle latency to transfer data from the DRAM data bus to the processor-memory bus, compute the latency for reading one 64-byte cache block. Now compute the peak data bandwidth for the memory interface, ignoring DRAM refresh cycles.

P3.32 Two page-table architectures are in common use today: multilevel forward page tables and hashed page tables. Write out a pseudocode function matching the following function declaration that searches a three-level forward page table and returns 1 on a hit and 0 on a miss, and assigns *realaddress on a hit.

```
int fptsearch(void *pagetablebase, void*
   virtualaddress, void** realaddress);
```

P3.33 As in Problem 3.32, write out a pseudocode function matching the following function declaration that searches a hashed page table and returns 1 on a hit and 0 on a miss, and assigns *realaddress on a hit:

```
int hptsearch(void *pagetablebase, void*
   virtualaddress, void** realaddress);
```

P3.34 Assume a single-platter disk drive with an average seek time of 4.5 ms, rotation speed of 7200 rpm, data transfer rate of 10 Mbytes/s per head, and controller overhead and queueing of 1 ms. What is the average access latency for a 4096-byte read?

P3.35 Recompute the average access latency for Problem34 assuming a rotation speed of 15K rpm, two platters, and an average seek time of 4.0 ms.

CHAPTER 4

Superscalar Organization

CHAPTER OUTLINE

4.1 Limitations of Scalar Pipelines
4.2 From Scalar to Superscalar Pipelines
4.3 Superscalar Pipeline Overview
4.4 Summary

References
Homework Problems

While pipelining has proved to be an extremely effective microarchitecture technique, the type of *scalar pipelines* presented in Chapter 2 have a number of shortcomings or limitations. Given the never-ending push for higher performance, these limitations must be overcome in order to continue to provide further speedup for existing programs. The solution is *superscalar pipelines* that are able to achieve performance levels beyond those possible with just scalar pipelines.

Superscalar machines go beyond just a single-instruction pipeline by being able to simultaneously advance multiple instructions through the pipeline stages. They incorporate multiple functional units to achieve greater concurrent processing of multiple instructions and higher instruction execution throughput. Another foundational attribute of superscalar processors is the ability to execute instructions in an order different from that specified by the original program. The sequential ordering of instructions in standard programs implies some unnecessary precedences between the instructions. The capability of executing instructions out of program order relieves this sequential imposition and allows more parallel processing of instructions without requiring modification of the original program. This and the following chapters attempt to codify the body of knowledge on superscalar processor design in a systematic fashion. This chapter focuses on issues related to the pipeline organization of superscalar machines. The techniques that address the dynamic interaction between the superscalar machine and the instructions being processed are presented in Chapter 5. Case studies of two commercial superscalar processors are presented in Chapters 6 and 7, while Chapter 8 provides a broad survey of historical and current designs.

4.1 Limitations of Scalar Pipelines

Scalar pipelines are characterized by a single-instruction pipeline of k stages. All instructions, regardless of type, traverse through the same set of pipeline stages. At most, one instruction can be resident in each pipeline stage at any one time, and the instructions advance through the pipeline stages in a lockstep fashion. Except for the pipeline stages that are stalled, each instruction stays in each pipeline stage for exactly one cycle and advances to the next stage in the next cycle. Such rigid scalar pipelines have three fundamental limitations:

1. The maximum throughput for a scalar pipeline is bounded by one instruction per cycle.

2. The unification of all instruction types into one pipeline can yield an inefficient design.

3. The stalling of a lockstep or rigid scalar pipeline induces unnecessary pipeline bubbles.

We elaborate on these limitations in Sections 4.1.1 to 4.1.3.

4.1.1 Upper Bound on Scalar Pipeline Throughput

As stated in Chapter 1 and as shown in Equation (4.1), processor performance can be increased either by increasing instructions per cycle (IPC) and/or frequency or by decreasing the total instruction count.

$$\text{Performance} = \frac{1}{\text{instruction count}} \times \frac{\text{instructions}}{\text{cycle}} \times \frac{1}{\text{cycle time}} = \frac{\text{IPC} \times \text{frequency}}{\text{instruction count}}$$

(4.1)

Frequency can be increased by employing a deeper pipeline. A deeper pipeline has fewer logic gate levels in each pipeline stage, which leads to a shorter cycle time and a higher frequency. However, there is a point of diminishing return due to the hardware overhead of pipelining. Furthermore, a deeper pipeline can potentially incur higher penalties, in terms of the number of penalty cycles, for dealing with inter-instruction dependences. The additional average cycles per instruction (CPI) overhead due to this higher penalty can possibly eradicate the benefit due to the reduction of cycle time.

Regardless of the pipeline depth, a scalar pipeline can only initiate the processing of at most one instruction in every machine cycle. Essentially, the average IPC for a scalar pipeline is fundamentally bounded by one. To get more instruction throughput, especially when deeper pipelining is no longer the most cost-effective way to get performance, the ability to initiate more than one instruction in every machine cycle is necessary. To achieve an IPC greater than one, a pipelined processor must be able to initiate the processing of more than one instruction in every machine cycle. This will require increasing the width of the pipeline to facilitate having more than one instruction resident in each pipeline stage at any one time. We identify such pipelines as *parallel pipelines*.

4.1.2 Inefficient Unification into a Single Pipeline

Recall that the second idealized assumption of pipelining is that all the repeated computations to be processed by the pipeline are identical. For instruction pipelines, this is clearly not the case. There are different instruction types that require different sets of subcomputations. In unifying these different requirements into one pipeline, difficulties and/or inefficiencies can result. Looking at the unification of different instruction types into the TYP pipeline in Chapter 2, we can observe that in the earlier pipeline stages (such as IF, ID, and RD stages) there is significant uniformity. However, in the execution stages (such as ALU and MEM stages) there is substantial diversity. In fact, in the TYP example, we have ignored floating-point instructions on purpose due to the difficulty of unifying them with the other instruction types. It is for this reason that at one point in time during the "RISC revolution," floating-point instructions were categorized as inherently CISC and considered to be violating RISC principles.

Certain instruction types make their unification into a single pipeline quite difficult. These include floating-point instructions and certain fixed-point instructions (such as multiply and divide instructions) that require multiple execution cycles. Instructions that require long and possibly variable latencies are difficult to unify with simple instructions that require only a single cycle latency. As the disparity between CPU and memory speeds continues to widen, the latency (in terms of number of machine cycles) of memory instructions will continue to increase. Other than latency differences, the hardware resources required to support the execution of these different instruction types are also quite different. With the continued push for faster hardware, more specialized execution units customized for specific instruction types will be required. This will also contribute to the need for greater diversity in the execution stages of the instruction pipeline.

Consequently, the forced unification of all the instruction types into a single pipeline becomes either impossible or extremely inefficient for future high-performance processors. For parallel pipelines there is a strong motivation not to unify all the execution hardware into one pipeline, but instead to implement multiple different execution units or subpipelines in the execution portion of parallel pipelines. We call such parallel pipelines *diversified pipelines*.

4.1.3 Performance Lost due to a Rigid Pipeline

Scalar pipelines are rigid in the sense that instructions advance through the pipeline stages in a lockstep fashion. Instructions enter a scalar pipeline according to program order, i.e., in order. When there are no stalls in the pipeline, all the instructions in the pipeline stages advance synchronously and the program order of instructions is maintained. When an instruction is stalled in a pipeline stage due to its dependence on a leading instruction, that instruction is held in the stalled pipeline stage while all leading instructions are allowed to proceed down the pipeline stages. Because of the rigid nature of a scalar pipeline, if a dependent instruction is stalled in pipeline stage i, then all earlier stages, i.e., stages $1, 2, \ldots, i-1$, containing trailing instructions are also stalled. All i stages of the pipeline are stalled until the instruction in stage i is forwarded its dependent operand. After the inter-instruction dependence is satisfied, then all i stalled instructions can again advance synchronously down the pipeline. For a rigid scalar pipeline, a stalled stage in the middle of the pipeline affects all earlier stages of the pipeline; essentially the stalling of stage i is propagated backward through all the preceding stages of the pipeline.

The backward propagation of stalling from a stalled stage in a scalar pipeline induces unnecessary pipeline bubbles or idling pipeline stages. While an instruction is stalled in stage i due to its dependence on a leading instruction, there may be another instruction trailing the stalled instruction which does not have a dependence on any leading instruction that would require its stalling. For example, this independent trailing instruction could be in stage $i-1$ and would be unnecessarily stalled due to the stalling of the instruction in stage i. According to program semantics, it is not necessary for this instruction to wait in stage $i-1$. If this instruction is allowed to bypass the stalled instruction and continue down the pipeline stages, an idling cycle of the pipeline can be eliminated, which effectively reduces the penalty due to the stalled instruction by one cycle; see Figure 4.1. If multiple instructions are able and allowed to bypass the stalled instruction, then multiple penalty cycles can be eliminated or "covered" in the sense that idling pipeline stages are given useful instructions to process. Potentially all the penalty cycles due to the stalled instruction can be covered. Allowing the bypassing of a stalled leading instruction by trailing instructions is referred to as an *out-of-order execution* of instructions. A rigid scalar pipeline does not allow out-of-order execution and hence can incur unnecessary penalty cycles in enforcing inter-instruction dependences. Parallel pipelines that support out-of-order execution are called *dynamic pipelines*.

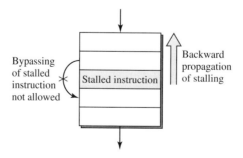

Figure 4.1

Unnecessary Stall Cycles Induced by Backward Propagation of Stalling in a Rigid Pipeline.

4.2 From Scalar to Superscalar Pipelines

Superscalar pipelines can be viewed as natural descendants of the scalar pipelines and involve extensions to alleviate the three limitations (see Section 4.1) with scalar pipelines. Superscalar pipelines are parallel pipelines, instead of scalar pipelines, in that they are able to initiate the processing of multiple instructions in every machine cycle. In addition, superscalar pipelines are diversified pipelines in employing multiple and heterogeneous functional units in their execution stage(s). Finally, superscalar pipelines can be implemented as dynamic pipelines in order to achieve the best possible performance without requiring reordering of instructions by the compiler. These three characterizing attributes of superscalar pipelines will be further elaborated in this section.

4.2.1 Parallel Pipelines

The degree of parallelism of a machine can be measured by the maximum number of instructions that can be concurrently in progress at any one time. A k-stage scalar pipeline can have k instructions concurrently resident in the machine and can potentially achieve a factor-of-k speedup over a nonpipelined machine. Alternatively, the same speedup can be achieved by employing k copies of the nonpipelined machine to process k instructions in parallel. These two forms of machine parallelism are illustrated in Figure 4.2(b) and (c), and they can be denoted *temporal machine parallelism* and *spatial machine parallelism*, respectively. Temporal and spatial parallelism of the same degree can yield about the same factor of potential speedup. Clearly, temporal parallelism via pipelining requires less hardware than spatial parallelism, which requires replication of the entire processing unit. Parallel pipelines can be viewed as employing both temporal and spatial machine parallelism, as illustrated in Figure 4.2(d), to achieve higher instruction processing throughput in an efficient manner.

The speedup of a scalar pipeline is measured with respect to a nonpipelined design and is primarily determined by the depth of the scalar pipeline. For parallel pipelines, or superscalar pipelines, the speedup now is usually measured with respect to a scalar pipeline and is primarily determined by the *width* of the parallel pipeline. A parallel pipeline with width s can concurrently process up to s instructions in each of its pipeline stages, which can lead to a potential speedup of s over a scalar pipeline. Figure 4.3 illustrates a parallel pipeline of width $s = 3$.

Significant additional hardware resources are required for implementing parallel pipelines. Each pipeline stage can potentially process and advance up to s instructions in every machine cycle. Hence, the logic complexity of each pipeline stage can increase by a factor of s. In the worst case, the circuitry for interstage interconnection can increase by a factor of s^2 if an $s \times s$ crossbar is used to connect all s instruction buffers from one stage to all s instruction buffers of the next stage. In order to support concurrent register file accesses by s instructions, the number of read and write ports of the register file must be increased by a factor of s. Similarly, additional I-cache and D-cache access ports must be provided.

As shown in Chapter 2, the Intel i486 is a five-stage scalar pipeline [Crawford, 1990]. The sequel to the i486 was the Pentium microprocessor from

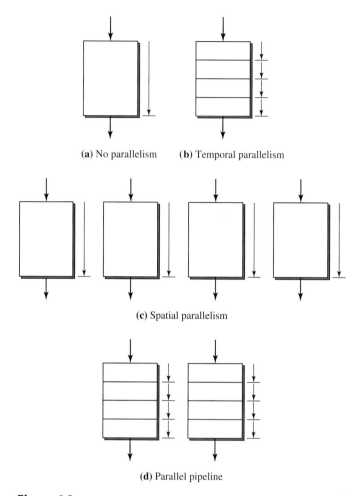

Figure 4.2
Machine Parallelism: (a) No Parallelism (Nonpipelined); (b) Temporal Parallelism (Pipelined); (c) Spatial Parallelism (Multiple Units); (d) Combined Temporal and Spatial Parallelism.

Intel [Intel Corp., 1993]. The Pentium microprocessor is a superscalar machine implementing a parallel pipeline of width $s = 2$. It essentially implements two i486 pipelines; see Figure 4.4. Multiple instructions can be fetched and decoded by the first two stages of the parallel pipeline in every machine cycle. In each cycle, potentially two instructions can be issued into the two execution pipelines, i.e., the U pipe and the V pipe. The goal is to maximize the number of dual-issue cycles. The superscalar Pentium microprocessor can achieve a peak execution rate of two instructions per machine cycle.

As compared to the scalar pipeline of i486, the Pentium parallel pipeline requires significant additional hardware resources. First, the five pipeline stages

SUPERSCALAR ORGANIZATION **183**

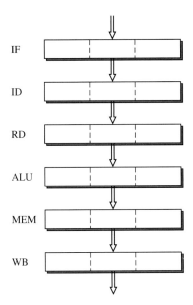

Figure 4.3
A Parallel Pipeline of Width $s = 3$.

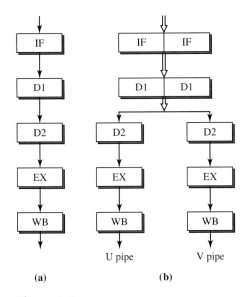

Figure 4.4
(a) The Five-Stage i486 Scalar Pipeline;
(b) The Five-Stage Pentium Parallel Pipeline
of Width $s = 2$.

have doubled in width. The two execution pipes can accommodate up to two instructions in each of the last three stages of the pipeline. The execute stage can perform an ALU operation or access the D-cache. Hence, additional ports to the register file must be provided to support the concurrent execution of two ALU operations in every cycle. If the two instructions in the execute stage are both load/store instructions, then the D-cache must provide dual access. A true dual-ported D-cache is expensive to implement. Instead, the Pentium D-cache is implemented as a single-ported D-cache with eight-way interleaving. Simultaneous accesses to two different banks by the two load/store instructions in the U and V pipes can be supported. If there is a bank conflict, i.e., both load/store instructions must access the same bank, then the two D-cache accesses are serialized.

4.2.2 Diversified Pipelines

The hardware resources required to support the execution of different instruction types can vary significantly. For a scalar pipeline, all the diverse requirements for the execution of all instruction types must be unified into a single pipeline. The resultant pipeline can be highly inefficient. Each instruction type only requires a subset of the execution stages, but it must traverse all the execution stages. Every instruction is idling as it traverses the unnecessary stages and incurs significant dynamic external fragmentation. The execution latency for all instruction types is equal to the total number of execution stages. This can result in unnecessary stalling of trailing instructions and/or require additional forwarding paths.

This inefficiency due to unification into one single pipeline is naturally addressed in parallel pipelines by employing multiple different functional units in the execution stage(s). Instead of implementing s identical pipes in an s-wide parallel pipeline, in the execution portion of the parallel pipeline, diversified execution pipes can be implemented; see Figure 4.5. In this example, four execution pipes, or functional units, of differing pipe depths are implemented. The RD stage dispatches instructions to the four execution pipes based on the instruction types.

There are a number of advantages in implementing diversified execution pipes. Each pipe can be customized for a particular instruction type, resulting in efficient hardware design. Each instruction type incurs only the necessary latency and makes use of all the stages of an execution pipe. This is certainly more efficient than implementing s identical copies of a universal execution pipe each of which can execute all instruction types. If all inter-instruction dependences between different instruction types are resolved prior to dispatching, then once instructions are issued into the individual execution pipes, no further stalling can occur due to instructions in other pipes. This allows the distributed and independent control of each execution pipe.

The design of a diversified parallel pipeline does require special considerations. One important consideration is the number and mix of functional units. Ideally the number of functional units should match the available instruction-level parallelism of the program, and the mix of functional units should match the dynamic mix of instruction types of the program. Most first-generation superscalar processors

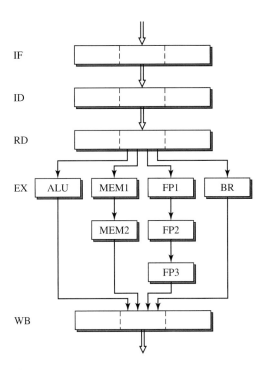

Figure 4.5
A Diversified Parallel Pipeline with Four Execution Pipes.

simply integrated a second execution pipe for processing floating-point instructions with the existing scalar pipe for processing non-floating-point instructions. As superscalar designs evolved from two-issue machines to four-issue machines, typically four functional units are implemented for executing integer, floating-point, load/store, and branch instructions. Some recent designs incorporate multiple integer units, some of which are dedicated to long-latency integer operations such as multiply and divide, and others are dedicated to the processing of special operations for image, graphics, and signal processing applications.

Similar to pipelining, the employment of a multiplicity of diversified functional units in the design of a high-performance CPU is not a recent invention. The CDC 6600 incorporates both pipelining and the use of multiple functional units [Thornton, 1964]. The CPU of the CDC 6600 employs 10 diversified functional units, as shown in Figure 4.6. The 10 functional units operate on data stored in 24 operating registers, which consist of 8 address registers (18 bits), 8 index registers (18 bits), and 8 floating-point registers (60 bits). The 10 functional units operate independently and consist of a fixed-point adder (18 bits), a floating-point adder (60 bits), two multiply units (60 bits), a divide unit (60 bits), a shift unit (60 bits), a boolean unit (60 bits), two increment units, and a branch unit. The CDC 6600 CPU is a pipelined processor with two decoding stages preceding the execution portion;

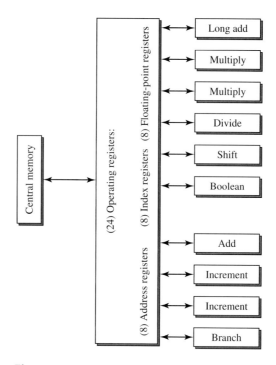

Figure 4.6
The CDC 6600 with 10 Diversified Functional Units in Its CPU.

however, the 10 functional units are not pipelined and have variable execution latencies. For example, a fixed-point add requires 3 cycles, and a floating-point multiply (divide) requires 10 (29) cycles. The goal of the CDC 6600 CPU is to sustain an issue rate of one instruction per machine cycle.

Another superscalar microprocessor employed a similar mix of functional units as the CDC 6600. Just prior to the formation of the PowerPC alliance with IBM and Apple, Motorola had developed a very clean design of a wide superscalar microprocessor called the 88110 [Diefendorf and Allen, 1992]. Interestingly, the 88110 also employs 10 functional units; see Figure 4.7. The 10 functional units consist of two integer units, a bit field unit, a floating-point add unit, a multiply unit, a divide unit, two graphic units, a load/store unit, and an instruction sequencing/branch unit. Most of the units have single-cycle latency. With the exception of the divide unit, the other units with multicycle latencies are all pipelined. In terms of the total number of functional units, the 88110 represents one of the wider superscalar designs.

4.2.3 Dynamic Pipelines

In any pipelined design, buffers are required between pipeline stages. In a scalar rigid pipeline, a single-entry buffer is placed between two consecutive pipeline stages

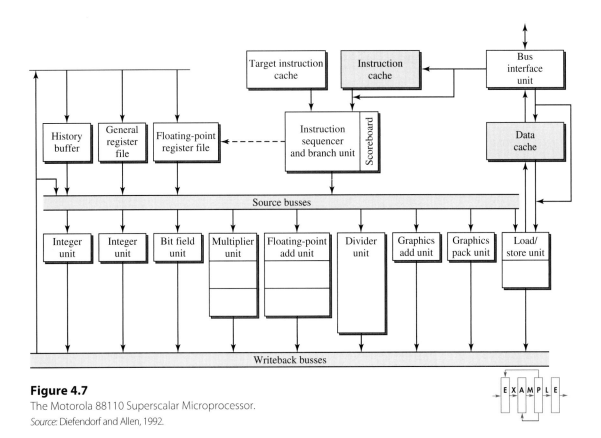

Figure 4.7
The Motorola 88110 Superscalar Microprocessor.
Source: Diefendorf and Allen, 1992.

(stages i and $i + 1$), as shown in Figure 4.8(a). The buffer holds all essential control and data bits for the instruction that has just traversed stage i of the pipeline and is ready to traverse stage $i + 1$ in the next machine cycle. Single-entry buffers are quite easy to control. In every machine cycle, the buffer's current content is used as input to stage $i + 1$; and at the end of the cycle, the buffer latches in the result produced by stage i. Essentially the buffer is clocked in every machine cycle. The exception occurs when the instruction in the buffer must be held back and prevented from traversing stage $i + 1$. In that case, the clocking of the buffer is disabled, and the instruction is stalled in the buffer. Clearly if this buffer is stalled in a scalar rigid pipeline, all stages preceding stage i must also be stalled. Hence, in a scalar rigid pipeline, if there is no stalling, then every instruction remains in each buffer for exactly one machine cycle and then advances to the next buffer. All the instructions enter and leave each buffer in exactly the same order as specified in the original sequential code.

In a parallel pipeline, multientry buffers are needed between two consecutive pipeline stages as shown in Figure 4.8(b). Multientry buffers can be viewed as simple extensions of the single-entry buffers. Multiple instructions can be latched into each multientry buffer in every machine cycle. In the next cycle, these instructions can then traverse the next pipeline stage. If all the instructions in a multientry buffer are required to advance simultaneously in a lockstep fashion, then the control of the

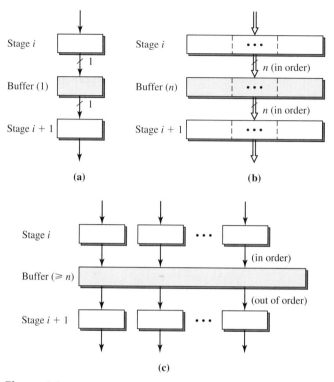

Figure 4.8
Interpipeline-Stage Buffers: (a) Single-Entry Buffer; (b) Multientry Buffer; (c) Multientry Buffer with Reordering.

multientry buffer is similar to that of the single-entry buffer. The entire multientry buffer is either clocked or stalled in each machine cycle. However, such operation of the parallel pipeline may induce unnecessary stalling of some of the instructions in a multientry buffer. For more efficient operation of a parallel pipeline, much more sophisticated multientry buffers are needed.

Each entry of the simple multientry buffer of Figure 4.8(b) is hardwired to one write port and one read port, and there is no interaction between the multiple entries. One enhancement to this simple buffer is to add connectivity between the entries to facilitate movement of data between entries. For example, the entries can be connected into a linear chain like a shift register and function as a FIFO queue. Another enhancement is to provide a mechanism for independent accessing of each entry in the buffer. This will require the ability to explicitly address each individual entry in the buffer and independently control the reading and writing of each entry. If each input/output port of the buffer is given the ability to access any entry in the buffer, then such a multientry buffer will effectively resemble a small multiported RAM. With such a buffer an instruction can remain in an entry of the buffer for many machine cycles and can be updated or modified while resident in that buffer. A further enhancement can incorporate associative accessing of the entries in the buffer. Instead of using conventional addressing to index into an entry in the buffer, the content of an entry can be used as an associative tag to index into that entry. With such accessing mechanism, the multientry buffer becomes a small associative cache memory.

Superscalar pipelines differ from (rigid) scalar pipelines in one key aspect, which is the use of complex multientry buffers for buffering instructions in flight. In order to minimize unnecessary stalling of instructions in a parallel pipeline, trailing instructions must be allowed to bypass a stalled leading instruction. Such bypassing can change the order of execution of instructions from the original sequential order of the static code. With out-of-order execution of instructions, there is the potential of approaching the data flow limit of instruction execution; i.e., instructions are executed as soon as their operands are available. A parallel pipeline that supports out-of-order execution of instructions is called a *dynamic pipeline*. A dynamic pipeline achieves out-of-order execution via the use of complex multientry buffers that allow instructions to enter and leave the buffers in different orders. Such a reordering multientry buffer is shown in Figure 4.8(c).

Figure 4.9 illustrates a parallel diversified pipeline of width $s = 3$ that is a dynamic pipeline. The execution portion of the pipeline, consisting of the four pipelined

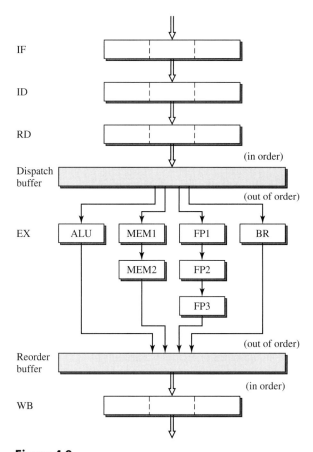

Figure 4.9
A Dynamic Pipeline of Width $s = 3$.

functional units, is bracketed by two reordering multientry buffers. The first buffer, called the *dispatch buffer,* is loaded with decoded instructions according to program order and then dispatches instructions to the functional units potentially in an order different from the program order. Hence instructions can leave the dispatch buffer in a different order than the order in which they enter the dispatch buffer. This pipeline also implements a set of diverse functional units with different latencies.

With potential out-of-order issuing into the functional units and/or the variable latencies of the functional units, instructions can finish execution out of order. To ensure that exceptions can be handled according to the original program order, the instructions must be completed (i.e., the machine state must be updated), in program order. When instructions finish execution out of order, another reordering multientry buffer is needed at the back end of the execution portion of the pipeline to ensure in-order completion. This buffer, called the *completion buffer,* buffers the instructions that may have finished execution out of order and retires the instructions in order by outputting instructions to the final writeback stage in program order. Such a dynamic pipeline facilitates the out-of-order execution of instructions in order to achieve the shortest possible execution time, and yet is able to provide precise exception by retiring the instructions and updating the machine state according to the program order.

4.3 Superscalar Pipeline Overview

This section presents an overview of the critical issues involved in the design of superscalar pipelines. The focus is on the organization, or structural design, of superscalar pipelines. Issues and techniques related to the dynamic interaction of machine organization and instruction semantics and the optimization of the resultant machine performance are covered in Chapter 5. Essentially this chapter focuses on the design of the machine organization, while Chapter 5 takes into account the interaction between the machine and the program.

Similar to the use of the six-stage TYP pipeline in Chapter 2 as a vehicle for presenting scalar pipeline design, we use the six-stage TEM superscalar pipeline shown in Figure 4.10 as a "template" for discussion on the organization of superscalar pipelines. Compared to scalar pipelines, there is far more variety and greater diversity in the implementation of superscalar pipelines. The TEM superscalar pipeline should not be viewed as an actual implementation of a typical or representative superscalar pipeline. The six stages of the TEM superscalar pipeline should be viewed as *logical* pipeline stages which may or may not correspond to six physical pipeline stages. The six stages of the TEM superscalar pipeline provide a nice framework or outline for discussing the six major portions of, or six major tasks performed by, most superscalar pipeline organizations.

The six stages of the TEM superscalar pipeline are *fetch, decode, dispatch, execute, complete,* and *retire*. The execute stage can include multiple (pipelined) functional units of different types with different execution latencies. This necessitates the dispatch stage to distribute instructions of different types to their corresponding functional units. With out-of-order execution of instructions in the execute stage, the complete stage is needed to reorder the instructions and ensure the in-order

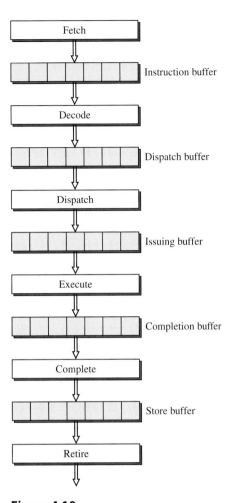

Figure 4.10
The Six-Stage Template (TEM) Superscalar Pipeline.

updating of the machine state. Note also that there are multientry buffers separating these six stages. The complexity of these buffers can vary depending on their functionality and location in the superscalar pipeline. These six stages and design issues related to them are now addressed in turn.

4.3.1 Instruction Fetching

Unlike a scalar pipeline, a superscalar pipeline, being a parallel pipeline, is capable of fetching more than one instruction from the I-cache in every machine cycle. Given a superscalar pipeline of width s, its fetch stage should be able to fetch s instructions from the I-cache in every machine cycle. This implies that the physical organization of the I-cache must be wide enough that each row of the I-cache array can store

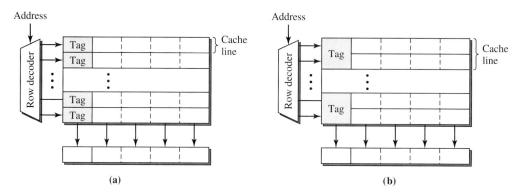

Figure 4.11
Organization of a Wide I-Cache: (a) One Cache Line is Equal to One Physical Row; (b) One Cache Line is Equal to Two Physical Rows.

s instructions and that an entire row can be accessed at one time. In our current discussion, we assume that the access latency of the I-cache is one cycle and that the fetch width is equal to the row width. Typically in such a wide cache organization, a cache line corresponds to a physical row in the cache array; it is also possible that a cache line can span several physical rows of the cache array, as illustrated in Figure 4.11.

The primary objective of the fetch stage is to maximize the instruction-fetching bandwidth. The sustained throughput achieved by the fetch stage will impact the overall throughput of the superscalar pipeline, because the throughput of all subsequent stages depends on and cannot possibly exceed the throughput of the fetch stage. Two primary impediments to achieving the maximum throughput of s instructions fetched per cycle are (1) the misalignment of the s instructions being fetched, called the *fetch group,* with respect to the row organization of the I-cache array; and (2) the presence of control-flow changing instructions in the fetch group.

In every machine cycle, the fetch stage uses the program counter (PC) to index into the I-cache to fetch the instruction pointed to by the PC along with the next $s - 1$ instructions, i.e., the s instructions of the fetch group. If the entire fetch group is stored in the same row of the cache array, then all s instructions can be fetched. On the other hand, if the fetch group crosses a row boundary, then not all s instructions can be fetched in that cycle (assuming that only one row of the I-cache can be accessed in each cycle). Hence, only those instructions in the first row can be fetched; the remaining instructions will require another cycle for their fetching. The fetch bandwidth is effectively reduced by one-half, for it now requires two cycles to fetch s instructions. This is due to the misalignment of the fetch group with respect to the row boundaries of the I-cache array, as illustrated in Figure 4.12. Such misalignments reduce the effective fetch bandwidth. In the case where each cache line corresponds to a physical row, as shown in Figure 4.11(a), then the crossing of a row boundary also corresponds to the crossing of a cache line boundary, which can incur additional problems. If a fetch group spans two cache lines, then it can induce an I-cache miss involving the second line even though the

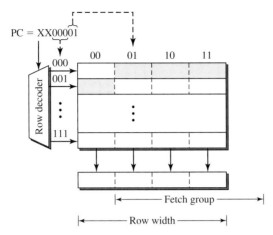

Figure 4.12
Misalignment of the Fetch Group Relative to the Row Boundaries of the I-Cache Array.

first line is resident. Even if both lines are resident in the I-cache, the physical accessing of multiple cache lines in one cycle is problematic.

There are two possible solutions to the misalignment problem. The first solution is a static technique employed at compile time. The compiler can be given information on the organization of the I-cache, e.g., its indexing scheme and row size. Based on this information, instructions can be appropriately placed in memory locations so as to ensure the aligning of fetch groups with physical rows. For example, every instruction that is the target of a branch can be placed in a memory location that is mapped to the first instruction of a row. This will increase the probability of fetching s instructions from the beginning of a row. Such techniques have been implemented and are reasonably effective. A problem with this solution is that the object code is tuned to a particular I-cache organization and may not be properly aligned for other I-cache organizations. Another problem is that the static code now can occupy a larger address range, which can potentially lead to a higher I-cache miss rate.

The second solution to the misalignment problem involves using hardware at run time. Alignment hardware can be incorporated to ensure that s instructions are fetched in every cycle even if the fetch group crosses a row boundary (but not a cache line boundary). Such alignment hardware is incorporated in the IBM RS/6000 design; we now briefly describe this design [Grohoski, 1990; Oehler and Groves, 1990].

The RS/6000 employs a two-way set-associative I-cache with a line size of 16 instructions (64 bytes). Each row of the I-cache array stores four associative sets (two per set) of instructions. Hence, each line of the I-cache spans four physical rows, as shown in Figure 4.13. The physical I-cache array is actually composed of four independent subarrays (denoted 0, 1, 2, and 3), which can be accessed in parallel. One instruction can be fetched from each subarray in every I-cache access. Which

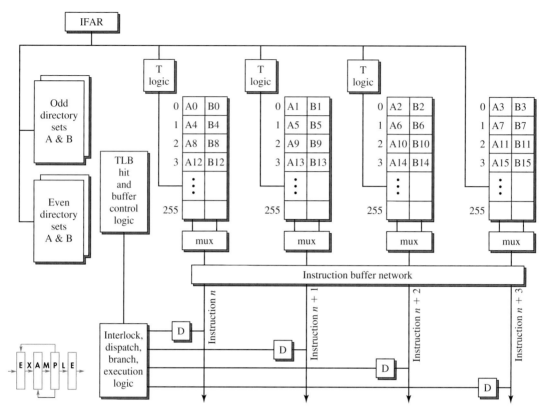

Figure 4.13
Organization of the RS/6000 Two-Way Set-Associative I-Cache with Auto-Realignment.

of the two instructions (either A or B) in the associative set is accessed depends on which of the two has a tag match with the address. The instruction addresses are allocated in an interleaved fashion across the four subarrays.

If the PC happens to point to the first subarray, i.e., subarray 0, then four consecutive instructions can be simultaneously fetched from the four subarrays. All four of these instructions reside in the same physical row of the I-cache, and all four subarrays are accessed using the same row address. On the other hand, if the PC indexes into the middle of the row, e.g., the first instruction of the fetch group resides in subarray 2, then the four consecutive instructions in the fetch group will span across two rows. The RS/6000 deals with this problem by detecting when the starting address points to a subarray other than subarray 0 and automatically incrementing the row address of the nonconsecutive subarrays. This is done by the "T-logic" hardware associated with each subarray. For example, if the PC indexes into subarray 2, then subarrays 2 and 3 will be accessed with the same row address presented to them. However the T-logic of subarrays 0 and 1 will detect this condition and automatically increment the row address presented to subarrays 0 and 1.

Consequently the two instructions fetched from subarrays 0 and 1 will actually be from the next physical row of the I-cache.

Therefore, regardless of the starting address and where that address points in an I-cache row, four consecutive instructions can always be fetched in every cycle as long as the fetch group does not cross a cache line boundary. When a fetch group crosses a cache line boundary, only instructions in the first cache line can be fetched in that cycle. Given the fact that the cache line of the RS/6000 consists of 16 instructions, and that there are 16 possible starting addresses of a word in a cache line, on the average the fetch bandwidth of this I-cache organization is $(13/16) \times 4 + (1/16) \times 3 + (1/16) \times 2 + (1/16) \times 1 = 3.625$ instructions per cycle.

Although the fetch group can begin in any one of the four subarrays, only subarrays 0, 1, and 2 require the T-logic hardware. The row address of subarray 3 never needs to be incremented regardless of the starting subarray of a fetch group. The *instruction buffer network* in the RS/6000 contains a rotating network which can rotate the four fetched instructions so as to present the four instructions, at its output, in original program order. This design of the I-cache is quite sophisticated and can ensure high fetch bandwidth even if the fetch group is misaligned with respect to the row organization of the I-cache. However, it is quite hardware intensive and was made feasible because the RS/6000 was implemented on multiple chips.

Other than the misalignment problem, the second impediment to sustaining the maximum fetch bandwidth of s instructions per cycle is the presence of control-flow changing instructions within the fetch group. If one of the instructions in the middle of the fetch group is a conditional branch, then the subsequent instructions in the fetch group will be discarded if the branch is taken. Consequently, when this happens, the fetch bandwidth is effectively reduced. This problem is fundamentally due to the presence of control dependences between instructions and is related to the handling of conditional branches. This topic, viewed as more related to the dynamic interaction between the machine and the program, is addressed in greater detail in Chapter 5, which covers techniques for dealing with control dependences and branch instructions.

4.3.2 Instruction Decoding

Instruction decoding involves the identification of the individual instructions, determination of the instruction types, and detection of inter-instruction dependences among the group of instructions that have been fetched but not yet dispatched. The complexity of the instruction decoding task is strongly influenced by two factors, namely, the ISA and the width of the parallel pipeline. For a typical RISC instruction set with fixed-length instructions and simple instruction formats, the decoding task is quite straightforward. No explicit effort is needed to determine the beginning and ending of each instruction. The relatively few different instruction formats and addressing modes make the distinguishing of instruction types reasonably easy. By simply decoding a small portion, e.g., one op code byte, of an instruction, the instruction type and the format used can be determined and the remaining fields of the instruction and their interpretation can be quickly determined. A RISC instruction set simplifies the instruction decoding task.

For a RISC scalar pipeline, instruction decoding is quite trivial. Frequently the decode stage is used for accessing the register operands and is merged with the register read stage. However, for a RISC parallel pipeline with multiple instructions being simultaneously decoded, the decode stage must identify dependences between these instructions and determine the independent instructions that can be dispatched in parallel. Furthermore, to support efficient instruction fetching, the decode stage must quickly identify control-flow changing branch instructions among the instructions being decoded in order to provide quick feedback to the fetch stage. These two tasks in conjunction with accessing many register operands can make the logic for the decode stage of a RISC parallel pipeline somewhat complex. A large number of comparators are needed for determining register dependences between instructions. The register files must be multiported and able to support many simultaneous accesses. Multiple busses are also needed to route the accessed operands to their appropriate destination buffers. It is possible that the decode stage can become the critical stage in the overall superscalar pipeline.

For a CISC parallel pipeline, the instruction decoding task can become even more complex and usually requires multiple pipeline stages. For such a parallel pipeline, the identification of individual instructions and their types is no longer trivial. Both the Intel Pentium and the AMD K5 employ two pipeline stages for decoding IA32 instructions. On the more deeply pipelined Intel Pentium Pro, a total of five machine cycles are required to access the I-cache and decode the IA32 instructions. The use of variable instruction lengths imposes an undesirable sequentiality to the instruction decoding task; the leading instruction must be decoded and have its length determined before the beginning of the next instruction can be identified. Consequently, the simultaneous parallel decoding of multiple instructions can become quite challenging. In the worst case, it must be assumed that a new instruction can begin anywhere within the fetch group, and a large number of decoders are used to simultaneously and "speculatively" decode instructions, starting at every byte boundary. This is extremely complex and can be quite inefficient.

There is an additional burden on the instruction decoder of a CISC parallel pipeline. The decoder must translate the architected instructions into internal low-level operations that can be directly executed by the hardware. Such a translation process was first described by Patt, Hwu, and Shebanow in their seminal paper on the high-performance substrate (HPS), which decomposed complex VAX CISC instructions into RISC-like primitives [Patt et al., 1985]. These internal operations resemble RISC instructions and can be viewed as vertical micro-instructions. In the AMD K5 these operations are called *RISC operations* or *ROPs* (pronounced "ar-ops"). In the Intel P6 these internal operations are identified as *micro-operations* or µ*ops* (pronounced "you-ops"). Each IA32 instruction is translated into one or more ROPs or µops. According to Intel, on average, one IA32 instruction is translated into 1.5 to 2.0 µops. In these CISC parallel pipelines, between the instruction decoding and instruction completion stages, all instructions in flight within the

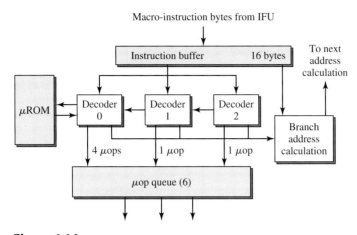

Figure 4.14
The Fetch/Decode Unit of the Intel P6 Superscalar Pipeline.

machine are these internal operations. In this book, for convenience we will adopt the Intel terminology and refer to these internal operations as μops.

The instruction decoder for the Intel Pentium Pro is presented as an illustrative example of instruction decoding for a CISC parallel pipeline. A diagram of the fetch/decode unit of the P6 is shown in Figure 4.14. In each machine cycle, the I-cache can deliver 16 aligned bytes to the instruction queue. Three parallel decoders simultaneously decode instruction bytes from the instruction queue. The first decoder at the front of the queue is capable of decoding all IA32 instructions, while the other two decoders have more limited capability and can only decode simple IA32 instructions such as register-to-register instructions.

The decoders translate IA32 instructions into the internal three-address μops. The μops employ the load/store model. Each IA32 instruction with complex addressing modes is translated into multiple μops. The first (generalized) decoder can generate up to four μops per cycle in response to the decoding of an IA32 instruction. Each of the other two (restricted) decoders can generate only one μop per cycle in response to the decoding of a simple IA32 instruction. In each machine cycle at least one IA32 instruction is decoded by the generalized decoder, leading to the generation of one or more μops. The goal is to go beyond this and have the other two restricted decoders also decode two simple IA32 instructions that trail the leading IA32 instruction in the same machine cycle. In the most ideal case the three parallel decoders can generate a total of six μops in one machine cycle. For those complex IA32 instructions that require more than four μops to translate, when they reach the front of the instruction queue, the generalized decoder will invoke a μops sequencer to emit microcode, which is simply a preprogrammed sequence of normal μops. These μops will require two or more machine cycles to generate. All the μops generated by the three parallel decoders are loaded into the reorder buffer (ROB), which has 40 entries to hold up to 40 μops, to await dispatching to the functional units.

For many superscalar processors, especially those that implement wide and/or CISC parallel pipelines, the instruction decoding hardware can be extremely complex and require partitioning into multiple pipeline stages. When the number of decoding stages is increased, the branch penalty, in terms of number of machine cycles, is also increased. Hence, it is not desirable to just keep increasing the depth of the decoding portion of the parallel pipeline. To help alleviate this complexity, a technique called *predecoding* has been proposed and implemented.

Predecoding moves a part of the decoding task to the other side, i.e., the input side, of the I-cache. When an I-cache miss occurs and a new cache line is being brought in from the memory, the instructions in that cache line are partially decoded by decoding hardware placed between the memory and the I-cache. The instructions and some additional decoded information are then stored in the I-cache. The decoded information, in the form of predecode bits, simplifies the instruction decoding task when the instructions are fetched from the I-cache. Hence, part of the decoding is performed only once when instructions are loaded into the I-cache, instead of every time when these instructions are fetched from the I-cache. With some of the decoding hardware having been moved to the input side of the I-cache, the instruction decoding complexity of the parallel pipeline can be simplified.

The AMD K5 is an example of a CISC superscalar pipeline that employs aggressive predecoding of IA32 instructions as they are fetched from memory and prior to their being loaded into the I-cache. In a single bus transaction a total of eight instruction bytes are fetched from memory. These bytes are predecoded, and five additional bits are generated by the predecoder for each of the instruction bytes. These five predecode bits contain information about the location of the start and end of an IA32 instruction, the number of μops (or ROPs) needed to translate that IA32 instruction, and the location of op codes and prefixes. These additional predecode bits are stored in the I-cache along with the original instruction's bytes. Consequently, the original I-cache line size of 128 bits (16 bytes) is increased by an additional 80 bits; see Figure 4.15. In each I-cache access, the 16 instruction bytes are fetched along with the 80 predecode bits. The predecode bits significantly simplify instruction decoding and allow the simultaneous decoding of multiple IA32 instructions by four identical decoders/translators that can generate up to four μops in each cycle.

There are two forms of overhead associated with predecoding. The I-cache miss penalty can be increased due to the necessity of predecoding the instruction bytes fetched from memory. This is not a serious problem if the I-cache miss rate is very low. The other overhead involves the storing of the predecode bits in the I-cache and the consequent increase of the I-cache size. For the K5 the size of the I-cache is increased by about 50%. There is clearly a tradeoff between the aggressiveness of predecoding and the I-cache size increase.

Predecoding is not just limited to alleviating the sequential bottleneck in parallel decoding of multiple CISC instructions in a CISC parallel pipeline. It can also be used to support RISC parallel pipelines. RISC instructions can be predecoded when they are being loaded into the I-cache. The predecode bits can be used to identify control-flow changing branch instructions within the fetch group and to explicitly

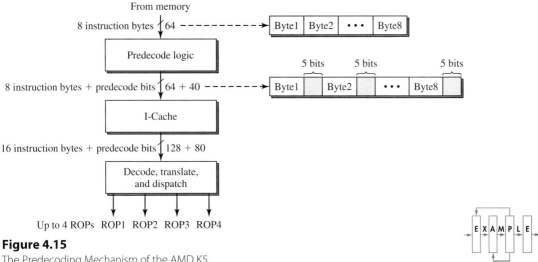

Figure 4.15
The Predecoding Mechanism of the AMD K5.

identify subgroups of independent instructions within the fetch group. For example, the PowerPC 620 employs 7 predecode bits for each instruction word in the I-cache. The UltraSPARC, MIPS R10000, and HP PA-8000 also employ either 4 or 5 predecode bits for each instruction.

As superscalar pipelines become wider and the number of instructions that must be simultaneously decoded increases, the instruction decoding task will become more of a bottleneck and more aggressive use of predecoding can be expected. The predecoder partially decodes the instructions, and effectively transforms the original undecoded instructions into a format that makes the final decoding task easier. One can view the predecoder as translating the instructions fetched from memory into different instructions that are then loaded into the I-cache. To expand this view, the possibility of enhancing the predecoder to do run-time object code translation between ISAs could be interesting.

4.3.3 Instruction Dispatching

Instruction dispatching is necessary for superscalar pipelines. In a scalar pipeline, all instructions regardless of their type flow through the same single pipeline. Superscalar pipelines are diversified pipelines that employ a multiplicity of heterogeneous functional units in their execution portion. Different types of instructions are executed by different functional units. Once the type of an instruction is identified in the decode stage, it must be routed to the appropriate functional unit for execution; this is the task of instruction dispatching.

Although superscalar pipelines are parallel pipelines, both the instruction fetching and instruction decoding tasks are usually carried out in a centralized fashion; i.e., all the instructions are managed by the same controller. Although multiple instructions are fetched in a cycle, all instructions must be fetched from the same I-cache. Hence all the instructions in the fetch group are accessed from the

I-cache at the same time, and they are all deposited into the same buffer. Instruction decoding is done in a centralized fashion because in the case of CISC instructions, all the bytes in the fetch group must be decoded collectively by a centralized decoder in order to identify the individual instructions. Even with RISC instructions, the decoder must identify inter-instruction dependences, which also requires centralized instruction decoding.

On the other hand, in a diversified pipeline all the functional units can operate independently in a distributed fashion in executing their own types of instructions once the inter-instruction dependences are resolved. Consequently, going from instruction decoding to instruction execution, there is a change from centralized processing of instructions to distributed processing of instructions. This change is carried out by, and is the reason for, the instruction dispatching stage in a superscalar pipeline. This is illustrated in Figure 4.16.

Another mechanism that is necessary between instruction decoding and instruction execution is the temporary buffering of instructions. Prior to its execution, an instruction must have all its operands. During decoding, register operands are fetched from the register files. In a superscalar pipeline it is possible that some of these operands are not yet ready because earlier instructions that update these registers have not finished their execution. When this situation occurs, an obvious solution is to stall the decoding stage until all register operands are ready. This solution seriously restricts the decoding throughput and is not desirable. A better solution is to fetch those register operands that are ready and go ahead and advance these

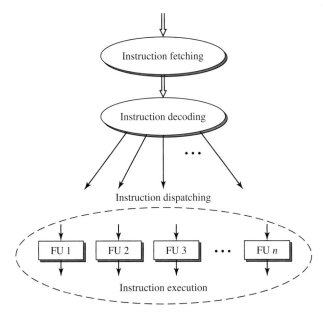

Figure 4.16
The Necessity of Instruction Dispatching in a Superscalar Pipeline.

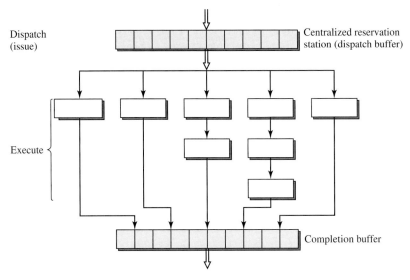

Figure 4.17
Centralized Reservation Station.

instructions into a separate buffer to await those register operands that are not ready. When all register operands are ready, those instructions can then exit this buffer and be issued into the functional units for execution. Borrowing the term used in the Tomasulo's algorithm employed in the IBM 360/91 [Tomasulo, 1967], we denote such a temporary instruction buffer as a *reservation station*. The use of a reservation station decouples instruction decoding and instruction execution and provides a buffer to take up the slack between decoding and execution stages due to the temporal variation of throughput rates in the two stages. This eliminates unnecessary stalling of the decoding stage and prevents unnecessary starvation of the execution stage.

Based on the placement of the reservation station relative to instruction dispatching, two types of reservation station implementations are possible. If a single buffer is used at the source side of dispatching, we identify this as a *centralized reservation station*. On the other hand, if multiple buffers are placed at the destination side of dispatching, they are identified as *distributed reservation stations*. Figures 4.17 and 4.18 illustrate the two ways of implementing reservation stations.

The Intel Pentium Pro implements a centralized reservation station. In such an implementation, one reservation station with many entries feeds all the functional units. Instructions are dispatched from this centralized reservation station directly to all the functional units to begin execution. On the other hand, the PowerPC 620 employs distributed reservation stations. In this implementation, each functional unit has its own reservation station on the input side of the unit. Instructions are dispatched to the individual reservation stations based on instruction type. These instructions remain in these reservation stations until they are ready to be issued into the functional units for execution. Of course, these two implementations of

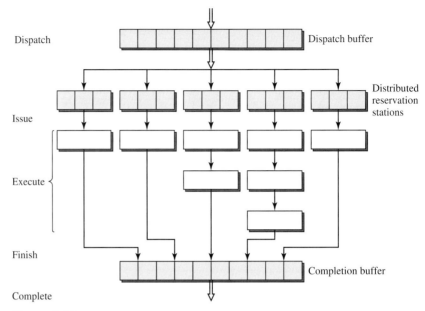

Figure 4.18
Distributed Reservation Stations.

reservation stations represent only the two extreme alternatives. Hybrids of these two approaches are also possible. For example, the MIPS R10000 employs one such hybrid implementation. We identified such hybrid implementations as *clustered reservation stations*. With clustered reservation stations, instructions are dispatched to multiple reservation stations, and each reservation station can feed or be shared by more than one functional unit. Typically the reservation stations and functional units are clustered based on instruction or data types.

Reservation station design involves certain tradeoffs. A centralized reservation station allows all instruction types to share the same reservation station and will likely achieve the best overall utilization of all the reservation station entries. However, a centralized implementation can incur the greatest complexity in its hardware design. It requires centralized control and a buffer that is highly multiported to allow multiple concurrent accesses. Distributed reservation stations can be single-ported buffers, each with only a small number of entries. However, each reservation station's idling entries cannot be used by instructions destined for execution in other functional units. The overall utilization of all the reservation station entries will be lower. It is also likely that one reservation station can saturate when all its entries are occupied and hence induce stalls in instruction dispatching.

With the different alternatives for implementing reservation stations, we need to clarify our use of certain terms. In this book the term *dispatching* implies the associating of instruction types with functional unit types after instructions have been decoded. On the other hand, the term *issuing* always means the initiation of execution in functional units. In a distributed reservation station design, these two events occur separately. Instructions are *dispatched* from the centralized decode/dispatch buffer to the individual reservation stations first, and when all their operands are available, then they are *issued* into the individual functional units for execution. With a centralized reservation station, the dispatching of instructions from the centralized reservation station does not occur until all their operands are ready. All instructions, regardless of type, are held in the centralized reservation station until they are ready to execute, at which time instructions are *dispatched* directly into the individual functional units to begin execution. Hence, in a machine with a centralized reservation station, the associating of instructions to individual functional units occurs at the same time as their execution is initiated. Therefore, with a centralized reservation station, instruction *dispatching* and instruction *issuing* occur at the same time, and these two terms become interchangeable. This is illustrated in Figure 4.17.

4.3.4 Instruction Execution

The instruction execution stage is the heart of a superscalar machine. The current trend in superscalar pipeline design is toward more parallel and more diversified pipelines. This translates into having more functional units and having these functional units be more specialized. By specializing them for executing specific instruction types, these functional units can be more performance efficient. Early scalar pipelined processors have essentially one functional unit. All instruction types (excluding floating-point instructions that are executed by a separate floating-point coprocessor chip) are executed by the same functional unit. In the TYP pipeline example, this functional unit is a two-stage pipelined unit consisting of the ALU and MEM stages of the TYP pipeline. Most first-generation superscalar processors are parallel pipelines with two diversified functional units, one executing integer instructions and the other executing floating-point instructions. These early superscalar processors simply integrated floating-point execution in the same instruction pipeline instead of employing a separate coprocessor unit.

Current superscalar processors can employ multiple integer units, and some have multiple floating-point units. These are the two most fundamental functional unit types. Some of these units are becoming quite sophisticated and are capable of executing more than one operation involving more than two source operands in each cycle. Figure 4.19(a) illustrates the integer execution unit of the TI SuperSPARC which contains a cascaded ALU configuration [Blanck and Krueger, 1992]. Three ALUs are included in this two-stage pipelined unit, and up to two integer operations can be issued into this unit in one cycle. If they are independent, then both operations are executed in the first stage using ALU0 and ALU2. If the second operation depends on the first, then the first one is executed in ALU2 during the first stage with the second one executed in ALUC in the second stage. Implementing such a functional unit allows more cycles in which two instructions are simultaneously issued.

The floating-point unit in the IBM RS/6000 is implemented as a two-stage pipelined multiply-add-fused (MAF) unit that takes three inputs (A, B, C) and performs $(A \times B) + C$. This is illustrated in Figure 4.19(b). The MAF unit is motivated by the most common use of floating-point multiplication to carry out the dot-product operation $D = (A \times B) + C$. If the compiler is able to merge many multiply-add

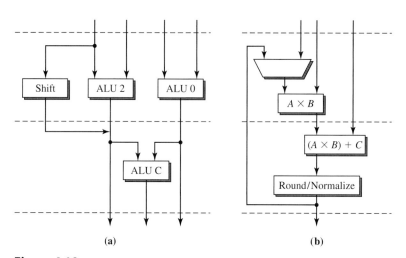

Figure 4.19
(a) Integer Functional Unit in the TI SuperSPARC; (b) Floating-Point Unit in the IBM RS/6000.

pairs of instructions into single MAF instructions, and the MAF unit can sustain the issuing of one MAF instruction in every cycle, then an effective throughput of two floating-point instructions per cycle can be achieved using only one MAF unit. The normal floating-point multiply instruction is actually executed by the MAF unit as $(A \times B) + 0$, while the floating-point add instruction is performed by the MAF unit as $(A \times 1) + C$. Since the MAF unit is pipelined, even without executing MAF instructions, it can still sustain an execution rate of one floating-point instruction per cycle.

In addition to executing integer ALU instructions, an integer unit can be used for generating memory addresses and executing branch and load/store instructions. However, in most recent designs separate branch and load/store units have been incorporated. The branch unit is responsible for updating the PC, while the load/store unit is directly connected to the D-cache. Other specialized functional units have emerged for supporting graphics and image processing applications. For example, in the Motorola 88110 there is a dedicated functional unit for bit manipulation and two functional units for supporting pixel processing. For many of the signal processing and multimedia applications, the common data type is a byte. Frequently 4 bytes are packed into a 32-bit word for simultaneous processing by specialized 32-bit functional units for increased throughput. In the TriMedia VLIW processor intended for such applications, such functional units are employed [Slavenburg et al., 1996]. For example, the TriMedia-1 processor can execute the quadavg instruction in one cycle. The quadavg instruction sums four rounded averages and is quite useful in MPEG decoding for decompressing compressed video images; it carries out the following computation.

$$\text{quadavg} = \frac{(a+e+1)}{2} + \frac{(b+f+1)}{2} + \frac{(c+g+1)}{2} + \frac{(d+h+1)}{2} \quad (4.2)$$

The eight variables denote 8-byte operands with *a*, *b*, *c*, and *d* stored as one 32-bit quantity and *e*, *f*, *g*, and *h* stored as another 32-bit quantity. The functional unit takes as input these two 32-bit operands and produces the quadavg result in one cycle. This single-cycle operation replaces numerous add and divide instructions that would have been required if the eight single-byte operands were manipulated individually. With the widespread deployment of multimedia applications, such specialized functional units that operate on special data types have emerged.

What is the best mix of functional units for a superscalar pipeline is an interesting question. The answer is dependent on the application domain. If we use the statistics from Chapter 2 of typical programs having 40% ALU instructions, 20% branches, and 40% load/store instructions, then we can have a 4-2-4 rule of thumb. For every four ALU units, we should have two branch units and four load/store units. Many of the current leading superscalar processors have four or more ALU-type functional units (including both integer and floating-point units). Most of them have only one branch unit but are able to speculate beyond one conditional branch instruction. However, most of these processors have only one load/store unit; some are able to process two load/store instructions in every cycle with some constraints. Clearly there seems be an imbalance in having too few load/store units. The reason is that implementing multiple load/store units that operate in parallel in accessing the same D-cache is a difficult task. It requires the D-cache to be multiported. Multiported memory modules involve very complex circuit design and can significantly slow down the memory speed.

In many designs multiple memory banks are used to simulate a truly multiported memory. A memory is partitioned into multiple banks. Each bank can perform a read/write operation in a machine cycle. If the effective addresses of two load/store instructions happen to reside on different banks, then both instructions can be carried out by the two different banks at the same time. However, if there is a bank conflict, then the two instructions must be serialized. Multibanked D-caches have been used to simulate multiported D-caches. For example, the Intel Pentium processor uses an eight-banked D-cache to simulate a two-ported D-cache [Intel Corp., 1993]. A truly multiported memory can guarantee conflict-free simultaneous accesses. Typically, more read ports than write ports are needed. Multiple read ports can be implemented by having multiple copies of the memory. All memory writes are broadcast to all the copies, with all the copies having identical content. Each copy can provide a small number of read ports with the total number of read ports being the sum of all the read ports on all the copies. For example, a memory with four read ports and two write ports can be implemented as two copies of simpler memory modules, each with only two write ports and two read ports. Implementing multiple, especially more than two, load/store units to operate in parallel can be a challenge in designing wide superscalar pipelines.

The amount of resource parallelism in the instruction execution portion is determined by the combination of *spatial* and *temporal* parallelism. Having multiple functional units is a form of spatial parallelism. Alternatively, parallelism can be obtained via pipelining of these functional units, which is a form of temporal parallelism. For example, instead of implementing a dual-ported D-cache, in some current designs D-cache access is pipelined into two pipeline stages so that two load/store instructions can be concurrently serviced by the D-cache. Currently, there is a general trend toward implementing deeper pipelines in order to reduce the cycle time and increase the clock speed. Spatial parallelism also tends to require greater hardware complexity and silicon real estate. Temporal parallelism makes more efficient use of hardware but does increase the overall instruction processing latency and potentially pipeline stall penalties due to inter-instruction dependences.

In real superscalar pipeline designs, we often see that the total number of functional units exceeds the actual width of the parallel pipeline. Typically the width of a superscalar pipeline is determined by the number of instructions that can be fetched, decoded, or completed in every machine cycle. However, because of the dynamic variation of instruction mix and the resultant nonuniform distribution of instruction mix during program execution on a cycle-by-cycle basis, there is a potential dynamic mismatch of instruction mix and functional unit mix. The former varies in time and the latter stays fixed. Because of the specialization and heterogeneity of the functional units the total number of functional units must exceed the width of the superscalar pipeline to avoid having the instruction execution portion become the bottleneck due to excessive structural dependences related to the unavailability of certain functional unit types. Some of the aggressive compiler back ends actually try to smooth out this dynamic variation of instruction mix to ensure a better sustained match with the functional unit mix. Of course, different application programs can exhibit a different inherent overall mix of instruction types. The compiler can only make localized adjustments to achieve some performance gain. Studies have been done in assessing the best number and mix of functional units based on SPEC benchmarks [Jourdan et al., 1995].

With a large number of functional units, there is additional hardware complexity other than the functional units themselves. Results from the outputs of functional units need to be forwarded to inputs of the functional units. A multiplicity of busses are required, and potentially logic for bus control and arbitration is needed. Usually a full crossbar interconnection network is too costly and not absolutely necessary. The mechanism for routing operands between functional units introduces another form of structural dependence. The interconnect mechanism also contributes to the latency of the execution stage(s) of the pipeline. In order to support data forwarding the reservation station(s) must monitor the busses for tag matches, indicating the availability of needed operands, and latch in the operands when they are broadcasted on the busses. Potentially the complexity of the instruction execution stage can grow at the rate of n^2, where n is the total number of functional units.

4.3.5 Instruction Completion and Retiring

An instruction is considered *completed* when it finishes execution and updates the machine state. An instruction finishes execution when it exits the functional unit and enters the completion buffer. Subsequently it exits the completion buffer and becomes completed. When an instruction finishes execution, its result may only reside in nonarchitected buffers. However, when it is completed, its result is written into an architecture register. With instructions that actually update memory locations, there can be a time period between when they are architecturally completed and when the memory locations are updated. For example, a store instruction can be architecturally completed when it exits the completion buffer and enters the store buffer to wait for the availability of a bus cycle in order to write to the D-cache. This store instruction is considered *retired* when it exits the store buffer and updates the D-cache. Hence, in this book instruction *completion* involves the updating of the machine state, whereas instruction *retiring* involves the updating of the memory state. For instructions that do not update the memory, retiring occurs at the same time as completion. So, in a distributed reservation station machine, an instruction can go through the following phases: *fetch, decode, dispatch, issue, execute, finish, complete,* and *retire*. Issuing and finishing simply refer to starting execution and ending execution, respectively. Some of the superscalar processor vendors use these terms in slightly different ways. Frequently, dispatching and issuing are used almost interchangeably, similar to completion and retiring. Sometimes completion is used to mean finishing execution, and retiring is used to mean updating the machine's architectural state. There is yet no standardization on the use of these terms.

During the execution of a program, interrupts and exceptions can occur that will disrupt the execution flow of a program. Superscalar processors employing dynamic pipelines that facilitate out-of-order execution must be able to deal with such disruptions of program execution. *Interrupts* are usually induced by the external environment such as I/O devices or the operating system. These occur in an asynchronous fashion with respect to the program execution. When an interrupt occurs, the program execution must be suspended to allow the operating system to service the interrupt. One way to do this is to stop fetching new instructions and allow the instructions that are already in the pipeline to finish execution, at which time the state of the machine can be saved. Once the interrupt has been serviced by the operating system, the saved machine state can be restored and the original program can resume execution.

Exceptions are induced by the execution of the instructions of the program. An instruction can induce an exception due to arithmetic operations, such as dividing by zero and floating-point overflow or underflow. When such exceptions occur, the results of the computation may no longer be valid and the operating system may need to intervene to log such exceptions. Exceptions can also occur due to the occurrence of page faults in a paging-based virtual memory system. Such exceptions can occur when instructions reference the memory. When such exceptions occur, a new page must be brought in from secondary storage, which can require on the order of thousands of machine cycles. Consequently, the execution of the program that induced the page fault is usually suspended, and the execution of a new program is initiated in the multiprogramming environment. After the page fault has been serviced, the original program can then resume execution.

It is important that the architectural state of the machine present at the time the excepting instruction is executed be saved so that the program can resume execution after the exception is serviced. Machines that are capable of supporting this suspension and resumption of execution of a program at the granularity of each individual instruction are said to have *precise exception*. Precise exception involves being able to checkpoint the state of the machine just prior to the execution of the excepting instruction and then resume execution by restoring the checkpointed state and restarting execution at the excepting instruction. In order to support precise exception, the superscalar processor must maintain its architectural state and evolve this machine state in such a way as if the instructions in the program are executed one at a time according to the original program order. The reason is that when an exception occurs, the state the machine is in at that time must reflect the condition that all instructions preceding the excepting instruction have completed while no instructions following the excepting instruction have completed. For a dynamic pipeline to have precise exception, this sequential evolving of the architectural state must be maintained even though instructions are actually executed out of program order.

In a dynamic pipeline, instructions are fetched and decoded in program order but are executed out of program order. Essentially, instructions can enter the reservation station(s) in order but exit the reservation station(s) out of order. They also finish execution out of order. To support precise exception, instruction completion must occur in program order so as to update the architectural state of the machine in program order. In order to accommodate out-of-order finishing of execution and in-order completion of instructions, a *reorder buffer* is needed in the instruction completion stage of the parallel pipeline. As instructions finish execution, they enter the reorder buffer out of order, but they exit the reorder buffer in program order. As they exit the reorder buffer, they are considered architecturally completed. This is illustrated in Figure 4.20 with the reservation station and the reorder buffer bounding the out-of-order region of the pipeline or essentially the instruction execution portion of the pipeline. The terms adopted in this book, referring to the various phases of instruction processing, are illustrated in Figure 4.20.

Precise exception is handled by the instruction completion stage using the reorder buffer. When an exception occurs, the excepting instruction is tagged in the reorder buffer. The completion stage checks each instruction before that instruction is completed. When a tagged instruction is detected, it is not allowed to be completed. All the instructions prior to the tagged instructions are allowed to be completed. The machine state is then checkpointed or saved. The machine state includes all the architected registers and the program counter. The remaining instructions in the pipeline, some of which may have already finished execution, are discarded. After the exception has been serviced, the checkpointed machine state is restored and execution resumes with the fetching of the instruction that triggered the original exception.

Early work on support for providing precise exceptions in a processor that supports out-of-order execution was conducted by Acosta et al. [1986], Sohi and Vajapeyam [1987], and Smith and Pleszkun [1988]. An early proposal describing the Metaflow processor, which was never completed, also provides interesting insights for the curious reader [Popescu et al., 1991].

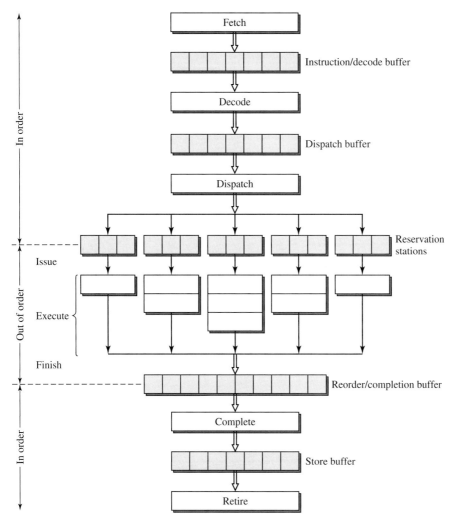

Figure 4.20
A Dynamic Pipeline with Reservation Station and Reorder Buffer.

4.4 Summary

Figure 4.20 represents an archetype of a contemporary out-of-order superscalar pipeline. It has an in-order front end, an out-of-order execution core, and an in-order back end. Both the front-end and back-end pipeline stages can advance multiple instructions per machine cycle. Instructions can remain in the reservation stations for one or more cycles while waiting for their source operands. Once the source operands are available, an instruction is issued from the reservation station into the execution unit. After execution, the instruction enters the reorder buffer (or completion buffer). Instructions in the reorder buffer are completed according to program order. In fact the reorder buffer is managed as a circular queue with instructions arranged according to the program order.

This chapter focuses on the superscalar pipeline organization and highlights the issues associated with the various pipeline stages. So far, we have addressed mostly the static structures of superscalar pipelines. Chapter 5 will get into the dynamic behavior of superscalar processors. We have chosen to present superscalar pipeline organization at a fairly high level, avoiding the implementation details. The main purpose of this chapter is to provide a bridge from scalar to superscalar pipelines and to convey a high-level framework for superscalar pipelines that will be a useful navigational aid when we get into the plethora of superscalar processor techniques in Chapter 5.

REFERENCES

Acosta, R., J. Kilestrup, and H. Torng: "An instruction issuing approach to enhancing performance in multiple functional unit processors," *IEEE Trans. on Computers,* C35, 9, 1986, pp. 815–828.

Blanck, G., and S. Krueger: "The SuperSPARC microprocessor," *Proc. IEEE COMPCON,* 1992, pp. 136–141.

Crawford, J.: "The execution pipeline of the Intel i486 CPU," *Proc. COMPCON Spring '90,* 1990, pp. 254–258.

Diefendorf, K., and M. Allen: "Organization of the Motorola 88110 superscalar RISC microprocessor," *IEEE MICRO,* 12, 2, 1992, pp. 40–63.

Grohoski, G.: "Machine organization of the IBM RISC System/6000 processor," *IBM Journal of Research and Development,* 34, 1, 1990, pp. 37–58.

Intel Corp.: *Pentium Processor User's Manual,* Vol. 3: *Architecture and Programming Manual.* Santa Clara, CA: Intel Corp., 1993.

Jourdan, S., P. Sainrat, and D. Litaize: "Exploring configurations of functional units in an out-of-order superscalar processor," *Proc. 22nd Annual Int. Symposium on Computer Architecture,* 1995, pp. 117–125.

Oehler, R. R., and R. D. Groves: "IBM RISC System/6000 processor architecture," *IBM Journal of Research and Development,* 34, 1, 1990, pp. 23–36.

Patt, Y., W. Hwu, and M. Shebanow: "HPS, a new microarchitecture: Introduction and rationale," *Proc. 18th Annual Workshop on Microprogramming (MICRO-18),* 1985, pp. 103–108.

Popescu, V., M. Schulz, J. Spracklen, G. Gibson, B. Lightner, and D. Isaman: "The Metaflow architecture," *IEEE Micro.,* June 1991, pp. 10–13, 63–73.

Slavenburg, G., S. Rathnam, and H. Dijkstra: "The TriMedia TM-1 PCI VLIW media processor," *Proc. Hot Chips 8,* 1996, pp. 171–178.

Smith, J., and A. Pleszkun: "Implementing precise interrupts in pipelined processors," *IEEE Trans. on Computers,* 37, 5, 1988, pp. 562–573.

Sohi, G., and S. Vajapeyam: "Instruction issue logic for high-performance, interruptible pipelined processors," *Proc. 14th Annual Int. Symposium on Computer Architecture,* 1987, pp. 27–34.

Thornton, J. E.: "Parallel operation in the Control Data 6600," *AFIPS Proc. FJCC* part 2, vol. 26, 1964, pp. 33–40.

Tomasulo, R.: "An efficient algorithm for exploiting multiple arithmetic units," *IBM Journal of Research and Development,* 11, 1967, pp. 25–33.

HOMEWORK PROBLEMS

P4.1 Is it reasonable to build a scalar pipeline that supports out-of-order execution? If so, describe a code execution scenario where such a pipeline would perform better than a conventional in-order scalar pipeline.

P4.2 Superscalar pipelines require replication of pipeline resources across each parallel pipeline, naively including the replication of cache ports. In practice, however, a two-wide superscalar pipeline may have two data cache ports but only a single instruction cache port. Explain why this is possible, but also discuss why a single instruction cache port can perform worse than two (replicated) instruction cache ports.

P4.3 Section 4.3.1 suggests that a compiler can generate object code where branch targets are aligned at the beginning of physical cache lines to increase the likelihood of fetching multiple instructions from the branch target in a single cycle. However, given a fixed number of instructions between taken branches, this approach may simply shift the unused fetch slots from before the branch target to after the branch that terminates sequential fetch at the target. For example, moving the code at `label0` so it aligns with a physical cache line will not improve fetch efficiency, since the wasted fetch slot shifts from the beginning of the physical line to the end.

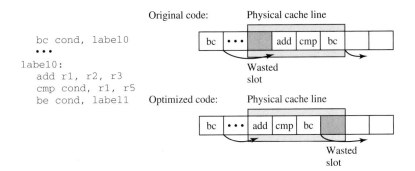

Discuss the relationship between fetch block size and the dynamic distance between taken branches. Describe how one affects the other, describe how important is branch target alignment for small vs. large fetch blocks and short vs. long dynamic distance, and describe how well static compiler-based target alignment might work in all cases.

P4.4 The auto-realigning instruction fetch hardware shown in Figure 4.13 still fails to achieve full-width fetch bandwidth (i.e., four instructions per cycle). Describe a more aggressive organization that is always able to fetch four instructions per cycle. Comment on the additional hardware such an organization implies.

P4.5 One idea to eliminate the branch misprediction penalty is to build a machine that executes both paths of a branch. In a two to three paragraph essay, explain why this may or may not be a good idea.

P4.6 Section 4.3.2 discusses adding predecode bits to the instruction cache to simplify the task of decoding instructions after they have been fetched. A logical extension of predecode bits is to simply store the instructions in decoded form in a decoded instruction cache; this is particularly attractive for processors like the Pentium Pro that dynamically translate fetched instructions into a sequence of simpler RISC-like instructions for the core to execute. Identify and describe at least one factor that complicates the building of decoded instruction caches for processors that translate from a complex instruction set to a simpler RISC-like instruction set.

P4.7 What is the most important advantage of a centralized reservation station over distributed reservation stations?

P4.8 In an in-order pipelined processor, pipeline latches are used to hold result operands from the time an execution unit computes them until they are written back to the register file during the writeback stage. In an out-of-order processor, rename registers are used for the same purpose. Given a four-wide out-of-order processor TYP pipeline, compute the minimum number of rename registers needed to prevent rename register starvation from limiting concurrency. What happens to this number if frequency demands force a designer to add five extra pipeline stages between dispatch and execute, and five more stages between execute and retire/writeback?

P4.9 A banked or interleaved cache can be an effective approach for allowing multiple loads or stores to be performed in one cycle. Sketch out the data flow for a two-way interleaved data cache attached to two load/store units. Now sketch the data flow for an eight-way interleaved data cache attached to four load/store units. Comment on how well interleaving scales or does not scale.

P4.10 The Pentium 4 processor operates its integer arithmetic units at double the nominal clock frequency of the rest of the processor. This is accomplished by pipelining the integer adder into two stages, computing the low-order 16 bits in the first cycle and the high-order 16 bits in the second cycle. Naively, this appears to increase ALU latency from one cycle to two cycles. However, assuming that two dependent instructions are both arithmetic instructions, it is possible to issue the second instruction in the cycle immediately following issue of the first instruction, since the low-order bits of the second instruction are dependent only on the low-order bits of the first instruction. Sketch out a pipeline diagram of such an ALU along with the additional bypass paths needed to handle this optimized case.

P4.11 Given the ALU configuration described in Problem 4.10, specify how many cycles a trailing dependent instruction of each of the following types must delay, following the issue of a leading arithmetic instruction: arithmetic, logical (and/or/xor), shift left, shift right.

P4.12 Explain why the kind of bit-slice pipelining described in Problem 4.10 cannot be usefully employed to pipeline dependent floating-point arithmetic instructions.

P4.13 Assume a four-wide superscalar processor that attempts to retire four instructions per cycle from the reorder buffer. Explain which data dependences need to be checked at this time, and sketch the dependence-checking hardware.

P4.14 Four-wide superscalar processors rarely sustain throughput much greater than one instruction per cycle (IPC). Despite this fact, explain why four-wide retirement is still useful in such a processor.

P4.15 Most general-purpose instruction sets have recently added multimedia extensions to support vector-like operations over arrays of small data types. For example, Intel IA32 has added the MMX and SSE instruction set extensions for this purpose. A single multimedia instruction will load, say, eight 8-bit operands into a 64-bit register in parallel, while arithmetic instructions will perform the same operation on all eight operands in single-instruction, multiple data (SIMD) fashion. Describe the changes you would have to make to the fetch, decode, dispatch, issue, execute, and retire logic of a typical superscalar processor to accommodate these instructions.

P4.16 The PowerPC instruction set provides support for a fused floating-point multiply-add operation that multiplies two of its input registers and adds the product to the third input register. Explain how the addition of such an instruction complicates the decode, dispatch, issue, and execute stages of a typical superscalar processor. What effect do you think these changes will have on the processor's cycle time?

P4.17 The semantics of the fused multiply-add instruction described in Problem 4.16 can be mimicked by issuing a separate floating-point add and floating-point multiply whenever such an instruction is decoded. In fact, the MIPS R10000 does just that; rather than supporting this instruction (which also exists in the MIPS instruction set) directly, the decoder simply inserts the add/multiply instruction pair into the execution window. Identify and discuss at least two reasons why this approach could reduce performance as measured in instructions per cycle.

P4.18 Does the ALU mix for the Motorola 88110 processor shown in Figure 4.7 agree with the IBM instruction mix provided in Section 2.2.4.3? If not, how would you change the ALU mix?

Terms and Buzzwords

These problems are similar to the "Jeopardy Game" on TV. The "answers" are given and you are to provide the best correct "questions." For each "answer" there may be more than one appropriate "question"; you need to provide the best one.

P4.19 A: A mechanism that tracks out-of-order execution and maintains speculative machine state.

Q: What is _____ ?

P4.20 A: It will significantly reduce the machine cycle time, but can increase the branch penalty.

Q: What is _____ ?

P4.21 A: Additional I-cache bits generated at cache refill time to ease the decoding/dispatching task.

Q: What are _____ ?

P4.22 A: A program attribute that causes inefficiencies in a superscalar fetch unit.

Q: What is _____ ?

P4.23 A: The internal RISC-like instruction executed by the Pentium Pro (P6) microarchitecture.

Q: What is _____ ?

P4.24 A: The logical pipeline stage that assigns an instruction to the appropriate execution unit.

Q: What is _____ ?

P4.25 A: An early processor design that incorporated 10 diverse functional units.

Q: What is _____ ?

P4.26 A: A new instruction that allows a scalar pipeline to achieve more than one floating-point operation per cycle.

Q: What is _____ ?

P4.27 A: An effective technique for allowing more than one memory operation to be performed per cycle.

   Q: What is _____ ?

P4.28 A: A useful architectural property that simplifies the task of writing low-level operating system code.

   Q: What is _____ ?

P4.29 A: The first research paper to describe run-time, hardware translation of one instruction set to another, simpler one.

   Q: What was _____ ?

P4.30 A: The first real processor to implement run-time, hardware translation of one instruction set to another, simpler one.

   Q: What was _____ ?

P4.31 A: This attribute of most RISC instruction sets substantially simplifies the task of decoding multiple instructions in parallel.

   Q: What was _____ ?

CHAPTER 5

Superscalar Techniques

CHAPTER OUTLINE

5.1 Instruction Flow Techniques
5.2 Register Data Flow Techniques
5.3 Memory Data Flow Techniques
5.4 Summary

References
Homework Problems

In Chapter 4 we focused on the structural, or organizational, design of the superscalar pipeline and dealt with issues that were somewhat independent of the specific types of instructions being processed. In this chapter we focus more on the dynamic behavior of a superscalar processor and consider techniques that deal with specific types of instructions. The ultimate performance goal of a superscalar pipeline is to achieve maximum throughput of instruction processing. It is convenient to view instruction processing as involving three component flows of instructions and/or data, namely, *instruction flow, register data flow,* and *memory data flow*. This partitioning into three flow paths is similar to that used in Mike Johnson's 1991 textbook entitled *Superscalar Microprocessor Design* [Johnson, 1991]. The overall performance objective is to maximize the volumes in all three of these flow paths. Of course, what makes this task interesting is that the three flow paths are not independent and their interactions are quite complex. This chapter classifies and presents superscalar microarchitecture techniques based on their association with the three flow paths.

The three flow paths correspond roughly to the processing of the three major types of instructions, namely, branch, ALU, and load/store instructions. Consequently, maximizing the throughput of the three flow paths corresponds to the minimizing of the branch, ALU, and load penalties.

1. *Instruction flow*. Branch instruction processing.

2. *Register data flow*. ALU instruction processing.

3. *Memory data flow*. Load/store instruction processing.

This chapter uses these three flow paths as a convenient framework for presenting the plethora of microarchitecture techniques for optimizing the performance of modern superscalar processors.

5.1 Instruction Flow Techniques

We present instruction flow techniques first because these deal with the early stages, e.g., the fetch and decode stages, of a superscalar pipeline. The throughput of the early pipeline stages will impose an upper bound on the throughput of all subsequent stages. For contemporary pipelined processors, the traditional partitioning of a processor into control path and data path is no longer clear or effective. Nevertheless, the early pipeline stages along with the branch execution unit can be viewed as corresponding to the traditional control path whose primary function is to enforce the control flow semantics of a program. The primary goal for all instruction flow techniques is to maximize the supply of instructions to the superscalar pipeline subject to the requirements of the control flow semantics of a program.

5.1.1 Program Control Flow and Control Dependences

The control flow semantics of a program are specified in the form of the control flow graph (CFG), in which the nodes represent basic blocks and the edges represent the transfer of control flow between basic blocks. Figure 5.1(a) illustrates a CFG with four basic blocks (dashed-line rectangles), each containing a number of instructions (ovals). The directed edges represent control flows between basic blocks. These edges are induced by conditional branch instructions (diamonds). The run-time execution of a program entails the dynamic traversal of the nodes and edges of its CFG. The actual path of traversal is dictated by the branch instructions and their branch conditions which can be dependent on run-time data.

The basic blocks, and their constituent instructions, of a CFG must be stored in sequential locations in the program memory. Hence the partial ordered basic blocks in a CFG must be arranged in a total order in the program memory. In mapping a CFG to linear consecutive memory locations, additional unconditional branch instructions must be added, as illustrated in Figure 5.1(b). The mapping of the CFG to a linear program memory facilitates an implied sequential flow of control along the sequential memory locations during program execution. However, the encounter of both conditional and unconditional branches at run time induces deviations from this implied sequential control flow and the consequent disruptions to the sequential fetching of instructions. Such disruptions cause stalls in the instruction fetch stage of the pipeline and reduce the overall instruction fetching bandwidth. Subroutine jump and return instructions also induce similar disruptions to the sequential fetching of instructions.

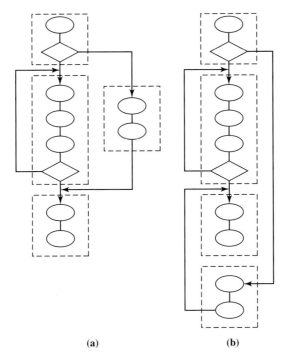

Figure 5.1
Program Control Flow: (a) The Control Flow Graph (CFG); (b) Mapping the CFG to Sequential Memory Locations.

5.1.2 Performance Degradation Due to Branches

A pipelined machine achieves its maximum throughput when it is in the streaming mode. For the fetch stage, streaming mode implies the continuous fetching of instructions from sequential locations in the program memory. Whenever the control flow of the program deviates from the sequential path, potential disruption to the streaming mode can occur. For unconditional branches, subsequent instructions cannot be fetched until the target address of the branch is determined. For conditional branches, the machine must wait for the resolution of the branch condition, and if the branch is to be taken, it must further wait until the target address is available. Figure 5.2 illustrates the disruption of the streaming mode by branch instructions. Branch instructions are executed by the branch functional unit. For a conditional branch, it is not until it exits the branch unit and when both the branch condition and the branch target address are known that the fetch stage can correctly fetch the next instruction.

As Figure 5.2 illustrates, this delay in processing conditional branches incurs a penalty of three cycles in fetching the next instruction, corresponding to the traversal of the decode, dispatch, and execute stages by the conditional branch. The actual lost-opportunity cost of three stalled cycles is not just three empty instruction

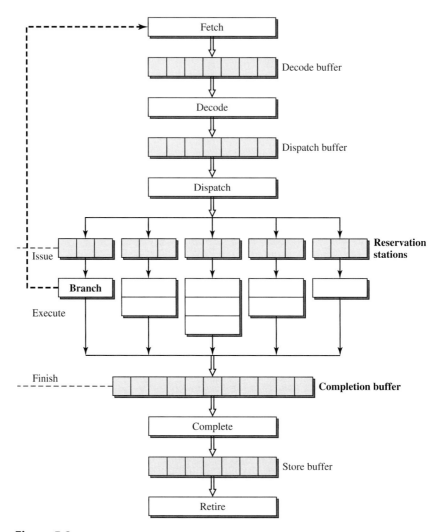

Figure 5.2
Disruption of Sequential Control Flow by Branch Instructions.

slots as in the scalar pipeline, but the number of empty instruction slots must be multiplied by the width of the machine. For example, for a four-wide machine the total penalty is 12 instruction "bubbles" in the superscalar pipeline. Also recall from Chapter 1, that such pipeline stall cycles effectively correspond to the *sequential bottleneck* of Amdahl's law and rapidly and significantly reduce the actual performance from the potential peak performance.

For conditional branches, the actual number of stalled or penalty cycles can be dictated by either target address generation or condition resolution. Figure 5.3 illustrates the potential cycles that can be incurred by target address generation. The

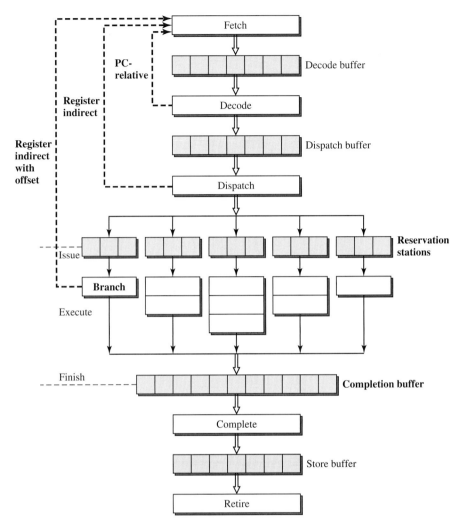

Figure 5.3
Branch Target Address Generation Penalties.

actual number of penalty cycles is determined by the addressing modes of the branch instructions. For the PC-relative addressing mode, the branch target address can be generated during the fetch stage, resulting in a penalty of one cycle. If the register indirect addressing mode is used, the branch instruction must traverse the decode stage to access the register. In this case a two-cycle penalty is incurred. For register indirect with an offset addressing mode, the offset must be added after register access and a total three-cycle penalty can result. For unconditional branches, only the penalty due to target address generation is of concern. For conditional branches, branch condition resolution latency must also be considered.

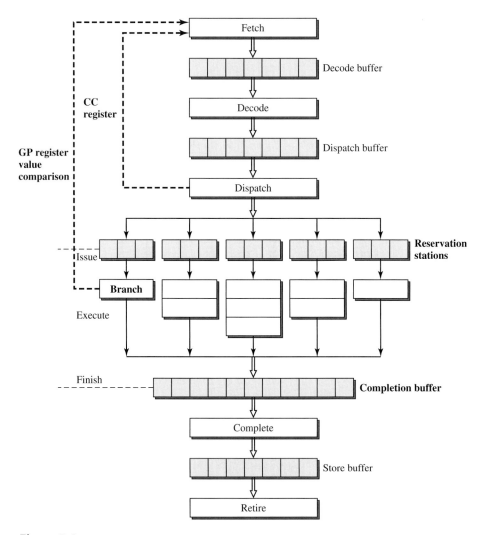

Figure 5.4
Branch Condition Resolution Penalties.

Different methods for performing condition resolution can also lead to different penalties. Figure 5.4 illustrates two possible penalties. If condition code registers are used, and assuming that the relevant condition code register is accessed during the dispatch stage, then a penalty of two cycles will result. If the ISA permits the comparison of two general-purpose registers to generate the branch condition, then one more cycle is needed to perform an ALU operation on the contents of the two registers. This will result in a penalty of three cycles. For a conditional branch, depending on the addressing mode and condition resolution method used, either one of the penalties may be the critical one. For example, even if the PC-relative addressing mode is used, a conditional branch that must access a condition code register will still incur a two-cycle penalty instead of the one-cycle penalty for target address generation.

Maximizing the volume of the instruction flow path is equivalent to maximizing the sustained instruction fetch bandwidth. To do this, the number of stall cycles in the fetch stage must be minimized. Recall that the total lost-opportunity cost is equal to the product of the number of penalty cycles and the width of a machine. For an n-wide machine each stalled cycle is equal to fetching n no-op instructions. The primary aim of instruction flow techniques is to minimize the number of such fetch stall cycles and/or to make use of these cycles to do potentially useful work. The current dominant approach to accomplishing this is via branch prediction which is the subject of Section 5.1.3.

5.1.3 Branch Prediction Techniques

Experimental studies have shown that the behavior of branch instructions is highly predictable. A key approach to minimizing branch penalty and maximizing instruction flow throughput is to speculate on both branch target addresses and branch conditions of branch instructions. As a static branch instruction is repeatedly executed at run time, its dynamic behavior can be tracked. Based on its past behavior, its future behavior can be effectively predicted. Two fundamental components of branch prediction are *branch target speculation* and *branch condition speculation*. With any speculative technique, there must be mechanisms to validate the prediction and to safely recover from any mispredictions. Branch misprediction recovery will be covered in Section 5.1.4.

Branch target speculation involves the use of a branch target buffer (BTB) to store previous branch target addresses. BTB is a small cache memory accessed during the instruction fetch stage using the instruction fetch address (PC). Each entry of the BTB contains two fields: the branch instruction address (BIA) and the branch target address (BTA). When a static branch instruction is executed for the first time, an entry in the BTB is allocated for it. Its instruction address is stored in the BIA field, and its target address is stored in the BTA field. Assuming the BTB is a fully associative cache, the BIA field is used for the associative access of the BTB. The BTB is accessed concurrently with the accessing of the I-cache. When the current PC matches the BIA of an entry in the BTB, a hit in the BTB results. This implies that the current instruction being fetched from the I-cache has been executed before and is a branch instruction. When a hit in the BTB occurs, the BTA field of the hit entry is accessed and can be used as the next instruction fetch address if that particular branch instruction is predicted to be taken; see Figure 5.5.

By accessing the BTB using the branch instruction address and retrieving the branch target address from the BTB all during the fetch stage, the speculative branch target address will be ready to be used in the next machine cycle as the new instruction fetch address if the branch instruction is predicted to be taken. If the branch instruction is predicted to be taken and this prediction turns out to be correct, then the branch instruction is effectively executed in the fetch stage, incurring no branch penalty. The nonspeculative execution of the branch instruction is still

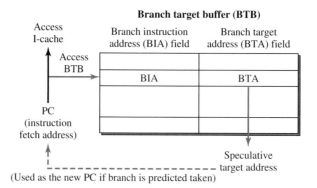

Figure 5.5
Branch Target Speculation Using a Branch Target Buffer.

performed for the purpose of validating the speculative execution. The branch instruction is still fetched from the I-cache and executed. The resultant target address and branch condition are compared with the speculative version. If they agree, then correct prediction was made; otherwise, misprediction has occurred and recovery must be initiated. The result from the nonspeculative execution is also used to update the content, i.e., the BTA field, of the BTB.

There are a number of ways to do branch condition speculation. The simplest form is to design the fetch hardware to be biased for not taken, i.e., to always predict not taken. When a branch instruction is encountered, prior to its resolution, the fetch stage continues fetching down the fall-through path without stalling. This form of minimal branch prediction is easy to implement but is not very effective. For example, many branches are used as loop closing instructions, which are mostly taken during execution except when exiting loops. Another form of prediction employs software support and can require ISA changes. For example, an extra bit can be allocated in the branch instruction format that is set by the compiler. This bit is used as a hint to the hardware to perform either predict not taken or predict taken depending on the value of this bit. The compiler can use branch instruction type and profiling information to determine the most appropriate value for this bit. This allows each static branch instruction to have its own specified prediction. However, this prediction is static in the sense that the same prediction is used for all dynamic executions of the branch. Such static software prediction technique is used in the Motorola 88110 [Diefendorf and Allen, 1992]. A more aggressive and dynamic form of prediction makes prediction based on the branch target address offset. This form of prediction first determines the relative offset between the address of the branch instruction and the address of the target instruction. A positive offset will trigger the hardware to predict not taken, whereas a negative offset, most likely indicating a loop closing branch, will trigger the hardware to predict taken. This branch offset-based technique is used in the original IBM RS/6000 design and has been adopted by other machines as well [Grohoski, 1990; Oehler and Groves, 1990]. The most common branch condition speculation technique

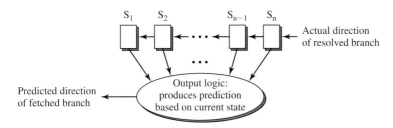

Figure 5.6
FSM Model for History-Based Branch Direction Predictors.

employed in contemporary superscalar machines is based on the history of previous branch executions.

History-based branch prediction makes a prediction of the branch direction, whether taken (T) or not taken (N), based on previously observed branch directions. This approach was first proposed by Jim Smith, who patented the technique on behalf of his employer, Control Data, and later published an important early study [Smith, 1981]. The assumption is that historical information on the direction that a static branch takes in previous executions can give helpful hints on the direction that it is likely to take in future executions. Design decisions for such type of branch prediction include how much history should be tracked and for each observed history pattern what prediction should be made. The specific algorithm for history-based branch direction prediction can be characterized by a finite state machine (FSM); see Figure 5.6. The n state variables encode the directions taken by the last n executions of that branch. Hence each state represents a particular history pattern in terms of a sequence of takens and not takens. The output logic generates a prediction based on the current state of the FSM. Essentially, a prediction is made based on the outcome of the previous n executions of that branch. When a predicted branch is finally executed, the actual outcome is used as an input to the FSM to trigger a state transition. The next state logic is trivial; it simply involves chaining the state variables into a shift register, which records the branch directions of the previous n executions of that branch instruction.

Figure 5.7(a) illustrates the FSM diagram of a typical 2-bit branch predictor that employs two history bits to track the outcome of two previous executions of the branch. The two history bits constitute the state variables of the FSM. The predictor can be in one of four states: NN, NT, TT, or TN, representing the directions taken in the previous two executions of the branch. The NN state can be designated as the initial state. An output value of either T or N is associated with each of the four states representing the prediction that would be made when a predictor is in that state. When a branch is executed, the actual direction taken is used as an input to the FSM, and a state transition occurs to update the branch history which will be used to do the next prediction.

The particular algorithm implemented in the predictor of Figure 5.7(a) is biased toward predicting branches to be taken; note that three of the four states

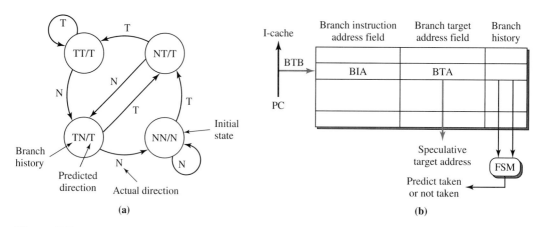

Figure 5.7
History-Based Branch Prediction: (a) A 2-Bit Branch Predictor Algorithm; (b) Branch Target Buffer with an Additional Field for Storing Branch History Bits.

predict the branch to be taken. It anticipates either long runs of N's (in the NN state) or long runs of T's (in the TT state). As long as at least one of the two previous executions was a taken branch, it will predict the next execution to be taken. The prediction will only be switched to not taken when it has encountered two consecutive N's in a row. This represents one particular branch direction prediction algorithm; clearly there are many possible designs for such history-based predictors, and many designs have been evaluated by researchers.

To support history-based branch direction predictors, the BTB can be augmented to include a history field for each of its entries. The width, in number of bits, of this field is determined by the number of history bits being tracked. When a PC address hits in the BTB, in addition to the speculative target address, the history bits are retrieved. These history bits are fed to the logic that implements the next-state and output functions of the branch predictor FSM. The retrieved history bits are used as the state variables of the FSM. Based on these history bits, the output logic produces the 1-bit output that indicates the predicted direction. If the prediction is a taken branch, then this output is used to steer the speculative target address to the PC to be used as the new instruction fetch address in the next machine cycle. If the prediction turns out to be correct, then effectively the branch instruction has been executed in the fetch stage without incurring any penalty or stalled cycle.

A classic experimental study on branch prediction was done by Lee and Smith [1984]. In this study, 26 programs from six different types of workloads for three different machines (IBM 370, DEC PDP-11, and CDC 6400) were used. Averaged across all the benchmarks, 67.6% of the branches were taken while 32.4% were not taken. Branches tend to be taken more than not taken by a ratio of 2 to 1. With static branch prediction based on the op-code type, the prediction accuracy ranged from 55% to 80% for the six workloads. Using only 1 bit of history, history-based dynamic branch prediction achieved prediction accuracies ranging from 79.7% to

96.5%. With 2 history bits, the accuracies for the six workloads ranged from 83.4% to 97.5%. Continued increase of the number of history bits brought additional incremental accuracy. However, beyond four history bits there is a very minimal increase in the prediction accuracy. They implemented a four-way set associative BTB that had 128 sets. The averaged BTB hit rate was 86.5%. Combining prediction accuracy with the BTB hit rate, the resultant average prediction effectiveness was approximately 80%.

Another experimental study was done in 1992 at IBM by Ravi Nair using the RS/6000 architecture and Systems Performance Evaluation Cooperative (SPEC) benchmarks [Nair, 1992]. This was a very comprehensive study of possible branch prediction algorithms. The goal for branch prediction is to overlap the execution of branch instructions with that of other instructions so as to achieve zero-cycle branches or accomplish *branch folding;* i.e., branches are folded out of the critical latency path of instruction execution. This study performed an exhaustive search for optimal 2-bit predictors. There are 2^{20} possible FSMs of 2-bit predictors. Nair determined that many of these machines are uninteresting and pruned the entire design space down to 5248 machines. Extensive simulations are performed to determine the optimal (achieves the best prediction accuracy) 2-bit predictor for each of the benchmarks. The list of SPEC benchmarks, their best prediction accuracies, and the associated optimal predictors are shown in Figure 5.8.

In Figure 5.8, the states denoted with bold circles represent states in which the branch is predicted taken; the nonbold circles represent states that predict not taken. Similarly the bold edges represent state transitions when the branch is actually taken; the nonbold edges represent transitions corresponding to the branch

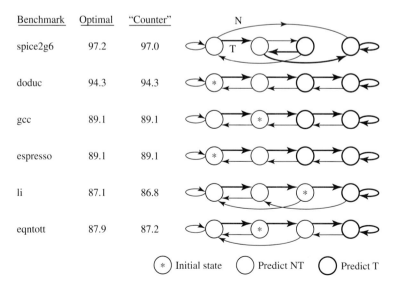

Figure 5.8
Optimal 2-Bit Branch Predictors for Six SPEC Benchmarks from the Nair Study.

actually not taken. The state denoted with an asterisk indicates the initial state. The prediction accuracies for the optimal predictors of these six benchmarks range from 87.1% to 97.2%. Notice that the optimal predictors for *doduc, gcc,* and *espresso* are identical (disregarding the different initial state of the *gcc* predictor) and exhibit the behavior of a 2-bit up/down saturating counter. We can label the four states from left to right as 0, 1, 2, and 3, representing the four count values of a 2-bit counter. Whenever a branch is resolved taken, the count is incremented; and it is decremented otherwise. The two lower-count states predict a branch to be not taken, while the two higher-count states predict a branch to be taken. Figure 5.8 also provides the prediction accuracies for the six benchmarks if the 2-bit saturating counter predictor is used for all six benchmarks. The prediction accuracies for *spice2g6, li,* and *eqntott* only decrease minimally from their optimal values, indicating that the 2-bit saturating counter is a good candidate for general use on all benchmarks. In fact, the 2-bit saturating counter, originally invented by Jim Smith, has become a popular prediction algorithm in real and experimental designs.

The same study by Nair also investigated the effectiveness of counter-based predictors. With a 1-bit counter as the predictor, i.e., remembering the direction taken last time and predicting the same direction for the next time, the prediction accuracies ranged from 82.5% to 96.2%. As we have seen in Figure 5.8, a 2-bit counter yields an accuracy range of 86.8% to 97.0%. If a 3-bit counter is used, the increase in accuracy is minimal; accuracies range from 88.3% to 97.0%. Based on this study, the 2-bit saturating counter appears to be a very good choice for a history-based predictor. Direct-mapped branch history tables are assumed in this study. While some programs, such as *gcc,* have more than 7000 conditional branches, for most programs, the branch penalty due to aliasing in finite-sized branch history tables levels out at about 1024 entries for the table size.

5.1.4 Branch Misprediction Recovery

Branch prediction is a speculative technique. Any speculative technique requires mechanisms for validating the speculation. Dynamic branch prediction can be viewed as consisting of two interacting engines. The leading engine performs speculation in the front-end stages of the pipeline, while a trailing engine performs validation in the later stages of the pipeline. In the case of misprediction the trailing engine also performs recovery. These two aspects of branch prediction are illustrated in Figure 5.9.

Branch speculation involves predicting the direction of a branch and then proceeding to fetch along the predicted path of control flow. While fetching from the predicted path, additional branch instructions may be encountered. Prediction of these additional branches can be similarly performed, potentially resulting in speculating past multiple conditional branches before the first speculated branch is resolved. Figure 5.9(a) illustrates speculating past three branches with the first and the third branches being predicted taken and the second one predicted not taken. When this occurs, instructions from three speculative basic blocks are now resident in the machine and must be appropriately identified. Instructions from each speculative basic block are given the same identifying tag. In the example of

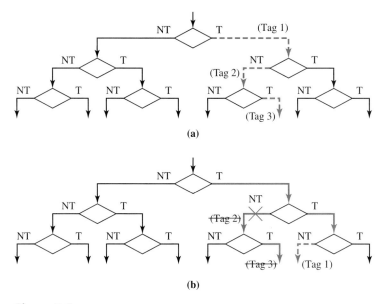

Figure 5.9
Two Aspects of Branch Prediction: (a) Branch Speculation; (b) Branch Validation/Recovery.

Figure 5.9(a), three distinct tags are used to identify the instructions from the three speculative basic blocks. A tagged instruction indicates that it is a speculative instruction, and the value of the tag identifies which basic block it belongs to. As a speculative instruction advances down the pipeline stages, the tag is also carried along. When speculating, the instruction addresses of all the speculated branch instructions (or the next sequential instructions) must be buffered in the event that recovery is required.

Branch validation occurs when the branch is executed and the actual direction of a branch is resolved. The correctness of the earlier prediction can then be determined. If the prediction turns out to be correct, the speculation tag is deallocated and all the instructions associated with that tag become nonspeculative and are allowed to complete. If a misprediction is detected, two actions are required; namely, the incorrect path must be terminated, and fetching from a new correct path must be initiated. To initiate a new path, the PC must be updated with a new instruction fetch address. If the incorrect prediction was a not-taken prediction, then the PC is updated with the computed branch target address. If the incorrect prediction was a taken prediction, then the PC is updated with the sequential (fall-through) instruction address, which is obtained from the previously buffered instruction address when the branch was predicted taken. Once the PC has been updated, fetching of instructions resumes along the new path, and branch prediction begins anew. To terminate the incorrect path, speculation tags are used. All the tags that are associated with the mispredicted branch are used to identify the instructions that must be eliminated. All such instructions that are still in the decode and dispatch buffers as well as those in reservation station entries are invalidated. Reorder buffer entries occupied by these instructions are deallocated. Figure 5.9(b) illustrates this validation/recovery task when the second of the three predictions is incorrect. The first branch is correctly predicted, and therefore instructions with Tag 1 become nonspeculative and are allowed to complete. The second prediction is incorrect, and all the instructions with Tag 2 and Tag 3 must be invalidated and their reorder buffer entries must be deallocated. After fetching down the correct path, branch prediction can begin once again, and Tag 1 is used again to denote the instructions in the first speculative basic block. During branch validation, the associated BTB entry is also updated.

We now use the PowerPC 604 superscalar microprocessor to illustrate the implementation of dynamic branch prediction in a real superscalar processor. The PowerPC 604 is a four-wide superscalar capable of fetching, decoding, and dispatching up to four instructions in every machine cycle [IBM Corp., 1994]. Instead of a single unified BTB, the PowerPC 604 employs two separate buffers to support branch prediction, namely, the branch target address cache (BTAC) and the branch history table (BHT); see Figure 5.10. The BTAC is a 64-entry fully associative cache that stores the branch target addresses, while the BHT, a 512-entry direct-mapped table, stores the history bits of branches. The reason for this separation will become clear shortly.

Both the BTAC and the BHT are accessed during the fetch stage using the current instruction fetch address in the PC. The BTAC responds in one cycle; however, the BHT requires two cycles to complete its access. If a hit occurs in the BTAC, indicating the presence of a branch instruction in the current fetch group, a predict taken occurs and the branch target address retrieved from the BTAC is used in the next fetch cycle. Since the PowerPC 604 fetches four instructions in a fetch cycle, there can be multiple branches in the fetch group. Hence, the BTAC entry indexed by the fetch address contains the branch target address of the first branch instruction in the fetch group that is predicted to be taken. In the second cycle, or during the decode stage, the history bits retrieved from the BHT are used to generate a history-based prediction on the same branch. If this prediction agrees with the taken prediction made by the BTAC, the earlier prediction is allowed to stand. On the other hand, if the BHT prediction disagrees with the BTAC prediction, the BTAC prediction is annulled and fetching from the fall-through path, corresponding to predict not taken, is initiated. In essence, the BHT prediction can overrule the BTAC prediction. As expected, in most cases the two predictions agree. In some cases, the BHT corrects the wrong prediction made by the BTAC. It is possible, however, for the BHT to erroneously change the correct prediction of the BTAC; this occurs very infrequently. When a branch is resolved, the BHT is updated; and based on its updated content the BHT in turn updates the BTAC by either leaving an entry in the BTAC if it is to be predicted taken the next time, or deleting an entry from the BTAC if that branch is to be predicted not taken the next time.

The PowerPC 604 has four entries in the reservation station that feeds the branch execution unit. Hence, it can speculate past up to four branches; i.e., there

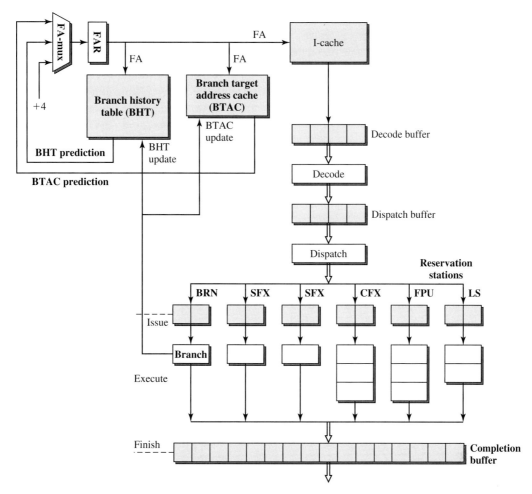

Figure 5.10
Branch Prediction in the PowerPC 604 Superscalar Microprocessor.

can be a maximum of four speculative branches present in the machine. To denote the four speculative basic blocks involved, a 2-bit tag is used to identify all speculative instructions. After a branch resolves, branch validation takes place and all speculative instructions either are made nonspeculative or are invalidated via the use of the 2-bit tag. Reorder buffer entries occupied by misspeculated instructions are deallocated. Again, this is performed using the 2-bit tag.

5.1.5 Advanced Branch Prediction Techniques

The dynamic branch prediction schemes discussed thus far have a number of limitations. Prediction for a branch is made based on the limited history of only that particular static branch instruction. The actual prediction algorithm does not take into account the dynamic context within which the branch is being executed. For example, it does not make use of any information on the particular control flow path taken in arriving at that branch. Furthermore the same fixed algorithm is used to make the prediction regardless of the dynamic context. It has been observed experimentally that the behavior of certain branches is strongly correlated with the behavior of other branches that precede them during execution. Consequently more accurate branch prediction can be achieved with algorithms that take into account the branch history of other correlated branches and that can adapt the prediction algorithm to the dynamic branching context.

In 1991, Yeh and Patt proposed a two-level adaptive branch prediction technique that can potentially achieve better than 95% prediction accuracy by having a highly flexible prediction algorithm that can adapt to changing dynamic contexts [Yeh and Patt, 1991]. In previous schemes, a single branch history table is used and indexed by the branch address. For each branch address there is only one relevant entry in the branch history table. In the two-level adaptive scheme, a set of history tables is used. These are identified as the pattern history table (PHT); see Figure 5.11. Each branch address indexes to a set of relevant entries; one of these entries is then selected based on the dynamic branching context. The context is determined by a specific pattern of recently executed branches stored in a branch history shift register (BHSR); see Figure 5.11. The content of the BHSR is used to index into the PHT to select one of the relevant entries. The content of this entry is then used as the state for the prediction algorithm FSM to produce a prediction. When a branch is resolved, the branch result is used to update both the BHSR and the selected entry in the PHT.

The two-level adaptive branch prediction technique actually specifies a framework within which many possible designs can be implemented. There are two options

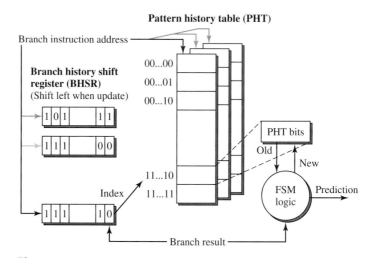

Figure 5.11
Two-Level Adaptive Branch Prediction of Yeh and Patt.
Source: Yeh and Patt, 1991.

to implementing the BHSR: *global* (G) and *individual* (P). The global implementation employs a single BHSR of *k* bits that tracks the branch directions of the last *k* dynamic branch instructions in program execution. These can involve any number (1 to *k*) of static branch instructions. The individual (called *per-branch* by Yeh and Patt) implementation employs a set of *k*-bit BHSRs as illustrated in Figure 5.11, one of which is selected based on the branch address. Essentially the global BHSR is shared by all static branches, whereas with individual BHSRs each BHSR is dedicated to each static branch or a subset of static branches if there is address aliasing when indexing into the set of BHSRs using the branch address. There are three options to implementing the PHT: *global* (g), *individual* (p), or *shared* (s). The global PHT uses a single table to support the prediction of all static branches. Alternatively, individual PHTs can be used in which each PHT is dedicated to each static branch (p) or a small subset of static branches (s) if there is address aliasing when indexing into the set of PHTs using the branch address. A third dimension to this design space involves the implementation of the actual prediction algorithm. When a history-based FSM is used to implement the prediction algorithm, Yeh and Patt identified such schemes as *adaptive* (A).

All possible implementations of the two-level adaptive branch prediction can be classified based on these three dimensions of design parameters. A given implementation can then be denoted using a three-letter notation; e.g., GAs represents a design that employs a single global BHSR, an adaptive prediction algorithm, and a set of PHTs with each being shared by a number of static branches. Yeh and Patt presented three specific implementations that are able to achieve a prediction accuracy of 97% for their given set of benchmarks:

- GAg: (1) BHSR of size 18 bits; (1) PHT of size $2^{18} \times 2$ bits.
- PAg: (512×4) BHSRs of size 12 bits; (1) PHT of size $2^{12} \times 2$ bits.
- PAs: (512×4) BHSRs of size 6 bits; (512) PHTs of size $2^{6} \times 2$ bits.

All three implementations use an adaptive (A) predictor that is a 2-bit FSM. The first implementation employs a global BHSR (G) of 18 bits and a global PHT (g) with 2^{18} entries indexed by the BHSR bits. The second implementation employs 512 sets (four-way set-associative) of 12-bit BHSRs (P) and a global PHT (g) with 2^{12} entries. The third implementation also employs 512 sets of four-way set-associative BHSRs (P), but each is only 6 bits wide. It also uses 512 PHTs (s), each having 2^{6} entries indexed by the BHSR bits. Both the 512 sets of BHSRs and the 512 PHTs are indexed using 9 bits of the branch address. Additional branch address bits are used for the set-associative access of the BHSRs. The 512 PHTs are direct-mapped, and there can be aliasing, i.e., multiple branch addresses sharing the same PHT. From experimental data, such aliasing had minimal impact on degrading the prediction accuracy. Achieving greater than 95% prediction accuracy by the two-level adaptive branch prediction schemes is quite impressive; the best traditional prediction techniques can only achieve about 90% prediction accuracy. The two-level adaptive branch prediction approach has been adopted by a number of real designs, including the Intel Pentium Pro and the AMD/NexGen Nx686.

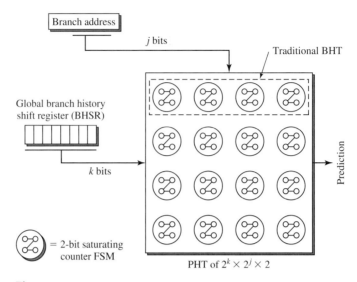

Figure 5.12
Correlated Branch Predictor with Global BHSR and Shared PHTs (GAs).

Following the original Yeh and Patt proposal, other studies by McFarling [1993], Young and Smith [1994], and Gloy et al. [1995] have gained further insights into two-level adaptive, or more recently called *correlated,* branch predictors. Figure 5.12 illustrates a correlated branch predictor with a global BHSR (G) and a shared PHT (s). The 2-bit saturating counter is used as the predictor FSM. The global BHSR tracks the directions of the last k dynamic branches and captures the dynamic control flow context. The PHT can be viewed as a single table containing a two-dimensional array, with 2^j columns and 2^k rows, of 2-bit predictors. If the branch address has n bits, a subset of j bits is used to index into the PHT to select one of the 2^j columns. Since j is less than n, some aliasing can occur where two different branch addresses can index into the same column of the PHT. Hence the designation of *shared* PHT. The k bits from the BHSR are used to select one of the 2^k entries in the selected column. The 2 history bits in the selected entry are used to make a history-based prediction. The traditional branch history table is equivalent to having only one row of the PHT that is indexed only by the j bits of the branch address, as illustrated in Figure 5.12 by the dashed rectangular block of 2-bit predictors in the first row of the PHT.

Figure 5.13 illustrates a correlated branch predictor with individual, or per-branch, BHSRs (P) and the same shared PHT (s). Similar to the GAs scheme, the PAs scheme also uses j bits of the branch address to select one of the 2^j columns of the PHT. However, i bits of the branch address, which can overlap with the j bits used to access the PHT, are used to index into a set of BHSRs. Depending on the branch address, one of the 2^i BHSRs is selected. Hence, each BHSR is associated with one particular branch address, or a set of branch addresses if there is aliasing. Essentially, instead of using a single BHSR to provide the dynamic control flow context for all static branches, multiple BHSRs are used to provide distinct dynamic control flow contexts for different subsets of static branches. This adds

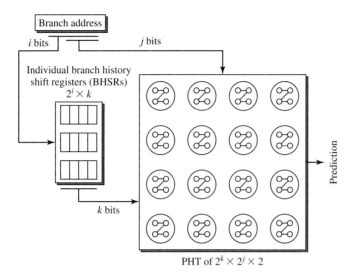

Figure 5.13
Correlated Branch Predictor with Individual BHSRs and Shared PHTs (PAs).

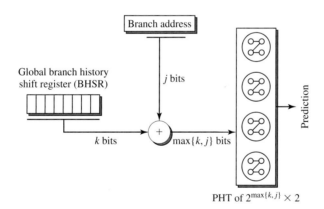

Figure 5.14
The *gshare* Correlated Branch Predictor of McFarling.
Source: McFarling, 1993.

flexibility in tracking and exploiting correlations between different branch instructions. Each BHSR tracks the directions of the last k dynamic branches belonging to the same subset of static branches. Both the GAs and the PAs schemes require a PHT of size $2^k \times 2^j \times 2$ bits. The GAs scheme has only one k-bit BHSR whereas the PAs scheme requires 2^i k-bit BHSRs.

A fairly efficient correlated branch predictor called *gshare* was proposed by Scott McFarling [1993]. In this scheme, j bits from the branch address are "hashed" (via bitwise XOR function) with the k bits from a global BHSR; see Figure 5.14. The resultant $\max\{k, j\}$ bits are used to index into a PHT of size $2^{\max\{k,j\}} \times 2$ bits to select one of the $2^{\max\{k,j\}}$ 2-bit branch predictors. The *gshare* scheme requires only one *k*-bit BHSR and a much smaller PHT, yet achieves comparable prediction accuracy to other correlated branch predictors. This scheme is used in the DEC Alpha 21264 4-way superscalar microprocessor [Keller, 1996].

5.1.6 Other Instruction Flow Techniques

The primary objective for instruction flow techniques is to supply as many useful instructions as possible to the execution core of the processor in every machine cycle. The two major challenges deal with conditional branches and taken branches. For a wide superscalar processor, to provide adequate conditional branch throughput, the processor must very accurately predict the outcomes and targets of multiple conditional branches in every machine cycle. For example, in a fetch group of four instructions, it is possible that all four instructions are conditional branches. Ideally one would like to use the addresses of all four instructions to index into a four-ported BTB to retrieve the history bits and target addresses of all four branches. A complex predictor can then make an overall prediction based on all the history bits. Speculative fetching can then proceed based on this prediction. Techniques for predicting multiple branches in every cycle have been proposed by Conte et al. [1995] as well as Rotenberg et al. [1996]. It is also important to ensure high accuracy in such predictions. Global branch history can be used in conjunction with per-branch history to achieve very accurate predictions. For those branches or sequences of branches that do not exhibit strongly biased branching behavior and therefore are not predictable, *dynamic eager execution* (DEE) has been proposed by Gus Uht [Uht and Sindagi, 1995]. DEE employs multiple PCs to simultaneously fetch from multiple addresses. Essentially the fetch stage pursues down multiple control flow paths until some branches are resolved, at which time some of the wrong paths are dynamically pruned by invalidating the instructions on those paths.

Taken branches are the second major obstacle to supplying enough useful instructions to the execution core. In a wide machine the fetch unit must be able to correctly process more than one taken branch per cycle, which involves predicting each branch's direction and target, and fetching, aligning, and merging instructions from multiple branch targets. An effective approach in alleviating this problem involves the use of a *trace cache* that was initially proposed by Eric Rotenberg [Rotenberg et al., 1996]. Since then, a form of trace caching has been implemented in Intel's most recent Pentium 4 superscalar microprocessor. Trace cache is a history-based fetch mechanism that stores dynamic instruction traces in a cache indexed by the fetch address and branch outcomes. These traces are assembled dynamically based on the dynamic branching behavior and can contain multiple nonconsecutive basic blocks. Whenever the fetch address hits in the trace cache, instructions are fetched from the trace cache rather than the instruction cache. Since a dynamic sequence of instructions in the trace cache can contain multiple taken branches but is stored sequentially, there is no need to fetch from multiple targets and no need for a multiported instruction cache or complex merging and aligning logic in the fetch stage. The trace cache can be viewed as doing dynamic basic block reordering according to the dominant execution paths taken by a program. The merging and aligning can be done at completion time, when nonconsecutive basic blocks on a dominant path are first executed, to assemble a trace, which is then stored in one line of the trace cache. The goal is that once the trace cache is warmed up, most of the fetching will come from the trace cache instead of the instruction cache. Since the reordered basic blocks in the trace cache better match the dynamic execution order, there will be fewer fetches from nonconsecutive locations in the trace cache, and there will be an effective increase in the overall throughput of taken branches.

5.2 Register Data Flow Techniques

Register data flow techniques concern the effective execution of ALU (or register-register) type instructions in the execution core of the processor. ALU instructions can be viewed as performing the "real" work specified by the program, with control flow and load/store instructions playing the supportive roles of providing the necessary instructions and the required data, respectively. In the most ideal machine, branch and load/store instructions, being "overhead" instructions, should take no time to execute and the computation latency should be strictly determined by the processing of ALU instructions. The effective processing of these instructions is foundational to achieving high performance.

Assuming a load/store architecture, ALU instructions specify operations to be performed on source operands stored in registers. Typically an ALU instruction specifies a binary operation, two source registers where operands are to be retrieved, and a destination register where the result is to be placed. $R_i \leftarrow F_n(R_j, R_k)$ specifies a typical ALU instruction, the execution of which requires the availability of (1) F_n, the functional unit; (2) R_j and R_k, the two source operand registers; and (3) R_i, the destination register. If the functional unit F_n is not available, then a structural dependence exists that can result in a structural hazard. If one or both of the source operands in R_j and R_k are not available, then a hazard due to true data dependence can occur. If the destination register R_i is not available, then a hazard due to anti- and output dependences can occur.

5.2.1 Register Reuse and False Data Dependences

The occurrence of anti- and output dependences, or false data dependences, is due to the reuse of registers. If registers are never reused to store operands, then such false data dependences will not occur. The reuse of registers is commonly referred to as *register recycling*. Register recycling occurs in two different forms, one static and the other dynamic. The static form is due to optimization performed by the compiler and is presented first. In a typical compiler, toward the back end of the compilation process two tasks are performed: *code generation* and *register allocation*. The code generation task is responsible for the actual emitting of machine instructions. Typically the code generator assumes the availability of an unlimited number of symbolic registers in which it stores all the temporary data. Each symbolic register is used to store one value and is only written once, producing what is commonly referred to as *single-assignment* code. However, an ISA has a limited number of architected registers, and hence the register allocation tool is used to map the unlimited number of symbolic registers to the limited and fixed number of architected registers. The register allocator attempts to keep as many of the temporary values in registers as possible to avoid having to move the data out to memory locations and reloading them later on. It accomplishes this by reusing registers. A register is written with a new value when the old value stored there is no longer needed; effectively each register is recycled to hold multiple values.

Writing of a register is referred to as the *definition* of a register and the reading of a register as the *use* of a register. After each definition, there can be one or more uses of that definition. The duration between the definition and the last use of a value is referred to as the *live range* of that value. After the last use of a live range, that register can be assigned to store another value and begin another live range. Register allocation procedures attempt to map nonoverlapping live ranges into the same architected register and maximize register reuse. In single-assignment code there is a one-to-one correspondence between symbolic registers and values. After register allocation, each architected register can receive multiple assignments, and the register becomes a variable that can take on multiple values. Consequently the one-to-one correspondence between registers and values is lost.

If the instructions are executed sequentially and a redefinition is never allowed to precede the previous definition or the last use of the previous definition, then the live ranges that share the same register will never overlap during execution and the recycling of registers does not induce any problem. Effectively, the one-to-one correspondence between values and registers can be maintained implicitly if all the instructions are processed in the original program order. However, in a superscalar machine, especially with out-of-order processing of instructions, register reading and writing operations can occur in an order different from the program order. Consequently the one-to-one correspondence between values and registers can potentially be perturbed; in order to ensure semantic correctness all anti- and output dependences must be detected and enforced. Out-of-order reading (writing) of registers can be permitted as long as all the anti- (output) dependences are enforced.

The dynamic form of register recycling occurs when a loop of instructions is repeatedly executed. With an aggressive superscalar machine capable of supporting many instructions in flight and a relatively small loop body being executed, multiple iterations of the loop can be simultaneously in flight in a machine. Hence, multiple copies of a register defining instruction from the multiple iterations can be simultaneously present in the machine, inducing the dynamic form of register recycling. Consequently anti- and output dependences can be induced among these dynamic instructions from the multiple iterations of a loop and must be detected and enforced to ensure semantic correctness of program execution.

One way to enforce anti- and output dependences is to simply stall the dependent instruction until the leading instruction has finished accessing the dependent register. If an anti- [write-after-read (WAR)] dependence exists between a pair of instructions, the trailing instruction (register updating instruction) must be stalled until the leading instruction has read the dependent register. If an output [write-after-write (WAW)] dependence exists between a pair of instructions, the trailing instruction (register updating instruction) must be stalled until the leading instruction has first updated the register. Such stalling of anti- and output dependent instructions can lead to significant performance loss and is not necessary. Recall that such false data dependences are induced by the recycling of the architected registers and are not intrinsic to the program semantics.

5.2.2 Register Renaming Techniques

A more aggressive way to deal with false data dependences is to dynamically assign different *names* to the multiple definitions of an architected register and, as a result, eliminate the presence of such false dependences. This is called *register renaming* and requires the use of hardware mechanisms at run time to undo the effects of register recycling by reproducing the one-to-one correspondence between registers and values for all the instructions that might be simultaneously in flight. By performing register renaming, single assignment is effectively recovered for the instructions that are in flight, and no anti- and output dependences can exist among these instructions. This will allow the instructions that originally had false dependences between them to be executed in parallel.

A common way to implement register renaming is to use a separate rename register file (RRF) in addition to the architected register file (ARF). A straightforward way to implement the RRF is to simply duplicate the ARF and use the RRF as a shadow version of the ARF. This will allow each architected register to be renamed once. However, this is not a very efficient way to use the registers in the RRF. Many existing designs implement an RRF with fewer entries than the ARF and allow each of the registers in the RRF to be flexibly used to rename any one of the architected registers. This facilitates the efficient use of the rename registers, but does require a mapping table to store the pointers to the entries in the RRF. The use of a separate RRF in conjunction with a mapping table to perform renaming of the ARF is illustrated in Figure 5.15.

When a separate RRF is used for register renaming, there are implementation choices in terms of where to place the RRF. One option is to implement a separate stand-alone structure similar to the ARF and perhaps adjacent to the ARF. This is shown in Figure 5.15(a). An alternative is to incorporate the RRF as part of the reorder buffer, as shown in Figure 5.15(b). In both options a busy field is added to the ARF along with a mapping table. If the busy bit of a selected entry of the ARF is set, indicating the architected register has been renamed, the corresponding entry of the map table is accessed to obtain the tag or the pointer to an RRF entry. In the former option, the tag specifies a rename register and is used to index into the RRF; whereas in the latter option, the tag specifies a reorder buffer entry and is used to index into the reorder buffer.

Based on the diagrams in Figure 5.15, the difference between the two options may seem artificial; however, there are important subtle differences. If the RRF is incorporated as part of the reorder buffer, every entry of the reorder buffer contains an additional field that functions as a rename register, and hence there is a rename register allocated for every instruction in flight. This is a design based on worst-case scenario and may be wasteful since not every instruction defines a register. For example, branch instructions do not update any architected register. On the

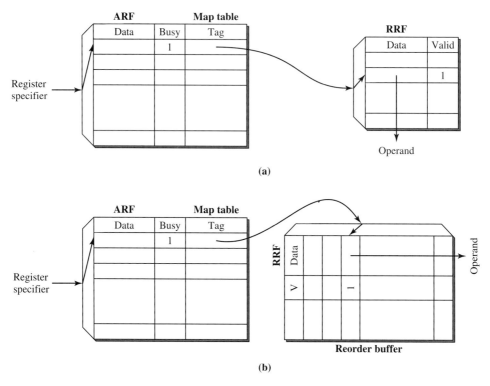

Figure 5.15
Rename Register File (RRF) Implementations: (a) Stand-Alone; (b) Attached to the Reorder Buffer.

other hand, a reorder buffer already contains ports to receive data from the functional units and to update the ARF at instruction completion time. When a separate stand-alone RRF is used, it introduces an additional structure that requires ports for receiving data from the functional units and for updating the ARF. The choice of which of the two options to implement involves design tradeoffs, and both options have been employed in real designs. We now focus on the stand-alone option to get a better feel of how register renaming actually works.

Register renaming involves three tasks: (1) *source read,* (2) *destination allocate,* and (3) *register update.* The first task of source read typically occurs during the decode (or possibly dispatch) stage and is for the purpose of fetching the register operands. When an instruction is decoded, its source register specifiers are used to index into a multiported ARF in order to fetch the register operands. Three possibilities can occur for each register operand fetch. First, if the busy bit is not set, indicating there is no pending write to the specified register and that the architected register contains the specified operand, the operand is fetched from the ARF. If the busy bit is set, indicating there is a pending write to that register and that the content of the architected register is stale, the corresponding entry of the map table is accessed to retrieve the rename tag. This rename tag specifies a

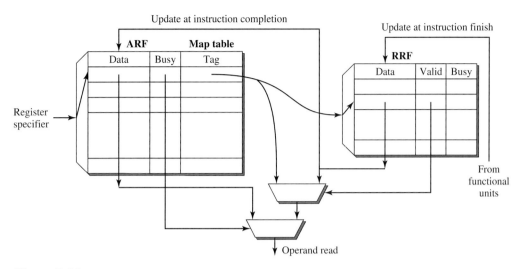

Figure 5.16
Register Renaming Tasks: Source Read, Destination Allocate, and Register Update.

rename register and is used to index into the RRF. Two possibilities can occur when indexing into the RRF. If the valid bit of the indexed entry is set, it indicates that the register-updating instruction has already finished execution, although it is still waiting to be completed. In this case, the source operand is available in the rename register and is retrieved from the indexed RRF entry. If the valid bit is not set, it indicates that the register-updating instruction still has not been executed and that the rename register has a pending update. In this case the tag, or the rename register specifier, from the map table is forwarded to the reservation station instead of to the source operand. This tag will be used later by the reservation station to obtain the operand when it becomes available. These three possibilities for source read are shown in Figure 5.16.

The task of destination allocate also occurs during the decode (or possibly dispatch) stage and has three subtasks, namely, *set busy bit, assign tag,* and *update map table*. When an instruction is decoded, its destination register specifier is used to index into the ARF. The selected architected register now has a pending write, and its busy bit must be set. The specified destination register must be mapped to a rename register. A particular unused (indicated by the busy bit) rename register must be selected. The busy bit of the selected RRF entry must be set, and the index of the selected RRF entry is used as a tag. This tag must then be written into the corresponding entry in the map table, to be used by subsequent dependent instructions for fetching their source operands.

While the task of register update takes place in the back end of the machine and is not part of the actual renaming activity of the decode/dispatch stage, it does have a direct impact on the operation of the RRF. Register update can occur in two separate steps; see Figure 5.16. When a register-updating instruction finishes execution, its result is written into the entry of the RRF indicated by the tag. Later on when this instruction is completed, its result is then copied from the RRF into the ARF. Hence, register update involves updating first an entry in the RRF and then an entry in the ARF. These two steps can occur in back-to-back cycles if the register-updating instruction is at the head of the reorder buffer, or they can be separated by many cycles if there are other unfinished instructions in the reorder buffer ahead of this instruction. Once a rename register is copied to its corresponding architected register, its busy bit is reset and it can be used again to rename another architected register.

So far we have assumed that register renaming implementation requires the use of two separate physical register files, namely the ARF and the RRF. However, this assumption is not necessary. The architected registers and the rename registers can be pooled together and implemented as a single physical register file with its number of entries equal to the sum of the ARF and RRF entry counts. Such a pooled register file does not rigidly designate some of the registers as architected registers and others as rename registers. Each physical register can be flexibly assigned to be an architected register or a rename register. Unlike a separate ARF and RRF implementation which must physically copy a result from the RRF to the ARF at instruction completion, the pooled register file only needs to change the designation of a register from being a rename register to an architected register. This will save the data transfer interconnect between the RRF and the ARF. The key disadvantage of the pooled register file is its hardware complexity. A secondary disadvantage is that at context swap time, when the machine state must be saved, the subset of registers constituting the architected state of the machine must be explicitly identified before state saving can begin.

The pooled register file approach is used in the floating-point unit of the original IBM RS/6000 design and is illustrated in Figure 5.17 [Grohoski, 1990; Oehler and Groves, 1990]. In this design, 40 physical registers are implemented for supporting an ISA that specifies 32 architected registers. A mapping table is implemented, based on whose content any subset of 32 of the 40 physical registers can be designated as the architected registers. The mapping table contains 32 entries indexed by the 5-bit architected register specifier. Each entry when indexed returns a 6-bit specifier indicating the physical register to which the architected register has been mapped.

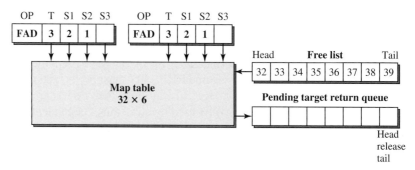

Figure 5.17
Floating-Point Unit (FPU) Register Renaming in the IBM RS/6000.

The floating-point unit (FPU) of the RS/6000 is a pipelined functional unit with the rename pipe stage preceding the decode pipe stage. The rename pipe stage contains the map table, two circular queues, and the associated control logic. The first queue is called the *free list* (FL) and contains physical registers that are available for new renaming. The second queue is called the *pending target return queue* (PTRQ) and contains those physical registers that have been used to rename architected registers that have been subsequently re-renamed in the map table. Physical registers in the PTRQ can be returned to the FL once the last use of that register has occurred. Two instructions can traverse the rename stage in every machine cycle. Because of the possibility of fused multiply-add (FMA) instructions that have three sources and one destination, each of the two instructions can contain up to four register specifiers. Hence, the map table must be eight-ported to support the simultaneous translation of the eight architected register specifiers. The map table is initialized with the identity mapping; i.e., architected register i is mapped to physical register i for $i = 0, 1, \ldots, 31$. At initialization, physical registers 32 to 39 are placed in the FL and the PTRQ is empty.

When an instruction traverses the rename stage, its architected register specifiers are used to index into the map table to obtain their translated physical register specifiers. The eight-ported map table has 32 entries, indexed by the 5-bit architected register specifier, with each entry containing 6 bits indicating the physical register to which the architected register is mapped. The content of the map table represents the latest mapping of architected registers to physical registers and specifies the subset of physical registers that currently represents the architected registers.

In the FPU of the RS/6000, by design only load instructions can trigger a new renaming. Such register renaming prevents the FPU from stalling while waiting for loads to execute in order to enforce anti- and output dependences. When a load instruction traverses the rename stage, its destination register specifier is used to index into the map table. The current content of that entry of the map table is pushed out to the PTRQ, and the next physical register in the FL is loaded into the map table. This effectively renames the redefinition of that destination register to a different physical register. All subsequent instructions that specify this architected register as a source operand will receive the new physical register specifier as the source register. Beyond the rename stage, i.e., in the decode and execute stages, the FPU uses only physical register specifiers, and all true register dependences are enforced using the physical register specifiers.

The map table approach represents the most aggressive and versatile implementation of register renaming. Every physical register can be used to represent any redefinition of any architected register. There is significant hardware complexity required to implement the multiported map table and the logic to control the two circular queues. The return of a register in the PTRQ to the FL is especially troublesome due to the difficulty in identifying the last-use instruction of a register. However, unlike approaches based on the use of separate rename registers, at instruction completion time no copying of the content of the rename registers to the architected registers is necessary. On the other hand, when interrupts occur and as part of context swap, the subset of physical registers that constitute the current architected machine state must be explicitly determined based on the map table contents.

Most contemporary superscalar microprocessors implement some form of register renaming to avoid having to stall for anti- and output register data dependences induced by the reuse of registers. Typically register renaming occurs during the instruction decoding time, and its implementation can become quite complex, especially for wide superscalar machines in which many register specifiers for multiple instructions must be simultaneously renamed. It's possible that multiple redefinitions of a register can occur within a fetch group. Implementing a register renaming mechanism for wide superscalars without seriously impacting machine cycle time is a real challenge. To achieve high performance the serialization constraints imposed by false register data dependences must be eliminated; hence, dynamic register renaming is absolutely essential.

5.2.3 True Data Dependences and the Data Flow Limit

A RAW dependence between two instructions is called a true data dependence due to the producer-consumer relationship between these two instructions. The trailing consumer instruction cannot obtain its source operand until the leading producer instruction produces its result. A true data dependence imposes a serialization constraint between the two dependent instructions; the leading instruction must finish execution before the trailing instruction can begin execution. Such true data dependences result from the semantics of the program and are usually represented by a data flow graph or data dependence graph (DDG).

Figure 5.18 illustrates a code fragment for a fast Fourier transform (FFT) implementation. Two source-level statements are compiled into 16 assembly instructions, including load and store instructions. The floating-point array variables

```
w[i+k].ip = z[i].rp + z[m+i].rp;
w[i+j].rp = e[k+1].rp * (z[i].rp - z[m+i].rp) - e[k+1].ip * (z[i].ip - z[m+i].ip);
```

(a)

```
i1:   f2    ← load, 4(r2)
i2:   f0    ← load, 4(r5)
i3:   f0    ← fadd, f2, f0
i4:   4(r6) ← store, f0
i5:   f14   ← laod, 8(r7)
i6:   f6    ← load, 0(r2)
i7:   f5    ← load, 0(r3)
i8:   f5    ← fsub, f6, f5
i9:   f4    ← fmul, f14, f5
i10:  f15   ← load, 12(r7)
i11:  f7    ← load, 4(r2)
i12:  f8    ← load, 4(r3)
i13:  f8    ← fsub, f7, f8
i14:  f8    ← fmul, f15, f8
i15:  f8    ← fsub, f4, f8
i16:  0(r8) ← store, f8
```

(b)

Figure 5.18
FFT Code Fragment: (a) Original Source Statements; (b) Compiled Assembly Instructions.

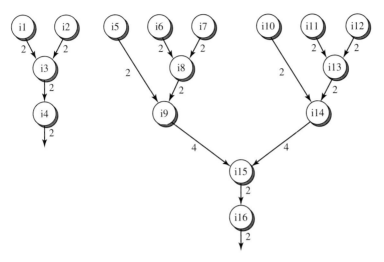

Figure 5.19
Data Flow Graph of the Code Fragment in Figure 5.18(b).

are stored in memory and must be first loaded before operations can be performed. After the computation, the results are stored back out to memory. Integer registers (ri) are used to hold addresses of arrays. Floating-point registers (fj) are used to hold temporary data. The DFG induced by the writing and reading of floating-point registers by the 16 instructions of Figure 5.18(b) is shown in Figure 5.19.

Each node in Figure 5.19 represents an instruction in Figure 5.18(b). A directed edge exists between two instructions if there exists a true data dependence between the two instructions. A dependent register can be identified for each of the dependence edges in the DFG. A latency can also be associated with each dependence edge. In Figure 5.19, each edge is labeled with the execution latency of the producer instruction. In this example, load, store, addition, and subtraction instructions are assumed to have two-cycle execution latency, while multiplication instructions require four cycles.

The latencies associated with dependence edges are cumulative. The longest dependence chain, measured in terms of total cumulative latency, is identified as the critical path of a DFG. Even assuming unlimited machine resources, a code fragment cannot be executed any faster than the length of its critical path. This is commonly referred to as the *data flow limit* to program execution and represents the best performance that can possibly be achieved. For the code fragment of Figure 5.19 the data flow limit is 12 cycles. The data flow limit is dictated by the true data dependences in the program. Traditionally, the *data flow execution model* stipulates that every instruction in a program begin execution immediately in the cycle following when all its operands become available. In effect, all existing register data flow techniques are attempts to approach the data flow limit.

5.2.4 The Classic Tomasulo Algorithm

The design of the IBM 360/91's floating-point unit, incorporating what has come to be known as *Tomasulo's algorithm,* laid the groundwork for modern superscalar processor designs [Tomasulo, 1967]. Key attributes of most contemporary register data flow techniques can be found in the classic Tomasulo algorithm, which deserves an in-depth examination. We first introduce the original design of the floating-point unit of the IBM 360, and then describe in detail the modified design of the FPU in the IBM 360/91 that incorporated Tomasulo's algorithm, and finally illustrate its operation and effectiveness in processing an example code sequence.

The original design of the IBM 360 floating-point unit is shown in Figure 5.20. The FPU contains two functional units: one floating-point add unit and one floating-point multiply/divide unit. There are three register files in the FPU: the floating-point registers (FLRs), the floating-point buffers (FLBs), and the store data buffers (SDBs). There are four FLR registers; these are the architected floating-point registers. Floating-point instructions with storage-register or storage-storage addressing modes are preprocessed. Address generation and memory

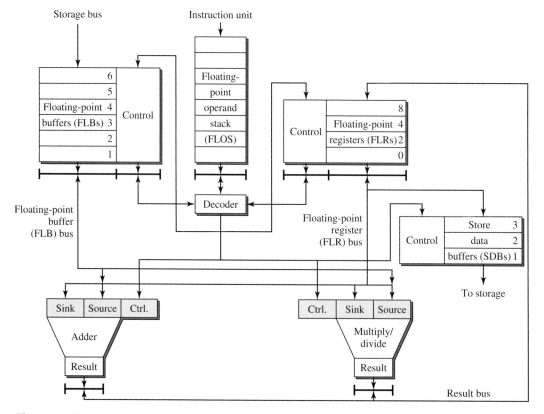

Figure 5.20
The Original Design of the IBM 360 Floating-Point Unit.

accessing are performed outside of the FPU. When the data are retrieved from the memory, they are loaded into one of the six FLB registers. Similarly if the destination of an instruction is a memory location, the result to be stored is placed in one of the three SDB registers and a separate unit accesses the SDBs to complete the storing of the result to a memory location. Using these two additional register files, the FLBs, and the SDBs, to support storage-register and storage-storage instructions, the FPU effectively functions as a register-register machine.

In the IBM 360/91, the instruction unit (IU) decodes all the instructions and passes all floating-point instructions (in order) to the floating-point operation stack (FLOS). In the FPU, floating-point instructions are then further decoded and issued in order from the FLOS to the two functional units. The two functional units are not pipelined and incur multiple-cycle latencies. The adder incurs 2 cycles for add instructions, while the multiply/divide unit incurs 3 cycles and 12 cycles for performing multiply and divide instructions, respectively.

In the mid-1960s, IBM began developing what eventually became Model 91 of the Systems 360 family. One of the goals was to achieve concurrent execution of multiple floating-point instructions and to sustain a throughput of one instruction per cycle in the instruction pipeline. This is quite aggressive considering the complex addressing modes of the 360 ISA and the multicycle latencies of the execution units. The end result is a modified FPU in the 360/91 that incorporated Tomasulo's algorithm; see Figure 5.21.

Tomasulo's algorithm consists of adding three new mechanisms to the original FPU design, namely, reservation stations, the common data bus, and register tags. In the original design, each functional unit has a single buffer on its input side to hold the instruction currently being executed. If a functional unit is busy, issuing of instructions by FLOS will stall whenever the next instruction to be issued requires the same functional unit. To alleviate this structural bottleneck, multiple buffers, called *reservation stations,* are attached to the input side of each functional unit. The adder unit has three reservation stations, while the multiply/divide unit has two. These reservation stations are viewed as virtual functional units; as long as there is a free reservation station, the FLOS can issue an instruction to that functional unit even if it is currently busy executing another instruction. Since the FLOS issues instructions in order, this will prevent unnecessary stalling due to unfortunate ordering of different floating-point instruction types.

With the availability of reservation stations, instructions can also be issued to the functional units by the FLOS even though not all their operands are yet available. These instructions can wait in the reservation station for their operands and only begin execution when they become available. The *common data bus* (CDB) connects the outputs of the two functional units to the reservation stations as well as the FLRs and SDB registers. Results produced by the functional units are broadcast into the CDB. Those instructions in the reservation stations needing the results as their operands will latch in the data from the CDB. Those registers in the FLR and SDB that are the destinations of these results also latch in the same data from the CDB. The CDB facilitates the forwarding of results directly from producer instructions to consumer instructions waiting in the reservation stations

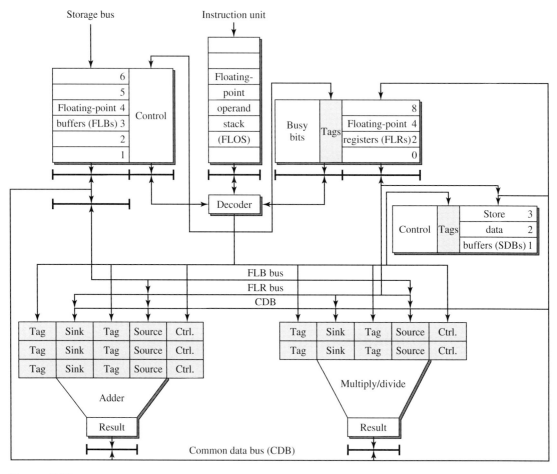

Figure 5.21
The Modified Design of the IBM 360/91 Floating-Point Unit with Tomasulo's Algorithm.

without having to go through the registers. Destination registers are updated simultaneously with the forwarding of results to dependent instructions. If an operand is coming from a memory location, it will be loaded into a FLB register once memory accessing is performed. Hence, the FLB can also output onto the CDB, allowing a waiting instruction in a reservation station to latch in its operand. Consequently, the two functional units and the FLBs can drive data onto the CDB, and the reservation station's FLRs and SDBs can latch in data from the CDB.

When the FLOS is dispatching an instruction to a functional unit, it allocates a reservation station and checks to see if the needed operands are available. If an operand is available in the FLRs, then the content of that register in the FLRs is copied to the reservation station; otherwise a tag is copied to the reservation station instead. The tag indicates where the pending operand is going to come from.

The pending operand can come from a producer instruction currently resident in one of the five reservation stations, or it can come from one of the six FLB registers. To uniquely identify one of these 11 possible sources for a pending operand, a 4-bit tag is required. If one of the two operand fields of a reservation station contains a tag instead of the actual operand, it indicates that this instruction is waiting for a pending operand. When that pending operand becomes available, the producer of that operand drives the tag along with the actual operand onto the CDB.

A waiting instruction in a reservation station uses its tag to monitor the CDB. When it detects a tag match on the CDB, it then latches in the associated operand. Essentially the producer of an operand broadcasts the tag and the operand on the CDB; all consumers of that operand monitor the CDB for that tag, and when the broadcasted tag matches their tag, they then latch in the associated operand from the CDB. Hence, all possible destinations of pending operands must carry a tag field and must monitor the CDB for a tag match. Each reservation station contains two operand fields, each of which must carry a tag field since each of the two operands can be pending. The four FLRs and the three registers in the SDB must also carry tag fields. This is a total of 17 tag fields representing 17 places that can monitor and receive operands; see Figure 5.22. The tag field at each potential consumer site is used in an associative fashion to monitor for possible matching of its content with the tag value being broadcasted on the CDB. When a tag match occurs, the consumer latches in the broadcasted operand.

The IBM 360 floating-point instructions use a two-address instruction format. Two source operands can be specified. The first operand specifier is called the *sink* because it also doubles as the destination specifier. The second operand specifier is called the *source*. Each reservation station has two operand fields, one for the sink and the other for the source. Each operand field is accompanied by a tag field. If an operand field contains real data, then its tag field is set to zero. Otherwise, its tag field identifies the source where the pending operand will be coming from, and is used to monitor the CDB for the availability of the pending operand. Whenever

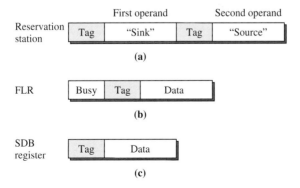

Figure 5.22

The Use of Tag Fields in (a) A Reservation Station, (b) A FLR, and (c) A SDB Register.

an instruction is dispatched by the FLOS to a reservation station, the data in the FLR corresponding to the sink operand are retrieved and copied to the reservation station. At the same time, the "busy" bit associated with this FLR is set, indicating that there is a pending update of that register, and the tag value that identifies the particular reservation station to which the instruction is being dispatched is written into the tag field of the same FLR. This clearly identifies which of the reservation stations will eventually produce the updated data for this FLR. Subsequently if a trailing instruction specifies this register as one of its source operands, when it is dispatched to a reservation station, only the tag field (called the *pseudo-operand*) will be copied to the corresponding tag field in the reservation station and not the actual data. When the busy bit is set, it indicates that the data in the FLR are stale and the tag represents the source from which the real data will come. Other than reservation stations and FLRs, SDB registers can also be destinations of pending operands and hence a tag field is required for each of the three SDB registers.

We now use an example sequence of instructions to illustrate the operation of Tomasulo's algorithm. We deviate from the actual IBM 360/91 design in several ways to help clarify the example. First, instead of the two-address format of the IBM 360 instructions, we will use three-address instructions to avoid potential confusion. The example sequence contains only register-register instructions. To reduce the number of machine cycles we have to trace, we will allow the FLOS to dispatch (in program order) up to two instructions in every cycle. We also assume that an instruction can begin execution in the same cycle that it is dispatched to a reservation station. We keep the same latencies of two and three cycles for add and multiply instructions, respectively. However, we allow an instruction to forward its result to dependent instructions during its last execution cycle, and a dependent instruction can begin execution in the next cycle. The tag values of 1, 2, and 3 are used to identify the three reservation stations of the adder functional unit, while 4 and 5 are used to identify the two reservation stations of the multiply/divide functional unit. These tag values are called the *IDs* of the reservation stations. The example sequence consists of the following four register-register instructions.

w: R4 ← R0 + R8

x: R2 ← R0 * R4

y: R4 ← R4 + R8

z: R8 ← R4 * R2

Figure 5.23 illustrates the first three cycles of execution. In cycle 1, instructions w and x are dispatched (in order) to reservation stations 1 and 4. The destination registers of instructions w and x are R4 and R2 (i.e., FLRs 4 and 2), respectively. The busy bits of these two registers are set. Since instruction w is dispatched to reservation station 1, the tag value of 1 is entered into the tag field of R4, indicating that the instruction in reservation station 1 will produce the result for updating R4. Similarly the tag value of 4 is entered into the tag field of R2. Both source operands of instruction w are available, so it begins execution immediately. Instruction x

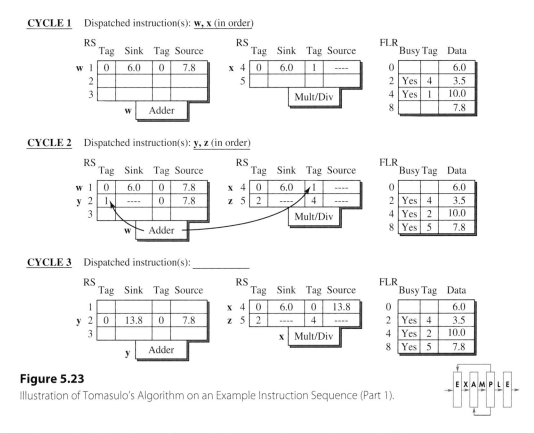

Figure 5.23
Illustration of Tomasulo's Algorithm on an Example Instruction Sequence (Part 1).

requires the result (R4) of instruction w for its second (source) operand. Hence when instruction x is dispatched to reservation station 4, the tag field of the second operand is written the tag value of 1, indicating that the instruction in reservation station 1 will produce the needed operand.

During cycle 2, instructions y and z are dispatched (in order) to reservation stations 2 and 5, respectively. Because it needs the result of instruction w for its first operand, instruction y, when it is dispatched to reservation station 2, receives the tag value of 1 in the tag field of the first operand. Similarly instruction z, dispatched to reservation station 5, receives the tag values of 2 and 4 in its two tag fields, indicating that reservation stations 2 and 4 will eventually produce the two operands it needs. Since R4 is the destination of instruction y, the tag field of R4 is updated with the new tag value of 2, indicating reservation station 2 (i.e., instruction y) is now responsible for the pending update of R4. The busy bit of R4 remains set. The busy bit of R8 is set when instruction z is dispatched to reservation station 5, and the tag field of R8 is set to 5. At the end of cycle 2, instruction w finishes execution and broadcasts its ID (reservation station 1) and its result onto the CDB. All the tag fields containing the tag value of 1 will trigger a tag match and latch in the broadcasted result. The first tag field of reservation station 2 (holding instruction y) and the second tag field of reservation station 4 (holding instruction x)

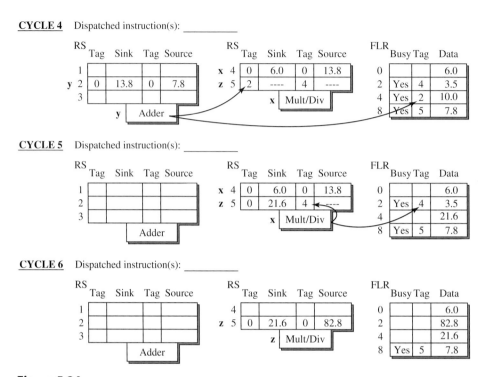

Figure 5.24
Illustration of Tomasulo's Algorithm on an Example Instruction Sequence (Part 2).

have such tag matches. Hence the result of instruction w is forwarded to dependent instructions x and y.

In cycle 3, instruction y begins execution in the adder unit, and instruction x begins execution in the multiply/divide unit. Instruction y finishes execution in cycle 4 (see Figure 5.24) and broadcasts its result on the CDB along with the tag value of 2 (its reservation station ID). The first tag field in reservation station 5 (holding instruction z) and the tag field of R4 have tag matches and pull in the result of instruction y. Instruction x finishes execution in cycle 5 and broadcasts its result on the CDB along with the tag value of 4. The second tag field in reservation station 5 (holding instruction z) and the tag field of R2 have tag matches and pull in the result of instruction x. In cycle 6, instruction z begins execution and finishes in cycle 8.

Figure 5.25(a) illustrates the data flow graph of this example sequence of four instructions. The four solid arcs represent the four true data dependences, while the other three arcs represent the anti- and output dependences. Instructions are dispatched in program order. Anti-dependences are resolved by copying an operand at dispatch time to the reservation station. Hence, it is not possible for a trailing instruction to overwrite a register before an earlier instruction has a chance to read that register. If the operand is still pending, the dispatched instruction will receive

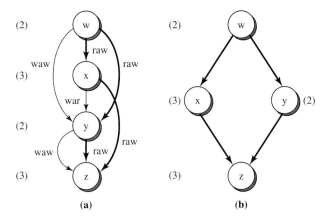

Figure 5.25
Data Flow Graphs of the Example Instruction Sequence: (a) All Data Dependences; (b) True Data Dependences.

the tag for that operand. When that operand becomes available, the instruction will receive that operand via a tag match in its reservation station.

As an instruction is dispatched, the tag field of its destination register is written with the reservation station ID of that instruction. When a subsequent instruction with the same destination register is dispatched, the same tag field will be updated with the reservation station ID of this new instruction. The tag field of a register always contains the reservation station ID of the latest updating instruction. If there are multiple instructions in flight that have the same destination register, only the latest instruction will be able to update that register. Output dependences are implicitly resolved by making it impossible for an earlier instruction to update a register after a later instruction has updated the same register. This does introduce the problem of not being able to support precise exception since the register file does not necessarily evolve through all its sequential states; i.e., a register can potentially miss an intermediate update. For example, in Figure 5.23, at the end of cycle 2, instruction w should have updated its destination register R4. However, instruction y has the same destination register, and when it was dispatched earlier in that cycle, the tag field of R4 was changed from 1 to 2 anticipating the update of R4 by instruction y. At the end of cycle 2 when instruction w broadcasts its tag value of 1, the tag field of R4 fails to trigger a tag match and does not pull in the result of instruction w. Eventually R4 will be updated by instruction y. However, if an exception is triggered by instruction x, precise exception will be impossible since the register file does not evolve through all its sequential states.

Tomasulo's algorithm resolves anti- and output dependences via a form of register renaming. Each definition of an FLR triggers the renaming of that register to a register tag. This tag is taken from the ID of the reservation station containing the instruction that redefines that register. This effectively removes false dependences from causing pipeline stalls. Hence, the data flow limit is strictly determined by the true data dependences. Figure 5.25(b) depicts the data flow graph involving only true data dependences. As shown in Figure 5.25(a) if all four instructions were required to execute sequentially to enforce all the data dependences, including anti- and output dependences, the total latency required for executing this sequence of instructions would be 10 cycles, given the latencies of 2 and 3 cycles for addition and multiplication instructions, respectively. When only true dependences are considered, Figure 5.25(b) reveals that the critical path is only 8 cycles, i.e., the path involving instructions w, x, and z. Hence, the data flow limit for this sequence of four instructions is 8 cycles. This limit is achieved by Tomasulo's algorithm as demonstrated in Figures 5.23 and 5.24.

5.2.5 Dynamic Execution Core

Most current state-of-the-art superscalar microprocessors consist of an out-of-order execution core sandwiched between an in-order front end, which fetches and dispatches instructions in program order, and an in-order back end, which completes and retires instructions also in program order. The out-of-order execution core (also referred to as the *dynamic execution* core), resembling a refinement of Tomasulo's algorithm, can be viewed as an embedded data flow, or *micro-dataflow,* engine that attempts to approach the data flow limit in instruction execution. The operation of such a dynamic execution core can be described according to the three phases in the pipeline, namely, *instruction dispatching, instruction execution,* and *instruction completion;* see Figure 5.26.

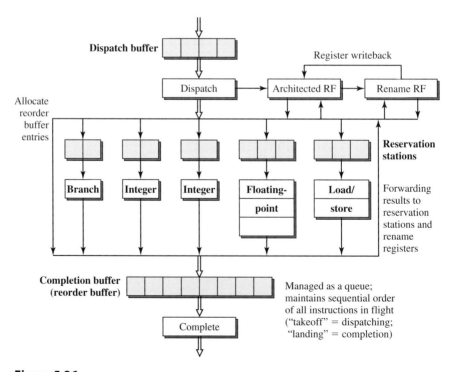

Figure 5.26
Micro-Dataflow Engine for Dynamic Execution.

The instruction dispatching phase consists of *renaming* of destination registers, *allocating* of reservation station and reorder buffer entries, and *advancing* instructions from the dispatch buffer to the reservation stations. For ease of presentation, we assume here that register renaming is performed in the dispatch stage. All redefinitions of architected registers are renamed to rename registers. Trailing uses of these redefinitions are assigned the corresponding rename register specifiers. This ensures that all producer-consumer relationships are properly identified and all false register dependences are removed.

Instructions in the dispatch buffer are then dispatched to the appropriate reservation stations based on instruction type. Here we assume the use of distributed reservation stations, and we use *reservation station* to refer to the (multientry) instruction buffer attached to each functional unit and *reservation station entry* to refer to one of the entries of this buffer. Simultaneous with the allocation of reservation station entries for the dispatched instructions is the allocation of entries in the reorder buffer for the same instructions. Reorder buffer entries are allocated according to program order.

Typically, for an instruction to be dispatched there must be the availability of a rename register, a reservation station entry, and a reorder buffer entry. If any one of these three is not available, instruction dispatching is stalled. The actual dispatching of instructions from the dispatch buffer entries to the reservation station entries is via a complex routing network. If the connectivity of this routing network is less than that of a full crossbar (this is frequently the case in real designs), then stalling can also occur due to resource contention in the routing network.

The instruction execution phase consists of *issuing* of ready instructions, *executing* the issued instructions, and *forwarding* of results. Each reservation station is responsible for identifying instructions that are ready to execute and for scheduling their execution. When an instruction is first dispatched to a reservation station, it may not have all its source operands and therefore must wait in the reservation station. Waiting instructions continually monitor the busses for tag matches. When a tag match is triggered, indicating the availability of the pending operand, the result being broadcasted is latched into the reservation station entry. When an instruction in a reservation station entry has all its operands, it becomes ready for execution and can be issued into the functional unit. In a given machine cycle if multiple instructions in a reservation station are ready, a scheduling algorithm is used (typically oldest first) to pick one of them for issuing into the functional unit to begin execution. If there is only one functional unit connected to a reservation station (as is the case for distributed reservation stations), then that reservation station can only issue one instruction per cycle.

Once issued into a functional unit, an instruction is executed. Functional units can vary in terms of their latency. Some have single-cycle latencies, others have fixed multiple-cycle latencies. Certain functional units can require a variable number of cycles, depending on the values of the operands and the operation being performed. Typically, even with function units that require multiple-cycle latencies, once an instruction begins execution in a pipelined functional unit, there is no further stalling of that instruction in the middle of the execution pipeline since all data dependences have been resolved prior to issuing and there shouldn't be any resource contention.

When an instruction finishes execution, it asserts its destination tag (i.e., the specifier of the rename register assigned for its destination) and the actual result onto a forwarding bus. All dependent instructions waiting in the reservation stations will trigger a tag match and latch in the broadcasted result. This is how an instruction forwards its result to other dependent instructions without requiring the intermediate steps of updating and then reading of the dependent register. Concurrent with result forwarding, the RRF uses the broadcasted tag as an index and loads the broadcasted result into the selected entry of the RRF.

Typically a reservation station entry is deallocated when its instruction is issued in order to allow another trailing instruction to be dispatched into it. Reservation station saturation can cause instruction dispatch to stall. Certain instructions whose execution can induce an exceptional condition may require rescheduling for execution in a future cycle. Frequently, for these instructions, their reservation station entries are not deallocated until they finish execution without triggering any exceptions. For example, a load instruction can potentially trigger a D-cache miss that may require many cycles to service. Instead of stalling the functional unit, such an excepting load can be reissued from the reservation station after the miss has been serviced.

In a dynamic execution core as described previously, a producer-consumer relationship is satisfied without having to wait for the writing and then the reading of the dependent register. The dependent operand is directly forwarded from the producer instruction to the consumer instruction to minimize the latency incurred due to the true data dependence. Assuming that an instruction can be issued in the same cycle that it receives its last pending operand via a forwarding bus, if there is no other instruction contending for the same functional unit, then this instruction should be able to begin execution in the cycle immediately following the availability of all its operands. Hence, if there are adequate resources such that no stalling due to structural dependences occurs, then the dynamic execution core should be able to approach the data flow limit.

5.2.6 Reservation Stations and Reorder Buffer

Other than the functional units, the critical components of the dynamic execution core are the *reservation stations* and the *reorder buffer*. The operations of these components dictate the function of the dynamic execution core. Here we present the issues associated with the implementation of the reservation station and the reorder buffer. We present their organization and behavior with special focus on loading and unloading of an entry of a reservation station and the reorder buffer.

There are three tasks associated with the operation of a reservation station: *dispatching, waiting,* and *issuing*. A typical reservation station is shown in Figure 5.27(b), and the various fields in an entry of a reservation station are illustrated in Figure 5.27(a). Each entry has a busy bit, indicating that the entry has been allocated, and a ready bit, indicating that the instruction in that entry has all its source operands. Dispatching involves loading an instruction from the dispatch buffer into an entry of the reservation station. Typically the dispatching of an instruction requires the following three steps: select a free, i.e., not busy, reservation

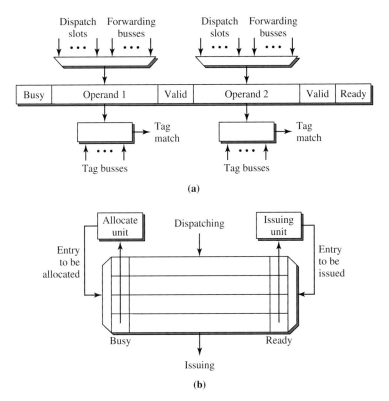

Figure 5.27
Reservation Station Mechanisms: (a) A Reservation Station Entry; (b) Dispatching into and Issuing from a Reservation Station.

station entry; load operands and/or tags into the selected entry; and set the busy bit of that entry. The selection of a free entry is based on the busy bits and is performed by the allocate unit. The allocate unit examines all the busy bits and selects one of the nonbusy entries to be the allocated entry. This can be implemented using a priority encoder. Once an entry is allocated, the operands and/or tags of the instruction are loaded into the entry. Each entry has two operand fields, each of which has an associated valid bit. If the operand field contains the actual operand, then the valid bit is set. If the field contains a tag, indicating a pending operand, then its valid bit is reset and it must wait for the operand to be forwarded. Once an entry is allocated, its busy bit must be set.

An instruction with a pending operand must wait in the reservation station. When a reservation station entry is waiting for a pending operand, it must continuously monitor the tag busses. When a tag match occurs, the operand field latches in the forwarded result and sets its valid bit. When both operand fields are valid, the ready bit is set, indicating that the instruction has all its source operands and is ready to be issued. This is usually referred to as *instruction wake up*.

The issuing step is responsible for selecting a ready instruction in the reservation station and issues it into the functional unit. This is usually referred to as *instruction select*. All the ready instructions are identified by their ready bits being set. The selecting of a ready instruction is performed by the issuing unit based on a scheduling heuristic; see Figure 5.27(b). The heuristic can be based on program order or how long each ready instruction has been waiting in the reservation station. Typically when an instruction is issued into the functional unit, its reservation station entry is deallocated by resetting the busy bit.

A large reservation station can be quite complex to implement. On its input side, it must support many possible sources, including all the dispatch slots and forwarding busses; see Figure 5.27(a). The data routing network on its input side can be quite complex. During the waiting step, all operand fields of a reservation station with pending operands must continuously compare their tags against potentially multiple tag busses. This is comparable to doing an associative search across all the reservation station entries involving multiple keys (tag busses). If the number of entries is small, this is quite feasible. However, as the number of entries increases, the increase in complexity is quite significant. This portion of the hardware is commonly referred to as the *wake-up logic*. When the entry count increases, it also complicates the issuing unit and the scheduling heuristic in selecting the best ready instruction to issue. This portion of the hardware is commonly referred to as the *select logic*. In any given machine cycle, there can be multiple ready instructions. The select logic must determine the best one to issue. For a superscalar machine, a reservation station can potentially support multiple instruction issues per cycle, in which case the select logic must pick the best subset of instructions to issue among all the ready instructions.

The reorder buffer contains all the instructions that are *in flight*, i.e., all the instructions that have been dispatched but not yet completed architecturally. These include all the instructions waiting in the reservation stations and executing in the functional units and those that have finished execution but are waiting to be completed in program order. The status of each instruction in the reorder buffer can be tracked using several bits in each entry of the reorder buffer. Each instruction can be in one of several states, i.e., waiting execution, in execution, and finished execution. These status bits are updated as an instruction traverses from one state to the next. An additional bit can also be used to indicate whether an instruction is speculative (in the predicted path) or not. If speculation can cross multiple branches, additional bits can be employed to identify which speculative basic block an instruction belongs to. When a branch is resolved, a speculative instruction can become nonspeculative (if the prediction is correct) or invalid (if the prediction is incorrect). Only finished and nonspeculative instructions can be completed. An instruction marked invalid is not architecturally completed when exiting the reorder buffer. Figure 5.28(a) illustrates the fields typically found in a reorder buffer entry; in this figure the rename register field is also included.

The reorder buffer is managed as a circular queue using a head pointer and a tail pointer; see Figure 5.28(b). The tail pointer is advanced when reorder buffer entries are allocated at instruction dispatch. The number of entries that can be

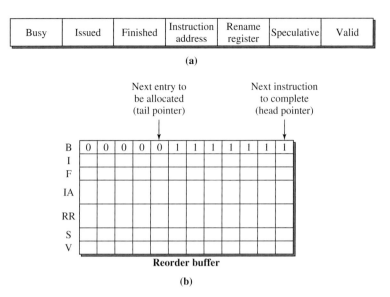

Figure 5.28
(a) Reorder Buffer Entry; (b) Reorder Buffer Organization.

allocated per cycle is limited by the dispatch bandwidth. Instructions are completed from the head of the queue. From the head of the queue as many instructions that have finished execution can be completed as the completion bandwidth allows. The completion bandwidth is determined by the capacity of another routing network and the ports available for register writeback. One of the critical issues is the number of write ports to the architected register file that are needed to support the transferring of data from the rename registers (or the reorder buffer entries if they are used as rename registers) to the architected registers. When an instruction is completed, its rename register and its reorder buffer entry are deallocated. The head pointer to the reorder buffer is also appropriately updated. In a way the reorder buffer can be viewed as the heart or the central control of the dynamic execution core because the status of all in-flight instructions is tracked by the reorder buffer.

It is possible to combine the reservation stations and the reorder buffer into one single structure, called the *instruction window,* that manages all the instructions in flight. Since at dispatch an entry in the reservation station and an entry in the reorder buffer must be allocated for each instruction, they can be combined as one entry in the instruction window. Hence, instructions are dispatched into the instruction window, entries of the instruction window monitor the tag busses for pending operands, results are forwarded into the instruction window, instructions are issued from the instruction window when ready, and instructions are completed from the instruction window. The size of the instruction window determines the maximum number of instructions that can be simultaneously in flight within the machine and consequently the degree of instruction-level parallelism that can be achieved by the machine.

5.2.7 Dynamic Instruction Scheduler

The dynamic instruction scheduler is the heart of a dynamic execution core. We use the term *dynamic instruction scheduler* to include the instruction window and its associated instruction wake-up and select logic. Currently there are two styles to the design of the dynamic instruction scheduler, namely, *with data capture* and *without data capture*.

Figure 5.29(a) illustrates a scheduler with data capture. With this style of scheduler design, when dispatching an instruction, those operands that are ready are copied from the register file (either architected or physical) into the instruction window; hence, we have the term *data captured*. For the operands that are not ready, tags are copied into the instruction window and used to latch in the operands when they are forwarded by the functional units. Results are forwarded to their waiting instructions in the instruction window. In effect, result forwarding and instruction wake up are combined, a la Tomasulo's algorithm. A separate forwarding path is needed to also update the register file so that subsequent dependent instructions can grab their source operands from the register file when they are being dispatched.

Some recent microprocessors have adopted a different style that does not employ data capture in the scheduler design; see Figure 5.29(b). In this style, register read is performed after the scheduler, as instructions are being issued to the functional units. At instruction dispatch there is no copying of operands into the instruction window; only tags (or pointers) for operands are loaded into the window. The scheduler still performs tag match to wake up ready instructions. However, results from functional units are only forwarded to the register file. All ready instructions that are issued obtain their operands directly from the register file just prior to execution. In effect, result forwarding and instruction wake up are decoupled. For instruction wake up only the tag needs to be forwarded to the scheduler. With the non-data-captured style of scheduler, the size (width) of the instruction window can be significantly reduced, and the much wider result-forwarding path to the scheduler is not needed.

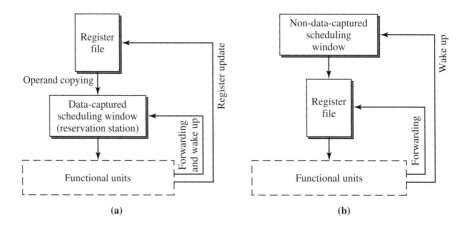

Figure 5.29
Dynamic Instruction Scheduler Design: (a) With Data Capture; (b) Without Data Capture.

There is a close relationship between register renaming and instruction scheduling. As stated earlier, one purpose for doing dynamic register renaming is to eliminate the false dependences induced by register recycling. Another purpose is to establish the producer-consumer relationship between two dependent instructions. A true data dependence is determined by the common rename register specifier in the producer and consumer instructions. The rename register specifier can function as the tag for result forwarding. In the non-data-captured scheduler of Figure 5.29(b) the register specifiers are used to access the register file for retrieving source operands; the (destination) register specifiers are used as tags for waking up dependent instructions in the scheduler. For the data-captured type of scheduler of Figure 5.29(a), the tags used for result forwarding and instruction wake up do not have to be actual register specifiers. The tags are mainly used to identify producer-consumer relationships between dependent instructions and can be assigned arbitrarily. For example, Tomasulo's algorithm uses reservation station IDs as the tags for forwarding results to dependent instructions as well as for updating architected registers. There is no explicit register renaming involving physical rename registers.

5.2.8 Other Register Data Flow Techniques

For many years the data flow limit has been assumed to be an absolute theoretical limit and the ultimate performance goal. Extensive research efforts on data flow architectures and data flow machines have been going on for over three decades. The data flow limit assumes that true data dependences are absolute and cannot possibly be overcome. Interestingly, in the late 1960s and the early 1970s a similar assumption was made concerning control dependences. It was generally thought that control dependences are absolute and that when encountering a conditional branch instruction there is no choice but to wait for that conditional branch to be executed before proceeding to the next instruction due to the uncertainty of the actual control flow. Since then, we have witnessed tremendous strides made in the area of branch prediction techniques. Conditional branches and associated control dependences are no longer absolute barriers and can frequently be overcome by speculating on the direction and the target address of the branch. What made such speculation possible is that frequently the outcome of a branch instruction is quite predictable. It wasn't until 1995 that researchers began to also question the absoluteness of true data dependences.

In 1996 several research papers appeared that proposed the concept of *value prediction*. The first paper by Lipasti, Wilkerson, and Shen focused on predicting load values based on the observation that frequently the values being loaded by a particular static load instruction are quite predictable [Lipasti et al., 1996]. Their second paper generalized the same basic idea for predicting the result of ALU instructions [Lipasti and Shen, 1996]. Experimental data based on real input data sets indicate that the results produced by many instructions are actually quite predictable. The notion of *value locality* indicates that certain instructions tend to repeatedly produce the same small set (sometimes one) of result values. By tracking the results produced by these instructions, future values can become predictable based on the historical values. Since these seminal papers, numerous papers have been published in recent years proposing various designs of value predictors [Mendelson and Gabbay, 1997; Sazeides and Smith, 1997; Calder et al., 1997; Gabbay and Mendelson, 1997;

1998a; 1998b; Calder et al., 1999]. In a recent study, it was shown that a hybrid value predictor can achieve prediction rates of up to 80% and a realistic design incorporating value prediction can achieve IPC improvements in the range of 8.6% to 23% for the SPEC benchmarks [Wang and Franklin, 1997].

When the result of an instruction is correctly predicted via value prediction, typically performed during the fetch stage, a subsequent dependent instruction can begin execution using this speculative result without having to wait for the actual decoding and execution of the leading instruction. This effectively removes the serialization constraint imposed by the true data dependence between these two instructions. In a way this particular dependence edge in the data flow graph is effectively removed when correct value prediction is performed. Hence, value prediction provides the potential to exceed the classical data flow limit. Of course, validation is still required to ensure that the prediction is correct and becomes the new limit on instruction execution throughput. Value prediction becomes effective in increasing machine performance if misprediction rarely occurs and the misprediction penalty is small (e.g., zero or one cycle) and if the validation latency is less than the average instruction execution latency. Clearly, efficient implementation of value prediction is crucial in ensuring its efficacy in improving performance.

Another recently proposed idea is called *dynamic instruction reuse* [Sodani and Sohi, 1997]. Similar to the concept of value locality, it has been observed through experiments with real programs that frequently the same sequence of instructions is repeatedly executed using the same set of input data. This results in redundant computation being performed by the machine. Dynamic instruction reuse techniques attempt to track such redundant computations, and when they are detected, the previous results are used without performing the redundant computations. These techniques are nonspeculative; hence, no validation is required. While value prediction can be viewed as the elimination of certain dependence edges in the data flow graph, dynamic instruction reuse techniques attempt to remove both nodes and edges of a subgraph from the data flow graph. A much earlier research effort had shown that such elimination of redundant computations can yield significant performance gains for programs written in functional languages [Harbison, 1980; 1982]. A more recent study also yields similar data on the presence of redundant computations in real programs [Richardson, 1992]. This is an area that is currently being actively researched, and new insightful results can be expected.

We will revisit these advanced register data flow techniques in Chapter 10 in greater detail.

5.3 Memory Data Flow Techniques

Memory instructions are responsible for moving data between the main memory and the register file, and they are essential for supporting the execution of ALU instructions. Register operands needed by ALU instructions must first be loaded from memory. With a limited number of registers, during the execution of a program not all the operands can be kept in the register file. The compiler generates *spill code* to temporarily place certain operands out to the main memory and to reload them when they are needed. Such spill code is implemented using store and load instructions. Typically, the compiler only allocates scalar variables into registers. Complex data structures, such as arrays and linked lists, that far exceed the size of the register file are usually kept in the main memory. To perform operations on such data structures, load and store instructions are required. The effective processing of load/store instructions can minimize the overhead of moving data between the main memory and the register file.

The processing of load/store instructions and the resultant memory data flow can become a bottleneck to overall machine performance due to the potential long latency for executing memory instructions. The long latency of load/store instructions results from the need to compute a memory address and the need to access a memory location. To support virtual memory, the computed memory address (called the *virtual address*) also needs to be translated into a physical address before the physical memory can be accessed. Cache memories are very effective in reducing the effective latency for accessing the main memory. Furthermore, various techniques have been developed to reduce the overall latency and increase the overall throughput for processing load/store instructions.

5.3.1 Memory Accessing Instructions

The execution of memory data flow instructions occurs in three steps: memory address generation, memory address translation, and data memory accessing. We first state the basis for these three steps and then describe the processing of load/store instructions in a superscalar pipeline.

The register file and the main memory are defined by the instruction set architecture for data storage. The main memory as defined in an instruction set architecture is a collection of 2^n memory locations with random access capability; i.e., every memory location is identified by an n-bit address and can be directly accessed with the same latency. Just like the architected register file, the main memory is an architected entity and is visible to the software instructions. However, unlike the register file, the address that identifies a particular memory location is usually not explicitly stored as part of the instruction format. Instead, a memory address is usually generated based on a register and an offset specified in the instruction. Hence, address generation is required and involves the accessing of the specified register and the adding of the offset value.

In addition to address generation, address translation is required when virtual memory is implemented in a system. The architected main memory constitutes the virtual address space of the program and is viewed by each program as its private address space. The physical memory that is implemented in a machine constitutes the physical address space, which may be smaller than the virtual address space and may even be shared by multiple programs. Virtual memory is a mechanism that maps the virtual address space of a program to the physical address space of the machine. With such address mapping, virtual memory is able to support the execution of a program with a virtual address space that is larger than the physical address space, and the multiprogramming paradigm by mapping multiple virtual address spaces to the same physical address space. This mapping mechanism involves the translation of the computed effective address, i.e., the virtual address, into a physical address that can be used to access the physical memory. This mechanism is usually implemented using a mapping table, and address translation is performed via a table lookup.

The third step in processing a load/store instruction is memory accessing. For load instructions data are read from a memory location and stored into a register, while for store instructions a register value is stored into a memory location. While the first two steps of address generation and address translation are performed in identical fashion for both loads and stores, the third step is performed differently for loads and stores by a superscalar pipeline.

In Figure 5.30, we illustrate these three steps as occurring in three pipeline stages. The first pipe stage performs effective address generation. We assume the typical addressing mode of register indirect with an offset for both load and store instructions. For a load instruction, as soon as the address register operand is available, it is issued into the pipelined functional unit and the effective address is generated by the first pipe stage. A store instruction must wait for the availability of both the address register and the data register operands before it is issued.

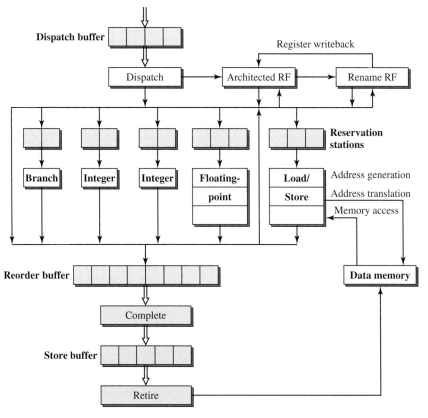

Figure 5.30
Processing of Load/Store Instructions.

After the first pipe stage generates the effective address, the second pipe stage translates this virtual address into a physical address. Typically, this is done by accessing the translation lookaside buffer (TLB), which is a hardware-controlled table containing the mapping of virtual to physical addresses. The TLB is essentially a cache of the page table that is stored in the main memory. (Section 3.6 provides more background material on the page table and the TLB.) It is possible that the virtual address being translated belongs to a page whose mapping is not currently resident in the TLB. This is called a *TLB miss*. If the particular mapping is present in the page table, then it can be retrieved by accessing the page table in the main memory. Once the missing mapping is retrieved and loaded into the TLB, the translation can be completed. It is also possible that the mapping is not resident even in the page table, meaning that the particular page being referenced has not been mapped and is not resident in the main memory. This will induce a *page fault* and require accessing disk storage to retrieve the missing page. This constitutes a program exception and will necessitate the suspension of the execution of the current program.

After successful address translation in the second pipe stage, a load instruction accesses the data memory during the third pipe stage. At the end of this machine cycle, the data are retrieved from the data memory and written into either the rename register or the reorder buffer. At this point the load instruction finishes execution. The updating of the architected register is not performed until this load instruction is completed from the reorder buffer. Here we assume that data memory access can be done in one machine cycle in the third pipe stage. This is possible if a data cache is employed. (Section 3.6 provides more background material on caches.) With a data cache, it is possible that the data being loaded are not resident in the data cache. This will result in a data *cache miss* and require the filling of the data cache from the main memory. Such cache misses can necessitate the stalling of the load/store pipeline.

Store instructions are processed somewhat differently than load instructions. Unlike a load instruction, a store instruction is considered as having *finished* execution at the end of the second pipe stage when there is a successful translation of the address. The register data to be stored to memory are kept in the reorder buffer. At the time when the store is being completed, these data are then written out to memory. The reason for this delayed access to memory is to prevent the premature and potentially erroneous update of the memory in case the store instruction may have to be flushed due to the occurrence of an exception or a branch misprediction. Since load instructions only read the memory, their flushing will not result in unwanted side effects to the memory state.

For a store instruction, instead of updating the memory at completion, it is possible to move the data to the store buffer at completion. The store buffer is a FIFO buffer that buffers architecturally completed store instructions. Each of these store instructions is then retired, i.e., updates the memory, when the memory bus is available. The purpose of the store buffer is to allow stores to be retired when the memory bus is not busy, thus giving priority to loads that need to access the memory bus. We use the term *completion* to refer to the updating of the CPU state and the term *retiring* to refer to the updating of the memory state. With the store buffer, a store instruction can be architecturally complete but not yet retired to memory.

When a program exception occurs, the instructions that follow the excepting instruction and that may have finished out of order, must be flushed from the reorder buffer; however, the store buffer must be drained, i.e., the store instructions in the store buffer must be retired, before the excepting program can be suspended.

We have assumed here that both address translation and memory accessing can be done in one machine cycle. Of course, this is only the case when both the TLB and the first level of the memory hierarchy return hits. An in-depth treatment of memory hierarchies that maximize the occurrence of cache hits by exploiting temporal and spatial locality and using various forms of caching is provided in Chapter 3. In Section 5.3.2, we will focus specifically on the additional complications that result from out-of-order execution of memory references and on some of the mechanisms used by modern problems to address these complications.

5.3.2 Ordering of Memory Accesses

A memory data dependence exists between two load/store instructions if they both reference the same memory location, i.e., there exists an *aliasing*, or collision, of the two memory addresses. A load instruction performs a read from a memory location, while a store instruction performs a write to a memory location. Similar to register data dependences, read-after-write (RAW), write-after-read (WAR), and write-after-write (WAW) dependences can exist between load and store instructions. A store (load) instruction followed by a load (store) instruction involving the same memory location will induce a RAW (WAR) memory data dependence. Two stores to the same memory location will induce a WAW dependence. These memory data dependences must be enforced in order to preserve the correct semantics of the program.

One way to enforce memory data dependences is to execute all load/store instructions in program order. Such total ordering of memory instructions is sufficient for enforcing memory data dependences but not necessary. It is conservative and can impose an unnecessary limitation on the performance of a program. We use the example in Figure 5.31 to illustrate this point. DAXPY is the name of a piece

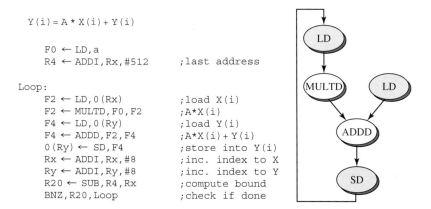

```
        Y(i) = A * X(i) + Y(i)

        F0 ← LD,a
        R4 ← ADDI,Rx,#512      ;last address

    Loop:
        F2 ← LD,0(Rx)          ;load X(i)
        F2 ← MULTD,F0,F2       ;A*X(i)
        F4 ← LD,0(Ry)          ;load Y(i)
        F4 ← ADDD,F2,F4        ;A*X(i)+Y(i)
        0(Ry) ← SD,F4          ;store into Y(i)
        Rx ← ADDI,Rx,#8        ;inc. index to X
        Ry ← ADDI,Ry,#8        ;inc. index to Y
        R20 ← SUB,R4,Rx        ;compute bound
        BNZ,R20,Loop           ;check if done
```

Figure 5.31
The DAXPY Example.

of code that multiplies an array by a coefficient and then adds the resultant array to another array. DAXPY (derived from "double precision A times X plus Y") is a kernel in the LINPAC routines and is commonly found in many numerical programs. Notice that all the iterations of this loop are data-independent and can be executed in parallel. However, if we impose the constraint that all load/store instructions be executed in total order, then the first load instruction of the second iteration cannot begin until the store instruction of the first iteration is performed. This constraint will effectively serialize the execution of all the iterations of this loop.

By allowing load/store instructions to execute out of order, without violating memory data dependences, performance gain can be achieved. Take the example of the DAXPY loop. The graph in Figure 5.31 represents the true data dependences involving the core instructions of the loop body. These dependences exist among the instructions of the same iteration of the loop. There are no data dependences between multiple iterations of the loop. The loop closing branch instruction is highly predictable; hence, the fetching of subsequent iterations can be done very quickly. The same architected registers specified by instructions from subsequent iterations are dynamically renamed by register renaming mechanisms; hence, there are no register dependences between the iterations due to the dynamic reuse of the same architected registers. Consequently if load/store instructions are allowed to execute out of order, the load instructions from a trailing iteration can begin before the execution of the store instruction from an earlier iteration. By overlapping the execution of multiple iterations of the loop, performance gain is achieved for the execution of this loop.

Memory models impose certain limitations on the out-of-order execution of load/store instructions by a processor. First, to facilitate recovery from exceptions, the sequential state of the memory must be preserved. In other words, the memory state must evolve according to the sequential execution of load/store instructions. Second, many shared-memory multiprocessor systems assume the sequential consistency memory model, which requires that the accessing of the shared memory by each processor be done according to program order [Lamport, 1979, Adve and Gharachorloo, 1996].[1] Both of these reasons effectively require that store instructions be executed in program order, or at least the memory must be updated as if stores are performed in program order. If stores are required to execute in program order, WAW and WAR memory data dependences are implicitly enforced and are not an issue. Hence, only RAW memory data dependences must be enforced.

5.3.3 Load Bypassing and Load Forwarding

Out-of-order execution of load instructions is the primary source for potential performance gain. As can be seen in the DAXPY example, load instructions are frequently at the beginning of dependence chains, and their early execution can facilitate the early execution of other dependent instructions. While relative to memory data dependences, load instructions are viewed as performing read operations on the memory locations, they are actually performing write operations to their destination registers. With loads being register-defining (DEF) instructions, they

---

[1]Consistency models are discussed in greater detail in Chapter 11.

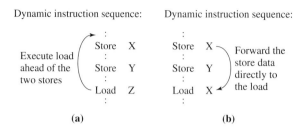

Figure 5.32
Early Execution of Load Instructions: (a) Load Bypassing; (b) Load Forwarding.

are typically followed immediately by other dependent register use (USE) instructions. The goal is to allow load instructions to begin execution as early as possible, possibly jumping ahead of other preceding store instructions, as long as RAW memory data dependences are not violated and that memory is updated according to the sequential memory consistency model.

Two specific techniques for early out-of-order execution of loads are *load bypassing* and *load forwarding*. As shown in Figure 5.32(a), load bypassing allows a trailing load to be executed earlier than preceding stores if the load address does not alias with the preceding stores; i.e., there is no memory data dependence between the stores and the load. On the other hand, if a trailing load aliases with a preceding store, i.e., there is a RAW dependence from the store to the load, load forwarding allows the load to receive its data directly from the store without having to access the data memory; see Figure 5.32(b). In both of these cases, earlier execution of a load instruction is achieved.

Before we discuss the issues that must be addressed in order to implement load bypassing and load forwarding, first we present the organization of the portion of the execution core responsible for processing load/store instructions. This organization, shown in Figure 5.33, is used as the vehicle for our discussion on load bypassing and load forwarding. There is one store unit (two pipe stages) and one load unit (three pipe stages); both are fed by a common reservation station. For now we assume that load and store instructions are issued from this shared reservation station in program order. The store unit is supported by a store buffer. The load unit and the store buffer can access the data cache.

Given the organization of Figure 5.33, a store instruction can be in one of several states while it is in flight. When a store instruction is dispatched to the reservation station, an entry in the reorder buffer is allocated for it. It remains in the reservation station until all its source operands become available and it is issued into the pipelined execution unit. Once the memory address is generated and successfully translated, it is considered to have finished execution and is placed into the finished portion of the store buffer (the reorder buffer is also updated). The store buffer operates as a queue and has two portions, *finished* and *completed*. The finished portion contains those stores that have finished execution but are not yet architecturally

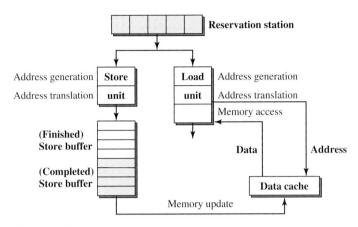

Figure 5.33

Mechanisms for Load/Store Processing: Separate Load and Store Units with In-Order Issuing from a Common Reservation Station.

completed. The completed portion of the store buffer contains those stores that are completed architecturally but waiting to update the memory. The identification of the two portions of the store buffer can be done via a pointer to the store buffer or a status bit in the store buffer entries. A store in the finished portion of the store buffer can potentially be speculative, and when a misspeculation is detected, it will need to be flushed from the store buffer. When a finished store is completed by the reorder buffer, it changes from the finished state to the completed state. This can be done by updating the store buffer pointer or flipping the status bit. When a completed store finally exits the store buffer and updates the memory, it is considered retired. Viewed from the perspective of the memory state, a store does not really finish its execution until it is retired. When an exception occurs, the stores in the completed portion of the store buffer must be drained in order to appropriately update the memory. So between being dispatched and retired, a store instruction can be in one of three states: *issued* (in the execution unit), *finished* (in the finished portion of the store buffer), or *completed* (in the completed portion of the store buffer).

One key issue in implementing load bypassing is the need to check for possible aliasing with preceding stores, i.e., those stores being bypassed. A load is considered to bypass a preceding store if the load reads from the memory before the store writes to the memory. Hence, before such a load is allowed to execute or read from the memory, it must be determined that it does not alias with all the preceding stores that are still in flight, i.e., those that have been issued but not retired. Assuming in-order issuing of load/store instructions from the load/store reservation station, all such stores should be sitting in the store buffer, including both the finished and the completed portions. The alias checking for possible dependence between the load and the preceding store can be done using the store buffer. A tag field containing the memory address of the store is incorporated with each entry of the store buffer. Once

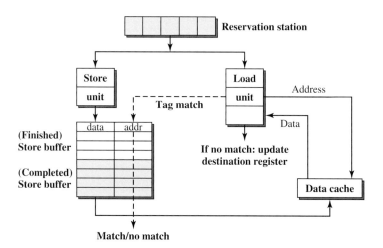

Figure 5.34
Illustration of Load Bypassing.

the memory address of the load is available, this address can be used to perform an associative search on the tag field of the store buffer entries. If a match occurs, then aliasing exists and the load is not allowed to execute out of order. Otherwise, the load is independent of the preceding stores in the store buffer and can be executed ahead of them. This associative search can be performed in the third pipe stage of the load unit concurrent with the accessing of the data cache; see Figure 5.34. If no aliasing is detected, the load is allowed to finish and the corresponding renamed destination register is updated with the data returned from the data cache. If aliasing is detected, the data returned by the data cache are discarded and the load is held back in the reservation station for future reissue.

Most of the complexity in implementing load bypassing lies in the store buffer and the associated associative search mechanism. To reduce the complexity, the tag field used for associative search can be reduced to contain only a subset of the address bits. Using only a subset of the address bits can reduce the width of the comparators needed for associative search. However, the result can be pessimistic. Potentially, an alias can be indicated by the narrower comparator when it really doesn't exist if the full-length address bits were used. Some of the load bypassing opportunities can be lost due to this compromise in the implementation. In general, the degradation of performance is minimal if enough address bits are used.

The load forwarding technique further enhances and complements the load bypassing technique. When a load is allowed to jump ahead of preceding stores, if it is determined that the load does alias with a preceding store, there is the potential to satisfy that load by forwarding the data directly from the aliased store. Essentially a memory RAW dependence exists between the leading store and the trailing load. The same associative search of the store buffer is needed. When aliasing is detected, instead of holding the load back for future reissue, the data from the aliased entry of the store buffer are forwarded to the renamed destination register

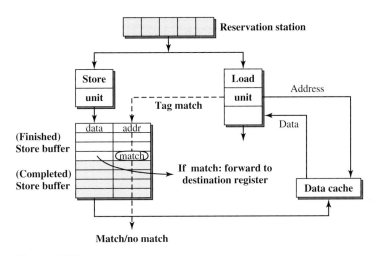

Figure 5.35
Illustration of Load Forwarding.

of the load instruction. This technique not only allows the load to be executed early, but also eliminates the need for the load to access the data cache. This can reduce the bandwidth pressure on the bus to the data cache.

To support load forwarding, added complexity to the store buffer is required; see Figure 5.35. First, the full-length address bits must be used for performing the associative search. When a subset of the address bits is used for supporting load bypassing, the only negative consequence is lost opportunity. For load forwarding the alias detection must be exact before forwarding of data can be performed; otherwise, it will lead to semantic incorrectness. Second, there can be multiple preceding stores in the store buffer that alias with the load. When such multiple matches occur during the associative search, there must be logic added to determine which of the aliased stores is the most recent. This will require additional priority encoding logic to identify the latest store on which the load is dependent before forwarding is performed. Third, an additional read port may be required for the store buffer. Prior to the incorporation of load forwarding, the store buffer has one write port that interfaces with the store unit and one read port that interfaces with the data cache. A new read port is now required that interfaces with the load unit; otherwise, port contention can occur between load forwarding and data cache update.

Significant performance improvement can be obtained with load bypassing and load forwarding. According to Mike Johnson, typically load bypassing can yield 11% to 19% performance gain, and load forwarding can yield another 1% to 4% of additional performance improvement [Johnson, 1991].

So far we have assumed that loads and stores share a common reservation station with instructions being issued from the reservation station to the store and the load units in program order. This in-order issuing assumption ensures that all the preceding stores to a load will be in the store buffer when the load is executed. This simplifies memory dependence checking; only an associative search of the store buffer is necessary. However, this in-order issuing assumption introduces an unnecessary limitation on the out-of-order execution of loads. A load instruction can be ready to be issued; however, a preceding store can hold up the issuing of the load even though the two memory instructions do not alias. Hence, allowing out-of-order issuing of loads and stores from the load/store reservation station can permit a greater degree of out-of-order and early execution of loads. This is especially beneficial if these loads are at the beginnings of critical dependence chains and their early execution can remove critical performance bottlenecks.

If out-of-order issuing from the load/store reservation station is allowed, a new problem must be solved. If a load is allowed to be issued out of order, then it is possible for some of the stores that precede it to still be in the reservation station or in the execution pipe stages, and not yet in the store buffer. Hence, simply performing an associative search on the entries of the store buffer is not adequate for checking for potential aliasing between the load and all its preceding stores. Worse yet, the memory addresses for these preceding stores that are still in the reservation station or in the execution pipe stages may not be available yet.

One approach is to allow the load to proceed, assuming no aliasing with the preceding stores that are not yet in the store buffer, and then validate this assumption later. With this approach, a load is allowed to issue out of order and be executed speculatively. If it does not alias with any of the stores in the store buffer, the load is allowed to finish execution. However, this load must be put into a new buffer called the finished load buffer; see Figure 5.36. The finished load buffer is managed in a similar fashion as the finished store buffer. A load is only resident in the finished load buffer after it finishes execution and before it is completed.

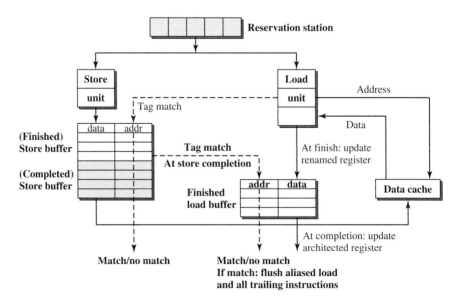

Figure 5.36
Fully Out-of-Order Issuing and Execution of Load and Store Instructions.

Whenever a store instruction is being completed, it must perform alias checking against the loads in the finished load buffer. If no aliasing is detected, the store is allowed to complete. If aliasing is detected, then it means that there is a trailing load that is dependent on the store, and that load has already finished execution. This implies that the speculative execution of that load must be invalidated and corrected by reissuing, or even refetching, that load and all subsequent instructions. This can require significant hardware complexity and performance penalty.

Aggressive early issuing of load instructions can lead to significant performance benefits. The ability to speculatively issue loads ahead of stores can lead to early execution of many dependent instructions, some of which can be other loads. This is important especially when there are cache misses. Early issuing of loads can lead to early triggering of cache misses which can in turn mask some or all of the cache miss penalty cycles. The downside with speculative issuing of loads is the potential overhead of having to recover from misspeculation. One way to reduce this overhead is to do alias or *dependence prediction*. In typical programs, the dependence relationship between a load and its preceding stores is quite predictable. A memory dependence predictor can be implemented to predict whether a load is likely to alias with its preceding stores. Such a predictor can be used to determine whether to speculatively issue a load or not. To obtain actual performance gain, aggressive speculative issuing of loads must be done very judiciously.

Moshovos [1998] proposed a memory dependence predictor that was used to directly bypass store data to dependent loads via memory cloaking. In subsequent work, Chrysos and Emer [1998] described a similar predictor called the store-set predictor.

5.3.4 Other Memory Data Flow Techniques

Other than load bypassing and load forwarding, there are other memory data flow techniques. These techniques all have the objectives of increasing the memory bandwidth and/or reducing the memory latency. As superscalar processors get wider, greater memory bandwidth capable of supporting multiple load/store instructions per cycle will be needed. As the disparity between processor speed and memory speed continues to increase, the latency of accessing memory, especially when cache misses occur, will become a serious bottleneck to machine performance. Sohi and Franklin [1991] studied high-bandwidth data memory systems.

One way to increase memory bandwidth is to employ multiple load/store units in the execution core, supported by a multiported data cache. In Section 5.3.3 we have assumed the presence of one store unit and one load unit supported by a single-ported data cache. The load unit has priority in accessing the data cache. Store instructions are queued in the store buffer and are retired from the store buffer to the data cache whenever the memory bus is not busy and the store buffer can gain access to the data cache. The overall data memory bandwidth is limited to one load/store instruction per cycle. This is a serious limitation, especially when there are bursts of load instructions. One way to alleviate this bottleneck is to provide two load units, as shown in Figure 5.37, and a *dual-ported data cache*. A dual-ported data cache is able to support two simultaneous cache accesses in

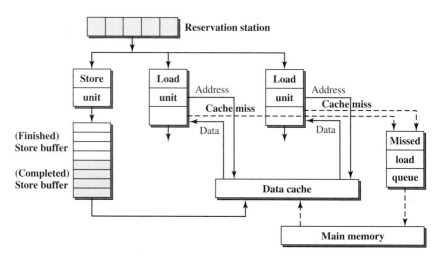

Figure 5.37
Dual-Ported and Nonblocking Data Cache.

every cycle. This will double the potential memory bandwidth. However, it comes with the cost of hardware complexity; a dual-ported cache can require doubling of the cache hardware. One way to alleviate this hardware cost is to implement interleaved data cache banks. With the data cache being implemented as multiple banks of memory, two simultaneous accesses to different banks can be supported in one cycle. If two accesses need to access the same bank, a bank conflict occurs and the two accesses must be serialized. From practical experience, a cache with eight banks can keep the frequency of bank conflicts down to acceptable levels.

The most common way to reduce memory latency is through the use of a cache. Caches are now widely employed. As the gap between processor speed and memory speed widens, multiple levels of caches are required. Most high-performance superscalar processors incorporate at least two levels of caches. The first level (L1) cache can usually keep up with the processor speed with access latency of one or very few cycles. Typically there are separate L1 caches for storing instructions and data. The second level (L2) cache typically supports the storing of both instructions and data, can be either on-chip or off-chip, and can be accessed in series (in case of a miss in the L1) or in parallel with the L1 cache. In some of the emerging designs a third level (L3) cache is used. It is likely that in future high-end designs a very large on-chip L3 cache will become commonplace. Other than the use of a cache or a hierarchy of caches, there are two other techniques for reducing the effective memory latency, namely, *nonblocking cache* and *prefetching cache*.

A nonblocking cache, first proposed by Kroft [1981], can reduce the effective memory latency by reducing the performance penalty due to cache misses. Traditionally, when a load instruction encounters a cache miss, it will stall the load unit pipeline and any further issuing of load instructions until the cache miss is serviced.

Such stalling is overly conservative and prevents subsequent and independent loads that may hit in the data cache from being issued. A nonblocking data cache alleviates this unnecessary penalty by putting aside a load that has missed in the cache into a *missed load queue* and allowing subsequent load instructions to issue; see Figure 5.37. A missed load sits in the missed load queue while the cache miss is serviced. When the missing block is fetched from the main memory, the missed load exits the missed load queue and finishes execution.

Essentially the cache miss penalty cycles are overlapped with, and masked by, the processing of subsequent independent instructions. Of course, if a subsequent instruction depends on the missed load, the issuing of that instruction is stalled. The number of penalty cycles that can be masked depends on the number of independent instructions following the missed load. A missed load queue can contain multiple entries, allowing multiple missed loads to be serviced concurrently. Potentially the cache penalty cycles of multiple missed loads can be overlapped to result in fewer total penalty cycles.

A number of issues must be considered when implementing nonblocking caches. Load misses can occur in bursts. The ability to support multiple misses and overlap their servicing is important. The interface to main memory, or a lower-level cache, must be able to support the overlapping or pipelining of multiple accesses. The filling of the cache triggered by the missed load may need to contend with the store buffer for the write port to the cache. There is one complication that can emerge with nonblocking caches. If the missed load is on a speculative path, i.e., the predicted path, there is the possibility that the speculation, i.e., branch prediction, will turn out to be incorrect. If a missed load is on a mispredicted path, the question is whether the cache miss should be serviced. In a machine with very aggressive branch prediction, the number of loads on the mispredicted path can be significant; servicing their misses speculatively can require significant memory bandwidth. Studies have shown that a nonblocking cache can reduce the amount of load miss penalty by about 15%.

Another way to reduce or mask the cache miss penalty is through the use of a prefetching cache. A prefetching cache anticipates future misses and triggers these misses early so as to overlap the miss penalty with the processing of instructions preceding the missing load. Figure 5.38 illustrates a prefetching data cache. Two structures are needed to implement a prefetching cache, namely, a *memory reference prediction table* and a *prefetch queue*. The memory reference prediction table stores information about previously executed loads in three different fields. The first field contains the instruction address of the load and is used as a tag field for selecting an entry of the table. The second field contains the previous data address of the load, while the third field contains a stride value that indicates the difference between the previous two data addresses used by that load. The memory reference prediction table is accessed via associative search using the fetch address produced by the branch predictor and the first field of the table. When there is a tag match, indicating a hit in the memory reference prediction table, the previous address is added to the stride value to produce a predicted memory address. This predicted address is then loaded into the prefetch queue. Entries in the prefetch queue are

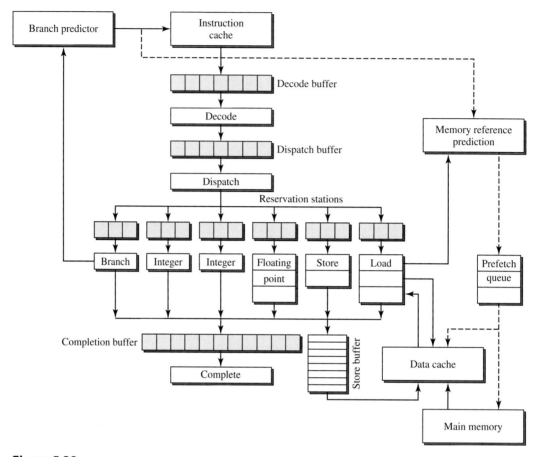

Figure 5.38
Prefetching Data Cache.

retrieved to speculatively access the data cache, and if a cache miss is triggered, the main memory or the next-level cache is accessed. The access to the data cache is in reality a cache touch operation; i.e., access to the cache is attempted in order to trigger a potential cache miss and not necessarily to actually retrieve the data from the cache. Early work on prefetching was done by Baer and Chen [1991] and Jouppi [1990].

The goal of a prefetching cache is to try to anticipate forthcoming cache misses and to trigger those misses early so as to hide the cache miss penalty by overlapping cache refill time with the processing of instructions preceding the missing load. When the anticipated missing load is executed, the data will be resident in the cache; hence, no cache miss is triggered, and no cache miss penalty is incurred. The actual effectiveness of prefetching depends on a number of factors. The prefetching distance, i.e., how far in advance the prefetching is being triggered, must be large enough to fully mask the miss penalty. This is the reason that the predicted instruction fetch address is used to access the memory reference prediction table, with the hope that the data prefetch will occur far enough in advance of the load. However, this makes prefetching effectiveness subject to the effectiveness of branch prediction. Furthermore, there is the potential of polluting the data cache with prefetches that are on the mispredicted path. Status or confidence bits can be added to each entry of the memory reference prediction table to modulate the aggressiveness of prefetching. Another problem can occur when the prefetching is performed too early so as to evict a useful block from the cache and induce an unnecessary miss. One more factor is the actual memory reference prediction algorithm used. Load address prediction based on stride is quite effective for loads that are stepping through an array. For other loads that are traversing linked list data structures, the stride prediction will not work very well. Prefetching for such memory references will require much more sophisticated prediction algorithms.

To enhance memory data flow, load instructions must be executed early and fast. Store instructions are less important because experimental data indicate that they occur less frequently than load instructions and they usually are not on the performance critical path. To speed up the execution of loads we must reduce the latency for processing load instructions. The overall latency for processing a load instruction includes four components: (1) pipeline front-end latency for fetching, decoding, and dispatching the load instruction; (2) reservation station latency of waiting for register data dependence to resolve; (3) execution pipeline latency for address generation and translation; and (4) the cache/memory access latency for retrieving the data from memory. Both nonblocking and prefetching caches address only the fourth component, which is a crucial component due to the slow memory speed. To achieve higher clocking rates, superscalar pipelines are quickly becoming deeper and deeper. Consequently the latencies, in terms of number of machine cycles, of the first three components are also becoming quite significant. A number of speculative techniques have been proposed to address the reduction of these latencies; they include load address prediction, load value prediction, and memory dependence prediction.

Recently *load address prediction* techniques have been proposed to address the latencies associated with the first three components [Austin and Sohi, 1995]. To deal with the latency associated with the first component, a load prediction table, similar to the memory reference prediction table, is proposed. This table is indexed with the predicted instruction fetch address, and a hit in this table indicates the presence of a load instruction in the upcoming fetch group. Hence, the prediction of the presence of a load instruction in the upcoming fetch group is performed during the fetch stage and without requiring the decode and dispatch stages. Each entry of this table contains the predicted effective address which is retrieved during the fetch stage, in effect eliminating the need for waiting in the reservation station for the availability of the base register value and the address generation stage of the execution pipeline. Consequently, data cache access can begin in the next cycle, and potentially data can be retrieved from the cache at the end of the decode stage. Such a form of load address prediction can effectively collapse the latencies of the first three components down to just two cycles, i.e., fetch and decode stages, if the address prediction is correct and there is a hit in the data cache.

While the hardware structures needed to support load address prediction are quite similar to those needed for memory prefetching, the two mechanisms have significant differences. Load address prediction is actually executing, though speculatively, the load instruction early, whereas memory prefetching is mainly trying to prefetch the needed data into the cache without actually executing the load instruction. With load address prediction, instructions that depend on the load can also execute early because their dependent data are available early. Since load address prediction is a speculative technique, it must be validated, and if misprediction is detected, recovery must be performed. The validation is performed by allowing the actual load instruction to be fetched from the instruction cache and executed in a normal fashion. The result from the speculative version is compared with that of the normal version. If the results concur, then the speculative result becomes nonspeculative and all the dependent instructions that were executed speculatively are also declared as nonspeculative. If the results do not agree, then the nonspeculative result is used and all dependent instructions must be reexecuted. If the load address prediction mechanism is quite accurate, mispredictions occur only infrequently, the misprediction penalty is minimal, and overall net performance gain can be achieved.

Even more aggressive than load address prediction is the technique of *load value prediction* [Lipasti et al., 1996]. Unlike load address prediction which attempts to predict the effective address of a load instruction, load value prediction actually attempts to predict the value of the data to be retrieved from memory. This is accomplished by extending the load prediction table to contain not just the predicted address, but also the predicted value for the destination register. Experimental studies have shown that many load instructions' destination values are quite predictable. For example, many loads actually load the same value as last time. Hence, by storing the last value loaded by a static load instruction in the load prediction table, this value can be used as the predicted value when the same static load is encountered again. As a result, the load prediction table can be accessed during the fetch stage, and at the end of that cycle, the actual destination value of a predicted load instruction can be available and used in the next cycle by a dependent instruction. This significantly reduces the latency required for processing a load instruction if the load value prediction is correct. Again, validation is required, and at times a misprediction penalty must be paid.

Other than load address prediction and load value prediction, a third speculative technique has been proposed called *memory dependence prediction* [Moshovos, 1998]. Recall from Section 5.3.3 that to perform load bypassing and load forwarding, memory dependence checking is required. For load bypassing, it must be determined that the load does not alias with any of the stores being bypassed. For load forwarding, the most recent aliased store must be identified. Memory dependence checking can become quite complex if a larger number of load/store instructions are involved and can potentially require an entire pipe stage. It would be nice to eliminate this latency. Experimental data have shown that the memory dependence relationship is quite predictable. It is possible to track the memory dependence relationship when load/store instructions are executed and use this information to make memory dependence prediction when the same load/store instructions are encountered again. Such memory dependence prediction can facilitate earlier execution of load bypassing and load forwarding. As with all speculative techniques, validation is needed and a recovery mechanism for misprediction must be provided.

5.4 Summary

In this chapter we attempt to cover all the fundamental microarchitecture techniques used in modern superscalar microprocessor design in a systematic and easy-to-digest way. We intentionally avoid inundating readers with lots of quantitative data and bar charts. We also focus on generic techniques instead of features of specific commercial products. In Chapters 6 and 7 we present detailed case studies of two actual products. Chapter 6 introduces the IBM/Motorola PowerPC 620 in great detail along with quantitative performance data. While not a successful commercial product, the PowerPC 620 represents one of the earlier and most aggressive out-of-order designs. Chapter 7 presents the Intel P6 microarchitecture, the first out-of-order implementation of the IA32 architecture. The Intel P6 is likely the most commercially successful microarchitecture. The fact that the P6 microarchitecture core provided the foundation for multiple generations of products, including the Pentium Pro, the Pentium II, the Pentium III, and the Pentium M, is a clear testimony to the effectiveness and elegance of its original design.

Superscalar microprocessor design is a rapidly evolving art which has been simultaneously harnessing the creative ideas of researchers and the insights and skills of architects and designers. Chapter 8 is a historical chronicle of the practice of this art form. Interesting and valuable lessons can be gleaned from this historical chronicle. The body of knowledge on superscalar microarchitecture techniques is constantly expanding. New innovative ideas from the research community as well as ideas from the traditional "macroarchitecture" domain are likely to find their way into future superscalar microprocessor designs. Chapters 9, 10, and 11 document some of these ideas.

REFERENCES

Adve, S. V., and K. Gharachorloo: "Shared memory consistency models: A tutorial," *IEEE Computer,* 29, 12, 1996, pp. 66–76.

Austin, T. M., and G. S. Sohi: "Zero-cycle loads: Microarchitecture support for reducing load latency," *Proc. 28th Annual ACM/IEEE Int. Symposium on Microarchitecture,* 1995, pp. 82–92.

Baer, J., and T. Chen: "An effective on-chip preloading scheme to reduce data access penalty," *Proc. Supercomputing '91,* 1991, pp. 176–186.

Calder, B., P. Feller, and A. Eustace: "Value profiling," *Proc. 30th Annual ACM/IEEE Int. Symposium on Microarchitecture,* 1997, pp. 259–269.

Calder, B., G. Reinman, and D. Tullsen: "Selective value prediction," *Proc. 26th Annual Int. Symposium on Computer Architecture (ISCA '99)*, vol. 27, 2 of *Computer Architecture News*, New York, N.Y.: ACM Press, 1999, pp. 64–75.

Chrysos, G., and J. Emer: "Memory dependence prediction using store sets," *Proc. 25th Int. Symposium on Computer Architecture*, 1998, pp. 142–153.

Conte, T., K. Menezes, P. Mills, and B. Patel: "Optimization of instruction fetch mechanisms for high issue rates," *Proc. 22nd Annual Int. Symposium on Computer Architecture*, 1995, pp. 333–344.

Diefendorf, K., and M. Allen: "Organization of the Motorola 88110 superscalar RISC microprocessor," *IEEE MICRO*, 12, 2, 1992, pp. 40–63.

Gabbay, F., and A. Mendelson: "Using value prediction to increase the power of speculative execution hardware," *ACM Transactions on Computer Systems*, 16, 3, 1988b, pp. 234–270.

Gabbay, F., and A. Mendelson: "Can program profiling support value prediction," *Proc. 30th Annual ACM/IEEE Int. Symposium on Microarchitecture*, 1997, pp. 270–280.

Gabbay, F., and A. Mendelson: "The effect of instruction fetch bandwidth on value prediction," *Proc. 25th Annual Int. Symposium on Computer Architecture*, Barcelona, Spain, 1998a, pp. 272–281.

Gloy, N. C., M. D. Smith, and C. Young: "Performance issues in correlated branch prediction schemes," *Proc. 27th Int. Symposium on Microarchitecture*, 1995, pp. 3–14.

Grohoski, G.: "Machine organization of the IBM RISC System/6000 processor," *IBM Journal of Research and Development*, 34, 1, 1990, pp. 37–58.

Harbison, S. P.: *A Computer Architecture for the Dynamic Optimization of High-Level Language Programs*. PhD thesis, Carnegie Mellon University, 1980.

Harbison, S. P.: "An architectural alternative to optimizing compilers," *Proc. Int. Conference on Architectural Support for Programming Languages and Operating Systems (ASPLOS)*, 1982, pp. 57–65.

IBM Corp.: *PowerPC 604 RISC Microprocessor User's Manual*. Essex Junction, VT: IBM Microelectronics Division, 1994.

Johnson, M.: *Superscalar Microprocessor Design*. Englewood Cliffs, NJ: Prentice Hall, 1991.

Jouppi, N. P.: "Improving direct-mapped cache performance by the addition of a small fully-associative cache and prefetch buffers," *Proc. of 17th Annual Int. Symposium on Computer Architecture*, 1990, pp. 364–373.

Kessler, R.: "The Alpha 21264 microprocessor," *IEEE MICRO*, 19, 2, 1999, pp. 24–36.

Kroft, D.: "Lockup-free instruction fetch/prefetch cache organization," *Proc. 8th Annual Symposium on Computer Architecture*, 1981, pp. 81–88.

Lamport, L.: "How to make a multiprocessor computer that correctly executes multiprocess programs," *IEEE Trans. on Computers*, C-28, 9, 1979, pp. 690–691.

Lee, J., and A. Smith: "Branch prediction strategies and branch target buffer design," *IEEE Computer*, 21, 7, 1984, pp. 6–22.

Lipasti, M. H., and J. P. Shen: "Exceeding the dataflow limit via value prediction," *Proc. 29th Annual ACM/IEEE Int. Symposium on Microarchitecture*, 1996, pp. 226–237.

Lipasti, M. H., C. B. Wilkerson, and J. P. Shen: "Value locality and load value prediction," *Proc. Seventh Int. Conference on Architectural Support for Programming Languages and Operating Systems (ASPLOS-VII)*, 1996, pp. 138–147.

McFarling, S.: "Combining branch predictors," Technical Report TN-36, Digital Equipment Corp. (**http://research.compaq.com/wrl/techreports/abstracts/TN-36.html**), 1993.

Mendelson, A., and F. Gabbay: "Speculative execution based on value prediction," Technical report, Technion (**http://www-ee.technion.ac.il/%7efredg**), 1997.

Moshovos, A.: "Memory Dependence Prediction," PhD thesis, University of Wisconsin, 1998.

Nair, R.: "Branch behavior on the IBM RS/6000," Technical report, IBM Computer Science, 1992.

Oehler, R. R., and R. D. Groves: "IBM RISC System/6000 processor architecture," *IBM Journal of Research and Development*, 34, 1, 1990, pp. 23–36.

Richardson, S. E.: "Caching function results: Faster arithmetic by avoiding unnecessary computation," Technical report, Sun Microsystems Laboratories, 1992.

Rotenberg, E., S. Bennett, and J. Smith: "Trace cache: a low latency approach to high bandwidth instruction fetching," *Proc. 29th Annual ACM/IEEE Int. Symposium on Microarchitecture*, 1996, pp. 24–35.

Sazeides, Y., and J. E. Smith: "The predictability of data values," *Proc. 30th Annual ACM/IEEE Int. Symposium on Microarchitecture*, 1997, pp. 248–258.

Smith, J. E.: "A study of branch prediction techniques," *Proc. 8th Annual Symposium on Computer Architecture*, 1981, pp. 135–147.

Sodani, A., and G. S. Sohi: "Dynamic instruction reuse," *Proc. 24th Annual Int. Symposium on Computer Architecture*, 1997, pp 194–205.

Sohi, G., and M. Franklin: "High-bandwidth data memory systems for superscalar processors," *Proc. 4th Int. Conference on Architectural Support for Programming Languages and Operating Systems*, 1991, pp. 53–62.

Tomasulo, R.: "An efficient algorithm for exploiting multiple arithmetic units," *IBM Journal of Research and Development*, 11, 1967, pp. 25–33.

Uht, A. K., and V. Sindagi: "Disjoint eager execution: An optimal form of speculative execution," *Proc. 28th Annual ACM/IEEE Int. Symposium on Microarchitecture*, 1995, pp. 313–325.

Wang, K., and M. Franklin: "Highly accurate data value prediction using hybrid predictors," *Proc. 30th Annual ACM/IEEE Int. Symposium on Microarchitecture*, 1997, pp. 281–290.

Yeh, T. Y., and Y. N. Patt: "Two-level adaptive training branch prediction," *Proc. 24th Annual Int. Symposium on Microarchitecture*, 1991, pp. 51–61.

Young, C., and M. D. Smith: "Improving the accuracy of static branch prediction using branch correlation," *Proc. 6th Int. Conference on Architectural Support for Programming Languages and Operating Systems (ASPLOS-VI)*, 1994, pp. 232–241.

HOMEWORK PROBLEMS

Problems 5.1 through 5.6

The displayed code that follows steps through the elements of two arrays (A[] and B[]) concurrently, and for each element, it puts the larger of the two values into the corresponding element of a third array (C[]). The three arrays are of length N.

The instruction set used for Problems 5.1 through 5.6 is as follows:

add	rd, rs, rt	rd ← rs + rt
addi	rd, rs, imm	rd ← rs + imm
lw	rd, offset(base)	rd ← MEM[offset+base] (offset = imm, base = reg)
sw	rs, offset(base)	MEM[offset+base] ← rs (offset = imm, base = reg)
bge	rs, rt, address	if (rs >= rt) PC ← address
blt	rs, rt, address	if (rs < rt) PC ← address
b	address	PC ← address

Note: r0 is hardwired to 0.

The benchmark code is as follows:

Static Inst#	Label	Assembly_Instruction		
	main:			
1		addi r2,	r0,	A
2		addi r3,	r0,	B
3		addi r4,	r0,	C
4		addi r5,	r0,	N
5		add r10,	r0,	r0
6		bge r10,	r5,	end
	loop:			
7		lw r20,	0(r2)	
8		lw r21,	0(r3)	
9		bge r20,	r21,	T1
10		sw r21,	0(r4)	
11		b T2		
	T1:			
12		sw r20,	0(r4)	
	T2:			
13		addi r10,	r10,	1
14		addi r2,	r2,	4
15		addi r3,	r3,	4
16		addi r4,	r4,	4
17		blt r10,	r5,	loop
	end:			

P5.1 Identify the basic blocks of this benchmark code by listing the static instructions belonging to each basic block in the following table. Number the basic blocks based on the lexical ordering of the code. *Note:* There may be more boxes than there are basic blocks.

	Basic Block No.								
	1	2	3	4	5	6	7	8	9
Instr. nos.									

P5.2 Draw the control flow graph for this benchmark.

P5.3 Now generate the instruction execution trace (i.e., the sequence of basic blocks executed). Use the following arrays as input to the program, and trace the code execution by recording the number of each basic block that is executed.

```
N = 5;
A[] = {8, 3, 2, 5, 9};
B[] = {4, 9, 8, 5, 1};
```

P5.4 Fill in Tables 5.1 and 5.2 based on the data you generated in Problem 5.3.

Table 5.1
Instruction mix

	Static		Dynamic	
Instr. Class	**Number**	**%**	**Number**	**%**
ALU				
Load/store				
Branch				

Table 5.2
Basic block/branch data*

	Static	Dynamic
Average basic block size (no. of instr.)		
Number of taken branches		
Number of not-taken branches		

*Count unconditional branches as taken branches.

P5.5 Given the branch profile information you collected in Problem 5.4, rearrange the basic blocks and reverse the sense of the branches in the program snippet to minimize the number of taken branches and to pack the code so that the frequently executed paths are placed together. Show the new program.

P5.6 Given the new program you wrote in Problem 5.5, recompute the branch statistics in the last two rows of Table 5.2.

Problems 5.7 through 5.13

Consider the following code segment within a loop body for Problems 5.7 through 5.13:

```
if (x is even) then                    ←(branch b1)
    increment a                        ←(b1 taken)
if (x is a multiple of 10) then        ←(branch b2)
    increment b                        ←(b2 taken)
```

Assume that the following list of nine values of x is to be processed by nine iterations of this loop.

$$8, 9, 10, 11, 12, 20, 29, 30, 31$$

Note: Assume that predictor entries are updated by each dynamic branch before the next dynamic branch accesses the predictor (i.e., there is no update delay).

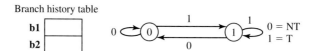

P5.7 Assume that a 1-bit (history bit) state machine (see above) is used as the prediction algorithm for predicting the execution of the two branches in this loop. Indicate the predicted and actual branch directions of the b1 and b2 branch instructions for each iteration of this loop. Assume an initial state of 0, i.e., NT, for the predictor.

	8	9	10	11	12	20	29	30	31
b1 predicted									
b1 actual									
b2 predicted									
b2 actual									

P5.8 What are the prediction accuracies for b1 and b2?

P5.9 What is the overall prediction accuracy?

P5.10 Assume a two-level branch prediction scheme is used. In addition to the 1-bit predictor, a 1-bit global register (g) is used. Register g stores the direction of the last branch executed (which may not be the same branch as the branch currently being predicted) and is used to index into two separate 1-bit branch history tables (BHTs) as shown in the following figure.

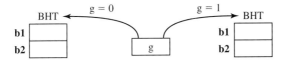

Depending on the value of g, one of the two BHTs is selected and used to do the normal 1-bit prediction. Again, fill in the predicted and actual branch directions of b1 and b2 for nine iterations of the loop. Assume the initial value of g = 0, i.e., NT. For each prediction, depending on the current value of g, only one of the two BHTs is accessed and updated. Hence, some of the entries in the following table should be empty.

Note: Assume that predictor entries are updated by each dynamic branch before the next dynamic branch accesses the predictor (i.e., there is no update delay).

	8	9	10	11	12	20	29	30	31
For g = 0:									
b1 predicted									
b1 actual									
b2 predicted									
b2 actual									
For g = 1:									
b1 predicted									
b1 actual									
b2 predicted									
b2 actual									

P5.11 What are the prediction accuracies for b1 and b2?

P5.12 What is the overall prediction accuracy?

P5.13 What is the prediction accuracy of b2 when g = 0? Explain why.

P5.14 The figure shows the control flow graph of a simple program. The CFG is annotated with three different execution trace paths. For each execution trace, circle which branch predictor (bimodal, local, or gselect) will *best* predict the branching behavior of the given trace. More

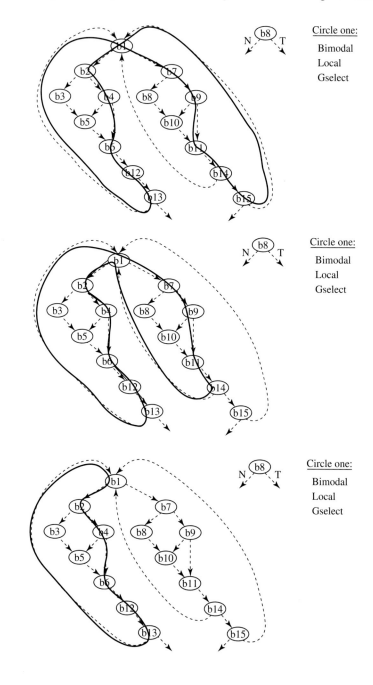

Circle one:

Bimodal
Local
Gselect

Circle one:

Bimodal
Local
Gselect

Circle one:

Bimodal
Local
Gselect

than one predictor may perform equally well on a particular trace. However, you are to use each of the three predictors *exactly once* in choosing the best predictors for the three traces. *Circle* your choice for each of the three traces and add. (Assume each trace is executed many times and every node in the CFG is a conditional branch. The branch history register for the local, global, and gselect predictors is limited to 4 bits.)

Problems 5.15 and 5.16: Combining Branch Prediction

Given a combining branch predictor with a two-entry direct-mapped bimodal branch direction predictor, a gshare predictor with a 1-bit BHR and two PHT entries, and a two-entry selector table, simulate a sequence of taken and not-taken branches as shown in the rows of the table in Problem 5.15, record the prediction made by the predictor before the branch is resolved as well as any change to the predictor entries after the branch resolves.

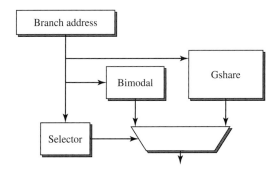

Use the following assumptions:

- Instructions are a fixed 4 bytes long; hence, the two low-order bits of the branch address should be shifted out when indexing into the predictor. Use the next lowest-order bit to index into the predictor.
- Each predictor and selector entry is a saturating up-down 2-bit Smith counter with the initial states shown.
- A taken branch (T) increments the predictor entry; a not-taken branch (N) decrements the predictor entry.
- A predictor entry less than 2 (0 or 1) results in a not-taken (N) prediction.
- A predictor entry greater than or equal to 2 (2 or 3) results in a taken (T) prediction.

- A selector value of 0 or 1 selects the bimodal predictor, while a selector value of 2 or 3 selects the gshare predictor.
- None of the predictors are tagged with the branch address.
- Avoid destructive interference by not updating the "wrong" predictor whenever the other predictor is right.

P5.15 Fill in the following table with the prediction outcomes, and the predictor state following resolution of each branch.

Branch Address	Branch Outcome (TNT)	Predicted Outcome (T/N)			Predictor State after Branch Is Resolved						
					Selector		Bimodal Predictor		Gshare Predictor		
		Bimodal	Gshare	Combined	PHT0	PHT1	PHT0	PHT1	BHR	PHT0	PHT1
Initial	N/A	N/A	N/A	N/A	2	0	0	2	0	2	1
0x654	N										
0x780	T										
0x78C	T										
0x990	T										
0xA04	N										
0x78C	N										

P5.16 Compute the overall branch prediction rates (number of correctly predicted branches / total number of predicted branches) for the bimodal, gshare, and final (combined) predictors.

P5.17 Branch predictions are resolved when a branch instruction executes. Early out-of-order processors like the PowerPC 604 simplified branch resolution by forcing branches to execute strictly in program order. Discuss why this simplifies branch redirect logic, and explain in detail how the microarchitecture must change to accommodate out-of-order branch resolution.

P5.18 Describe a scenario in which out-of-order branch resolution would be important for performance. State your hypothesis and describe a set of experiments to validate your hypothesis. Optionally, modify a timing simulator and conduct these experiments.

Problems 5.19 and 5.20: Register Renaming

Given the DAXPY kernel shown in Figure 5.31 and the IBM RS/6000 (RIOS-I) floating-point load renaming scheme also discussed in class (both are shown in the following figure), simulate the execution of two iterations of the DAXPY loop and show the state of the floating-point map table, the pending target return queue, and the free list.

- Assume the initial state shown in the table for Problem 5.19.
- Note the table only contains columns for the registers that are referenced in the DAXPY loop.
- As in the RS/6000 implementation discussed, assume only a single load instruction is renamed per cycle and that only a single floating-point instruction can complete per cycle.
- Only floating-point load, multiply, and add instructions are shown in the table, since only these are relevant to the renaming scheme.
- Remember that only load destination registers are renamed.
- The first load from the loop prologue is filled in for you.

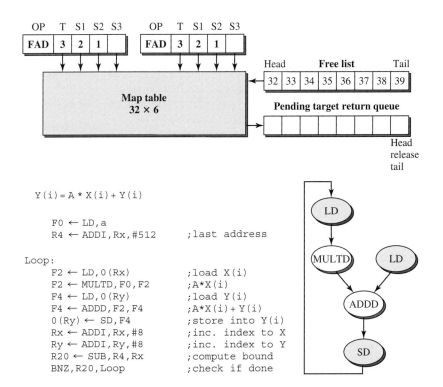

```
Y(i) = A * X(i) + Y(i)

        F0  ← LD,a
        R4  ← ADDI,Rx,#512      ;last address

Loop:
        F2  ← LD,0(Rx)          ;load X(i)
        F2  ← MULTD,F0,F2       ;A*X(i)
        F4  ← LD,0(Ry)          ;load Y(i)
        F4  ← ADDD,F2,F4        ;A*X(i)+Y(i)
        0(Ry) ← SD,F4           ;store into Y(i)
        Rx  ← ADDI,Rx,#8        ;inc. index to X
        Ry  ← ADDI,Ry,#8        ;inc. index to Y
        R20 ← SUB,R4,Rx         ;compute bound
        BNZ,R20,Loop            ;check if done
```

P5.19 Fill in the remaining rows in the following table with the map table state and pending target return queue state after the instruction is renamed, and the free list state after the instruction completes.

Floating-Point Instruction	Pending Target Return Queue after Instruction Renamed	Free List after Instruction Completes	Map Table Subset		
			F0	F2	F4
Initial state		32,33,34,35,36,37,38,39	0	2	4
F0 ⇐ LD, a	0	33,34,35,36,37,38,39	32	2	4
F2 ⇐ LD, 0(Rx)					
F2 ⇐ MULTD, F0, F2					
F4 ⇐ LD, 0(Ry)					
F4 ⇐ ADDD, F2, F4					
F2 ⇐ LD, 0(Rx)					
F2 ⇐ MULTD, F0, F2					
F4 ⇐ LD, 0(Ry)					
F4 ⇐ ADDD, F2, F4					

P5.20 Given that the RS/6000 can rename a floating-point load in parallel with a floating-point arithmetic instruction (mult/add), and assuming the map table is a write-before-read structure, is any internal bypassing needed within the map table? Explain why or why not.

P5.21 Simulate the execution of the following code snippet using Tomasulo's algorithm. Show the contents of the reservation station entries, register file busy, tag (the tag is the RS ID number), and data fields for each cycle (make a copy of the table shown on the next page for each cycle that you simulate). Indicate which instruction is executing in each functional unit in each cycle. Also indicate any result forwarding across a common data bus by circling the producer and consumer and connecting them with an arrow.

i: $R4 \leftarrow R0 + R8$
j: $R2 \leftarrow R0 * R4$
k: $R4 \leftarrow R4 + R8$
l: $R8 \leftarrow R4 * R2$

Assume dual dispatch and a dual common data bus (CDB). Add latency is two cycles, and multiply latency is three cycles. An instruction can begin execution in the same cycle that it is dispatched, assuming all dependences are satisfied.

P5.22 Determine whether or not the code executes at the data-flow limit for Problem 5.21. Explain why or why not. Show your work.

CYCLE #: _____

ID	Tag	Sink	Tag	Source
1				
2				
3				

Adder

ID	Tag	Sink	Tag	Source
4				
5				

Mult/Div

	Busy	Tag	Data
0			6.0
2			3.5
4			10.0
8			7.8

DISPATCHED INSTRUCTION(S): _____

P5.23 As presented in this chapter, load bypassing is a technique for enhancing memory data flow. With load bypassing, load instructions are allowed to jump ahead of earlier store instructions. Once address generation is done, a store instruction can be completed architecturally and can then enter the store buffer to await an available bus cycle for writing to memory. Trailing loads are allowed to bypass these stores in the store buffer if there is no address aliasing.

In this problem you are to simulate such load bypassing (there is no load forwarding). You are given a sequence of load/store instructions and their addresses (symbolic). The number to the left of each instruction indicates the cycle in which that instruction is dispatched to the reservation station; it can begin execution in that same cycle. Each store instruction will have an additional number to its right, indicating the cycle in which it is ready to retire, i.e., exit the store buffer and write to the memory.

Use the following assumptions:

- All operands needed for address calculation are available at dispatch.
- One load and one store can have their addresses calculated per cycle.
- One load or one store can be executed, i.e., allowed to access the cache, per cycle.
- The reservation station entry is deallocated the cycle after address calculation and issue.
- The store buffer entry is deallocated when the cache is accessed.
- A store instruction can access the cache the cycle after it is ready to retire.
- Instructions are issued in order from the reservation stations.
- Assume 100% cache hits.

292 MODERN PROCESSOR DESIGN

Example:

Dispatch cycle	Instruction	Retire cycle
1	Load A	
1	Load B	
1	Store C	5

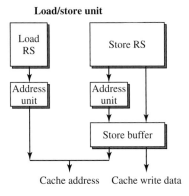

Cycle	Load Reservation Station	Store Reservation Station	Store Buffer	Cache Address	Cache Write Data
1	Ld A Ld B	St C			
2	Ld B		St C	Ld A	
3			St C	Ld B	
4			St C		
5			St C		
6				St C	data

Code:

Dispatch cycle	Instruction	Retire cycle
1	Store A	6
2	Load B	
3	Load A	
4	Store D	10
4	Load E	
4	Load A	
5	Load D	

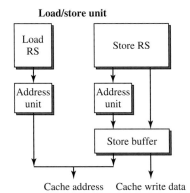

Cycle	Load Reservation Station	Store Reservation Station	Store Buffer	Cache Address	Cache Write Data
1					
2					
3					
4					
5					
6					
7					
8					
9					
10					
11					
12					
13					
14					
15					

P5.24 In one or two sentences compare and contrast load forwarding with load bypassing.

P5.25 Would load forwarding improve the performance of the code sequence from Problem 5.23? Why or why not?

Problems 5.26 through 5.28

The goal of lockup-free cache designs is to increase the amount of concurrency or parallelism in the processor by overlapping cache miss handling with other processing or other cache misses. For this problem, assume the following simple workload running on a processor with a primary cache with 64-byte lines and a 16-byte memory bus. Assume that t_{miss} is 5 cycles, and $t_{transfer}$ for each 16-byte sub-block is 1 cycle. Hence, for a simple blocking cache, a miss will take $t_{miss} + (64/16) \times t_{transfer} = 5 + 4 \times 1 = 9$ cycles. Here is the workload:

```
for(i=1;i<10000;++i)
        a += A[i] + B[i];
```

In RISC assembly language, assuming r3 points to A[0] and r4 points to B[0]:

```
        li      r2,9999     # load iteration count into r2
loop:   lfdu    r5,8(r3)    # load A[i], incr. pointer in r3
        lfdu    r6,8(r4)    # load B[i], incr. pointer in r4
```

```
add     r7,r7,r5    # add A[i] to a
add     r7,r7,r6    # add B[i] to a
bdnz    r2,loop     # decrement r2, branch if not zero
```

Here is a timing diagram for the loop body assuming neither array hits in the cache and the cache is a simple blocking design (m = miss latency, t = transfer latency, A = load from A, B = load from B, a = add, and b = branch):

Cycle	0 1 2 3 4 5 6 7 8 9	1 0 1 2 3 4 5 6 7 8 9	2 0 1 2 3 4 5 6 7 8 9
Array A	m m m m m t t t t		
Array B		m m m m m t t t t	
Execution	A	B	a a b A B a a b A B

For Problems 5.26 through 5.28 assume that each instruction takes a single cycle to execute and that all subsequent instructions are stalled on a cache miss until the requested data are returned by the cache (as shown in the timing diagram). Furthermore, assume that no portion of either array (A or B) is in the cache initially, but must be fetched on demand misses. Also, assume there are enough loop iterations to fill all the table entries provided.

P5.26 Assume a blocking cache design with critical word forwarding (i.e., the requested word is forwarded as soon as it has been transferred), but support for only a single outstanding miss. Fill in the timing diagram and explain your work.

Cycle	0 1 2 3 4 5 6 7 8 9	1 0 1 2 3 4 5 6 7 8 9	2 0 1 2 3 4 5 6 7 8 9
Array A	m m m m m		
Array B			
Execution	A		

Cycle	3 0 1 2 3 4 5 6 7 8 9	4 0 1 2 3 4 5 6 7 8 9	5 0 1 2 3 4 5 6 7 8 9
Array A			
Array B			
Execution			

P5.27 Now fill in the timing diagram for a lockup-free cache that supports multiple outstanding misses, and explain your work.

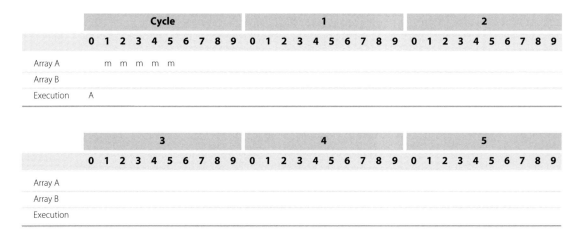

P5.28 Instead of a 64-byte cache line, assume a 32-byte line size for the cache, and fill in the timing diagram as in Problem 5.27. Explain your work

Problems 5.29 through 5.30

The goal of prefetching and nonblocking cache designs is to increase the amount of concurrency or parallelism in the processor by overlapping cache miss handling with other processing or other cache misses. For this problem, assume the same simple workload from Problem 5.26 running on a processor with a primary cache with 16-byte lines and an 4-byte memory bus. Assume that t_{miss} is 2 cycles and

$t_{transfer}$ for each 4-byte subblock is 1 cycle. Hence, a miss will take $t_{miss} + (16/4) \times t_{transfer} = 2 + 4 \times 1 = 6$ cycles.

For Problems 5.29 and 5.30, assume that each instruction takes a single cycle to execute and that all subsequent instructions are stalled on a cache miss until the requested data are returned by the cache (as shown in the table in Problem 5.29). Furthermore, assume that no portion of either array (A or B) is in the cache initially, but must be fetched on demand misses. Also, assume there are enough loop iterations to fill all the table entries provided.

Further assume a stride-based hardware prefetch mechanism that can track up to two independent strided address streams, and issues a stride prefetch the cycle after it has observed the same stride twice, from observing three strided misses (e.g., misses to A, A + 32, A + 64 triggers a prefetch for A + 96). The prefetcher will issue its next strided prefetch in the cycle following a demand reference to its previous prefetch, but no sooner (to avoid overwhelming the memory subsystem). Assume that a demand reference will always get priority over a prefetch for any shared resource.

P5.29 Fill in the following table to indicate miss (m) and transfer (t) cycles for both demand misses and prefetch requests for the assumed workload. Annotate each prefetch with a symbolic address (e.g., A + 64). The first 30 cycles are filled in for you.

	Cycle										1										2									
	0	1	2	3	4	5	6	7	8	9	0	1	2	3	4	5	6	7	8	9	0	1	2	3	4	5	6	7	8	9
Array A		m	m	t	t	t	t														m	m	t	t	t	t				
Array B						m	m	t	t	t		t	t														m	m	t	t	t	t
Prefetch 1																														
Prefetch 2																														
Execution	A						B				a	a	b	A	B	a	a	b	A					B					a	a

	3										4										5									
	0	1	2	3	4	5	6	7	8	9	0	1	2	3	4	5	6	7	8	9	0	1	2	3	4	5	6	7	8	9
Array A																														
Array B																														
Prefetch 1																														
Prefetch 2																														
Execution																														

	6										7										8									
	0	1	2	3	4	5	6	7	8	9	0	1	2	3	4	5	6	7	8	9	0	1	2	3	4	5	6	7	8	9
Array A																														
Array B																														
Prefetch 1																														
Prefetch 2																														
Execution																														

P5.30 Report the overall miss rate for the portion of execution shown in Problem 5.29, as well as the coverage of the prefetches generated (coverage is defined as: (number of misses eliminated by prefetching)/(number of misses without prefetching).

P5.31 A victim cache is used to augment a direct-mapped cache to reduce conflict misses. For additional background on this problem, read Jouppi's paper on victim caches [Jouppi, 1990]. Please fill in the following table to reflect the state of each cache line in a four-entry direct-mapped cache and a two-entry fully associative victim cache following each memory reference shown. Also, record whether the reference was a cache hit or a cache miss. The reference addresses are shown in hexadecimal format. Assume the direct-mapped cache is indexed with the low-order bits above the 16-byte line offset (e.g., address 40 maps to set 0, address 50 maps to set 1). Use a dash (—) to indicate an invalid line and the address of the line to indicate a valid line. Assume LRU policy for the victim cache and mark the LRU line as such in the table.

		Direct-Mapped Cache				Victim Cache	
Reference Address	Hit/Miss	Line 0	Line 1	Line 2	Line 3	Line 0	Line 1
[init state]		—	110	—	FF0	1F0	210/LRU
80							
A0							
200							
80							
B0							
E0							
200							
80							
200							

P5.32 Given your results from Problem 5.31, and excluding the data provided to the processor to supply the requested instruction words, compute the total number of bytes that were transferred into and out of the direct-mapped cache array. Assume this is a read-only instruction cache; hence, there are no writebacks of dirty lines. Show your work.

P5.33 Fill in the details for the block diagram of the read-only victim I-cache design shown in the following figure (based on the victim cache outlined in Problem 5.31). For additional background on this problem, read Jouppi's paper on victim caches [Jouppi, 1990]. Show data and address paths, identify which address bits are used for indexing and tag comparison, and show the logic for generating a hit/miss signal as well as control logic for any multiplexers that are included in your design. Don't forget the data paths for "swapping" cache lines between the victim cache and the DM cache. Assume that the arrays are read in the first half of each clock cycle and written in the second half of each clock cycle. Do not include LRU update or control or data paths for handling misses.

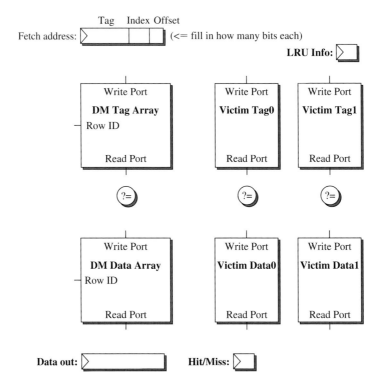

Problems 5.34 through 5.36: Simplescalar Simulation

Problems 5.34 through 5.36 require you to use and modify the Simplescalar 3.0 simulator suite, available from **http://www.simplescalar.com.** Use the Simplescalar simulators to execute the four benchmarks in the instructional benchmark suite from **http://www.simplescalar.com.**

P5.34 First use the sim-outorder simulator with the default machine parameters to simulate an out-of-order processor that performs both right-path and wrong-path speculative references against the instruction cache, data cache, and unified level 2 cache. Report the instruction cache, data cache, and l2 cache miss rates (misses per reference) as well as the total number of references and the total number of misses for each of the caches.

P5.35 Now simulate the exact same memory hierarchy as in Problem 5.34 but using the sim-cache simulator. Note that you will have to determine the parameters to use when you invoke sim-cache. Report the same statistics as in Problem 5.34, and compute the increase (or decrease) in each statistic.

P5.36 Now modify sim-outorder.c to inhibit cache misses caused by wrong-path references. This is very easy to do for instruction fetches in sim-outorder, since the global variable "spec_mode" is set whenever the processor begins to execute instructions from an incorrect branch path. You can use this global flag to inhibit instruction cache misses from wrong path instruction fetches. For data cache misses, you can check the "spec_mode" flag within each RUU entry and inhibit data cache misses for any such instruction. "Inhibiting misses" means don't even bother to check the cache hierarchy; simply treat these references as if they hit the cache. You can find the places where the cache is accessed by searching for calls to the function "cache_access" within sim-outorder.c. Now recollect the statistics from Problem 5.34 in your modified simulator and compare your results to Problem 5.35. If your results still differ from Problem 5.35, explain why that might be the case.

CHAPTER 6

Trung A. Diep

The PowerPC 620

CHAPTER OUTLINE

6.1   Introduction
6.2   Experimental Framework
6.3   Instruction Fetching
6.4   Instruction Dispatching
6.5   Instruction Execution
6.6   Instruction Completion
6.7   Conclusions and Observations
6.8   Bridging to the IBM POWER3 and POWER4
6.9   Summary

      References
      Homework Problems

The PowerPC family of microprocessors includes the 64-bit PowerPC 620 microprocessor. The 620 was the first 64-bit superscalar processor to employ true out-of-order execution, aggressive branch prediction, distributed multientry reservation stations, dynamic renaming for all register files, six pipelined execution units, and a completion buffer to ensure precise exceptions. Most of these features had not been previously implemented in a single-chip microprocessor. Their actual effectiveness is of great interest to both academic researchers as well as industry designers. This chapter presents an instruction-level, or machine-cycle level, performance evaluation of the 620 microarchitecture using a VMW-generated performance simulator of the 620 (VMW is the Visualization-based Microarchitecture Workbench from Carnegie Mellon University) [Levitan et al., 1995; Diep et al., 1995].

We also describe the IBM POWER3 and POWER4 designs, and we highlight how they differ from the predecessor PowerPC 620. While they are fundamentally similar in that they aggressively extract instruction-level parallelism from sequential code, the differences between the 620, the POWER3, and the POWER4 designs help to highlight recent trends in processor implementation: increased memory bandwidth through aggressive cache hierarchies, better branch prediction, more execution resources, and deeper pipelining.

6.1 Introduction

The PowerPC Architecture is the result of the PowerPC alliance among IBM, Motorola, and Apple [May et al., 1994]. It is based on the Performance Optimized with Enhanced RISC (POWER) Architecture, designed to facilitate parallel instruction execution and to scale well with advancing technology. The PowerPC alliance has released and announced a number of chips. The first, which provided a transition from the POWER Architecture to the PowerPC Architecture, was the PowerPC 601 microprocessor [IBM Corp., 1993]. The second, a low-power chip, was the PowerPC 603 microprocessor [Motorola, Inc., 2002]. Subsequently, a more advanced chip for desktop systems, the PowerPC 604 microprocessor, has been shipped [IBM Corp., 1994]. The fourth chip was the 64-bit 620 [Levitan et al., 1995; Diep et al., 1995].

More recently, Motorola and IBM have pursued independent development of general-purpose PowerPC-compatible parts. Motorola has focused on 32-bit desktop chips for Apple, while IBM has concentrated on server parts for its Unix (AIX) and business (OS/400) systems. Recent 32-bit Motorola designs, not detailed here, are the PowerPC G3 and G4 designs [Motorola, Inc., 2001; 2003]. These are 32-bit parts derived from the PowerPC 603, with short pipelines, limited execution resources, but very low cost. IBM's server parts have included the in-order multithreaded Star series (Northstar, Pulsar, S-Star [Storino et al., 1998]), as well as the out-of-order POWER3 [O'Connell and White, 2000] and POWER4 [Tendler et al., 2001]. In addition, both Motorola and IBM have developed various PowerPC cores for the embedded marketplace. Our focus in this chapter is on the PowerPC 620 and its heirs at the high-performance end of the marketplace, the POWER3 and the POWER4.

The PowerPC Architecture has 32 general-purpose registers (GPRs) and 32 floating-point registers (FPRs). It also has a condition register which can be addressed as one 32-bit register (CR), as a register file of 8 four-bit fields (CRFs), or as 32 single-bit fields. The architecture has a count register (CTR) and a link register (LR), both primarily used for branch instructions, and an integer exception register (XER) and a floating-point status and control register (FPSCR), which are used to record the exception status of the appropriate instruction types. The PowerPC instructions are typical RISC instructions, with the addition of floating-point fused multiply-add (FMA) instructions, load/store instructions with addressing modes that update the effective address, and instructions to set, manipulate, and branch off of the condition register bits.

The 620 is a four-wide superscalar machine. It uses aggressive branch prediction to fetch instructions as early as possible and a dispatch policy to distribute those

THE POWERPC 620

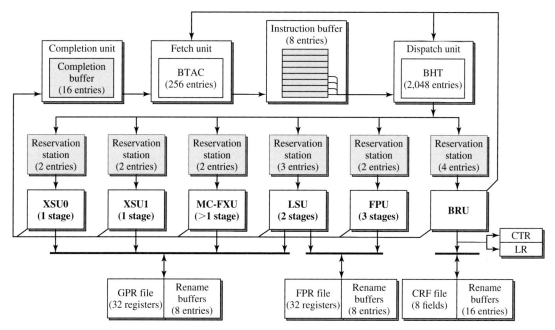

Figure 6.1
Block Diagram of the PowerPC 620 Microprocessor.

instructions to the execution units. The 620 uses six parallel execution units: two simple (single-cycle) integer units, one complex (multicycle) integer unit, one floating-point unit (three stages), one load/store unit (two stages), and a branch unit. The 620 uses distributed reservation stations and register renaming to implement out-of-order execution. The block diagram of the 620 is shown in Figure 6.1.

The 620 processes instructions in five major stages, namely the fetch, dispatch, execute, complete, and writeback stages. Some of these stages are separated by buffers to take up slack in the dynamic variation of available parallelism. These buffers are the instruction buffer, the reservation stations, and the completion buffer. The pipeline stages and their buffers are shown in Figure 6.2. Some of the units in the execute stage are actually multistage pipelines.

Fetch Stage. The fetch unit accesses the instruction cache to fetch up to four instructions per cycle into the instruction buffer. The end of a cache line or a taken branch can prevent the fetch unit from fetching four useful instructions in a cycle. A mispredicted branch can waste cycles while fetching from the wrong path. During the fetch stage, a preliminary branch prediction is made using the branch target address cache (BTAC) to obtain the target address for fetching in the next cycle.

Instruction Buffer. The instruction buffer holds instructions between the fetch and dispatch stages. If the dispatch unit cannot keep up with the fetch unit, instructions are buffered until the dispatch unit can process them. A maximum of eight

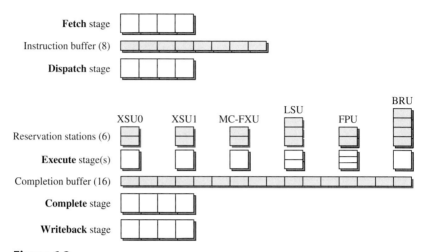

Figure 6.2
Instruction Pipeline of the PowerPC 620 Microprocessor.

instructions can be buffered at a time. Instructions are buffered and shifted in groups of two to simplify the logic.

Dispatch Stage. The dispatch unit decodes instructions in the instruction buffer and checks whether they can be dispatched to the reservation stations. If all dispatch conditions are fulfilled for an instruction, the dispatch stage will allocate a reservation station entry, a completion buffer entry, and an entry in the rename buffer for the destination, if needed. Each of the six execution units can accept at most one instruction per cycle. Certain infrequent serialization constraints can also stall instruction dispatch. Up to four instructions can be dispatched in program order per cycle.

There are eight integer register rename buffers, eight floating-point register rename buffers, and 16 condition register field rename buffers. The count register and the link register have one shadow register each, which is used for renaming. During dispatch, the appropriate buffers are allocated. Any source operands which have been renamed by previous instructions are marked with the tags of the associated rename buffers. If the source operand is not available when the instruction is dispatched, the appropriate result busses for forwarding results are watched to obtain the operand data. Source operands which have not been renamed by previous instructions are read from the architected register files.

If a branch is being dispatched, resolution of the branch is attempted immediately. If resolution is still pending, that is, the branch depends on an operand that is not yet available, it is predicted using the branch history table (BHT). If the prediction made by the BHT disagrees with the prediction made earlier by the BTAC in the fetch stage, the BTAC-based prediction is discarded and fetching proceeds along the direction predicted by the BHT.

Reservation Stations. Each execution unit in the execute stage has an associated reservation station. Each execution unit's reservation station holds those instructions waiting to execute there. A reservation station can hold two to four instruction entries, depending on the execution unit. Each dispatched instruction waits in a reservation station until all its source operands have been read or forwarded and the execution unit is available. Instructions can leave reservation stations and be issued into the execution units out of order [except for FPU and branch unit (BRU)].

Execute Stage. This major stage can require multiple cycles to produce its results, depending on the type of instruction being executed. The load/store unit is a two-stage pipeline, and the floating-point unit is a three-stage pipeline. At the end of execution, the instruction results are sent to the destination rename buffers and forwarded to any waiting instructions.

Completion Buffer. The 16-entry completion buffer records the state of the in-flight instructions until they are architecturally complete. An entry is allocated for each instruction during the dispatch stage. The execute stage then marks an instruction as finished when the unit is done executing the instruction. Once an instruction is finished, it is eligible for completion.

Complete Stage. During the completion stage, finished instructions are removed from the completion buffer in order, up to four at a time, and passed to the writeback stage. Fewer instructions will complete in a cycle if there are an insufficient number of write ports to the architected register files. By holding instructions in the completion buffer until writeback, the 620 guarantees that the architected registers hold the correct state up to the most recently completed instruction. Hence, precise exception is maintained even with aggressive out-of-order execution.

Writeback Stage. During this stage, the writeback logic retires those instructions completed in the previous cycle by committing their results from the rename buffers to the architected register files.

6.2 Experimental Framework

The performance simulator for the 620 was implemented using the VMW framework developed at Carnegie Mellon University. The five machine specification files for the 620 were generated based on design documents provided and periodically updated by the 620 design team. Correct interpretation of the design documents was checked by a member of the design team through a series of refinement cycles as the 620 design was finalized.

Instruction and data traces are generated on an existing PowerPC 601 microprocessor via software instrumentation. Traces for several SPEC 92 benchmarks, four integer and three floating-point, are generated. The benchmarks and their dynamic instruction mixes are shown in Table 6.1. Most integer benchmarks have similar instruction mixes; *li* contains more multicycle instructions than the rest. Most of these instructions move values to and from special-purpose registers. There is greater diversity among the floating-point benchmarks. *Hydro2d* uses more nonpipelined floating-point instructions. These instructions are all floating-point divides, which require 18 cycles on the 620.

Table 6.1
Dynamic instruction mix of the benchmark set*

Instruction Mix	Integer Benchmarks (SPECInt92)				Floating-Point Benchmarks (SPECfp92)		
	compress	eqntott	espresso	li	alvinn	hydro2d	tomcatv
Integer							
Arithmetic (single cycle)	42.73	48.79	48.30	29.54	37.50	26.25	19.93
Arithmetic (multicycle)	0.89	1.26	1.25	5.14	0.29	1.19	0.05
Load	25.39	23.21	24.34	28.48	0.25	0.46	0.31
Store	16.49	6.26	8.29	18.60	0.20	0.19	0.29
Floating-point							
Arithmetic (pipelined)	0.00	0.00	0.00	0.00	12.27	26.99	37.82
Arithmetic (nonpipelined)	0.00	0.00	0.00	0.00	0.08	1.87	0.70
Load	0.00	0.00	0.00	0.01	26.85	22.53	27.84
Store	0.00	0.00	0.00	0.01	12.02	7.74	9.09
Branch							
Unconditional	1.90	1.87	1.52	3.26	0.15	0.10	0.01
Conditional	12.15	17.43	15.26	12.01	10.37	12.50	3.92
Conditional to count register	0.00	0.44	0.10	0.39	0.00	0.16	0.05
Conditional to link register	4.44	0.74	0.94	2.55	0.03	0.01	0.00

*Values given are percentages.

Trace-driven performance simulation is used. With trace-driven simulation, instructions with variable latency such as integer multiply/divide and floating-point divide cannot be simulated accurately. For these instructions, we assume the minimum latency. The frequency of these operations and the amount of variance in the latencies are both quite low. Furthermore, the traces only contain those instructions that are actually executed. No speculative instructions that are later discarded due to misprediction are included in the simulation runs. Both I-cache and D-cache activities are included in the simulation. The caches are 32K bytes and 8-way set-associative. The D-cache is two-way interleaved. Cache miss latency of eight cycles and a perfect unified L2 cache are also assumed.

Table 6.2
Summary of benchmark performance

Benchmarks	Dynamic Instructions	Execution Cycles	IPC
compress	6,884,247	6,062,494	1.14
eqntott	3,147,233	2,188,331	1.44
espresso	4,615,085	3,412,653	1.35
li	3,376,415	3,399,293	0.99
alvinn	4,861,138	2,744,098	1.77
hydro2d	4,114,602	4,293,230	0.96
tomcatv	6,858,619	6,494,912	1.06

Table 6.2 presents the total number of instructions simulated for each benchmark and the total number of 620 machine cycles required. The sustained average number of instructions per cycle (IPC) achieved by the 620 for each benchmark is also shown. The IPC rating reflects the overall degree of instruction-level parallelism achieved by the 620 microarchitecture, the detailed analysis of which is presented in Sections 6.3 to 6.6.

6.3 Instruction Fetching

Provided that the instruction buffer is not saturated, the 620's fetch unit is capable of fetching four instructions in every cycle. If the fetch unit were to wait for branch resolution before continuing to fetch nonspeculatively, or if it were to bias naively for branch-not-taken, machine execution would be drastically slowed by the bottleneck in fetching down taken branches. Hence, accurate branch prediction is crucial in keeping a wide superscalar processor busy.

6.3.1 Branch Prediction

Branch prediction in the 620 takes place in two phases. The first prediction, done in the fetch stage, uses the BTAC to provide a preliminary guess of the target address when a branch is encountered during instruction fetch. The second, and more accurate, prediction is done in the dispatch stage using the BHT, which contains branch history and makes predictions based on the two history bits.

During the dispatch stage, the 620 attempts to resolve immediately a branch based on available information. If the branch is unconditional, or if the condition register has the appropriate bits ready, then no branch prediction is necessary. The branch is executed immediately. On the other hand, if the source condition register bits are unavailable because the instruction generating them is not finished, then branch prediction is made using the BHT. The BHT contains two history bits per entry that are accessed during the dispatch stage to predict whether the branch will be taken or not taken. Upon resolution of the predicted branch, the actual direction of the branch is updated to the BHT. The 2048-entry BHT is a direct-mapped table, unlike the BTAC, which is an associative cache. There is no concept of a hit or a miss. If two branches that update the BHT are an exact multiple of 2048 instructions apart, i.e., aliased, they will affect each other's predictions.

The 620 can resolve or predict a branch at the dispatch stage, but even that can incur one cycle delay until the new target of the branch can be fetched. For this reason, the 620 makes a preliminary prediction during the fetch stage, based solely on the address of the instruction that it is currently fetching. If one of these addresses hits in the BTAC, the target address stored in the BTAC is used as the fetch address in the next cycle. The BTAC, which is smaller than the BHT, has 256 entries and is two-way set-associative. It holds only the targets of those branches that are predicted taken. Branches that are predicted not taken (fall through) are not stored in the BTAC. Only unconditional and PC-relative conditional branches use the BTAC. Branches to the count register or the link register have unpredictable target addresses and are never stored in the BTAC. Effectively, these branches are always predicted not taken by the BTAC in the fetch stage. A link register stack, which stores the addresses of subroutine returns, is used for predicting conditional return instructions. The link register stack is not modeled in the simulator.

There are four possible cases in the BTAC prediction: a BTAC miss for which the branch is not taken (correct prediction), a BTAC miss for which the branch is taken (incorrect prediction), a BTAC hit for a taken branch (correct prediction), and a BTAC hit for a not-taken branch (incorrect prediction). The BTAC can never hit on a taken branch and get the wrong target address; only PC-relative branches can hit in the BTAC and therefore must always use the same target address. Two predictions are made for each branch, once by the BTAC in the fetch stage, and another by the BHT in the dispatch stage. If the BHT prediction disagrees with the BTAC prediction, the BHT prediction is used, while the BTAC prediction is discarded. If the predictions agree and are correct, all instructions that are speculatively fetched are used and no penalty is incurred.

In combining the possible predictions and resolutions of the BHT and BTAC, there are six possible outcomes. In general, the predictions made by the BTAC and BHT are strongly correlated. There is a small fraction of the time that the wrong prediction made by the BTAC is corrected by the right prediction of the BHT. There is the unusual possibility of the correct prediction made by the BTAC being undone by the incorrect prediction of the BHT. However, such cases are quite rare; see Table 6.3. The BTAC makes an early prediction without using branch history. A hit in the BTAC effectively implies that the branch is predicted taken. A miss in the BTAC implicitly means a not-taken prediction. The BHT prediction is based on branch history and is more accurate but can potentially incur a one-cycle penalty if its prediction differs from that made by the BTAC. The BHT tracks the branch history and updates the entries in the BTAC. This is the reason for the strong correlation between the two predictions.

Table 6.3 summarizes the branch prediction statistics for the benchmarks. The BTAC prediction accuracy for the integer benchmarks ranges from 75% to 84%. For the floating-point benchmarks it ranges from 88% to 94%. For these correct predictions by the BTAC, no branch penalty is incurred if they are likewise predicted

Table 6.3
Branch prediction data*

Branch Processing	compress	eqntott	espresso	li	alvinn	hydro2d	tomcatv
Branch resolution							
Not taken	40.35	31.84	40.05	33.09	6.38	17.51	6.12
Taken	59.65	68.16	59.95	66.91	93.62	82.49	93.88
BTAC prediction							
Correct	84.10	82.64	81.99	74.70	94.49	88.31	93.31
Incorrect	15.90	17.36	18.01	25.30	5.51	11.69	6.69
BHT prediction							
Resolved	19.71	18.30	17.09	28.83	17.49	26.18	45.39
Correct	68.86	72.16	72.27	62.45	81.58	68.00	52.56
Incorrect	11.43	9.54	10.64	8.72	0.92	5.82	2.05
BTAC incorrect and BHT correct	0.01	0.79	1.13	7.78	0.07	0.19	0.00
BTAC correct and BHT incorrect	0.00	0.12	0.37	0.26	0.00	0.08	0.00
Overall branch prediction accuracy	88.57	90.46	89.36	91.28	99.07	94.18	97.95

*Values given are percentages.

correctly by the BHT. The overall branch prediction accuracy is determined by the BHT. For the integer benchmarks, about 17% to 29% of the branches are resolved by the time they reach the dispatch stage. For the floating-point benchmarks, this range is 17% to 45%. The overall misprediction rate for the integer benchmarks ranges from 8.7% to 11.4%; whereas for the floating-point benchmarks it ranges from 0.9% to 5.8%. The existing branch prediction mechanisms work quite well for the floating-point benchmarks. There is still room for improvement in the integer benchmarks.

6.3.2 Fetching and Speculation

The main purpose for branch prediction is to sustain a high instruction fetch bandwidth, which in turn keeps the rest of the superscalar machine busy. Misprediction translates into wasted fetch cycles and reduces the effective instruction fetch bandwidth. Another source of fetch bandwidth loss is due to I-cache misses. The effects of these two impediments on fetch bandwidth for the benchmarks are shown in Table 6.4. Again, for the integer benchmarks, significant percentages (6.7% to 11.8%) of the fetch cycles are lost due to misprediction. For all the benchmarks, the I-cache misses resulted in the loss of less than 1% of the fetch cycles.

Table 6.4
Zero bandwidth fetch cycles*

Benchmarks	Misprediction	I-Cache Miss
compress	6.65	0.01
eqntott	11.78	0.08
espresso	10.84	0.52
li	8.92	0.09
alvinn	0.39	0.02
hydro2d	5.24	0.12
tomcatv	0.68	0.01

*Values given are percentages.

Table 6.5
Distribution and average number of branches bypassed*

Benchmarks	Number of Bypassed Branches					Average
	0	1	2	3	4	
compress	66.42	27.38	5.40	0.78	0.02	0.41
eqntott	48.96	28.27	20.93	1.82	0.02	0.76
espresso	53.39	29.98	11.97	4.63	0.03	0.68
li	63.48	25.67	7.75	2.66	0.45	0.51
alvinn	83.92	15.95	0.13	0.00	0.00	0.16
hydro2d	68.79	16.90	10.32	3.32	0.67	0.50
tomcatv	92.07	2.30	3.68	1.95	0.00	0.16

*Columns 0–4 show percentage of cycles.

Branch prediction is a form of speculation. When speculation is done effectively, it can increase the performance of the machine by alleviating the constraints imposed by control dependences. The 620 can speculate past up to four predicted branches before stalling the fifth branch at the dispatch stage. Speculative instructions are allowed to move down the pipeline stages until the branches are resolved, at which time if the speculation proves to be incorrect, the speculated instructions are canceled. Speculative instructions can potentially finish execution and reach the completion stage prior to branch resolution. However, they are not allowed to complete until the resolution of the branch.

Table 6.5 displays the frequency of bypassing specific numbers of branches, which reflects the degree of speculation sustained. The average number of branches bypassed is determined by obtaining the number of correctly predicted branches that are bypassed in each cycle. Once a branch is determined to be mispredicted, speculation of instructions beyond that branch is not simulated. For the integer benchmarks, in 34% to 51% of the cycles, the 620 is speculatively executing beyond one or more branches. For floating-point benchmarks, the degree of speculation is lower. The frequency of misprediction, shown in Table 6.4, is related to the combination of the average number of branches bypassed, provided in Table 6.5, and the prediction accuracy, provided in Table 6.3.

6.4 Instruction Dispatching

The primary objective of the dispatch stage is to advance instructions from the instruction buffer to the reservation stations. The 620 uses an in-order dispatch policy.

6.4.1 Instruction Buffer

The eight-entry instruction buffer sits between the fetch stage and the dispatch stage. The fetch stage is responsible for filling the instruction buffer. The dispatch stage examines the first four entries of the instruction buffer and attempts to dispatch them to the reservation stations. As instructions are dispatched, the remaining instructions in the instruction buffer are shifted in groups of two to fill the vacated entries.

Figure 6.3(a) shows the utilization of the instruction buffer by profiling the frequencies of having specific numbers of instructions in the instruction buffer. The instruction buffer decouples the fetch stage and the dispatch stage and moderates the temporal variations of and differences between the fetching and dispatching parallelisms. The frequency of having zero instructions in the instruction buffer is significantly lower in the floating-point benchmarks than in the integer benchmarks. This frequency is directly related to the misprediction frequency shown in Table 6.4. At the other end of the spectrum, instruction buffer saturation can cause fetch stalls.

6.4.2 Dispatch Stalls

The 620 dispatches instructions by checking in parallel for all conditions that can cause dispatch to stall. This list of conditions is described in the following in greater detail. During simulation, the conditions in the list are checked one at a time and in the order listed. Once a condition that causes the dispatch of an instruction to stall is identified, checking of the rest of the conditions is aborted, and only that condition is identified as the source of the stall.

Serialization Constraints. Certain instructions cause *single-instruction serialization*. All previously dispatched instructions must complete before the serializing instruction can begin execution, and all subsequent instructions must wait until the serializing instruction is finished before they can dispatch. This condition, though extremely disruptive to performance, is quite rare.

Branch Wait for mtspr. Some forms of branch instructions access the count register during the dispatch stage. A move to special-purpose register (mtspr) instruction that writes to the count register will cause subsequent dependent branch instructions to delay dispatching until it is finished. This condition is also rare.

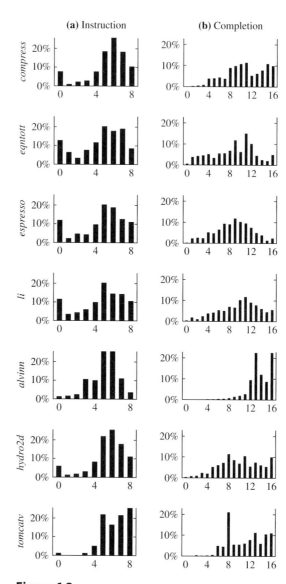

Figure 6.3
Profiles of the (a) Instruction Buffer and (b) Completion Buffer Utilizations.

Register Read Port Saturation. There are seven read ports for the general-purpose register file and four read ports for the floating-point register file. Occasionally, saturation of the read ports occurs when a read port is needed but none is available. There are enough condition register field read ports (three) that saturation cannot occur.

Reservation Station Saturation. As an instruction is dispatched, the instruction is placed into the reservation station of the instruction's associated execution unit. The instruction remains in the reservation station until it is issued. There is one reservation station per execution unit, and each reservation station has multiple entries, depending on the execution unit. Reservation station saturation occurs when an instruction can be dispatched to a reservation station but that reservation station has no more empty entries.

Rename Buffer Saturation. As each instruction is dispatched, its destination register is renamed into the appropriate rename buffer files. There are three rename buffer files, for general-purpose registers, floating-point registers, and condition register fields. Both the general-purpose register file and the floating-point register file have eight rename buffers. The condition register field file has 16 rename buffers.

Completion Buffer Saturation. Completion buffer entries are also allocated during the dispatch stage. They are kept until the instruction has completed. The 620 has 16 completion buffer entries; no more than 16 instructions can be in flight at the same time. Attempted dispatch beyond 16 in-flight instructions will cause a stall. Figure 6.3(b) illustrates the utilization profiles of the completion buffer for the benchmarks.

Another Dispatched to Same Unit. Although a reservation station has multiple entries, each reservation station can receive at most one instruction per cycle even when there are multiple available entries in a reservation station. Essentially, this constraint is due to the fact that each of the reservation stations has only one write port.

6.4.3 Dispatch Effectiveness

The average utilization of all the buffers is provided in Table 6.6. Utilization of the load/store unit's three reservation station entries averages 1.36 to 1.73 entries for integer benchmarks and 0.98 to 2.26 entries for floating-point benchmarks. Unlike the other execution units, the load/store unit does not deallocate a reservation station entry as soon as an instruction is issued. The reservation station entry is held until the instruction is finished, usually two cycles after the instruction is issued. This is due to the potential miss in the D-cache or the TLB. The reservation station entries in the floating-point unit are more utilized than those in the integer units. The in-order issue constraint of the floating-point unit and the nonpipelining of some floating-point instructions prevent some ready instructions from issuing. The average utilization of the completion buffer ranges from 9 to 14 for the benchmarks and corresponds with the average number of instructions that are in flight.

Sources of dispatch stalls are summarized in Table 6.7 for all benchmarks. The data in the table are percentages of all the cycles executed by each of the benchmarks. For example, in 24.35% of the *compress* execution cycles, no dispatch stalls occurred; i.e., all instructions in the dispatch buffer (first four entries of the instruction buffer) are dispatched. A common and significant source of bottleneck

Table 6.6
Summary of average number of buffers used

Buffer Usage	compress	eqntott	espresso	li	alvinn	hydro2d	tomcatv
Instruction buffers (8)	5.41	4.43	4.72	4.65	5.13	5.44	6.42
Dispatch buffers (4)	3.21	2.75	2.89	2.85	3.40	3.10	3.53
XSU0 RS entries (2)	0.37	0.66	0.68	0.36	0.48	0.23	0.11
XSU1 RS entries (2)	0.42	0.51	0.65	0.32	0.24	0.17	0.10
MC-FXU RS entries (2)	0.04	0.07	0.09	0.28	0.01	0.10	0.00
FPU RS entries (2)	0.00	0.00	0.00	0.00	0.70	1.04	0.89
LSU RS entries (3)	1.69	1.36	1.60	1.73	2.26	0.98	1.23
BRU RS entries (4)	0.45	0.84	0.75	0.59	0.19	0.54	0.17
GPR rename buffers (8)	2.73	3.70	3.25	2.77	3.79	1.83	1.97
FPR rename buffers (8)	0.00	0.00	0.00	0.00	5.03	2.85	3.23
CR rename buffers (16)	1.25	1.32	1.19	0.98	1.27	1.20	0.42
Completion buffers (16)	10.75	8.83	8.75	9.87	13.91	10.10	11.16

Table 6.7
Frequency of dispatch stall cycles*

Sources of Dispatch Stalls	compress	eqntott	espresso	li	alvinn	hydro2d	tomcatv
Serialization	0.00	0.00	0.00	0.00	0.00	0.00	0.00
Move to special register constraint	0.00	4.49	0.94	3.44	0.00	0.95	0.08
Read port saturation	0.26	0.00	0.02	0.00	0.32	2.23	6.73
Reservation station saturation	36.07	22.36	31.50	34.40	22.81	42.70	36.51
Rename buffer saturation	24.06	7.60	13.93	17.26	1.36	16.98	34.13
Completion buffer saturation	5.54	3.64	2.02	4.27	21.12	7.80	9.03
Another to same unit	9.72	20.51	18.31	10.57	24.30	12.01	7.17
No dispatch stalls	24.35	41.40	33.28	30.06	30.09	17.33	6.35

*Values given are percentages.

for all the benchmarks is the saturation of reservation stations, especially in the load/store unit. For the other sources of dispatch stalls, the degrees of various bottlenecks vary among the different benchmarks. Saturation of the rename buffers is significant for *compress* and *tomcatv*, even though on average their rename buffers are less than one-half utilized. Completion buffer saturation is highest in *alvinn*, which has the highest frequency of having all 16 entries utilized; see Figure 6.3(b). Contention for the single write port to each reservation station is also a serious bottleneck for many benchmarks.

Figure 6.4(a) displays the distribution of dispatching parallelism (the number of instructions dispatched per cycle). The number of instructions dispatched in each cycle can range from 0 to 4. The distribution indicates the frequency (averaged across the entire trace) of dispatching n instructions in a cycle, where $n = 0, 1, 2, 3, 4$. In all benchmarks, at least one instruction is dispatched per cycle for over one-half of the execution cycles.

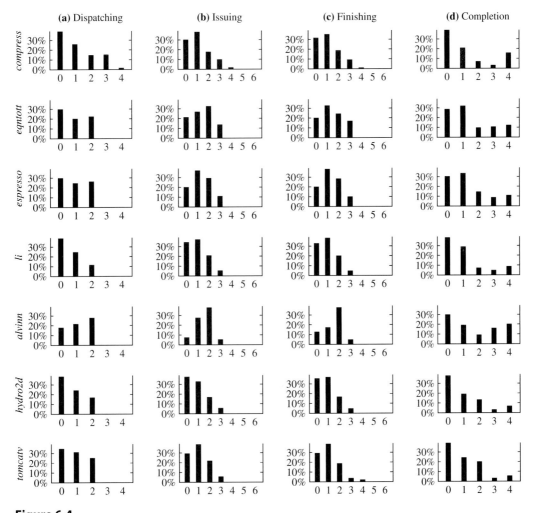

Figure 6.4

Distribution of the Instruction (a) Dispatching, (b) Issuing, (c) Finishing, and (d) Completion Parallelisms.

6.5 Instruction Execution

The 620 widens in the execute stage. While the fetch, dispatch, complete, and writeback stages all are four wide, i.e., can advance up to four instructions per cycle, the execute stage contains six execution units and can issue and finish up to six instructions per cycle. Furthermore, unlike the other stages, the execute stage processes instructions out of order to achieve maximum throughput.

6.5.1 Issue Stalls

Once instructions have been dispatched to reservation stations, they must wait for their source operands to become available, and then begin execution. There are a few other constraints, however. The full list of issuing hazards is described here.

Out of Order Disallowed. Although out-of-order execution is usually allowed from reservation stations, it is sometimes the case that certain instructions may not proceed past a prior instruction in the reservation station. This is the case in the branch unit and the floating-point unit, where instructions must be issued in order.

Serialization Constraints. Instructions which read or write non-renamed registers (such as XER), which read or write renamed registers in a non-renamed fashion (such as load/store multiple instructions), or which change or synchronize machine state (such as the *eieio* instruction, which enforces in-order execution of I/O) must wait for all prior instructions to complete before executing. These instructions stall in the reservation stations until their serialization constraints are satisfied.

Waiting for Source Operand. The primary purpose of reservation stations is to hold instructions until all their source operands are ready. If an instruction requires a source that is not available, it must stall here until the operand is forwarded to it.

Waiting for Execution Unit. Occasionally, two or more instructions will be ready to begin execution in the same cycle. In this case, the first will be issued, but the second must wait. This condition also applies when an instruction is executing in the MC-FXU (a nonpipelined unit) or when a floating-point divide instruction puts the FPU into nonpipelined mode.

The frequency of occurrence for each of the four issue stall types is summarized in Table 6.8. The data are tabulated for all execution units except the branch unit. Thus, the in-order issuing restriction only concerns the floating-point unit. The number of issue serialization stalls is roughly proportional to the number of multicycle integer instructions in the benchmark's instruction mix. Most of these multicycle instructions access the special-purpose registers or the entire condition register as a non-renamed unit, which requires serialization. Most issue stalls due to waiting for an execution unit occur in the load/store unit. More load/store instructions are ready to execute than the load/store execution unit can accommodate. Across all benchmarks, in significant percentages of the cycles, no issuing stalls are encountered.

Table 6.8
Frequency of issue stall cycles*

Sources of Issue Stalls	compress	eqntott	espresso	li	alvinn	hydro2d	tomcatv
Out of order disallowed	0.00	0.00	0.00	0.00	0.72	11.03	1.53
Serialization	1.69	1.81	3.21	10.81	0.03	4.47	0.01
Waiting for source	21.97	29.30	37.79	32.03	17.74	22.71	3.52
Waiting for execution unit	13.67	3.28	7.06	11.01	2.81	1.50	1.30
No issue stalls	62.67	65.61	51.94	46.15	78.70	60.29	93.64

*Values given are percentages.

6.5.2 Execution Parallelism

Here we examine the propagation of instructions through issuing, execution, and finish. Figure 6.4(b) shows the distribution of issuing parallelism (the number of instructions issued per cycle). The maximum number of instructions that can be issued in each cycle is six, the number of execution units. Although the issuing parallelism and the dispatch parallelism distributions must have the same average value, issuing is less centralized and has fewer constraints and can therefore achieve a more consistent rate of issuing. In most cycles, the number of issued instructions is close to the overall sustained IPC, while the dispatch parallelism has more extremes in its distribution.

We expected the distribution of finishing parallelism, shown in Figure 6.4(c), to look like the distribution of issuing parallelism because an instruction after it is issued must finish a certain number of cycles later. Yet this is not always the case as can be seen in the issuing and finishing parallelism distributions of the *eqntott, alvinn,* and *hydro2d* benchmarks. The difference comes from the high frequency of load/store and floating-point instructions. Since these instructions do not take the same amount of time to finish after issuing as the integer instructions, they tend to shift the issuing parallelism distribution. The integer benchmarks, with their more consistent instruction execution latencies, generally have more similarity between their issuing and finishing parallelism distributions.

6.5.3 Execution Latency

It is of interest to examine the average latency encountered by an instruction from dispatch until finish. If all the issuing constraints are satisfied and the execution unit is available, an instruction can be dispatched from the dispatch buffer to a reservation station, and then issued into the execution unit in the next cycle. This is the best case. Frequently, instructions must wait in the reservation stations. Hence, the overall execution latency includes the waiting time in the reservation station and the actual latency of the execution units. Table 6.9 shows the average overall execution latency encountered by the benchmarks in each of the six execution units.

Table 6.9
Average execution latency (in cycles) in each of the six execution units for the benchmarks

Execution Units	compress	eqntott	espresso	li	alvinn	hydro2d	tomcatv
XSU0 (1 stage)	1.53	1.62	1.89	2.28	1.05	1.48	1.01
XSU1 (1 stage)	1.72	1.73	2.23	2.39	1.13	1.78	1.03
MC-FXU (>1 stage)	4.35	4.82	6.18	5.64	3.48	9.61	1.64
FPU (3 stages)	—*	—*	—*	—*	5.29	6.74	4.45
LSU (2 stages)	3.56	2.35	2.87	3.22	2.39	2.92	2.75
BRU (≥1 stage)	2.71	2.86	3.11	3.28	1.04	4.42	4.14

*Very few instructions are executed in this unit for these benchmarks.

6.6 Instruction Completion

Once instructions finish execution, they enter the completion buffer for in-order completion and writeback. The completion buffer functions as a reorder buffer to reorder the out-of-order execution in the execute stage back to the sequential order for in-order retiring of instructions.

6.6.1 Completion Parallelism

The distributions of completion parallelism for all the benchmarks are shown in Figure 6.4(d). Again, similar to dispatching, up to four instructions can be completed per cycle. An average value can be computed for each of the parallelism distributions. In fact, for each benchmark, the average completion parallelism should be exactly equal to the average dispatching, issuing, and finishing parallelisms, which are all equal to the sustained IPC for the benchmark. In the case of instruction completion, while instructions are allowed to finish out of order, they can only complete in order. This means that occasionally the completion buffer will have to wait for one slow instruction to finish, but then will be able to retire its maximum of four instructions at once. On some occasions, the completion buffer can saturate and cause the stalling at the dispatch stage; see Figure 6.3(b).

The integer benchmarks with their more consistent execution latencies usually have one instruction completed per cycle. *Hydro2d* completes zero instructions in a large percentage of cycles because it must wait for floating-point divide instructions to finish. Usually, instructions cannot complete because they, or instructions preceding them, are not finished yet. However, occasionally there are other reasons. The 620 has four integer and two floating-point writeback ports. It is rare to run out of integer register file write ports. However, floating-point write port saturation occurs occasionally.

6.6.2 Cache Effects

The D-cache behavior has a direct impact on the CPU performance. Cache misses can cause additional stall cycles in the execute and complete stages. The D-cache in the 620 is interleaved in two banks, each with an address port. A load or store instruction can use either port. The cache can service at most one load and one store at the same time. A load instruction and a store instruction can access the cache in the same cycle if the accesses are made to different banks. The cache is nonblocking, only for the port with a load access. When a load cache miss is encountered and while a cache line is being filled, a subsequent load instruction can proceed to access the cache. If this access results in a cache hit, the instruction can proceed without being blocked by the earlier miss. Otherwise, the instruction is returned to the reservation station. The multiple entries in the load/store reservation station and the out-of-order issuing of instructions allow the servicing of a load with a cache hit past up to three outstanding load cache misses.

The sequential consistency model for main memory imposes the constraint that all memory instructions must appear to execute in order. However, if all memory instructions are to execute in sequential order, a significant amount of performance can be lost. The 620 executes all store instructions, which access the cache after the complete stage (using the physical address), in order; however, it allows load instructions, which access the cache in the execute stage (using the virtual address), to bypass store instructions. Such relaxation is possible due to the weak memory consistency model specified by the PowerPC ISA [May et al., 1994]. When a store is being completed, aliasing of its address with that of loads that have bypassed and finished is checked. If aliasing is detected, the machine is flushed when the next load instruction is examined for completion, and refetching of that load instruction is carried out. No forwarding of data is made from a pending store instruction to a dependent load instruction. The weak ordering of memory accesses can eliminate some unnecessary stall cycles. Most load instructions are at the beginning of dependence chains, and their earliest possible execution can make available other instructions for earlier execution.

Table 6.10 summarizes the nonblocking cache effect and the weak ordering of load/store instructions. The first line in the table gives the D-cache hit rate for all the benchmarks. The hit rate ranges from 94.2% to 99.9%. Because of the nonblocking feature of the cache, a load can bypass another load if the trailing load is a cache hit at the time that the leading load is being serviced for a cache miss. The percentage (as percentage of all load instructions) of all such trailing loads that actually bypass a missed load is given in the second line of Table 6.10. When a store is completed, it enters the complete store queue, waits there until the store writes to the cache, and then exits the queue. During the time that a pending store is in the queue, a load can potentially access the cache and bypass the store. The third line of Table 6.10 gives the percentage of all loads that, at the time of the load cache access, bypass at least one pending store. Some of these loads have addresses that alias with the addresses of the pending stores. The percentage of all loads that bypass a pending store and alias with any of the pending store addresses is given in the fourth line of the table. Most of the benchmarks have an insignificant number of aliasing occurrences. The fifth line of the table gives the average number of pending stores, or the number of stores in the store complete queue, in each cycle.

Table 6.10
Cache effect data*

Cache Effects	compress	eqntott	espresso	li	alvinn	hydro2d	tomcatv
Loads/stores with cache hit	94.17	99.57	99.92	99.74	99.99	94.58	96.24
Loads that bypass a missed load	8.45	0.53	0.11	0.14	0.01	4.82	5.45
Loads that bypass a pending store	58.85	21.05	27.17	48.49	98.33	58.26	43.23
Load that aliased with a pending store	0.00	0.31	0.77	2.59	0.27	0.21	0.29
Average number of pending stores per cycle	1.96	0.83	0.97	2.11	1.30	1.01	1.38

*Values given are percentages, except for the average number of pending stores per cycle.

6.7 Conclusions and Observations

The most interesting parts of the 620 microarchitecture are the branch prediction mechanisms, the out-of-order execution engine, and the weak ordering of memory accesses. The 620 does reasonably well on branch prediction. For the floating-point benchmarks, about 94% to 99% of the branches are resolved or correctly predicted, incurring little or no penalty cycles. Integer benchmarks yield another story. The range drops down to 89% to 91%. More sophisticated prediction algorithms, for example, those using more history information, can increase prediction accuracy. It is also clear that floating-point and integer benchmarks exhibit significantly different branching behaviors. Perhaps separate and different branch prediction schemes can be employed for dealing with the two types of benchmarks.

Even with having to support precise exceptions, the out-of-order execution engine in the 620 is still able to achieve a reasonable degree of instruction-level parallelism, with sustained IPC ranging from 0.99 to 1.44 for integer benchmarks and from 0.96 to 1.77 for floating-point benchmarks. One hot spot is the load/store unit. The number of load/store reservation station entries and/or the number of load/store units needs to be increased. Although the difficulties of designing a system with multiple load/store units are myriad, the load/store bottleneck in the 620 is evident. Having only one floating-point unit for three integer units is also a source of bottleneck. The integer benchmarks rarely stall on the integer units, but the floating-point benchmarks do stall while waiting for floating-point resources. The single dispatch to each reservation station in a cycle is also a source of dispatch stalls, which can reduce the number of instructions available for out-of-order execution. One interesting tradeoff involves the choice of implementing distributed reservation stations, as in the 620, versus one centralized reservation station, as in the Intel P6. The former approach permits simpler hardware since there are only single-ported reservation stations. However, the latter can share the multiple ports and the reservation station entries among different instruction types.

Allowing weak-ordering memory accesses is essential in achieving high performance in modern wide superscalar processors. The 620 allows loads to bypass stores and other loads; however, it does not provide forwarding from pending stores to dependent loads. The 620 allows loads to bypass stores but does not check for aliasing until completing the store. This store-centric approach makes forwarding difficult and requires that the machine be flushed from the point of the dependent load when aliasing occurs. The 620 does implement the D-cache as two interleaved banks, and permits the concurrent processing of one load and one store in the same cycle if there is no bank conflict. Using the standard rule of thumb for dynamic instruction mix, there is a clear imbalance with the processing of load/store instructions in current superscalar processors. Increasing the throughput of load/store instructions is currently the most critical challenge. As future superscalar processors get wider and their clock speeds increase, the memory bottleneck problem will be further exacerbated. Furthermore, commercial applications such as transaction processing (not characterized by the SPEC benchmarks) put even greater pressure on the memory and chip I/O bottleneck.

It is interesting to examine superscalar introductions contemporaneous to the 620 by different companies, and how different microprocessor families have evolved; see Figure 6.5. The PowerPC microprocessors and the Alpha AXP microprocessors represent two different approaches to achieving high performance in superscalar machines. The two approaches have been respectively dubbed "brainiacs vs. speed demons." The PowerPC microprocessors attempt to achieve

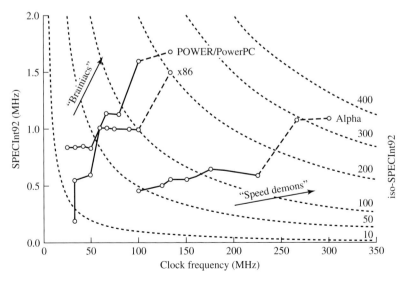

Figure 6.5
Evolution of Superscalar Families.

the highest level of IPC possible without overly compromising the clock speed. On the other hand, the Alpha AXP microprocessors go for the highest possible clock speed while achieving a reasonable level of IPC. Of course, future versions of PowerPC chips will get faster and future Alpha AXP chips will achieve higher IPC. The key issue is, which should take precedence, IPC or MHz? Which approach will yield an easier path to get to the next level of performance? Although these versions of the PowerPC 620 microprocessor and the Alpha AXP 21164 microprocessor seem to indicate that the speed demons are winning, there is no strong consensus on the answer for the future. In an interesting way, the two rivaling approaches resemble, and perhaps are a reincarnation of, the CISC vs. RISC debate of a decade earlier [Colwell et al., 1985].

The announcement of the P6 from Intel presents another interesting case. The P6 is comparable to the 620 in terms of its microarchitectural aggressiveness in achieving high IPC. On the other hand the P6 is somewhat similar to the 21164 in that they both are more "superpipelined" than the 620. The P6 represents yet a third, and perhaps hybrid, approach to achieving high performance. Figure 6.5 reflects the landscape that existed circa the mid-1990s. Since then the landscape has shifted significantly to the right. Today we are no longer using SPECInt92 benchmarks but SPECInt2000, and we are dealing with frequencies in the multiple-gigahertz range.

6.8 Bridging to the IBM POWER3 and POWER4

The PowerPC 620 was intended as the initial high-end 64-bit implementation of the PowerPC architecture that would satisfy the needs of the server and high-performance workstation market. However, because of numerous difficulties in finishing the design in a timely fashion, the part was delayed by several years and ended up only being used in a few server systems developed by Groupe Bull. In the meantime, IBM was able to satisfy its need in the server product line with the Star series and POWER2 microprocessors, which were developed by independent design teams and differed substantially from the PowerPC 620.

However, the IBM POWER3 processor, released in 1998, was heavily influenced by the PowerPC 620 design and reused its overall pipeline structure and many of its functional blocks [O'Connell and White, 2000]. Table 6.11 summarizes some of the key differences between the 620 and the POWER3 processors. Design optimization combined with several years of advances in semiconductor technology resulted in nearly tripling the processor frequency, even with a similar pipeline structure, resulting in noticeable performance improvement.

The POWER3 addressed some of the shortcomings of the PowerPC 620 design by substantially improving both instruction execution bandwidth as well as memory bandwidth. Although the front and back ends of the pipeline remained the same width, the POWER3 increased the peak issue rate to eight instructions per cycle by providing two load/store units and two fully pipelined floating-point units. The effective window size was also doubled by increasing the completion buffer to 32 entries and by doubling the number of integer rename registers and tripling the number of floating-point rename registers. Memory bandwidth was further enhanced with a novel 128-way set-associative cache design that embeds

Table 6.11
PowerPC 620 versus IBM POWER3 and POWER4

Attribute	620	POWER3	POWER4
Frequency	172 MHz	450 MHz	1.3 GHz
Pipeline length	5+	5+	15+
Branch prediction	Bimodal BHT/BTAC	Same as 620	3 × 16K combining
Fetch/issue/completion width	4/6/4	4/8/4	4/8/5
Rename/physical registers	8 Int, 8 FP	16 Int, 24 FP	80 Int, 72 FP
In-flight instructions	16	32	Up to 200
Floating-point units	1	2	2
Load/store units	1	2	2
Instruction cache	32K 8-way SA	32K 128-way SA	64K DM
Data cache	32K 8-way SA	64K 128-way SA	32K 2-way SA
L2/L3 size	4 Mbytes	16 Mbytes	~1.5 Mbytes/32 Mbytes
L2 bandwidth	1 Gbytes/s	6.4 Gbytes/s	100+ Gbytes/s
Store queue entries	6 × 8 bytes	16 × 8 bytes	12 × 64 bytes
MSHRs	I:1/D:1	I:2/D:4	I:2/D:8
Hardware prefetch	None	4 streams	8 streams

SA—set-associative   DM—direct mapped

tag match hardware directly into the tag arrays of both L1 caches, significantly reducing the miss rates, and by doubling the overall size of the data cache. The L2 cache size also increased substantially, as did available bandwidth to the off-chip L2 cache. Memory latency was also effectively decreased by incorporating an aggressive hardware prefetch engine that can detect up to four independent reference streams and prefetch them from memory. This prefetching scheme works extremely well for floating-point workloads with regular, predictable access patterns. Finally, support for multiple outstanding cache misses was added by providing two miss-status handling registers (MSHRs) for the instruction cache and four MSHRs for the data cache.

The next new high-performance processor in the PowerPC family was the POWER4 processor, introduced in 2001 [Tendler et al., 2001]. Key attributes of this entirely new core design are summarized in Table 6.11. IBM achieved yet another tripling of processor frequency, this time by employing a substantially deeper pipeline in conjunction with major advances in process technology (i.e., reduced feature sizes, copper interconnects, and silicon-on-insulator technology). The POWER4 pipeline is illustrated in Figure 6.6 and extends to 15 stages for the best case of single-cycle integer ALU instructions. To keep this pipeline fed with useful instructions, the POWER4 employs an advanced combining branch predictor that uses a 16K entry selector table to choose between a 16K entry *bimodal* predictor and a 16K entry *gshare* predictor. Each entry in each of the three tables is only 1 bit,

324 MODERN PROCESSOR DESIGN

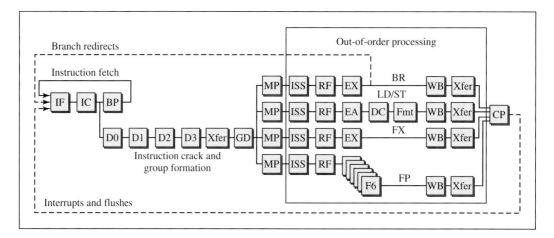

Figure 6.6
POWER4 Pipeline Structure.
Source: Tendler et al., 2001.

rather than a 2-bit up-down counter, since studies showed that 16K 1-bit entries performed better than 8K 2-bit entries. This indicates that for the server workloads the POWER4 is optimized for, branch predictor capacity misses are more important than the hysteresis provided by 2-bit counters.

The POWER4 matches the POWER3 in execution bandwidth, but provides substantially more rename registers (now in the form of a single physical register file) and supports up to 200 in-flight instructions in its pipeline. As in the POWER3, memory bandwidth and latency were important considerations, and multiple load/store units, support for up to eight outstanding cache misses, a very-high-bandwidth interface to the on-chip L2, and support for a massive off-chip L3 cache, all play an integral role in improving overall performance. The POWER4 also packs two complete processor cores sharing an L2 cache on a single chip in a *chip multiprocessor* configuration. More details on this arrangement are discussed in Chapter 11.

6.9 Summary

The PowerPC 620 is an interesting first-generation out-of-order superscalar processor design that exemplifies the short pipelines, aggressive support for extracting instruction-level parallelism, and support for weak ordering of memory references that are typical for other processors of a similar vintage (for example, the HP PA-8000 [Gwennap, 1994] and the MIPS R10000 [Yeager, 1996]). Its evolution into the IBM POWER3 part illustrates the natural extension of execution resources to extract even greater parallelism while also tackling the memory bandwidth and latency bottlenecks. Finally, the recent POWER4 design highlights the seemingly heroic efforts of microprocessors today to tolerate memory bandwidth and latency with aggressive on- and off-chip cache hierarchies, stream-based hardware prefetching, and very large instruction windows. At the same time, the POWER4 illustrates the trend toward higher and higher clock frequency through extremely deep pipelining, which can only be sustained as a result of increasingly accurate branch predictors that keep such pipelines filled with useful instructions.

REFERENCES

Colwell, R., C. Hitchcock, E. Jensen, H. B. Sprunt, and C. Kollar: "Instructions sets and beyond: Computers, complexity, and controversy," *IEEE Computer,* 18, 9, 1985, pp. 8–19.

Diep, T. A., C. Nelson, and J. P. Shen: "Performance evaluation of the PowerPC 620 microarchitecture," *Proc. 22nd Int. Symposium on Computer Architecture,* Santa Margherita Ligure, Italy, 1995.

Gwennap, L.: "PA-8000 combines complexity and speed," *Microprocessor Report,* 8, 15, 1994, pp. 6–9.

IBM Corp.: *PowerPC 601 RISC Microprocessor User's Manual.* IBM Microelectronics Division, 1993.

IBM Corp.: *PowerPC 604 RISC Microprocessor User's Manual.* IBM Microelectronics Division, 1994.

Levitan, D., T. Thomas, and P. Tu: "The PowerPC 620 microprocessor: A high performance superscalar RISC processor," *Proc. of COMPCON 95,* 1995, pp. 285–291.

May, C., E. Silha, R. Simpson, and H. Warren: *The PowerPC Architecture: A Specification for a New Family of RISC Processors,* 2nd ed. San Francisco, CA, Morgan Kauffman, 1994.

Motorola, Inc.: *MPC750 RISC Microprocessor Family User's Manual.* Motorola, Inc., 2001.

Motorola, Inc.: *MPC603e RISC Microprocessor User's Manual.* Motorola, Inc., 2002.

Motorola, Inc.: *MPC7450 RISC Microprocessor Family User's Manual.* Motorola, Inc., 2003.

O'Connell, F., and S. White: "POWER3: the next generation of PowerPC processors," *IBM Journal of Research and Development,* 44, 6, 2000, pp. 873–884.

Storino, S., A. Aipperspach, J. Borkenhagen, R. Eickemeyer, S. Kunkel, S. Levenstein, and G. Uhlmann: "A commercial multi-threaded RISC processor," *Int. Solid-State Circuits Conference,* 1998.

Tendler, J. M., S. Dodson, S. Fields, and B. Sinharoy: "IBM eserver POWER4 system microarchitecture," IBM Whitepaper, 2001.

Yeager, K.: "The MIPS R10000 superscalar microprocessor," *IEEE Micro,* 16, 2, 1996, pp. 28–40.

HOMEWORK PROBLEMS

P6.1 Assume the IBM instruction mix from Chapter 2, and consider whether or not the PowerPC 620 completion buffer versus integer rename buffer design is reasonably balanced. Assume that load and ALU instructions need an integer rename buffer, while other instructions do not. If the 620 design is not balanced, how many rename buffers should there be?

P6.2 Assuming instruction mixes in Table 6.2, which benchmarks are likely to be rename buffer constrained? That is, which ones run out of rename buffers before they run out of completion buffers and vice versa?

P6.3 Given the dispatch and retirement bandwidth specified, how many integer architected register file (ARF) read and write ports are needed to sustain peak throughput? Given instruction mixes in Table 6.2, also compute average ports needed for each benchmark. Explain why you would not just build for the average case. Given the actual number of read and write ports specified, how likely is it that dispatch will be port-limited? How likely is it that retirement will be port-limited?

P6.4 Given the dispatch and retirement bandwidth specified, how many integer rename buffer read and write ports are needed? Given instruction mixes in Table 6.2, also compute average ports needed for each benchmark. Explain why you would not just build for the average case.

P6.5 Compare the PowerPC 620 BTAC design to the next-line/next-set predictor in the Alpha 21264 as described in the *Alpha 21264 Microprocessor Hardware Reference Manual* (available from **www.compaq.com**). What are the key differences and similarities between the two techniques?

P6.6 How would you expect the results in Table 6.5 to change for a more recent design with a deeper pipeline (e.g., 20 stages, like the Pentium 4)?

P6.7 Judging from Table 6.7, the PowerPC 620 appears reservation station–starved. If you were to double the number of reservation stations, how much performance improvement would you expect for each of the benchmarks? Justify your answer.

P6.8 One of the most obvious bottlenecks of the 620 design is the single load/store unit. The IBM POWER3, a subsequent design based heavily on the 620 microarchitecture, added a second load/store unit along with a second floating-point multiply/add unit. Compare the SPECInt2000 and SPECFP2000 score of the IBM POWER3 (as reported on **www.spec.org**) with another modern processor, the Alpha AXP 21264. Normalized to frequency, which processor scores higher? Why is it not fair to normalize to frequency?

P6.9 The data in Table 6.10 seems to support the 620 designers' decision to not implement load/store forwarding in the 620 processor. Discuss how this tradeoff changes as pipeline depth increases and relative memory latency (as measured in processor cycles) increases.

P6.10 Given the data in Table 6.9, present a hypothesis for why XSU1 appears to have consistently longer execution latency than XSU0. Describe an experiment you might conduct to verify your hypothesis.

P6.11 The IBM POWER3 can detect up to four regular access streams and issue prefetches for future references. Construct an address reference trace that will utilize all four streams.

P6.12 The IBM POWER4 can detect up to eight regular access streams and issue prefetches for future references. Construct an address reference trace that will utilize all eight streams.

P6.13 The stream prefetching of the POWER3 and POWER4 processors is done outside the processor core, using the physical addresses of cache lines that miss the L1 cache. Explain why large virtual memory page sizes can improve the efficiency of such a prefetch scheme.

P6.14 Assume that a program is streaming sequentially through a 1-Gbyte array by reading each aligned 8-byte floating-point double in the 1-Gbyte array. Further assume that the prefetch engine will start prefetching after it has seen three consecutive cache line references that miss the L1 cache (i.e., the fourth cache line in a sequential stream will be prefetched). Assuming 4K page sizes and given the cache line sizes for the POWER3 and POWER4, compute the overall miss rate for this program for each of the following three assumptions: no prefetching, physical-address prefetching, and virtual-address prefetching. Report the miss rate per L1 D-cache reference, assuming that there is a single reference to every 8-byte word.

P6.15 Download and install the sim-outorder simulator from the Simplescalar simulator suite (available from **www.simplescalar.com**). Configure the simulator to match (as closely as possible) the microarchitecture of the PowerPC 620. Now collect branch prediction and cache hit data using the instructional benchmarks available from the Simplescalar website. Compare your results to Tables 6.3 and 6.10 and provide some reasons for differences you might observe.

Robert P. Colwell
Dave B. Papworth
Glenn J. Hinton
Mike A. Fetterman
Andy F. Glew

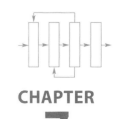

CHAPTER
7

Intel's P6 Microarchitecture

CHAPTER OUTLINE

7.1 Introduction
7.2 Pipelining
7.3 The In-Order Front End
7.4 The Out-of-Order Core
7.5 Retirement
7.6 Memory Subsystem
7.7 Summary
7.8 Acknowledgments

   References
   Homework Problems

In 1990, Intel began development of a new 32-bit Intel Architecture (IA32) microarchitecture core known as the P6. Introduced as a product in 1995 [Colwell and Steck, 1995], it was named the Pentium Pro processor and became very popular in workstation and server systems. A desktop proliferation of the P6 core, the Pentium II processor, was launched in May 1997, which added the MMX instructions to the basic P6 engine. The P6-based Pentium III processor followed in 1998, which included MMX and SSE instructions. This chapter refers to the core as the P6 and to the products by their respective product names.

The P6 microarchitecture is a 32-bit Intel Architecture-compatible, high-performance, superpipelined dynamic execution engine. It is order-3 superscalar and uses out-of-order and speculative execution techniques around a micro-dataflow execution core. P6 includes nonblocking caches and a transactions-based snooping bus. This chapter describes the various components of the design and how they combine to deliver extraordinary performance on an economical die size.

329

7.1 Introduction

The basic block diagram of the P6 microarchitecture is shown in Figure 7.1. There are three basic sections to the microarchitecture: the in-order front end, an out-of-order middle, and an in-order back-end "retirement" process. To be Intel Architecture–compatible, the machine must obey certain conventions on execution of its program code. But to achieve high performance, it must relax other conventions, such as execution of the program's operators strictly in the order implied by the program itself. True data dependences must be observed, but beyond that, only certain memory ordering constraints and the precise faulting semantics of the IA32 architecture must be guaranteed.

To maintain precise faulting semantics, the processor must ensure that asynchronous events such as interrupts and synchronous but awkward events such as faults and traps will be handled in exactly the same way as they would have in an i486 system.[1] This implies an in-order retirement process that reimposes the original program ordering to the commitment of instruction results to permanent architectural

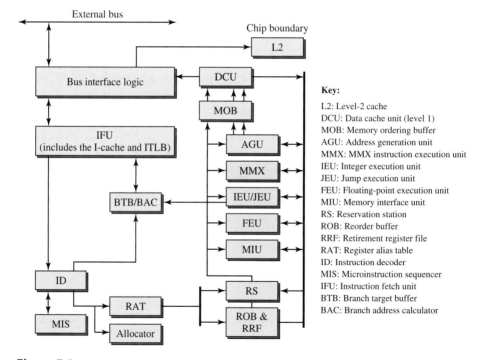

Figure 7.1
P6 Microarchitecture Block Diagram.

---

[1] Branches and faults use the same mechanism to recover state. However, for performance reasons, branches clear and restart the front end as early as possible. Page faults are handled speculatively, but floating-point faults are handled only when the machine is sure the faulted instructions were on the path of certain execution.

machine state. With these in-order mechanisms at both ends of the execution pipeline, the actual execution of the instructions can proceed unconstrained by any artifacts other than true data dependences and machine resources. We will explore the details of all three sections of the machine in the remainder of this chapter.

There are many novel aspects to this microarchitecture. For instance, it is almost universal that processors have a central controller unit somewhere that monitors and controls the overall pipeline. This controller "understands" the state of the instructions flowing through the pipeline, and it governs and coordinates the changes of state that constitute the computation process. The P6 microarchitecture purposely avoids having such a centralized resource. To simplify the hardware in the rest of the machine, this microarchitecture translates the Intel Architecture instructions into simple, stylized atomic units of computation called *micro-operations* (micro-ops or μops). All that the microarchitecture knows about the state of a program's execution, and the only way it can change its machine state, is through the manipulation of these μops.

The P6 microarchitecture is very deeply pipelined, relative to competitive designs of its era. This deep pipeline is implemented as several short pipe segments connected by queues. This approach affords a much higher clock rate, and the negative effects of deep pipelining are ameliorated by an advanced branch predictor, very fast high-bandwidth access to the L2 cache, and a much higher clock rate (for a given semiconductor process technology).

This microarchitecture is a speculative, out-of-order engine. Any engine that speculates can also misspeculate and must provide means to detect and recover from that condition. To ensure that this fundamental microarchitecture feature would be implemented as error-free as humanly possible, we designed the recovery mechanism to be extremely simple. Taking advantage of that simplicity, and the fact that this mechanism would be very heavily validated, we mapped the machine's event-handling events (faults, traps, interrupts, breakpoints) onto the same set of protocols and mechanisms.

The front-side bus is a change from Intel's Pentium processor family. It is transaction-oriented and designed for high-performance multiprocessing systems. Figure 7.2 shows how up to four Pentium Pro processors can be connected to a single shared bus. The chipset provides for main memory, through the data

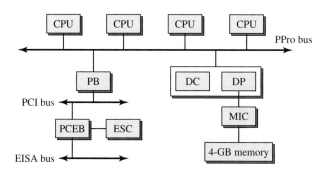

Figure 7.2
P6 Pentium Pro System Block Diagram.

Figure 7.3
P6 Product Packaging.

path (DP) and data controller (DC) parts, and then through the memory interface controller (MIC) chip. I/O is provided via the industry standard PCI bus, with a bridge to the older Extended Industry Standard Architecture (EISA) standard bus.

Various proliferations of the P6 had chipsets and platform designs that were optimized for multiple market segments including the (1) high-volume, (2) workstation, (3) server, and (4) mobile market segments. These differed mainly in the amount and type of memory that could be accommodated, the number of CPUs that could be supported in a single platform, and L2 cache design and placement. The Pentium II processor does not use the two-die-in-a-package approach of the original Pentium Pro; that approach yielded a very fast system product but was expensive to manufacture due to the special ceramic dual-cavity package and the unusual manufacturing steps required. P6-based CPUs were packaged in several formats (see Figure 7.3):

- Slot 1 brought the P6 microarchitecture to volume price points, by combining one P6 CPU with two commodity cache RAMs and a tag chip on one FR4 fiberglass substrate cartridge. This substrate has all its electrical contacts contained in one edge connector, and a heat sink attached to one side of the packaged substrate. Slot 1's L2 cache runs at one-half the clock frequency of the CPU.

- Slot 2 is physically larger than Slot 1, to allow up to four custom SRAMs to form the very large caches required by the high-performance workstation and server markets. Slot 2 cartridges are carefully designed so that, despite the higher number of loads on the L2 bus, they can access the large L2 cache at the full clock frequency of the CPU.

- With improved silicon process technology, in 1998 Intel returned to pin-grid-array packaging on the Celeron processor, with the L2 caches contained on the CPU die itself. This obviated the need for the Pentium Pro's two-die-in-a-package or the Slot 1/Slot 2 cartridges.

7.1.1 Basics of the P6 Microarchitecture

In subsequent sections, the operation of the various components of the microarchitecture will be examined. But first, it may be helpful to consider the overall machine organization at a higher level.

A useful way to view the P6 microarchitecture is as a dataflow engine, fed by an aggressive front end, constrained by implementation and code compatibility. It is not difficult to design microarchitectures that are capable of expressing instruction-level parallelism; adding multiple execution units is trivial. Keeping those execution units gainfully employed is what is hard.

The P6 solution to this problem is to

- Extract useful work via deep speculation on the front end (instruction cache, decoder, and register renaming).
- Provide enough temporary storage that a lot of work can be "kept in the air."
- Allow instructions that are ready to execute to pass others that are not (in the out-of-order middle section).
- Include enough memory bandwidth to keep up with all the work in progress.

When speculation is proceeding down the right path, the µops generated in the front end flow smoothly into the reservation station (RS), execute when all their data operands have become available (often in an order other than that implied by the source program), take their place in the retirement line in the reorder buffer (ROB), and retire when it is their turn.

Micro-ops carry along with them all the information required for their scheduling, dispatch, execution, and retirement. Micro-ops have two source references, one destination, and an operation-type field. These logical references are renamed in the register alias table (RAT) to physical registers residing in the ROB.

When the inevitable misprediction occurs,[2] a very simple protocol is exercised within the out-of-order core. This protocol ensures that the out-of-order core flushes the speculative state that is now known to be bogus, while keeping any other work that is not yet known to be good or bogus. This same protocol directs the front end to drop what it was doing and start over at the mispredicted target's correct address.

Memory operations are a special category of µop. Because the IA32[3] instruction set architecture has so few registers, IA32 programs must access memory frequently. This means that the dependency chains that characterize a program generally start with a memory load, and this in turn means that it is important that loads be speculative. (If they were not, all the rest of the speculative engine would be starved while waiting for loads to go in order.) But not all loads can be speculative; consider a load in a memory-mapped I/O system where the load has a nonrecoverable side effect. Section 7.6 will cover some of these special cases. Stores are never speculative, there being no way to "put back the old data" if a misspeculated store were later found to have been in error. However, for performance reasons, store data can be forwarded from the store buffer (SB), before the data have actually

---

[2]For instance, on the first encounter with a branch, the branch predictor does not "know" there is a branch at all, much less which way the branch might go.

[3]This chapter uses IA32 to refer to the standard 32-bit Intel Architecture, as embodied in processors such as the Intel 486, or the Pentium, Pentium II, Pentium III, and Pentium 4 processors.

appeared in any caches or memory, to allow dependent loads (and their progeny) to proceed. This is closely analogous to a writeback cache, where data can be loaded from a cache that has not yet written its data to main memory.

Because of IA32 coding semantics, it is important to carefully control the transfer of information from the out-of-order, speculative engine to the permanent machine state that is saved and restored in program context switches. We call this "retirement." Essentially, all the machine's activity up until this point can be undone. Retirement is the act of irrevocably committing changes to a program's state. The P6 microarchitecture can retire up to three µops per clock cycle, and therefore can retire as many as three IA32 instructions' worth of changes to the permanent state. (If more than three ops are needed to express a given IA32 instruction, the retirement process makes sure the necessary all-or-none atomicity is obeyed.)

7.2 Pipelining

We will examine the individual elements of the P6 micro-architecture in this chapter, but before we look at the pieces, it may help to see how they all fit together. The pipeline diagram of Figure 7.4 may help put the pieces into perspective. The first thing to note about the figure is that it appears to show many separate pipelines, rather than one single pipeline. This is intentional; it reflects both the philosophy and the design of the microarchitecture.

The pipeline segments are in three clusters. First is the in-order front end, second is the out-of-order core, and third is the retirement. For reasons that should become clear, it is essential that these pipeline segments be separable operationally. For example, when recovering from a mispredicted branch, the front end of the machine will immediately flush the bogus information it had been processing from the mispredicted target address and will refetch the corrected stream from the corrected branch target, and all the while the out-of-order core continues working on previously fetched instructions (up until the mispredicted branch).

It is also important to separate the overall pipeline into independent segments so that when the various queues happen to fill up, and thus require their suppliers to stall until the tables have drained, only as little of the overall machine as necessary stalls, not the entire machine.

7.2.1 In-Order Front-End Pipeline

The first stage (pipe stage 11[4]) of the in-order front-end pipeline is used by the branch target buffer (BTB) to generate a pointer for the instruction cache (I-cache) to use, in accessing what we hope will be the right set of instruction bytes. Remember that the machine is always speculating, and this guess can be wrong; if it is, the error will be recognized at one of several places in the machine and a misprediction recovery sequence will be initiated at that time.

The second pipe stage in the in-order pipe (stage 12) initiates an I-cache fetch at the address that the BTB generated in pipe stage 11. The third pipe stage (stage 13)

---
[4]Why is the first stage numbered 11 instead of 1? We don't remember. We think it was arbitrary.

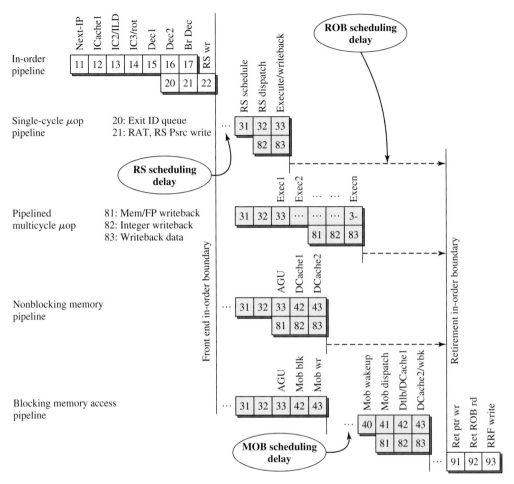

Figure 7.4
P6 Pipelining.

continues the I-cache access. The fourth pipe stage (stage 14) completes the I-cache fetch and transfers the newly fetched cache line to the instruction decoder (ID) so it can commence decoding.

Pipe stages 15 and 16 are used by the ID to align the instruction bytes, identify the ends of up to three IA32 instructions, and break these instructions down into sequences of their constituent μops.

Pipe stage 17 is the stage where part of the ID can detect branches in the instructions it has just decoded. Under certain conditions (e.g., an unpredicted but unconditional branch), the ID can notice that a branch went unpredicted by the BTB (probably because the BTB had never seen that particular branch before) and can flush the in-order pipe and refetch from the branch target, without having to wait until the branch actually tries to retire many cycles in the future.

Pipe stage 17 is synonymous with pipe stage 21,[5] which is the rename stage. Here the register alias table (RAT) renames μop destination/source linkages to a large set of physical registers in the reorder buffer. In pipe stage 22, the RAT transfers the μops (three at a time, since the P6 microarchitecture is an order-3 superscalar) to the out-of-order core. Pipe stage 22 marks the transition from the in-order section to the out-of-order section of the machine. The μops making this transition are written into both the reservation station (where they will wait until they can execute in the appropriate execution unit) and the reorder buffer (where they "take a place in line," so that eventually they can commit their changes to the permanent machine state in the order implied by the original user program).

7.2.2 Out-of-Order Core Pipeline

Once the in-order front end has written a new set of (up to) three μops into the reservation station, these μops become possible candidates for execution. The RS takes several factors into account when deciding which μops are ready for execution: The μop must have all its operands available; the execution unit (EU) needed by that μop must be available; a writeback bus must be ready in the cycle in which the EU will complete that μop's execution; and the RS must not have other μops that it thinks (for whatever reason) are more important to overall performance than the μop under discussion. Remember that this is the out-of-order core of the machine. The RS does not know, and does not care about, the original program order. It only observes data dependences and tries to maximize overall performance while doing so.

One implication of this is that any given μop can wait from zero to dozens or even hundreds of clock cycles after having been written into the RS. That is the point of the RS scheduling delay label on Figure 7.4. The scheduling delay can be as low as zero, if the machine is recovering from a mispredicted branch, and these are the first "known good" μops from the new instruction stream. (There would be no point to writing the μops into the RS, only to have the RS "discover" that they are data-ready two cycles later. The first μop to issue from the in-order section is guaranteed to be dependent on no speculative state, because there is no more speculative state at that point!)

Normally, however, the μops do get written into the RS, and there they stay until the RS notices that they are data-ready (and the other constraints previously listed are satisfied). It takes the RS two cycles to notice, and then *dispatch*, μops to the execution units. These are pipe stages 31 and 32.

Simple, single-cycle-execution μops such as logical operators or simple arithmetic operations execute in pipe stage 33. More complex operations, such as integer multiply, or floating-point operations, take as many cycles as needed.

One-cycle operators provide their results to the writeback bus at the end of pipe stage 33. The writeback busses are a shared resource, managed by the RS, so the RS must ensure that there will be a writeback bus available in some future cycle for a given μop at the time the μop is dispatched. Writeback bus scheduling occurs in pipe stage 82, with the writeback itself in the execution cycle, 83 (which is synonymous with pipe stage 33).

---

[5] Why do some pipe stages have more than one name? Because the pipe segments are independent. Sometimes part of one pipe segment lines up with one stage of a different pipe segment, and sometimes with another.

Memory operations are a bit more complicated. All memory operations must first generate the effective address, per the usual IA32 methods of combining segment base, offset, base, and index. The μop that generates a memory address executes in the address generation unit (AGU) in pipe stage 33. The data cache (DCache) is accessed in the next two cycles, pipe stages 42 and 43. If the access is a cache hit, the accessed data return to the RS and become available as a source to other μops.

If the DCache reference was a miss, the machine tries the L2 cache. If that misses, the load μop is suspended (no sense trying it again any time soon; we must refill the miss from main memory, which is a slow operation). The memory ordering buffer (MOB) maintains the list of active memory operations and will keep this load μop suspended until its cache line refill has arrived. This conserves cache access bandwidth for other μop sequences that may be independent of the suspended μop; these other μops can go around the suspended load and continue making forward progress. Pipe stage 40 is used by the MOB to identify and "wake up" the suspended load. Pipe stage 41 re-dispatches the load to the DCache, and (as earlier) pipe stages 42 and 43 are used by the DCache in accessing the line. This MOB scheduling delay is labeled in Figure 7.4.

7.2.3 Retirement Pipeline

Retirement is the act of transferring the speculative state into the permanent, irrevocable architectural machine state. For instance, the speculative out-of-order core may have a μop that wrote 0xFA as its instruction result into the appropriate field of ROB entry 14. Eventually, if no mispredicted branch is found in the interim, it will become that μop's turn to retire next, when it has become the oldest μop in the machine. At that point, the μop's original intention (to write 0xFA into, e.g., the EAX register) is realized by transferring the 0xFA data in ROB Slot 14 to the retirement register file's (RRF) EAX register.

There are several complicating factors to this simple idea. First, what the ROB is actually retiring should not be viewed as just a sequence of μops, but rather a series of IA32 instructions. Since it is architecturally illegal to retire only part of an IA32 instruction, then either all μops comprising an IA32 instruction retire, or none do. This atomicity requirement generally demands that the partially modified architectural state never be visible to the world outside the processor. So part of what the ROB must do is to detect the beginning and end of a given IA32 instruction and to make sure the atomicity rule is strictly obeyed. The ROB does this by observing some marks left on the μops by the instruction decoder (ID): some μops are marked as the first μop in an IA32 instruction, and others are marked as last. (Obviously, others are not marked at all, implying they are somewhere in the middle of an IA32 μop sequence.)

While retiring a sequence of μops that comprise an IA32 instruction, no external events can be handled. Those simply have to wait, just as they do in previous generations of the Intel Architecture (i486 and Pentium processors, for instance). But between two IA32 instructions, the machine must be capable of taking interrupts, breakpoints, traps, handling faults, and so on. The reorder buffer makes sure that these events are only possible at the right times and that multiple pending events are serviced in the priority order implied by the Intel instruction set architecture.

The processor must have the capability to stop a microcode flow partway through, switch to a microcode assist routine, perform some number of μops, and then resume the flow at the point of interruption, however. In that sense, an instruction may be considered to be partially executed at the time the trap is taken. The first part of the instruction cannot be discarded and restarted, because this would prevent forward progress. This kind of behavior occurs for TLB updates, some kinds of floating-point assists, and more.

The reorder buffer is implemented as a circular list of μops, with one retirement pointer and one new-entry pointer. The reorder buffer writes the results of a just-executed μop into its array in pipe stage 22. The μop results from the ROB are read in pipe stage 82 and committed to the permanent machine state in the RRF in pipe stage 93.

7.3 The In-Order Front End

The primary responsibility of the front end is to keep the execution engine full of useful work to do. On every clock cycle, the front end makes a new guess as to the best I-cache address from which to fetch a new line, and it sends the cache line guessed from the last clock to the decoders so they can get started. This guess can, of course, be discovered to have been incorrect, whereupon the front end will later be redirected to where the fetch really should have been from. A substantial performance penalty occurs when a mispredicted branch is discovered, and a key challenge for a microarchitecture such as this one is to ensure that branches are predicted correctly as often as possible, and to minimize the recovery time when they are found to have been mispredicted. This will be discussed in greater detail shortly.

The decoders convert up to three IA instructions into their corresponding μops (or μop flows, if the IA instructions are complex enough) and push these into a queue. A register renamer assigns new physical register designators to the source and destination references of these μops, and from there the μops issue to the out-of-order core of the machine.

As implied by the pipelining diagram in Figure 7.4, the in-order front end of the P6 microarchitecture runs independently from the rest of the machine. When a mispredicted branch is detected in the out-of-order core [in the jump execution unit (JEU)], the out-of-order core continues to retire μops older than the mispredicted branch μop, but flushes everything younger. Refer to Figure 7.5. Meanwhile, the front end is immediately flushed and begins to refetch and decode instructions starting at the correct branch target (supplied by the JEU). To simplify the handoff between the in-order front end and the out-of-order core, the new μops from the corrected branch target are strictly quarantined from whatever μops remain in the out-of-order core, until the out-of-order section has drained. Statistically, the out-of-order core will usually have drained by the time the new FE2 μops get through the in-order front end.

7.3.1 Instruction Cache and ITLB

The on-chip instruction cache (I-cache) performs the usual function of serving as a repository of recently used instructions. Figure 7.6 shows the four pipe stages of the instruction fetch unit (IFU). In its first pipe stage (pipe stage 11), the IFU

INTEL'S P6 MICROARCHITECTURE

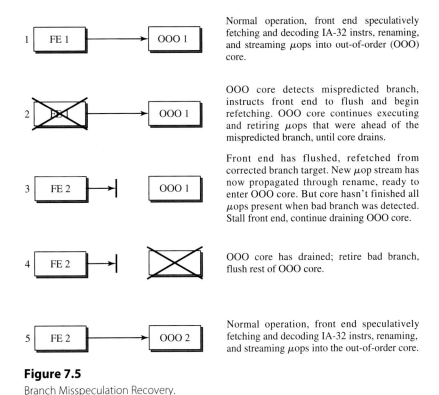

Figure 7.5
Branch Misspeculation Recovery.

selects the address of the next cache access. This address is selected from a number of competing fetch requests that arrive at the IFU from (among others) the BTB and branch address calculator (BAC). The IFU picks the request with the highest priority and schedules it for service by the second pipe stage (pipe stage 12). In the second pipe stage, the IFU accesses its many caches and buffers using the fetch address selected by the previous stage. Among the caches and buffers accessed are the instruction cache and the instruction streaming buffer. If there is a hit in any of these caches or buffers, instructions are read out and forwarded to the third pipe stage. If there is a miss in all these buffers, an external fetch is initiated by sending a request to the external bus logic (EBL).

Two other caches are also accessed in pipe stage 12 using the same fetch address: the ITLB in the IFU and the branch target buffer (BTB). The ITLB access obtains the physical address and memory type of the fetch, and the BTB access obtains a branch prediction. The BTB takes two cycles to complete one access. In the third pipe stage (13), the IFU marks the instructions received from the previous stage (12). Marking is the process of determining instruction boundaries. Additional marks for predicted branches are delivered by the BTB by the end of pipe stage 13. Finally, in the fourth pipe stage (14), the instructions and their marks are written into the instruction buffer and optionally steered to the ID, if the instruction buffer is empty.

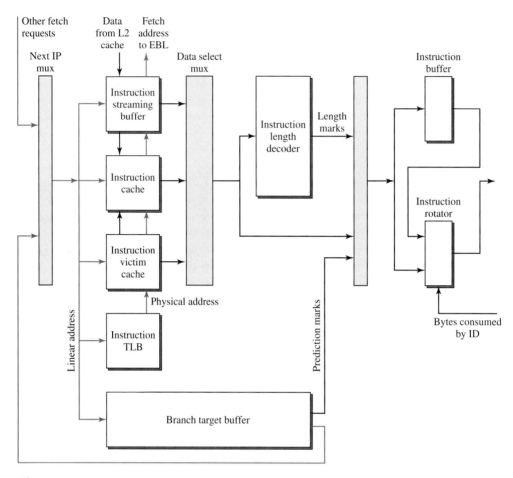

Figure 7.6
Front-End Pipe Staging.

The fetch address selected by pipe stage 11 for service in pipe stage 12 is a *linear* address, not a *virtual* or *physical* address. In fact, the IFU is oblivious to virtual addresses and indeed all segmentation. This allows the IFU to ignore segment boundaries while delaying the checking of segmentation-related violations to units downstream from the IFU in the P6 pipeline. The IFU does, however, deal with paging. When paging is turned off, the linear fetch address selected by pipe stage 11 is identical to the physical address and is directly used to search all caches and buffers in pipe stage 12. However, when paging is turned on, the linear address must be translated by the ITLB into a physical address. The virtual to linear to physical sequence is shown in Figure 7.7.

The IFU caches and buffers that require a physical address (actually, untranslated bits, with a match on physical address) for access are the instruction cache, the instruction streaming buffer, and the instruction victim cache. The branch target

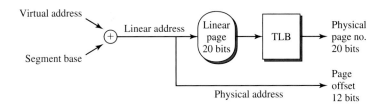

Figure 7.7
Virtual to Linear to Physical Addresses.

buffer is accessed using the linear fetch address. A block diagram of the front end, and the BTB's place in it, is shown in Figure 7.6.

7.3.2 Branch Prediction

The branch target buffer has two major functions: to predict branch direction and to predict branch targets. The BTB must operate early in the instruction pipeline to prevent the machine from executing down a wrong program stream. (In a speculative engine such as the P6, executing down the wrong stream is, of course, a performance issue, not a correctness issue. The machine will always execute correctly, the only question is how quickly.)

The branch *decision* (taken or not taken) is known when the jump execution unit (JEU) resolves the branch (pipe stage 33). Cycles would be wasted were the machine to wait until the branch is resolved to start fetching the instructions after the branch. To avoid this delay, the BTB predicts the decision of the branch as the IFU fetches it (pipe stage 12). This prediction can be wrong. The machine is able to detect this case and recover. All predictions made by the BTB are verified downstream by either the branch address calculator (pipe stage 17) or the JEU (pipe stage 33).

The BTB takes the starting linear address of the instructions being fetched and produces the prediction and target address of the branch instructions being fetched. This information (prediction and target address) is sent to the IFU, and the next cache line fetch will be redirected if a branch is predicted taken. A branch's entry in the BTB is updated or allocated in the BTB cache only when the JEU resolves it. A branch update is sometimes too late to help the next instance of the branch in the instruction stream. To overcome this delayed update problem, branches are also speculatively updated (in a separately maintained BTB state) when the BTB makes a prediction (pipe stage 13).

7.3.2.1 Branch Prediction Algorithm.
Dynamic branch prediction in the P6 BTB is related to the two-level adaptive training algorithm proposed by Yeh and Patt [1991]. This algorithm uses two levels of branch history information to make predictions. The first level is the history of the branches. The second level is the branch behavior for a specific pattern of branch history. For each branch, the BTB keeps N bits of "real" branch history (i.e., the branch decision for the last N dynamic occurrences). This history is called the branch history register (BHR).

The pattern in the BHR indexes into a 2^N entry table of states, the pattern table (PT). The state for a given pattern is used to predict how the branch will act the next time it is seen. The states in the pattern table are updated using Lee and Smith's [1984] saturating up-down counter.

The BTB uses a 4-bit semilocal pattern table per set. This means 4 bits of history are kept for each entry, and all entries in a set use the same pattern table (the four branches in a set share the same pattern table). This has equivalent performance to a 10-bit global table, with less hardware complexity and a smaller die area. A speculative copy of the BHR is updated in pipe stage 13, and the real one is updated upon branch resolution in pipe stage 83. But the pattern table is updated only for conditional branches, as they are computed in the jump execution unit.

To obtain the prediction of a branch, the *decision* of the branch (taken or not taken) is shifted into the old history pattern of the branch, and this field is used to index the pattern table. The most significant bit of the state in the pattern table indicates the prediction used the next time it is seen. The old state indexed by the old history pattern is updated using the Lee and Smith state machine.

An example of how the algorithm works is shown in Figure 7.8. The history of the entry to be updated is *0010,* and the branch *decision* was taken. The new

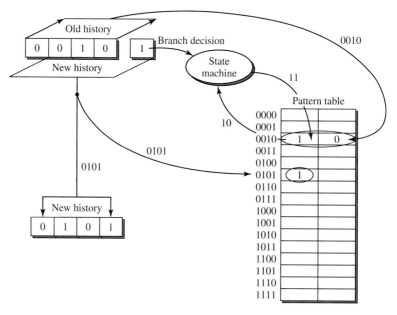

Two processes occur in parallel:

1. The new history is used to access the pattern table to get the new prediction bit. This prediction bit is written into the BTB in the next phase.

2. The old history is used to access the pattern table to get the state that has to be updated. The updated state is then written back to the pattern table.

Figure 7.8
Yeh's Algorithm.

INTEL'S P6 MICROARCHITECTURE 343

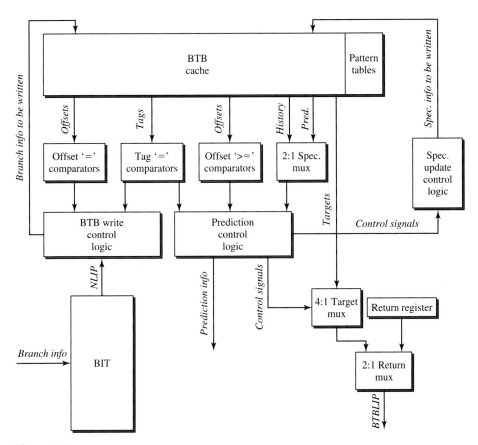

Figure 7.9
Simplified BTB Block Diagram.

history *0101* is used to index into the pattern table and the new prediction *1* for the branch (the most significant bit of the state) is obtained. The old history *0010* is used to index the pattern table to get the old state *10*. The old state *10* is sent to the state machine along with the branch *decision*, and the new state *11* is written back into the pattern table.

The BTB also maintains a 16-deep return stack buffer to help predict returns. For circuit speed reasons, BTB accesses require two clocks. This causes predicted-taken branches to insert a one-clock fetch "bubble" into the front end. The double-buffered fetch lines into the instruction decoder and the ID's output queue help eliminate most of these bubbles in normal execution. A block diagram of the BTB is shown in Figure 7.9.

7.3.3 Instruction Decoder

The first stage of the ID is known as the instruction steering block (ISB) and is responsible for latching instruction bytes from the IFU, picking off individual instructions in order, and steering them to each of the three decoders. The ISB

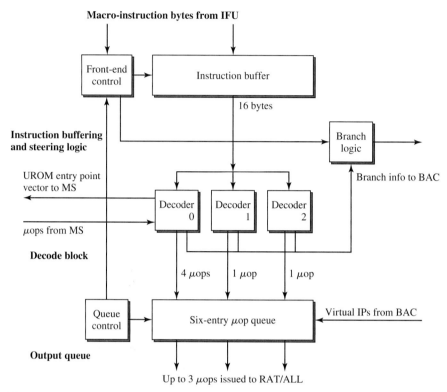

Figure 7.10
ID Block Diagram.

quickly detects how many instructions are decoded each clock to make a fast determination of whether or not the instruction buffer is empty. If empty, it enables the latch to receive more instruction bytes from the IFU (refer to Figure 7.10).

There are other miscellaneous functions performed by this logic as the "front end" of the ID. It detects and generates the correct sequencing of predicted branches. In addition, the ID front end generates the valid bits for the µops produced by the decode PLAs and detects stall conditions in the ID.

Next, the instruction buffer loads 16 bytes at a time from the IFU. These data are aligned such that the first byte in the buffer is guaranteed to be the first byte of a complete instruction. The average instruction length is 2.7 to 3.1 bytes. This means that on average five to six complete instructions will be loaded into the buffer. Loading a new batch of instruction bytes is enabled under any of the following conditions:

- A processor front-end reset occurs due to branch misprediction.
- All complete instructions currently in the buffer are successfully decoded.
- A BTB predicted-taken branch is successfully decoded in any of the three decoders.

Steering three properly aligned macro-instructions to three decoders in one clock is complicated due to the variable length of IA32 instructions. Even determining the length of one instruction itself is not straightforward, as the first bytes of an instruction must be decoded in order to interpret the bytes that follow. Since the process of steering three variable-length instructions is inherently serial, it is helpful to know beforehand the location of each macro-instruction's boundaries. The instruction length decoder (ILD), which resides in the IFU, performs this pre-decode function. It scans the bytes of the macro-instruction stream locating instruction boundaries and marking the first opcode and end-bytes of each. In addition, the IFU marks the bytes to indicate BTB branch predictions and code breakpoints.

There may be from 1 to 16 instructions loaded into the instruction buffer during each load. Each of the first three instructions is steered to one of three decoders. If the instruction buffer does not contain three complete instructions, then as many as possible are steered to the decoders. The steering logic uses the *first opcode markers* to align and steer the instructions in parallel.

Since there may be up to 16 instructions in the instruction buffer, it may take several clocks to decode all of them. The starting byte location of the three instructions steered in a given clock may lie anywhere in the buffer. Hardware aligns three instructions and steers them to the three decoders.

Even though three instructions may be steered to the decoders in one cycle, all three may not get successfully decoded. When an instruction is not successfully decoded, then that specific decoder is flushed and all μops resulting from that decode attempt will be invalidated. It can take multiple cycles to consume (decode) all the instructions in the buffer. The following situations result in the invalidation of μops and the resteering of their corresponding macro-instructions to another decoder during a subsequent cycle:

- If a complex macro-instruction is detected on decoder 0, requiring assistance from the microcode sequencer (MS) microcode read-only memory (UROM), then the μops from all subsequent decoders are invalidated. When the MS has completed sequencing the rest of the *flow*,[6] subsequent macro-instructions are decoded.

- If a macro-instruction is steered to a limited-functionality decoder (which is not able to decode it), then the macro-instructions and all subsequent macro-instructions are resteered to other decoders in the next cycle. All μops produced by this and subsequent decoders are invalidated.

- If a branch is encountered, then all μops produced by subsequent decoders are invalidated. Only one branch can be decoded per cycle.

Note that the number of macro-instructions that can be decoded simultaneously does not directly relate to the number of μops that the ID can issue because the decoder queue can store μops and issue them later.

---

[6]*Flow* refers to a sequence of μops emitted by the microcode ROM. Such sequences are commonly used by the microcode to express IA32 instructions, or microarchitectural housekeeping.

7.3.3.1 Complex Instructions.
Complex instructions are those requiring the MS to sequence μops from the UROM. Only decoder 0 can handle these instructions. There are two ways in which the MS microcode will be invoked:

- Long flows where decoder 0 generates up to the first four μops of the flow and the MS sequences the remaining μops.
- Low-performance instructions where decoder 0 issues no μops but transfers control to the MS to sequence from the UROM.

Decoders 1 and 2 cannot decode complex instructions, a design tradeoff that reflects both the silicon expense of implementation as well as the statistics of dynamic IA32 code execution. Complex instructions will be resteered to the next-lower available decoder during subsequent clocks until they reach decoder 0. The MS receives a UROM entry point vector from decoder 0 and begins sequencing μops until the end of the microcode flow is encountered.

7.3.3.2 Decoder Branch Prediction.
When the macro-instruction buffer is loaded from the IFU, the ID looks at the prediction byte marks to see if there are any predicted-taken branches (predicted dynamically by the BTB) in the set of complete instructions in the buffer. A proper prediction will be found on the byte corresponding to the last byte of a branch instruction. If a predicted-taken branch is found anywhere in the buffer, the ID indicates to the IFU that the ID has "grabbed" the predicted branch. The IFU can now let the 16-byte block, fetched at the target address of the branch, enter the buffer at the input of its rotator. The rotator then aligns the instruction at the branch target so that it will be the next instruction loaded into the ID's instruction buffer. The ID may decode the predicted branch immediately, or it may take several cycles (due to decoding all the instructions ahead of it). After the branch is finally decoded, the ID will latch the instructions at the branch target in the next clock cycle.

Static branch prediction (prediction made without reference to run-time history) is made by the branch address calculator (BAC). If the BAC decides to take a branch, it gives the IFU a target IP where the IFU should start fetching instructions. The ID must not, of course, issue any μops of instructions after the branch, until it decodes the branch target instruction. The BAC will make a static branch prediction under two conditions: It sees an absolute branch that the BTB did not make a prediction on, or it sees a conditional branch with a target address whose direction is "backward" (which suggests it is the return edge of a loop).

7.3.4 Register Alias Table
The register alias table (RAT) provides register renaming of integer and floating-point registers and flags to make available a larger register set than is explicitly provided in the Intel Architecture. As μops are presented to the RAT, their logical sources and destination are mapped to the corresponding physical ROB addresses where the data are found. The mapping arrays are then updated with new physical destination addresses granted by the allocator for each new μop.

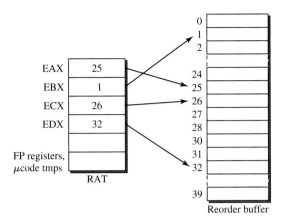

Figure 7.11
Basic RAT Register Renaming.

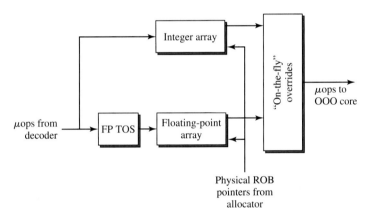

Figure 7.12
RAT Block Diagram.

Refer to Figures 7.11 and 7.12. In each clock cycle, the RAT must look up the physical ROB locations corresponding to the logical source references of each μop. These physical designators become part of the μop's overall state and travel with the μop from this point on. Any machine state that will be modified by the μop (its "destination" reference) is also renamed, via information provided by the allocator. This physical destination reference becomes part of the μop's overall state and is written into the RAT for use by subsequent μops whose sources refer to the same logical destination. Because the physical destination value is unique to each μop, it is used as an identifier for the μop throughout the out-of-order section. All checks and references to a μop are performed by using this physical destination (PDst) as its name.

Since the P6 is a superscalar design, multiple μops must be renamed in a given clock cycle. If there is a true dependency chain through these three μops, say,

μop0: ADD EAX, EBX; src 1 = EBX, src 2 = EAX, dst = EAX

μop1: ADD EAX, ECX;

μop2: ADD EAX, EDX;

then the RAT must supply the renamed source locations "on the fly," via logic, rather than just looking up the destination, as it does for dependences tracked across clock cycles. Bypass logic will directly supply μop1's source register, src 2, EAX, to avoid having to wait for μop0's EAX destination to be written into the RAT and then read as μop1's src.

The state in the RAT is speculative, because the RAT is constantly updating its array entries per the μop destinations flowing by. When the inevitable branch misprediction occurs, the RAT must flush the bogus state it has collected and revert to logical-to-physical mappings that will work with the next set of μops. The P6's branch misprediction recovery scheme guarantees that the RAT will have to do no new renamings until the out-of-order core has flushed all its bogus misspeculated state. That is useful, because it means that register references will now reside in the retirement register file until new speculative μops begin to appear. Therefore, to recover from a branch misprediction, all the RAT needs to do is to revert all its integer pointers to point directly to their counterparts in the RRF.

7.3.4.1 RAT Implementation Details.

The IA32 architecture allows partial-width reads and writes to the general-purpose integer registers (i.e., EAX, AX, AH, AL), which presents a problem for register renaming. The problem occurs when a partial-width write is followed by a larger-width read. In this case, the data required by the larger-width read must be an assimilation of multiple previous writes to different pieces of the register.

The P6 solution to the problem requires that the RAT remember the width of each integer array entry. This is done by maintaining a 2-bit *size* field for each entry in the integer low and high banks. The 2-bit encoding will distinguish between the three register write sizes of 32, 16, and 8 bits. The RAT uses the register size information to determine if a larger register value is needed than has previously been written. In this case, the RAT must generate a partial-write stall.

Another case, common in 16-bit code, is the *independent* use of the 8-bit registers. If only one alias were maintained for all three sizes of an integer register access, then independent use of the 8-bit subsets of the registers would cause a tremendous number of false dependences. Take, for example, the following series of μops:

μop0: MOV AL,#DATA1

μop1: MOV AH,#DATA2

μop2: ADD AL,#DATA3

μop3: ADD AH,#DATA4

Micro-ops 0 and 1 move independent data into AL and AH. Micro-ops 3 and 4 source AL and AH for the addition. If only one alias were available for the "A" register, then µop1's pointer to AH would overwrite µop0's pointer to AL. Then when µop2 tried to read AL, the RAT would not know the correct pointer and would have to stall until µop1 retired. Then µop3's AH source would again be lost due to µop2's write to AL. The CPU would essentially be serialized, and performance would be diminished.

To prevent this, two integer register banks are maintained in the RAT. For 32-bit and 16-bit RAT accesses, data are *read* only from the low bank, but data are *written* into both banks simultaneously. For 8-bit RAT accesses, however, only the appropriate high or low bank is read or written, according to whether it was a high byte or low byte access. Thus, the high and low byte registers use different rename entries, and both can be renamed independently. Note that the high bank only has four array entries because four of the integer registers (namely, EBP, ESP, EDI, ESI) cannot have 8-bit accesses, per the Intel Architecture specification.

The RAT physical source (PSrc) designators point to locations in the ROB array where data may currently be found. Data do not actually appear in the ROB until after the µop generating the data has executed and written back on one of the writeback busses. Until execution writeback of a PSrc, the ROB entry contains junk.

Each RAT entry has an RRF bit to select one of two address spaces, the RRF or the ROB. If the RRF bit is set, then the data are found in the real register file; the physical address bits are set to the appropriate entry of the RRF. If the RRF bit is clear, then the data are found in the ROB, and the physical address points to the correct position in the ROB. The 6-bit physical address field can access any of the ROB entries. If the RRF bit is set, the entry points to the real register file; its physical address field contains the pointer to the appropriate RRF register. The busses are arranged such that the RRF can source data in the same way that the ROB can.

7.3.4.2 Basic RAT Operation. To rename logical sources (LSrc's), the six sources from the three ID-issued µops are used as the indices into the RAT's integer array. Each entry in the array has six read ports to allow all six LSrc's to each read any logical entry in the array.

After the read phase has been completed, the array must be updated with new physical destinations (PDst's) from the allocator associated with the destinations of the current µops being processed. Because of possible intracycle destination dependences, a priority write scheme is employed to guarantee that the correct PDst is written to each array destination.

The priority write mechanism gives priority in the following manner:

Highest: Current µop2's physical destination
           Current µop1's physical destination
           Current µop0's physical destination
Lowest: Any of the retiring µops physical destinations

Retirement is the act of removing a completed µop from the ROB and committing its state to the appropriate permanent architectural state in the machine. The ROB informs the RAT that the retiring µop's destination can no longer be found in the reorder buffer but must (from now on) be taken from the real register file (RRF). If the retiring PDst is found in the array, the matching entry (or entries) is reset to point to the RRF.

The retirement mechanism requires the RAT to do three associative matches of each array PSrc against all three retirement pointers that are valid in the current cycle. For all matches found, the corresponding array entries are reset to point to the RRF. Retirement has lowest priority in the priority writeback mechanism; logically, retirement should happen before any new µops write back. Therefore, if any µops want to write back concurrently with a retirement reset, then the PDst writeback would happen last.

Resetting the floating-point register rename apparatus is more complicated, due to the Intel Architecture FP register stack organization. Special hardware is provided to remove the top-of-stack (TOS) offset from FP register references. In addition, a retirement FP RAT (RfRAT) table is maintained, which contains nonspeculative alias information for the floating-point stack registers. It is updated only upon µop retirement. Each RfRAT entry is 4 bits wide: a 1-bit retired stack valid and a 3-bit RRF pointer. In addition, the RfRAT maintains its own nonspeculative TOS pointer. The reason for the RfRAT's existence is to be able to recover from mispredicted branches and other events in the presence of the FXCH instruction.

The FXCH macro-op swaps the floating-point TOS register entry with any stack entry (including itself, oddly enough). FXCH could have been implemented as three MOV µops, using a temporary register. But the Pentium processor-optimized floating-point code uses FXCH extensively to arrange data for its dual execution units. Using three µops for the FXCH would be a heavy performance hit for the P6 processors on Pentium processor-optimized FP code, hence the motivation to implement FXCH as a single µop.

P6 processors handle the FXCH operation by having the FP part of the RAT (fRAT) merely swap its array pointers for the two source registers. This requires extra write ports in the fRAT but obviates having to swap 80+ bits of data between any two stack registers in the RRF. In addition, since the pointer swap operation would not require the resources of an execution unit, the FXCH is marked as "completed" in the ROB as soon as the ROB receives it from the RAT. So the FXCH effectively takes no RS resources and executes in zero cycles.

Because of any number of previous FXCH operations, the fRAT may speculatively swap any number of its entries before a mispredicted branch occurs. At this point, all instructions issued down this branch are stopped. Sometime later, a signal will be asserted by the ROB indicating that all µops up to and including the branching µop have retired. This means that all arrays in the CPU have been reset, and macroarchitectural state must be restored to the machine state existing at the time of the mispredicted branch. The trick is to be able to correctly undo the effects of the speculative FXCHs. The fRAT entries cannot simply be reset to constant RRF values, as integer rename references are, because any number of retired FXCHs may have occurred, and the fRAT must forevermore remember the retired FXCH mappings. This is the purpose of the retirement fRAT: to "know" what to reset the FP entries to when the front end must be flushed.

7.3.4.3 Integer Retirement Overrides.
When a retiring μop's PDst is still being referenced in the RAT, then at retirement that RAT entry reverts to pointing into the retirement register file. This implies that the retirement of μops must take precedence over the table read. This operation is performed as a bypass *after* the table read in hardware. This way, the data read from the table will be overridden by the most current μop retirement information.

The integer retirement override mechanism requires doing an associative match of the integer arrays' PSrc entries against all retirement pointers that are valid in the current cycle. For all matches found, the corresponding array entries are reset to point to the RRF.

Retirement overrides must occur, because retiring PSrc's read from the RAT will no longer point to the correct data. The ROB array entries that are retiring during the current cycle cannot be referenced by any current μop (because the data will now be found in the RRF).

7.3.4.4 New PDst Overrides.
Micro-op logical source references are used as indices into the RAT's multiported integer array, and physical sources are output by the array. These sources are then subject to retirement overrides. At this time, the RAT also receives newly allocated physical destinations (PDst's) from the allocator. Priority comparisons of logical sources and destinations from the ID are used to gate out either PSrc's from the integer array or PDst's from the allocator as the actual renamed μop physical sources. Notice that source 0 is never overridden because it has no previous μop in the cycle on which to be dependent. A block diagram of the RAT's override hardware is shown in Figure 7.13.

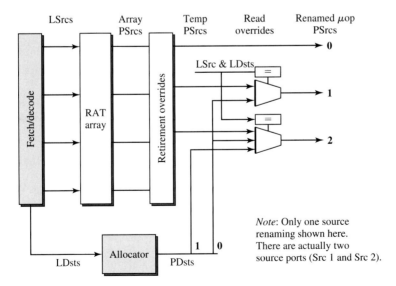

Figure 7.13
RAT New PDst Overrides.

Suppose that the following µops are being processed:

µop0: r1 + r3 → r3

µop1: r3 + r2 → r3

µop2: r3 + r4 → r5

Notice that a µop1 source relies on the destination reference of µop0. This means that the data required by µop1 are not found in the register pointed to by the RAT, but rather are found at the new location provided by the allocator. The PSrc information in the RAT is made stale by the allocator PDst of µop0 and must be overridden before the renamed µop physical sources are output to the RS and to the ROB. Also notice that a µop2 source uses the same register as was written by both µop0 and µop1. The *new PDst override* control must indicate that the PDst of µop1 (not µop0) is the appropriate pointer to use as the override for µop2's source.

Note that the µop groups can be a mixture of both integer and floating-point operations. Although there are two separate control blocks to perform integer and FP overrides, comparison of the logical register names sufficiently isolates the two classes of µops. It is naturally the case that only like types of sources and destinations can override each other. (For example, an FP destination cannot override an integer source.) Therefore, differences in the floating-point overrides can be handled independently of the integer mechanism.

The need for floating-point overrides is the same as for the integer overrides. Retirement and concurrent issue of µops prevent the array from being updated with the newest information before those concurrent µops read the array. Therefore, PSrc information read from the RAT arrays must be overridden by both retirement overrides and new PDst overrides.

Floating-point retirement overrides are identical to integer retirement overrides except that the value to which a PSrc is overridden is not determined by the logical register source name as in the integer case. Rather, the retiring logical register destination reads the RfRAT for the reset value. Depending on which retirement µop content addressable memory (CAM) matched with this array read, the retirement override control must choose between one of the three RfRAT reset values. These reset values must have been modified by any concurrent retiring FXCHs as well.

7.3.4.5 RAT Stalls.
The RAT can stall in two ways, internally and externally. The RAT generates an internal stall if it is unable to completely process the current set of µops, due to a partial register write, a flag mismatch, or other microarchitectural conditions. The allocator may also be unable to process all µops due to an RS or ROB table overflow; this is an external stall to the RAT.

Partial Write Stalls. When a partial-width write (e.g., AX, AL, AH) is followed by a larger-width read (e.g., EAX), the RAT must stall until the last partial-width write of the desired register has retired. At this point, all portions of the register have been reassembled in the RRF, and a single PSrc can be specified for the required data.

The RAT performs this function by maintaining the size information (8, 16, or 32 bits) for each register alias. To handle the independent use of 8-bit registers, two entries and aliases (H and L) are maintained in the integer array for each of the registers EAX, EBX, ECX, and EDX. (The other macroregisters cannot be partially written, as per the Intel Architecture specification.) When 16- or 32-bit writes occur, both entries are updated. When 8-bit writes occur, only the corresponding entry (H or L, not both) is updated.

Thus when an entry is targeted by a logical source, the size information read from the array is compared to the requested size information specified by the µop. If the size needed is greater than the size available (read from array), then the RAT stalls both the instruction decoder and the allocator. In addition, the RAT clears the "valid bits" on the µop causing the stall (and any µops younger than it is) until the partial write retires; this is the in-order pipe, and subsequent µops cannot be allowed to pass the stalling µop here.

Mismatch Stalls. Since reading and writing the flags are common occurrences and are therefore performance-critical, they are renamed just as the registers are. There are two alias entries for flags, one for arithmetic flags and one for floating-point condition code flags, that are maintained in much the same fashion as the other integer array entries. When a µop is known to write the flags, the PDst granted for the µop is written into the corresponding flag entry (as well as the destination register entry). When subsequent µops use the flags as a source, the appropriate flag entry is read to find the PDst where the flags live.

In addition to the general renaming scheme, each µop emitted by the ID has associated flag information, in the form of masks, that tell the RAT which flags the µop touches and which flags the µop needs as input. In the event a previous but not yet retired µop did not touch all the flags that a current µop needs as input, the RAT stalls the in-order machine. This informs the ID and allocator that no new µops can be driven to the RAT because one or more of the current µops cannot be issued until a previous flag write retires.

7.3.5 Allocator

For each clock cycle, the allocator assumes that it will have to allocate three reorder buffer, reservation station, and load buffer entries and two store buffer entries. The allocator generates pointers to these entries and decodes the µops coming from the ID unit to determine how many entries of each resource are really needed and which RS dispatch port they will be dispatched on.

Based on the µop decoding and valid bits, the allocator will determine whether or not resource needs have been met. If not, then a stall is asserted and µop issue is frozen until sufficient resources become available through retirement of previous µops.

The first step in allocation is the decoding of the µops that are delivered by the ID. Some µops need an LB or SB entry; all µops need an ROB entry.

7.3.5.1 ROB Allocation.
The ROB entry addresses are the physical destinations or PDst's which were assigned by the allocator. The PDst's are used to directly

address the ROB. This means if the ROB is full, the allocator must assert a stall signal early enough to prevent overwriting valid ROB data.

The ROB buffer is treated as a circular buffer by the allocator. In other words, entry addresses are assigned sequentially from 0 until the highest address, and then wraparound back to 0. A three-or-none allocation policy is used: every cycle, at least three ROB entries must be available or the allocator will stall. This means ROB allocation is independent of the type of μop and does not even depend on the μop's validity. The three-or-none policy simplifies allocation.

At the end of the cycle, the address of the last ROB entry allocated is preserved and becomes the new allocation starting point. Note that this does depend on the real number of valid μops. The ROB also uses the number of valid μops to determine where to stop retiring.

7.3.5.2 MOB Allocation.
All μops have a load buffer ID and a store buffer ID (together known as a MOB ID, or MBID) stored with them. Load μops will have a newly allocated LB address and the last SB address that was allocated. Nonload μops (store or any other μop) have MBID with LBID = 0 and the SBID (or store color) of the last store allocated.

The LB and SB are treated as circular buffers, as is the ROB. However, the allocation policy is slightly different. Since every μop does not need an LB or SB entry, it would be a big performance hit to use a three-or-none policy (or two-or-none for SB) and stall whenever the LB or SB has less than three free entries. Instead we use an all-or-none policy. This means that stalling will occur only when not all the valid MOB μops can be allocated.

Another important part of MOB allocation is the handling of entries containing *senior* stores. These are stores that have been committed or retired by the CPU but are still actually awaiting completion of execution to memory. These store buffer entries cannot be deallocated until the store is actually performed to memory.

7.3.5.3 RS Allocation.
The allocator also generates write enable bits which are used by the RS directly for its entry enables. If the RS is full, a stall indication must be given early in order to prevent the overwrite of valid RS data. In fact if the RS is full, the enable bits will all be cleared and thus no entry will be enabled for writing. If the RS is not full but a stall occurs due to some other resource conflict, the RS invalidates data written to any RS entry in that cycle (i.e., data get written but are marked as invalid).

The RS allocation works differently from the ROB or MOB circular buffer model. Since the RS dispatches μops out of order (as they become data-ready), its free entries are typically interspersed with used or allocated entries, and so a circular buffer model does not work. Instead, a bitmap scheme is used where each RS entry maps to a bit of the RS allocation pool. In this way, entries may be drawn or replaced from the pool in any order. The RS searches for free entries by scanning from location 0 until the first three free entries are found.

Some μops can dispatch to more than one port, and the act of committing a given μop to a given port is called binding. The binding of μops to the RS functional unit interfaces is done at allocation. The allocator has a load-balancing algorithm that knows how many μops in the RS are waiting to be executed on a given interface. This algorithm is only used for μops that can execute on more than one EU. This is referred to as a *static binding with load balancing* of ready μops to an execution interface.

7.4 The Out-of-Order Core

7.4.1 Reservation Station

The reservation station (RS) is basically a place for μops to wait until their operands have all become ready and the appropriate execution unit has become available. In each cycle, the RS determines execution unit availability and source data validity, performs out-of-order scheduling, dispatches μops to execution units, and controls data bypassing to RS array and execution units. All entries of the RS are identical and can hold any kind of μops.

The RS has 20 entries. The control portion of an entry (μop, entry valid, etc.) can be written from one of three ports (there are three ports because the P6 microarchitecture is of superscalar order 3.). This information comes from the allocator and RAT. The data portion of an entry can be written from one of six ports (three ROB and three execution unit writebacks). CAMs control the snarfing of valid writeback data into μop Src fields and data bypassing at the execution unit (EU) interfaces. The CAMs, EU arbitration, and control information are used to determine data validity and EU availability for each entry (ready bit generation). The scheduler logic uses this ready information to schedule up to five μops. The entries that have been scheduled for dispatch are then read out of the array and driven to the execution unit.

During pipe stage 31, the RS determines which entries are, or will be, ready for dispatch in stage 32. To do this, it is necessary to know the availability of data and execution resources (EU/AGU units). This ready information is sent to the scheduler.

7.4.1.1 Scheduling. The basic function of the scheduler is to enable the dispatching of up to five μops per clock from the RS. The RS has five schedulers, one for each execution unit interface. Figure 7.14 shows the mapping of the functional units to their RS ports.

The RS uses a priority pointer to specify where the scheduler should begin its scan of the 20 entries. The priority pointer will change according to a pseudo-FIFO algorithm. This is used to reduce stale entry effects and increase performance in the RS.

7.4.1.2 Dispatch. The RS can dispatch up to five μops per clock. There are two EU and two AGU interfaces and one store data (STD) interface. Figure 7.14 shows the connections of the execution units to the RS ports. Before instruction dispatch time, the RS determines whether or not all the resources needed for a particular μop to execute are available, and then the ready entries are scheduled. The RS then dispatches all the necessary μop information to the scheduled functional unit. Once a μop has been dispatched to a functional unit and no cancellation has

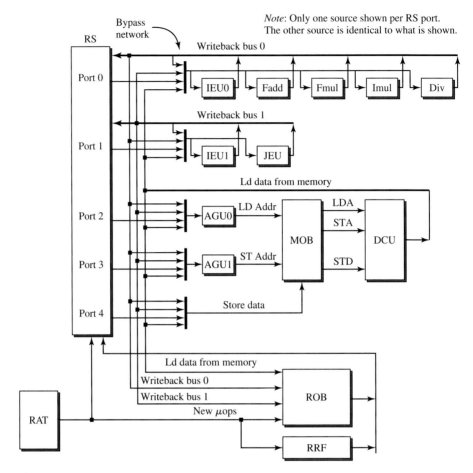

Figure 7.14
Execution Unit Data Paths.

occurred due to a cache miss, the entry can be deallocated for use by a new μop. Every cycle, deallocation pointers are used to signal the allocator about the availability of all 20 entries in the RS.

7.4.1.3 Data Writeback. It is possible that source data will not be valid at the time the RS entry is initially written. The μop must then remain in the RS until all its sources are valid. The content addressable memories (CAMs) are used to compare the writeback physical destination (PDst) with the stored physical sources (PSrc). When a match occurs, the corresponding write enables are asserted to snarf the needed writeback data into the appropriate source in the array.

7.4.1.4 Cancellation. Cancellation is the inhibiting of a μop from being scheduled, dispatched, or executed due to a cache miss or possible future resource conflict.

All canceled μops will be rescheduled at a later time unless the out-of-order machine is reset.

There are times when writeback data are invalid, e.g., when the memory unit detects a cache miss. In this case, dispatching μops that are dependent on the writeback data need to be canceled and rescheduled at a later time. This can happen because the RS pipeline assumes cache accesses will be hits, and schedules dependent μops based on that assumption.

7.5 Retirement

7.5.1 The Reorder Buffer

The reorder buffer (ROB) participates in three fundamental aspects of the P6 microarchitecture: speculative execution, register renaming, and out-of-order execution. In some ways, the ROB is similar to the register file in an in-order machine, but with additional functionality to support retirement of speculative operations and register renaming.

The ROB supports speculative execution by buffering the results of the execution units (EUs) before committing them to architecturally visible state. This allows most of the microengine to fetch and execute instructions at a maximum rate by assuming that branches are properly predicted and that no exceptions occur. If a branch is mispredicted or if an exception occurs in executing an instruction, the microengine can *recover* simply by discarding the speculative results stored in the ROB. The microengine can also *restart* at the proper instruction by examining the committed architectural state in the ROB. A key function of the ROB is to control retirement or completion of μops.

The buffer storage for EU results is also used to support register renaming. The EUs write result data *only* into the renamed register in the ROB. The retirement logic in the ROB updates the architectural registers based upon the contents of each renamed instance of the architectural registers. Micro-ops which source an architectural register obtain either the contents of the actual architectural register or the contents of the renamed register. Since the P6 microarchitecture is superscalar, different μops in the same clock which use the same architectural register may in fact access different physical registers.

The ROB supports out-of-order execution by allowing EUs to complete their μops and write back the results without regard to other μops which are executing simultaneously. Therefore, as far as the execution units are concerned, μops complete out of order. The ROB retirement logic *reorders* the completed μops into the original sequence issued by the instruction decoder as it updates the architectural state during retirement.

The ROB is active in three separate parts of the processor pipeline (refer to Figure 7.4): the rename and register read stages, the execute/writeback stage, and the retirement stages.

The placement of the ROB relative to other units in the P6 is shown in the block diagram in Figure 7.1. The ROB is closely tied to the allocator (ALL) and register alias table (RAT) units. The allocator manages ROB physical registers to support speculative operations and register renaming. The actual renaming of architectural registers in the ROB is managed by the RAT. Both the allocator and the RAT function within the in-order part of the P6 pipeline. Thus, the rename and register read (or ROB read) functions are performed in the same sequence as in the program flow.

The ROB interface with the reservation station (RS) and the EUs in the out-of-order part of the machine is loosely coupled in nature. The data read from the ROB during the register read pipe stage consist of operand sources for the μop. These operands are stored in the RS until the μop is dispatched to an execution unit. The EUs write back μop results to the ROB through the five writeback ports (three full writeback ports, two partial writebacks for STD and STA). The result writeback is out of order with respect to μops issued by the instruction decoder. Because the results from the EUs are speculative, any exceptions that were detected by the EUs may or may not be "real." Such exceptions are written into a special field of the μop. If it turns out that the μop was misspeculated, then the exception was not "real" and will be flushed along with the rest of the μop. Otherwise, the ROB will notice the exceptional condition during retirement of the μop and will cause the appropriate exception-handling action to be invoked then, before making the decision to commit that μop's result to architectural state.

The ROB retirement logic has important interfaces to the micro-instruction sequencer (MS) and the memory ordering buffer (MOB). The ROB/MS interface allows the ROB to signal an exception to the MS, forcing the micro-instruction sequencer to jump to a particular exception handler microcode routine. Again, the ROB must force the control flow change because the EUs report events out of order with respect to program flow. The ROB/MOB interface allows the MOB to commit memory state from stores when the store μop is committed to the machine state.

7.5.1.1 ROB Stages in the Pipeline.
The ROB is active in both the in-order and out-of-order sections of the P6 pipeline. The ROB is used in the in-order pipe in pipe stages 21 and 22. Entries in the reorder buffer which will hold the results of the speculative μops are allocated in pipe stage 21. The reorder buffer is managed by the allocator and the retirement logic as a circular buffer. If there are unused entries in the reorder buffer, the allocator will use them for the μops being issued in the clock. The entries used are signaled to the RAT, allowing it to update its renaming or alias tables. The addresses of the entries used (PDst's) are also written into the RS for each μop. The PDst is the key token used by the out-of-order section of the machine to identify μops in execution; it is the actual slot number in the ROB. As the entries in the ROB are allocated, certain fields in them are written with data from fields in the μops. This information can be written either at allocation time or with the results written back by the EUs. To reduce the width of the RS entries as well as to reduce the amount of information which must be circulated to the EUs or memory subsystem, any μop information required to retire a μop which is determined strictly at decode time is written into the ROB at allocation time.

In pipe stage 22, immediately following entry allocation, the sources for the μops are read from the ROB. The physical source addresses, PSrc's, are delivered by the RAT based upon the alias table update performed in pipe stage 21. A source may reside in one of three places: in the committed architectural state (retirement register file), in the reorder buffer, or from a writeback bus. (The RRF contains both architectural state and microcode visible state. Subsequent references to RRF state will call them macrocode and microcode visible state). Source operands read from the RRF are always valid, ready for execution unit use. Sources read from the ROB may or may not be valid, depending on the timing of the source read with respect to writebacks of previous µops which updated the entries read. If the source operand delivered by the ROB is invalid, the RS will wait until an EU writes back to a PDst which matches the physical source address for a source operand in order to capture (or bypass at the EU) the valid source operand for a given µop.

An EU writes back destination data into the entry allocated for the µop, along with any event information, in pipe stage 83. (*Event* refers to exceptions, interrupts, microcode assists, and so on.) The writeback pipe stage is decoupled from the rename and register read pipe stages because the µops are issued out of order from the RS. Arbitration for use of writeback busses is determined by the EUs along with the RS. The ROB is simply the terminus for each of the writeback busses and stores whatever data are on the busses into the writeback PDst's signaled by the EUs.

The ROB retirement logic commits macrocode and microcode visible state in pipe stages 92 and 93. The retirement pipe stages are decoupled from the writeback pipe stage because the writebacks are out of order with respect to the program or microcode order. Retirement effectively *reorders* the out-of-order completion of µops by the EUs into an in-order completion of µops by the machine as a whole. Retirement is a two-clock operation, but the retirement stages are pipelined. If there are allocated entries in the reorder buffer, the retirement logic will attempt to deallocate or retire them. Retirement treats the reorder buffer as FIFO in deallocating the entries, since the µops were originally allocated in a sequential FIFO order earlier in the pipeline. This ensures that retirement follows the original program source order, in terms of allowing the architectural state to be modified.

The ROB contains all the P6 macrocode and microcode state which may be modified without serialization of the machine. (Serialization limits to one the number of µops which may flow through the out-of-order section of the machine, effectively making them execute in order.) Much of this state is updated directly from the speculative state in the reorder buffer. The extended instruction pointer (EIP) is the one architectural register which is an exception to this norm. The EIP requires a significant amount of hardware in the ROB for each update. The reason is the number of µops which may retire in a clock varies from zero to three.

The ROB is implemented as a multiported register file with separate ports for allocation time writes of µop fields needed at retirement, EU writebacks, ROB reads of sources for the RS, and retirement logic reads of speculative result data.

The ROB has 40 entries. Each entry is 157 bits wide. The allocator and retirement logic manage the register file as FIFO. Both source read and destination writeback functions treat the reorder buffer as a register file.

Table 7.1
Registers in the RRF

Qty.	Register Name(s)	Size (bits)	Description
8	i486 general registers	32	EAX, ECX, EDX, EBX, EBP, ESP, ESI, EDI
8	i486 FP stack registers	86	FST(0-7)
12	General microcode temp. registers	86	For storing both integer and FP values
4	Integer microcode temp. registers	32	For storing integer values
1	EFLAGS	32	The i486 system flags register
1	ArithFlags	8	The i486 flags which are renamed
2	FCC	4	The FP condition codes
1	EIP	32	The architectural instruction pointer
1	FIP	32	The architectural FP instruction pointer
1	EventUIP	12	The micro-instruction reporting an event
2	FSW	16	The FP status word

The RRF contains both the macrocode and microcode visible state. Not all such processor state is located in the RRF, but any state which may be renamed is there. Table 7.1 gives a listing of the registers in the RRF.

Retirement logic generates the addresses for the retirement reads performed in each clock. The retirement logic also computes the retirement valid signals indicating which entries with valid writeback data may be retired.

The IP calculation block produces the architectural instruction pointer as well as several other macro- and micro-instruction pointers. The macro-instruction pointer is generated based on the lengths of all the macro-instructions which may retire, as well as any branch target addresses which may be delivered by the jump execution unit.

When the ROB has determined that the processor has started to execute operations down the wrong path of a branch, any operations in that path must not be allowed to retire. The ROB accomplishes this by asserting a "clear" signal at the point just before the first of these operations would have retired. All speculative operations are then flushed from the machine. When the ROB retires an operation that faults, it clears both the in-order and out-of-order sections of the machine in pipe stages 93 and 94.

7.5.1.2 Event Detection. Events include faults, traps, assists, and interrupts. Every entry in the reorder buffer has an event information field. The execution units write back into this field. During retirement the retirement logic looks at this field for the three entries that are candidates for retirement. The event information field tells the retirement logic whether there is an exception and whether it is a fault or a trap or an assist. Interrupts are signaled directly by the interrupt unit. The jump unit marks the event information field in case of taken or mispredicted branches.

If an event is detected, the ROB clears the machine of all μops and forces the MS to jump to a microcode event handler. Event records are saved to allow the microcode handler to properly repair the result or invoke the correct macrocode handler. Macro- and micro-instruction pointers are also saved to allow program resumption upon termination of the event handler.

7.6 Memory Subsystem

The memory ordering buffer (MOB) is a part of the memory subsystem of the P6. The MOB interfaces the processor's out-of-order engine to the memory subsystem. The MOB contains two main buffers, the load buffer (LB) and the store address buffer (SAB). Both of these buffers are circular queues with each entry within the buffer representing either a load or a store micro-operation, respectively. The SAB works in unison with the memory interface unit's (MIU) store data buffer (SDB) and the DCache's physical address buffer (PAB) to effectively manage a processor store operation. The SAB, SDB, and PAB can be viewed as one buffer, the store buffer (SB).

The LB contains 16 buffer entries, holding up to 16 loads. The LB queues up load operations that were unable to complete when originally dispatched by the reservation station (RS). The queued operations are redispatched when the conflict has been removed. The LB maintains processor ordering for loads by snooping external writes against completed loads. A second processor's write to a speculatively read memory location forces the out-of-order engine to clear and restart the load operation (as well as any younger μops).

The SB contains 12 entries, holding up to 12 store operations. The SB is used to queue up all store operations before they dispatch to memory. These stores are then dispatched in original program order, when the OOO engine signals that their state is no longer speculative. The SAB also checks all loads for store address conflicts. This checking keeps loads consistent with previously executed stores still in the SB.

The MOB resources are allocated by the allocator when a load or store operation is issued into the reservation station. A load operation decodes into one μop and a store operation is decoded into two μops: store data (STD) and store address (STA). At allocation time, the operation is tagged with its eventual location in the LB or SB, collectively referred to as the MOB ID (MBID). Splitting stores into two distinct μops allows any possible concurrency between generation of the address and data to be stored to be expressed.

The MOB receives speculative LD and STA operations from the reservation station. The RS provides the opcode, while the address generation unit (AGU) calculates and provides the linear address for the access. The DCache either executes these operations immediately, or they are dispatched later by the MOB. In either case they are written into one of the MOB arrays. During memory operations, the data translation lookaside buffer (DTLB) converts the linear address to a physical address or signals a page miss to the page miss handler (PMH). The MOB will also perform numerous checks on the linear address and data size to determine if the operation can continue or if it must block.

In the case of a load, the data cache unit is expected to return the data to the core. In parallel, the MOB writes address and status bits into the LB, to signal the operation's completion. In the case of a STA, the MOB completes the operation by writing a valid bit (AddressDone) into the SAB array and to the reorder buffer. This indicates that the address portion of the store has completed. The data portion of the store is executed by the SDB. The SDB will signal the ROB and SAB when the data have been received and written into the buffer. The MOB will retain the store information until the ROB indicates that the store operation is retired and committed to the processor state. It will then dispatch from the MOB to the data cache unit to commit the store to the system state. Once completed, the MOB signals deallocation of SAB resources for reuse by the allocator. Stores are executed by the memory subsystem in program order.

7.6.1 Memory Access Ordering

Micro-op register operand dependences are tracked explicitly, based on the register references in the original program instructions. Unfortunately, memory operations have implicit dependences, with load operations having a dependency on any previous store that has address overlap with the load. These operations are often speculative inside the MOB, both the stores and loads, so that system memory access may return stale data and produce incorrect results.

To maintain self-consistency between loads and stores, the P6 employs a concept termed *store coloring*. Each load operation is tagged with the store buffer ID (SBID) of the store previous to it. This ID represents the relative location of the load compared to all stores in the execution sequence. When the load executes in the memory subsystem, the MOB will use this SBID as a beginning point for analyzing the load against all older stores in the buffer, while also allowing the MOB to ignore younger stores.

Store coloring is used to maintain ordering consistency between loads and stores of the same processor. A similar problem occurs between processors of a multiprocessing system. If loads execute out of order, they can effectively make another processor's store operations appear out of order. This results from a younger load passing an older load that has not been performed yet. This younger load reads old data, while the older load, once performed, has the chance of reading new data written by another processor. If allowed to commit to state, these loads would violate *processor ordering*. To prevent this violation, the LB watches (snoops) all data writes on the bus. If another processor writes a location that was speculatively read, the speculatively completed load and subsequent operations will be cleared and re-executed to get the correct data.

7.6.2 Load Memory Operations

Load operations issue to the RS from the allocator and register allocation table (RAT). The allocator assigns a new load buffer ID (LBID) to each load that issues into the RS. The allocator also assigns a store color to the load, which is the SBID of the last store previously allocated. The load waits in the RS for its data operands to become available. Once available, the RS dispatches the load on port 2 to the AGU and LB. Assuming no other dispatches are waiting for this port, the LB bypasses this operation for immediate execution by the memory subsystem. The AGU generates the linear address to be used by the DTLB, MOB, and DCU. As the DTLB does its translation to the physical address, the DCU does an initial data lookup using the lower-order 12 bits. Likewise, the SAB uses the lower-order 12 bits along with the store color SBID to check potential conflicting addresses of previous stores (previous in program order, not time order). Assuming a DTLB page hit and no SAB conflicts, the DCU uses the physical address to do a final tag match and return the correct data (assuming no miss or block). This completes the load operation, and the RS, ROB, and MOB write their completion status.

If the SAB noticed an address match, the SAB would cause the SDB to forward SDB data, ignoring the DCU data. If a SAB conflict existed but the addresses did not match (a false conflict detection), then the load would be blocked and written into the LB. The load will wait until the conflicting store has left the store buffer.

7.6.3 Basic Store Memory Operations

Store operations are split into two micro-ops, store data (STD) followed by a store address (STA). Since a store is represented by the combination of these operations, the allocator allocates a store buffer entry only when the STD is issued into the RS. The allocation of a store buffer entry reserves the same location in the SAB, the SDB, and the PAB. When the store's source data become available, the RS dispatches the STD on port 4 to the MOB for writing into the SDB. As the STA address source data become available, the RS dispatches the STA on port 3 to the AGU and SAB. The AGU generates the linear address for translation by the DTLB and for writing into the SAB. Assuming a DTLB page hit, the physical address is written into the PAB. This completes the STA operation, and the MOB and ROB update their completion status.

Assuming no faults or mispredicted branches, the ROB retires both the STD and STA. Monitoring this retirement, the SAB marks the store (STD/STA pair) as the committed, or *senior,* processor state. Once senior, the MOB dispatches these operations by sending the opcode, SBID, and lower 12 address bits to the DCU. The DCU and MIU use the SBID to access the physical address in the PAB and store data in the SDB, respectively, to complete the final store operation.

7.6.4 Deferring Memory Operations

In general, most memory operations are expected to complete three cycles after dispatch from the RS (which is only two clocks longer than an ALU operation). However, memory operations are not totally predictable as to their translation and availability from the L1 cache. In cases such as these, the operations require other resources, e.g., DCU fill buffers on a pending cache miss, that may not be available. Thus, the operations must be deferred until the resource becomes available.

The MOB load buffer employs a general mechanism of blocking load memory operation until a later wakeup is received. The blocking information associated with each entry of the load buffer contains two fields: a blocking code or type and a blocking identifier. The block code identifies the source of the block (e.g., address block, PMH resource block). The block identifier refers to a specific ID of a resource associated with the block code. When a wakeup signal is received, all deferred memory operations that match the blocking code and identifier are marked "ready for dispatch." The load buffer then schedules and dispatches one of these ready operations in a manner that is very similar to RS dispatching.

The MOB store buffer uses a restricted mechanism for blocking STA memory operations. The operations remain blocked until the ROB retirement pointers indicate that STA µop is the oldest nonretired operation in the machine. This operation will then dispatch at retirement with the write to the DCU occurring simultaneously with the dispatch of the STA. This simplified mechanism for stores was used because STAs are rarely blocked.

7.6.5 Page Faults

The DTLB translates the linear addresses to physical addresses for all memory load and store address µops. The DTLB does the address translation by performing a lookup in a cache array for the physical address of the page being accessed. The DTLB also caches page attributes with the physical address. The DTLB uses this information to check for page protection faults and other paging-related exceptions.

The DTLB stores physical addresses for only a subset of all possible memory pages. If an address lookup fails, the DTLB signals a miss to the PMH. The PMH executes a *page walk* to fetch the physical address from the page tables located in physical memory. The PMH then looks up the effective memory type for the physical address from its on-chip memory type range registers and supplies both the physical address and the effective memory type to the DTLB to store in its cache array. (These memory type range registers are usually configured at processor boot time.)

Finally, the DTLB performs the fault detection and writeback for various types of faults including page faults, assists, and machine check architecture errors for the DCU. This is true for data and instruction pages. The DTLB also checks for I/O and data breakpoint traps, and either writes back (for store address µops) or passes (for loads and I/O µops) the results to the DCU which is responsible for supplying the data for the ROB writeback.

7.7 Summary

The design described in this chapter began as the brainchild of the authors of this chapter, but also reflects the myriad contributions of hundreds of designers, microcoders, validators, and performance analysts. Subject only to the economics that rule Intel's approach to business, we tried at all times to obey the prime directive: Make choices that maximize delivered performance, and quantify those choices wherever possible. The out-of-order, speculative execution, superpipelined, superscalar, micro-dataflow, register-renaming, glueless multiprocessing design that we described here was the result. Intel has shipped approximately one billion P6-based microprocessors as of 2002, and many of the fundamental ideas described in this chapter have been reused for the Pentium 4 processor generation.

Further details on the P6 microarchitecture can be found in Colwell and Steck [1995] and Papworth [1996].

7.8 Acknowledgments

The design of the P6 microarchitecture was a collaborative effort among a large group of architects, designers, validators, and others. The microarchitecture described here benefited enormously from contributions from these extraordinarily talented people. They also contributed some of the text descriptions found in this chapter. Thank you, one and all.

We would also like to thank Darrell Boggs for his careful proofreading of a draft of this chapter.

REFERENCES

Colwell, Robert P., and Randy Steck: "A 0.6 µm BiCMOS microprocessor with dynamic execution," *Proc. Int. Solid State Circuits Conference,* San Francisco, CA 1995, pp. 176–177.

Lee, J., and A. J. Smith: "Branch predictions and branch target buffer design," *IEEE Computer,* January 1984, 21, 7, pp. 6–22.

Papworth, David B.: "Tuning the Pentium Pro microarchitecture," *IEEE Micro,* August 1996, pp. 8–15.

Yeh, T.-Y., and Y. N. Patt: "Two-level adaptive branch prediction," *The 24th ACM/IEEE Int. Symposium and Workshop on Microarchitecture,* November 1991, pp. 51–61.

HOMEWORK PROBLEMS

P7.1 The PowerPC 620 does not implement load/store forwarding, while the Pentium Pro does. Explain why both design teams are likely to have made the right design tradeoff.

P7.2 The P6 recovers from branch mispredictions in a somewhat coarse-grained manner, as illustrated in Figure 7.5. Explain how this simplifies the misprediction recovery logic that manages the reorder buffer (ROB) as well as the register alias table (RAT).

P7.3 AMD's Athlon (K7) processor takes a somewhat different approach to dynamic translation from IA32 macro-instructions to the machine instructions it actually executes. For example, an ALU instruction with one memory operand (e.g., add eax,[eax]) would translate into two µops in the Pentium Pro: a load that writes a temporary register followed by a register-to-register add instruction. In contrast, the Athlon would simply dispatch the original instruction into the issue queue as a macro-op, but it would issue from the queue twice: once as a load and again as an ALU operation. Identify and discuss at least two microarchitectural benefits that accrue from this "macro-op" approach to instruction-set translation.

P7.4 The P6 two-level branch predictor has a speculative and nonspeculative branch history register stored at each entry. Describe when and how each branch history register is updated, and provide some reasoning that justifies this design decision.

P7.5 The P6 dynamic branch predictor is backed up by a static predictor that is able to predict branch instructions that for some reason were not predicted by the dynamic predictor. The static prediction occurs in pipe stage 17 (refer to Figure 7.4). One scenario in which the static prediction is used occurs when the BTB reports a tag mismatch, reflecting the fact that it has no branch history information for this particular branch. Assume the static branch prediction turns out to be correct. One possible optimization would be to avoid installing such a branch (one that is statically predictable) in the BTB, since it might displace another branch that needs dynamic prediction. Discuss at least one reason why this might be a bad idea.

P7.6 Early in the P6 development, the design had two different ROBs, one for integers and another for floating-point. To save die size, these were combined into one during the Pentium Pro development. Explain the advantages and disadvantages of the separate integer ROB and floating-point ROB versus the unified ROB.

P7.7 From the timing diagrams you can see that the P6 retirement process takes three clock cycles. Suppose you knew a way to implement the ROB so that retirement only took two clock cycles. Would you expect a substantial performance boost? Explain.

P7.8 Section 7.3.4.5 describes a mismatch stall that occurs when condition flags are only partially written by an in-flight μop. Suggest a solution that would prevent the mismatch stall from occurring in the renaming process.

P7.9 Section 7.3.5.3 describes the RS allocation policy of the Pentium Pro. Based on this description, would you call the P6 a centralized RS design or a distributed RS design? Justify your answer.

P7.10 If the P6 microarchitecture had to support an instruction set that included predication, what effect would that have on the register renaming process?

P7.11 As described in the text, the P6 microarchitecture splits store operations into a STA and STD pair for handling address generation and data movement. Explain why this makes sense from a microarchitectural implementation perspective.

P7.12 Following up on Problem 7.11, would there be a performance benefit (measured in instructions per cycle) if stores were not split? Explain why or why not?

P7.13 What changes would one have to make to the P6 microarchitecture to accommodate stores that are not split into separate STA and STD operations? What would be the likely effect on cycle time?

P7.14 AMD has recently announced the x86-64 extensions to the Intel IA32 architecture that add support for 64-bit registers and addressing. Investigate these extensions (more information is available from **www.amd.com**) and outline the changes you would need to make to the P6 architecture to accommodate these additional instructions.

CHAPTER 8

Mark Smotherman

Survey of Superscalar Processors

CHAPTER OUTLINE

8.1 Development of Superscalar Processors
8.2 A Classification of Recent Designs
8.3 Processor Descriptions
8.4 Verification of Superscalar Processors
8.5 Acknowledgments

References
Homework Problems

The 1990s was the decade in which superscalar processor design blossomed. However, the idea of decoding and issuing multiple instructions per cycle from a single instruction stream dates back 25 years before that. In this chapter we review the history of superscalar design and examine a number of selected designs.

8.1 Development of Superscalar Processors

This section reviews the history of superscalar design, beginning with the IBM Stretch and its direct superscalar descendant, the Advanced Computer System (ACS), and follows developments up through current processors.

8.1.1 Early Advances in Uniprocessor Parallelism: The IBM Stretch

The first efforts at what we now call superscalar instruction issue started with an IBM machine directly descended from the IBM Stretch. Because of its use of aggressive implementation techniques (such as pre-decoding, out-of-order execution, speculative execution, branch misprediction recovery, and precise exceptions) and because it was a precursor to the IBM ACS in the 1960s (and the RS/6000 POWER

369

architecture in the 1990s), it is appropriate to review the Stretch, also known as the IBM 7030 [Buchholz, 1962].

The Stretch design started in 1955 when IBM lost a bid on a high-performance decimal computer system for the University of California Radiation Laboratory (Livermore Lab). Univac, IBM's competitor and the dominant computer manufacturer at the time, won the contract to build the Livermore Automatic Research Computer (LARC) by promising delivery of the requested machine in 29 months [Bashe et al., 1986]. IBM had been more aggressive, and its bid was based on a renegotiation clause for a machine that was four to five times faster than requested and cost $3.5 million rather than the requested $2.5 million.

In the following year, IBM bid a binary computer of "speed at least 100 times greater than that of existing machines" to the Los Alamos Scientific Laboratory and won a contract for what would become the Stretch. Delivery was slated for 1960. Stephen Dunwell was chosen to head the project, and among those he recruited for the design effort were Gerrit Blaauw, Fred Brooks, John Cocke, and Harwood Kolsky. While Blaauw and Brooks investigated instruction set design ideas, which would later serve them as they worked on the IBM S/360, Cocke and Kolsky constructed a crucial simulator that would help the team explore organization options. Erich Bloch, later to become chief scientist at IBM, was named engineering manager in 1958 and led the implementation efforts on prototype units in that year and on an engineering model in 1959.

Five test programs were selected for the simulation to help determine machine parameters: a hydrodynamics mesh problem, a Monte Carlo neutron-diffusion code, the inner loop of a second neutron diffusion code, a polynomial evaluation routine, and the inner loop of a matrix inversion routine. Several Stretch instructions intended for scientific computation of this kind, such as a branch-on-count and multiply-and-add (called *cumulative multiply* in Stretch and later known as *fused multiply and add*), would become important to RS/6000 performance some 30 years later.

Instructions in Stretch flowed through two processing elements: an indexing and instruction unit that fetched, pre-decoded, and partially executed the instruction stream, and an arithmetic unit that executed the remainder of the instructions. Stretch also partitioned its registers according to this organization; a set of sixteen 64-bit index registers was associated with the indexing and instruction unit, and a set of 64-bit accumulators and other registers was associated with the arithmetic unit. Partitioned register sets also appear on the ACS and the RS/6000.

The indexing and instruction unit (see Figure 8.1) of Stretch fetched 64-bit memory words into a two-word instruction buffer. Instructions could be either 32 or 64 bits in length, so up to four instructions could be buffered. The indexing and instruction unit directly executed indexing instructions and *prepared* arithmetic instructions by calculating effective addresses (i.e., adding index register contents to address fields) and starting memory operand fetches. The unit was itself pipelined and decoded instructions in parallel with execution. One interesting feature of the instruction fetch logic was the addition of pre-decoding bits to all instructions; this was done one word at a time, so two half-word instructions could be pre-decoded in parallel.

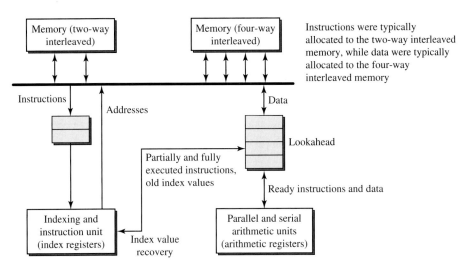

Figure 8.1
IBM Stretch Block Diagram.

Unconditional branches and conditional branches that depended on the state of the index registers, such as the branch-on-count instruction, could be fully executed in the indexing and instruction unit (compare with the branch unit on RS/6000). Conditional branches that depended on the state of the arithmetic registers were predicted untaken, and the untaken path was speculatively executed.

All instructions, either fully executed or prepared, were placed into a novel form of buffering called a *lookahead* unit, which was at that time also called a *virtual memory* but which we would view today as a combination of a completion buffer and a history buffer. A fully executed indexing instruction would be placed into one of the four levels of lookahead along with its instruction address and the previous value of any index register that had been modified. This history of old values provided a way for the lookahead levels to be rolled back and thus restore the contents of index registers on a mispredicted branch, interrupt, or exception. A prepared arithmetic instruction would also be placed into a lookahead level along with its instruction address, and there it would wait for the completion of its memory operand fetch. A feature that foreshadows many current processors is that some of the more complex Stretch instructions had to be broken down into separate parts and stored into multiple lookahead levels.

An arithmetic instruction would be executed by the arithmetic unit whenever its lookahead level became the oldest and its memory operand was available. Arithmetic exceptions were made precise by causing a rollback of the lookahead levels, just as would be done in the case of a mispredicted branch. A store instruction was also executed when its lookahead level became the oldest. While the store was in the lookahead, store forwarding was implemented by checking the memory address of each subsequent load placed in the lookahead. If the address to be read matched the address to be stored, the load was canceled, and the store value was directly copied into the buffer reserved for the load value (called *short-circuiting*). Only one outstanding store at a time was allowed in the lookahead. Also, because of potential instruction modification, the store address was compared to each of the instruction addresses in the lookahead levels.

Stretch was implemented with 169,100 transistors and 96K 64-bit words of core memory. The clock cycle time was 300 ns (up from the initial estimates of 100 ns) for the indexing unit and lookahead unit, while the clock cycle time for the variable-length field unit and parallel arithmetic unit was 600 ns. Twenty-three levels of logic were allowed in a path, and a connection of approximately 15 feet (ft) was counted as one-half level. The parallel arithmetic unit performed one floating-point add each 1.5 μs and one floating-point multiply every 2.4 μs. The processing units dissipated 21 kilowatts (kW). The CPU alone (without its memory banks) measured 30 ft by 6 ft by 5 ft.

As the clock cycle change indicates, Stretch did not live up to its initial performance promises, which had ranged from 60 to 100 times the performance of a 704. In 1960, product planners set a price of $13.5 million for the commercial form of Stretch, the 7030. They estimated that its performance would be eight times the performance of a 7090, which was itself eight times the performance of a 704. This estimation was heavily based on arithmetic operation timings.

When Stretch became operational in 1961, benchmarks indicated that it was only four times faster than a 7090. This difference was in large part due to the store latency and the branch misprediction recovery time, since both cases stalled the arithmetic unit. Even though Stretch was the fastest computer in the world (and remained so until the introduction of the CDC 6600 in 1964), the performance shortfall caused considerable embarrassment for IBM. In May 1961, Tom Watson announced a price cut of the 7030s under negotiation to $7.78 million and immediately withdrew the product from further sales.

While Stretch turned out to be slower than expected and was delivered a year later than planned, it provided IBM with enormous advances in transistor design and computer organization principles. Work on Stretch circuits allowed IBM to deliver the first of the popular 7090 series 13 months after the initial contract in 1958; and, multiprogramming, memory protection, generalized interrupts, the 8-bit byte, and other ideas that originated in Stretch were subsequently used in the very successful S/360. Stretch also pioneered techniques in uniprocessor parallelism, including decoupled access-execute execution, speculative execution, branch misprediction recovery, and precise exceptions. It was also the first machine to use memory interleaving and the first to buffer store values and provide forwarding to subsequent loads. Stretch provided a wonderful training ground for John Cocke and others who would later propose and investigate the idea of parallel decoding of multiple instructions in follow-on designs.

8.1.2 First Superscalar Design: The IBM Advanced Computer System

In 1961, IBM started planning for two high-performance projects to exceed the capabilities of Stretch. *Project X* had a goal of 10 to 30 times the performance of Stretch, and this led to the announcement of the IBM S/360 Model 91 in 1964 and its delivery in 1967. The Model 91's floating-point unit is famous for executing instructions out-of-order, according to an algorithm devised by Robert Tomasulo. The initial cycle time goal for Project X was 50 ns, and the Model 91 shipped at a 60-ns cycle time. Mike Flynn was the project manager for the IBM S/360 Model 91 up until he left IBM in 1966.

The second project, named *Project Y,* had a goal of building a machine that was 100 times faster than Stretch. Project Y started in 1961 at IBM Watson Research Center. However, because of Watson's overly critical assessment of Stretch, Project Y languished until the 1963 announcement of the CDC 6600 (which combined scalar instruction issue with out-of-order instruction execution among its 10 execution units and ran with a 100-ns cycle time; see Figure 4.6). Project Y was then assigned to Jack Bertram's experimental computers and programming group; and John Cocke, Brian Randell, and Herb Schorr began playing major roles in defining the circuit technology, instruction set, and compiler technology.

In late 1964, sales of the CDC 6600 and the announcement of a 25-ns cycle time 6800 (later redesigned and renamed the 7600) added urgency to the Project Y effort. Watson decided to "go for broke on a very advanced machine" (memo dated May 17, 1965 [Pugh, 1991]), and in May 1965, a supercomputer laboratory was established in Menlo Park, California, under the direction of Max Paley and Jack Bertram. The architecture team was led by Herb Schorr, the circuits team by Bob Domenico, the compiler team by Fran Allen, and the engineering team by Russ Robelen. John Cocke arrived in California to work on the compilers in 1966. The design became known as the Advanced Computer System 1 (ACS-1) [Sussenguth, 1990].

The initial clock cycle time goal for ACS-1 was 10 ns, and a more aggressive goal was embraced of 1000 times the performance of a 7090. To reach the cycle time goal, the ACS-1 pipeline was designed with a target of five gate levels of logic per stage. The overall plan was ambitious and included an optimizing compiler as well as a new operating system, streamlined I/O channels, and multi-headed disks as integral parts of the system. Delivery was at first anticipated for 1968 to expected customers such as Livermore and Los Alamos. However, in late 1965, the target introduction date was moved back to the 1970 time frame.

Like the CDC 6600 and modern RISC architectures, most ACS-1 instructions were defined with three register specifiers. There were thirty-one 24-bit index registers and thirty-one 48-bit arithmetic registers. Because it was targeted to number-crunching at the national labs, the single-precision floating-point data used a 48-bit format and double-precision data used 96 bits. The ACS-1 also used 31 *backup registers,* each one being paired with a corresponding arithmetic register. This provided a form of register renaming, so that a load or writeback could occur to the backup register whenever a dependency on the previous register value was still outstanding.

Parallel decoding of multiple instructions and dispatch to two reservation stations, one of which provided out-of-order issue, were proposed for the processor (see Figure 8.2). Schorr wrote in his 1971 paper on the ACS-1 that "multiple decoding was a new function examined by this project." Cocke in a 1994 interview stated that he arrived at the idea of multiple instruction decoding for ACS-1 in response to an IBM internal report written by Gene Amdahl in the early 1960s

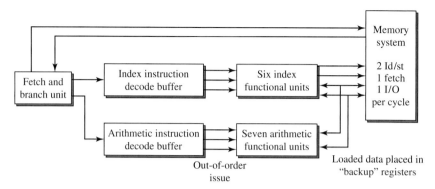

Figure 8.2
IBM ACS Block Diagram.

in which Amdahl postulated one instruction decode per cycle as one of the fundamental limits on obtainable performance. Cocke wanted to test each supposed fundamental limitation and decided that multiple decoding was feasible. (See also Flynn [1966] for a discussion of this limit and the difficulty of multiple decoding.)

Although Cocke had made some early proposals for methods of multiple instruction issue, in late 1965 Lynn Conway made the contribution of the generalized scheme for dynamic instruction scheduling that was used in the design. She described a *contender stack* that scheduled instructions in terms of source and destination *scheduling matrices* and a *busy vector*. Instruction decoding and filling of the matrices would stop on the appearance of a conditional branch and resume only when that branch was resolved. The matrices were also scanned in reverse order to give priority to the issue of the conditional branch.

The resulting ACS-1 processor design had six function units for index operations: compare, shift, add, branch address calculation, and two effective address adders. It had seven function units for arithmetic operations: compare, shift, logic, add, divide/integer multiply, floating-point add, and floating-point multiply. Up to seven instructions could be issued per cycle: three index operations (two of which could be load/stores), three arithmetic operations, and one branch. The eight-entry load/store/index instruction buffer could issue up to three instructions in order. The eight-entry arithmetic instruction buffer would search for up to three ready instructions and could issue these instructions out of order. (See U.S. Patent 3,718,912.) Loads were sent to both instruction buffers to maintain instruction ordering.

Recognizing that they could lose half or more of the design's performance on branching, the designers adopted several aggressive techniques to reduce the number of branches and to speed up the processing of those branches and other transfers of control that remained:

- Ed Sussenguth and Herb Schorr divided the actions of a conditional branch into three separate categories: branch target address calculation, taken/untaken determination, and PC update. The ACS-1 combined the first two actions in a *prepare-to-branch* instruction and used an *exit* instruction to perform the last action. This allowed a variable number of branch delay slots to be filled (called *anticipating a branch*); but, more importantly, it provided for multiway branch specification. That is, multiple prepare-to-branch instructions could be executed and thereby set up an internal table of multiple branch conditions and associated target addresses, only one of which (the first one that evaluated to true) would be used by the exit instruction. Thus only one redirection of the instruction fetch stream would be required. (See U.S. Patent 3,577,189.)

- A set of 24 condition code registers allowed precalculation of branch conditions and also allowed a single prepare-to-branch instruction to specify a logical expression involving any two of the condition codes. This is similar in concept to the eight independent condition codes in the RS/6000.

- To handle the case of a forward conditional branch with small displacement, a conditional bit was added to each instruction format (i.e., a form of predication). A special form of the prepare-to-branch instruction was used as a conditional *skip*. At the point of condition resolution, if the condition in the skip instruction was true, any instructions marked as conditional were removed from the instruction queues. If the condition was resolved to be false, then the marked instructions were unmarked and allowed to execute. (See U.S. Patent 3,577,190.)

- Dynamic branch prediction with 1-bit histories provided for instruction prefetch into the decoder, but speculative execution was ruled out because of the previous performance problems with Stretch. A 12-entry target instruction cache with eight instructions per entry was also proposed by Ed Sussenguth to provide the initial target instructions and thus eliminate the four-cycle penalty for taken branches. (See U.S. Patent 3,559,183.)

- Up to 50 instructions could be in some stage of execution at any given time, so interrupts and exceptions could be costly. Most external interrupts were converted by the hardware into specially marked branches to the appropriate interrupt handler routines and then inserted into the instruction stream to allow the previously issued instructions to complete. (These were called *soft interrupts*.) Arithmetic exceptions were handled by having two modes: one for multiple issue with imprecise interrupts and one for serialized issue. This approach was used for the S/360 Model 91 and for the RS/6000.

Main memory was 16-way interleaved, and a store buffer provided for load bypassing, as done in Stretch. Cache memory was introduced within IBM in 1965, leading to the announcement of the S/360 Model 85 in 1968. The ACS-1 adopted the cache memory approach and proposed a 64K-word unified instruction and data cache. The ACS-1 cache was to be two-way set-associative with a line size of 32 words and last-recently-used (LRU) replacement; a block of up to eight 24-bit instructions could be fetched each cycle. A cache hit would require five cycles, and a cache miss would require 40 cycles. I/O was to be performed to and from the cache.

The ACS-1 processor design called for 240,000 circuits: 50% of these were for floating-point, 24% were for indexing, and the remaining 26% were for instruction sequencing. Up to 40 circuits were to be included on an integrated-circuit die. At approximately 30 mW per circuit, the total power dissipation of the processor was greater than 7 kW.

An optimizing compiler with instruction scheduling, register allocation, and global code motion was developed in parallel with the machine design by Fran Allen and John Cocke [Allen, 1981]. Simulation demonstrated that the compiler could produce better code than careful hand optimization in several instances. In her article, Fran Allen credits the ACS-1 work as providing the foundations for program analysis and machine independent/dependent optimization.

Special emphasis was given to six benchmark kernels by both the instruction set design and compiler groups. One of these was double-precision floating-point inner product. Cocke estimated that the machine could reach five to six instructions per cycle on linear algebra codes of this type [1998]. Schorr [1971] and Sussenguth [1990] contain performance comparisons between the IBM 7090, CDC 6600, S/360 Model 91, and ACS-1 for a simple loop [Lagrangian hydrodynamics calculation (LHC)] and a very complex loop [Newtonian diffusion (ND)], and these comparisons are given in Table 8.1.

An analysis of several machines was also performed that normalized relative performance with respect to the number of circuits and the circuit speed of each machine. The result was called a relative *architectural factor*, although average memory access time (affected by the presence or absence of cache memory) affected the results. Based on this analysis, and using the 7090 for normalization, Stretch had a factor of 1.2; both the CDC 6600 and the Model 91 had factors of 1.1; and the Model 195 (with cache) had a factor of 1.7. The ACS-1 had a factor of 5.2.

The ACS-1 was in competition with other projects within IBM, and by the late 1960s, a design that was incompatible with the S/360 architecture was losing support within the company. Gene Amdahl, having become an IBM Fellow

Table 8.1

ACS-1 performance comparison

	7090	CDC 6600	S/360 M91	ACS-1
Relative performance on LHC	1	50	110	2500
Relative performance on ND	1	21	72	1608
Sustained IPC on ND	0.26	0.27	0.4	1.8
Performance limiter on ND	Sequential nature of the machine	Inst. fetch	Branches	Arithmetic

in 1965 and having come to California as a consultant to Paley, began working with John Earle on a proposal to redesign the ACS to provide S/360 compatibility. In early 1968, persuaded by increased sales forecasts, IBM management accepted the Amdahl-Earle plan. However, the overall project was thrown into a state of disarray by this decision, and approximately one-half of the design team left.

Because of the constraint of architectural compatibility, the ACS-360 had to discard the innovative branching and predication schemes, and it also had to provide a strongly ordered memory model as well as precise interrupts. Compatibility also meant that an extra gate level of logic was required in the execution stage, with consequent loss of clock frequency. One ACS-360 instruction set innovation that later made it into the S/370 was start I/O fast release (SIOF), so that the processor would not be unduly slowed by the initiation of I/O channels.

Unfortunately, with the design no longer targeted to number-crunching, the ACS-360 had to compete with other IBM S/360 projects on the basis of benchmarks that included commercial data processing. The result was that the IPC of the ACS-360 was less than one. In 1968, a second instruction counter and a second set of registers were added to the simulator to make the ACS-360 the first simultaneous multithreaded design. Instructions were tagged with an additional *red/blue bit* to designate the instruction stream and register set; and, as project members had expected, the utilization of the function units increased.

However, it was too late. By 1969, emitter coupled logic (ECL) circuit design problems, coupled with the performance achievements of the cache-based S/360 Model 85, a slowdown in the national economy, and East Coast/West Coast tensions within the company, led to the cancellation of the ACS-360 [Pugh et al., 1991]. Amdahl left shortly thereafter to start his own company. Further work was done at IBM on superscalar S/370s up through the 1990s. However, IBM never produced a superscalar mainframe, with the notable exception of a processor announced 25 years later, the ES/9000 Model 520 [Liptay, 1992].

8.1.3 Instruction-Level Parallelism Studies

In the early 1970s two important studies on multiple instruction decoding and issue were published: one by Gary Tjaden and Mike Flynn [1970] and one by Ed Riseman and Caxton Foster [1972]. Flynn remembers being skeptical of the idea of multiple decoding, but later, with his student Tjaden, he examined some of the inherent problems of interlocking and control in the context of a multiple-issue 7094. Flynn also published what appears to be the first open-literature reference to multiple decoding as part of his classic SISD/SIMD/MIMD paper [Flynn, 1966].

While Tjaden and Flynn concentrated on the decoding logic for a multiple-issue IBM 7094, Riseman and Foster examined the effect of branches in CDC 3600 programs. Both groups reported small amounts of available parallelism in the benchmarks they studied (1.86 and 1.72 instructions per cycle, respectively); however, Riseman and Foster found increasing levels of parallelism as the number of branches were eliminated by knowing which paths were executed.

The results of these papers were taken as quite negative and dampened general enthusiasm for fine-grain, single-program parallelism (see Section 1.4.2). It would be the early 1980s before Josh Fisher and Bob Rau's VLIW efforts [Fisher, 1983; Rau et al., 1982] and Tilak Agerwala and John Cocke's superscalar efforts (see the following) would convince designers of the feasibility of multiple instruction issue and thus inspire numerous design efforts.

8.1.4 By-Products of DAE: The First Multiple-Decoding Implementations

In the early 1980s, work by Jim Smith appeared on decoupled access-execute (DAE) architectures [Smith, 1982; 1984; Smith and Kaminski, 1982; Smith et al., 1986]. Smith was a veteran of Control Data Corporation (CDC) design efforts and was now teaching at the University of Wisconsin. In his 1982 International Symposium on Computer Architecture (ISCA) paper he gives credit to the IBM Stretch as the first machine to decouple access and execution, thereby allowing memory loads to start as early as possible. Smith's design efforts included architecturally visible queues on which the loads and stores operated. Computational instructions referenced either registers or loaded-data queues. His ideas led to the design and development of the dual-issue Astronautics ZS-1 in the mid-1980s [Smith et al., 1987].

As shown in Figure 8.3, the ZS-1 fetched 64-bit words from memory into an instruction *splitter*. Instructions could be either 32 or 64 bits in length, so the splitter could fetch up to two instructions per cycle. Branches were 64 bits and were fully executed in the splitter and removed from the instruction stream;

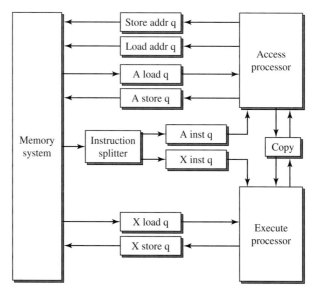

Figure 8.3
Astronautics ZS-1 Block Diagram.

unresolved conditional branches stalled the splitter. Access (A) instructions were placed in a four-entry A instruction queue, and execute (X) instructions were placed in a 24-entry X instruction queue. In-order issue occurs from these instruction queues; issue requires that there be no dependences or conflicts, and operands are fetched at that time from registers and/or load queues, as specified in the instruction. The access processor included three execution units: integer ALU, shift, and integer multiply/divide; and the execute processor included four execution units: logical, floating-point adder, floating-point multiplier, and reciprocal approximation unit. In this manner, up to two instructions could be issued per cycle.

In the 1982 ISCA paper, Smith also cites the CSPI MAP 200 array processor as an example of decoupling access and execution [Cohler and Storer, 1981]. The MAP 200 had separate access and execute processors coupled by FIFO buffers, but each processor had its own program memory. It was up to the programmer to ensure correct coordination of the processors.

In 1986 Glen Culler announced a dual-issue DAE machine, the Culler-7, a multiprocessor with an M68010-based kernel processor and up to four user processors [Lichtenstein, 1986]. Each user processor was a combination of an A machine, used to control program sequencing and data memory addressing and access, and a microcoded X machine, used for floating-point computations and which could run in parallel with the A machine. The A and X machines were coupled by a four-entry input FIFO buffer and a single-entry output buffer. A program memory contained sequences of X instructions, sometimes paired with and then trailed by some number of A instructions.

The X instructions were lookups into a control store of microcode routines; these routines were sequences of horizontal micro-instructions that specified operations for a floating-point adder and multiplier, two 4K-entry scratch pad memories, and various registers and busses. Single-precision floating-point operations were single-cycle, while double-precision operations took two cycles. User-microcoded routines could also be placed in the control store.

X and A instruction pairs were fetched, decoded, and executed together when available. A common sequence was a single X instruction, which would start a microcoded routine, followed by a series of A instructions to provide the necessary memory accesses. The first pair would be fetched and executed together, and the remaining A instructions would be fetched and executed in an overlapping manner with the multicycle X instruction. The input and output buffers between the X and A machines were interlocked, but the programmer/compiler was responsible for deadlock avoidance (e.g., omission of a required A instruction before the next X instruction).

The ZS-1 and Culler-7, developed without knowledge of each other, represent the first commercially sold processors in which multiple instructions from a single instruction stream were fetched, decoded, and issued in parallel. This dual issue of access and execute instructions will appear several times in later designs (albeit without the FIFO buffers) in which an integer unit will have responsibility for both integer instructions and memory loads and stores and can issue these in parallel with floating-point computation instructions on a floating-point unit.

8.1.5 IBM Cheetah, Panther, and America

Tilak Agerwala at IBM Research started a dual-issue project, code-named Cheetah, in the early 1980s with the urging and support of John Cocke. This design incorporated ACS ideas, such as backup registers, as well as ideas from the IBM 801 RISC experimental machine, another John Cocke project (circa 1974 to 1978). The three logical unit types seen in the RS/6000, i.e., branch, fixed-point (integer), and floating-point, were first proposed in Cheetah. A member of the Cheetah group, Pradip Bose, published a compiler research paper at the 1986 Fall Joint Computer Conference describing dual-issue machines such as the Astronautics ZS-1 and the IBM design.

In invited talks at several universities during 1983 to 1984, Agerwala first publicly used the term he had coined for ACS and Cheetah-like machines: *superscalar*. This name helped describe the potential performance of multiple-decoding machines, especially as compared to vector processors. These talks, some of which were available on videotape, and a related IBM technical report were influential in rekindling interest in multiple-decoding designs. By the time Jouppi and Wall presented their paper on available instruction-level parallelism at ASPLOS-III [1989] and Smith, Johnson, and Horowitz presented their paper on the limits on multiple instruction issue [1989], also at ASPLOS-III, superscalar and VLIW processors were hot topics.

Further development of the Cheetah/Panther design occurred in 1985 to 1986 and led to a four-way issue design called America [Special issue, *IBM Journal of Research and Development,* 1990]. The design team was led by Greg Grohoski and included Marc Auslander, Al Chang, Marty Hopkins, Peter Markstein, Vicky Markstein, Mark Mergen, Bob Montoye, and Dan Prener. In this design, a generalized register renaming facility for floating-point loads replaced the use of backup registers, and a more aggressive branch-folding approach replaced the Cheetah's delayed branching scheme. In 1986 the IBM Austin development lab adopted the America design and began refining it into the RS/6000 architecture (also known as RIOS and POWER).

8.1.6 Decoupled Microarchitectures

In the middle 1980s, Yale Patt and his students at the University of California, Berkeley, including Wen-Mei Hwu, Steve Melvin, and Mike Shebanow, proposed a generalization of the Tomasulo floating-point unit of the IBM S/360 Model 91, which they called *restricted data flow*. The key idea was that a sequential instruction stream could be dynamically converted into a partial data flow graph and executed in a data flow manner. The results of decoding the instruction stream would be stored in a decoded instruction cache (DIC), and this buffer area *decouples* the instruction decoding engine from the execution engine.

8.1.6.1 Instruction Fission.
In their work on the *high-performance substrate* (HPS), Patt and his students determined that regardless of the complexity of the target instruction set, the nodes of the partial dataflow graph stored in the DIC could be RISC-like micro-instructions. They applied this idea to the VAX ISA and found that an average of four HPS micro-instructions were needed per VAX instruction and that a restricted dataflow implementation could reduce the CPI of a VAX instruction stream from the then-current 6 to 2 [Patt et al., 1986; Wilson et al., 1987].

The translation of CISC instruction streams into dynamically scheduled, RISC-like micro-instruction streams was the basis of a number of IA32 processors, including the NexGen Nx586, the AMD K5, and the Intel Pentium Pro. The recent Pentium 4 caches the translated micro-instruction stream in its trace cache, similar to the decoded instruction cache of HPS. This fission-like approach is also used by some nominally reduced instruction set computer processors. One example is the recent POWER4, which *cracks* some of the more complex PowerPC instructions into multiple internal operations.

8.1.6.2 Instruction Fusion. Another approach to a decoupled microarchitecture is to fetch instructions and then allow the decoding logic to fuse compatible instructions together, rather than break each apart into smaller micro-operations. The resulting instruction group traverses the execution engine as a unit, in almost the same manner as a VLIW instruction.

One early effort along this line was undertaken at AT&T Bell Labs in the middle 1980s to design a decoupled scalar pipeline as part of the C Machine Project. The result was the CRISP microprocessor, described in 1987 [Ditzel and McLellan, 1987; Ditzel et al., 1987]. CRISP translated variable-length instructions into fixed-length formats, including next-address fields, during traversal of a three-stage decode pipeline. The resulting decoded instructions were placed into a 32-entry DIC, and a three-stage execution pipeline fetched and executed these decoded entries. By collapsing computation instructions and branches in this manner, CRISP could run simple instruction sequences at a rate of greater than one instruction per cycle. The Motorola 68060 draws heavily from this design.

Another effort at fusing instructions was the National Semiconductor Swordfish. The design, led by Don Alpert, began in Israel in the late 1980s and featured dual integer pipelines (A and B) and a multiple-unit floating-point coprocessor. A decoded instruction cache was organized into instruction pair entries. An instruction cache miss started a fetch and pre-decode process, called *instruction loading*. This process examined the instructions, precalculated branch target addresses, and checked opcodes and register dependences for dual issue. If dual issue was possible, a special bit in the cache entry was set. Regardless of dual issue, the first instruction in a cache entry was always sent to pipeline A, and the second instruction was supplied to both pipeline B and the floating-point pipeline. Program-sequencing instructions could only be executed by pipeline B. Loads could be performed on either pipeline, and thus they could issue on A in parallel with branches or floating-point operations on B. Pipeline B operated in lockstep with the floating-point pipeline; and in cases where a floating-point operation could trap, pipeline B cycled twice in its memory stage so that it and the floating-point pipeline would enter their writeback stages simultaneously. This provided in-order completion and thus made floating-point exceptions precise.

Other designs using instruction fusion include the Transputer T9000, introduced in 1991 and the TI SuperSPARC, introduced in 1992. Within the T9000, up to four instructions could be fetched per cycle, but an instruction *grouper* could build groups of up to eight instructions that would flow through the five-stage pipeline together [May et al., 1991]. The SuperSPARC had a similar grouping stage that combined up to three instructions.

Some recent processors use the idea of grouping instructions into larger units as a way to gain efficiency for reservation station slot allocation, reorder buffer allocation, and retirement actions, e.g., the Alpha 21264, AMD Athlon, Intel Pentium 4, and IBM POWER4. However, in these cases the instructions or micro-operations are not truly fused together but are independently executed within the execution engine.

8.1.7 Other Efforts in the 1980s

There were several efforts at multiple-instruction issue undertaken in the 1980s. H. C. Torng at Cornell University examined multiple-instruction issue for Cray-like machines and developed an out-of-order multiple-issue mechanism called the *dispatch stack* [Acosta et al., 1986]. Introduced in 1986 was the Stellar GS-1000 graphics supercomputer workstation [Sporer et al., 1988]. The GS-1000 used a four-way multithreaded, 12-stage pipelined processor in which two adjacent instructions in an instruction stream could be *packetized* and executed in a single cycle.

The Apollo DN10000 and the Intel i860 were dual-issue processors introduced in the late 1980s, but in each case the compile-time marking of dual issue makes these machines better understood as long-instruction-word architectures rather than as superscalars. In particular, the Apollo design used a bit in the integer instruction format to indicate whether a *companion* floating-point instruction (immediately following the integer instruction, with the pair being double-word aligned) should be dual-issued. The i860 used a bit in the floating-point instruction format to indicate *dual operation mode* in which aligned pairs of integer and floating-point instructions would be fetched and executed together. Because of pipelining, the effect of the bit in the i860 governed dual issue of the next pair of instructions.

8.1.8 Wide Acceptance of Superscalar

In 1989 Intel announced the first single-chip superscalar, the i960CA, which was a triple-issue implementation of the i960 embedded processor architecture [Hinton, 1989]. Also 1989 saw the announcement of the IBM RS/6000 as the first superscalar workstation; a special session with three RS/6000 papers was presented at the International Conference on Computer Design that October. Appearing in 1990 was the aggressively microcoded Tandom Cyclone, which executed special dual-instruction-execution microprograms whenever possible [Horst et al., 1990], and Motorola introduced the dual-issue 88110 in 1991. Mainframe manufacturers were also experimenting with superscalar designs; Univac announced the A19 in 1991, and in the following year Liptay [1992] described the IBM ES/9000 Model 520. A flurry of announcements occurred in the early 1990s, including the dual-issue Intel Pentium and the triple-issue PowerPC 601 for personal computers. And 1995 saw the introduction of five major processor cores that, with various tweaks, have powered computer systems for the past several years: HP 8000, Intel P6 (basis for Pentium Pro/II/III) MIPS R10000, HaL SPARC64, and UltraSPARC-I. Intel recently introduced the Pentium 4 with a redesigned core, and Sun has introduced the redesigned UltraSPARC-III. AMD has been actively involved in multiple superscalar designs since the K5 in 1995 through the current Athlon (K7) and Opteron (K8). IBM and Motorola have also introduced multiple designs in the POWER and PowerPC families. However, several system manufacturers, such as Compaq and MIPS (SGI), have trimmed or canceled their superscalar processor design plans in anticipation of adopting processors from the Intel Itanium processor family, a new explicitly parallel instruction computing (EPIC) architecture. For example, the Alpha line of processors began with the introduction of the dual-issue 21064 (EV5) in 1992 and continued until the cancellation of the eight-issue 21464 (EV9) design in 2001.

Figure 8.4 presents a time line of the designs, papers, and commercially available processors that have been important in the development of superscalar

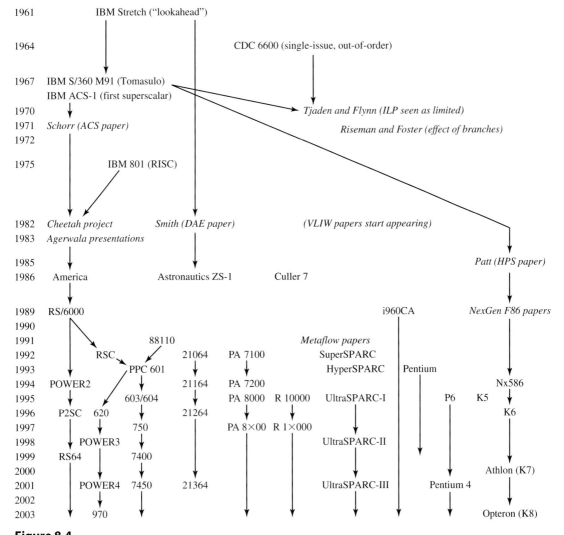

Figure 8.4
Time Line of Superscalar Development.

techniques. (There are many more superscalar processors available today than can fit in the figure, so your favorite one may not be listed.)

8.2 A Classification of Recent Designs

This section presents a classification of superscalar designs. We distinguish among various techniques and levels of sophistication that were used to provide multiple issue for pipelined ISAs, and we compare superscalar processors developed for the fastest clock cycle times (*speed demons*) and those developed for high issue rates (*brainiacs*). Of course, designers pick and choose from among the different design techniques and a given processor may exhibit characteristics from multiple categories.

8.2.1 RISC and CISC Retrofits

Many manufacturers chose to compete at the level of performance introduced by the IBM RS/6000 in 1989 by retrofitting superscalar techniques onto their 1980s-era RISC architectures, which were typically optimized for a single integer pipeline, or onto legacy complex instruction set computer (CISC) architectures. Six subcategories, or levels of sophistication, of retrofit are evident (these are adapted from Shen and Wolfe [1993]). These levels are design points rather than being strictly chronological developments. For example, in 1996, QED chose to use the first design point for the 200-MHz MIPS R5000 and obtained impressive SPEC95 numbers: 70% of the SPECint95 performance and 85% of the SPECfp95 performance of a contemporary 200-MHz Pentium Pro (a level-6 design style).

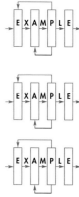

1. Floating-point coprocessor style
   - These processors cannot issue multiple integer instructions, or even an integer instruction and a branch in the same cycle; instead, the issue logic allows the dual issue of an integer instruction and a floating-point instruction. This is the easiest extension of a pipelined RISC. Performance is gained on floating-point codes by allowing the integer unit to execute the necessary loads and stores of floating-point values, as well as index register updates and branching.
   - Examples: Hewlett-Packard PA-RISC 7100 and MIPS R5000.
2. Integer with branch
   - This type of processor allows combined issue of integer instructions and branches. Thus performance on integer codes is improved.
   - Examples: Intel i960CA and HyperSPARC.
3. Multiple integer issue
   - These processors include multiple integer units and allow dual issue of multiple integer and/or memory instructions.
   - Examples: Hewlett-Packard PA-RISC 7100LC, Intel i960MM, and Intel Pentium.

4. Dependent integer issue
   - This type of processor uses cascaded or three-input ALUs to allow multiple issue of dependent integer instructions. A related technique is to double-pump the ALU each clock cycle.
   - Examples: SuperSPARC and Motorola 68060.

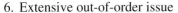

5. Multiple function units with precise exceptions
   - This type of processor emphasizes a precise exception model with sophisticated recovery mechanisms and includes a large number of function units with few, if any, issue restrictions. Restricted forms of out-of-order execution using distributed reservation stations are possible (i.e., interunit slip).
   - Example: Motorola 88110.

6. Extensive out-of-order issue
   - This type of processor provides complete out-of-order issue for all instructions. In addition to the normal pipeline stages, there is an identifiable dispatch stage, in which instructions are placed into a centralized reservation station or a set of distributed reservation stations, and an identifiable retirement stage, at which point the instructions are allowed to change the architectural register state and stores are allowed to change the state of the data cache.
   - Examples: Pentium Pro and HaL SPARC64.

Table 8.2 illustrates the variety of buffering choices found in level-5 and -6 designs.

Table 8.2
Out-of-order organization

Processor	Reservation Station Structure (Number of Entries)	Operand Copies	Entries in Reorder Buffer	Result Copies
Alpha 21264	Queues (15, 20)	No	20 × 4 insts. each	No
HP PA 8000	Queues (28, 28)	No	Combined w/RS	No
AMD K5	Decentralized (1, 2, 2, 2, 2, 2)	Yes	16	Yes
AMD K7	Schedulers (6 × 3, 12 × 3)	Yes, no	24 × 3 macroOps each	Yes, no
Pentium Pro	Centralized (20)	Yes	40	Yes
Pentium 4	Queues and schedulers	No	128	No
MIPS R10000	Queues (16, 16, 16)	No	32 (active list)	No
PPC 604	Decentralized (2, 2, 2, 2, 2, 2)	Yes	16	No
PPC 750	Decentralized (1, 1, 1, 1, 2, 2)	Yes	6	No
PPC 620	Decentralized (2, 2, 2, 2, 3, 4)	Yes	16	No
POWER3	Queues (3, 4, 6, 6, 8)	No	32	No
POWER4	Queues (10, 10, 10, 12, 18, 18)	No	20 × 5 IOPs each	No
SPARC64	Decentralized (8, 8, 8, 12)	Yes	64 (A-ring)	No

8.2.2 Speed Demons: Emphasis on Clock Cycle Time

High clock rate is the primary goal for a speed demon design. Such designs are characterized by deep pipelines, and designers will typically trade off lower issue rates and longer load-use and branch misprediction penalties for clock rate. Section 2.3 discusses these tradeoffs in more detail.

The initial DEC Alpha implementation, the 21064, illustrates the speed demon approach. The 21064 combined superpipelining and two-way superscalar issue and used seven stages in its integer pipeline, whereas most contemporary designs in 1992 used five or at most six stages. However, the tradeoff is that the 21064 would be classified only at level 1 of the retrofit categories given earlier. This is because only one integer instruction could be issued per cycle and could not be paired with an integer branch.

An alternative view of a speed demon processor is to consider it without the superpipelining exposed, that is, to look at what is accomplished in every two clock cycles. This is again illustrated by the 21064 since its clock rate was typically two or more times the clock rates of other contemporary chips. With this view, the 21064 is a four-way issue design with dependent instructions allowed with only a mild ordering constraint (i.e., the dependent instructions cannot be in the same doubleword); thus it is at level 4 of the retrofit categories.

A high clock rate often dictates a full custom logic design. Bailey gives a brief overview of the clocking, latching, and choices between static and dynamic logic used in the first three Alpha designs [1998]; he claims that full custom design is neither as difficult nor as time-consuming as is generally thought. Grundmann et al. [1997] also discusses the full-custom philosophy used in the Alpha designs.

8.2.3 Brainiacs: Emphasis on IPC

A separate design philosophy, the brainiac approach, is based on getting the most work done per clock cycle. This can involve instruction set design decisions as well as implementation decisions. Designers from this school of thought will trade off large reservation stations, complex dynamic scheduling logic, and lower clock cycle times for higher IPC. Other characteristics of this approach include emphasis on low load-use penalties and special support for dependent instruction execution.

The brainiac approach to architecture and implementation is illustrated by the IBM POWER (performance optimized with enhanced RISC). *Enhanced* instructions, such as fused multiply-add, load-multiple/store-multiple, string operations, and automatic index register updates for load/stores, were included in order to reduce the number of instructions that needed to be fetched and executed. The instruction cache was specially designed to avoid alignment constraints for full-width fetches, and the instruction distribution crossbar and front ends of the execution pipelines were designed to accept as many instructions as possible so that branches could be fetched and handled as quickly as possible. IBM also emphasized time to market and, for many components, used a standard-cell design approach that left the circuit design relatively unoptimized. This was especially true for the POWER2. Thus, for example, in 1996, the fastest clock rates on POWER and POWER2 implementations were 62.5 and 71.5 MHz, respectively, while the Alpha 21064A and 21264A ran at 300 and 500 MHz, respectively. Smith and Weiss [1994] offer an interesting comparison of the DEC and IBM design philosophies. (See also Section 6.7 and Figure 6.5.)

The brainiac approach to implementation can be seen in levels 4 and 6 of the retrofit categories.

8.3 Processor Descriptions

This section presents brief descriptions of several superscalar processors. The descriptions are ordered alphabetically according to manufacturer and/or architecture family (e.g., AMD and Cyrix are described with the Intel IA32 processors). The descriptions are not intended to be complete but rather to give brief overviews and highlight interesting or unusual design choices. More information on each design can be obtained from the references cited. *Microprocessor Reports* is also an excellent source of descriptive articles on the microarchitectural features of processors; these descriptions are often derived from manufacturer presentations at the annual Microprocessor Forum. The annual IEEE International Solid-State Circuits Conference typically holds one or more sessions with short papers on the circuit design techniques used in the newest processors.

8.3.1 Compaq / DEC Alpha

The DEC Alpha was designed as a 64-bit replacement for the 32-bit VAX architecture. Alpha architects Richard Sites and Rich Witek paid special attention to multiprocessor support, operating system independence, and multiple issue [Sites, 1993]. They explicitly rejected what they saw as scalar RISC implementation artifacts found in contemporary instruction sets, such as delayed branches and single-copy resources like multiplier-quotient and string registers. They also spurned mode bits, condition codes, and strict memory ordering.

In contrast to most other recent superscalar designs, the Alpha architects chose to allow imprecise arithmetic exceptions and, furthermore, not to provide a mode bit to change to a precise-exception mode. Instead, they defined a trap barrier instruction (TRAPB, and the almost identical EXCB) that will serialize any implementation so that pending exceptions will be forced to occur. Precise floating-point exceptions can then be provided in a naive way by inserting a TRAPB after each floating-point operation. A more efficient approach is to ensure that the compiler's register allocation will not allow instructions to overwrite source registers within a basic block or smaller region (e.g., the code block corresponding to a single high-level language statement); this constraint allows precise exceptions to be provided with one TRAPB per basic block (or smaller region) since the exception handler can then completely determine the correct values for all destination registers.

The Alpha architects also rejected byte and 16-bit word load/store operations, since they require a shift and mask network and a read-modify-write sequencing mechanism between memory and the processor. Instead, short instruction sequences were developed to perform byte and word operations in software. However, this turned out to be a design mistake, particularly painful when emulating IA32 programs on the Alpha; and, in 1995, byte and short loads and stores were introduced into the Alpha architecture and then supported on the 21164A.

8.3.1.1 Alpha 21064 (EV4) / 1992.
The 21064 was the first implementation of the Alpha architecture, and the design team was led by Alpha architect Rich Witek. The instruction fetch/issue unit could fetch two instructions per cycle on an aligned doubleword boundary. These two instructions could be issued together according to some complex rules, which were direct consequences of the allocation of register file ports and instruction issue paths within the design. The decoder was unaggressive; that is, if only the first instruction of the pair could be issued, no other instructions were fetched or examined until the second instruction of the pair had also been issued and removed from the decoder. However, a pipe stage was dedicated to swapping the instruction pair into appropriate issue slots to eliminate some ordering constraints in the issue rules. This simple approach to instruction issue was one of the many tradeoffs made in the design to support the highest clock rate possible. The pipeline is illustrated in Figure 8.5.

The 8K-byte instruction cache contained a 1-bit dynamic branch predictor for each instruction (2 bits on the 21064A); however, by appropriately setting a control register, static prediction based on the sign of the displacement could instead be selected. A four-entry subroutine address prediction stack was also included in the 21064, but hint bits had to be explicitly set within the jump instructions to push, pop, or ignore this stack.

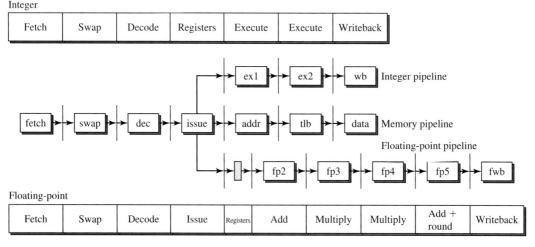

Figure 8.5
Alpha 21064 Pipeline Stages.

The 21064 had three function units: integer, load/store, and floating-point. The integer unit was pipelined in two stages for longer-executing instructions such as shifts; however, adds and subtracts finished in the first stage. The load/store unit interfaced to an 8K-byte data cache and a four-entry write buffer. Each entry was cache-line sized even though the cache was writethrough; this sizing provided for write merging. Load bypass was provided, and up to three outstanding data cache misses were supported.

For more information, see the special issue of *Digital Technical Journal* [1992], Montanaro [1992], Sites [1993], and McLellan [1993].

8.3.1.2 Alpha 21164 (EV5) / 1995. The Alpha 21164 was an aggressive second implementation of the Alpha architecture. John Edmondson was the lead architect during design, and Jim Keller was lead architect during advanced development. Pete Bannon was a contributor and also led the design of a follow-on chip, the 21164PC. The 21164 integrated four function units, three separate caches, and an L3 cache controller on chip. The function units were integer unit 0, which also performed integer shift and load and store; integer unit 1, which also performed integer multiply, integer branch, and load (but not store); floating-point unit 0, which performed floating-point add, subtract, compare, and branch and which controlled a floating-point divider; and floating-point unit 1, which performed floating-point multiply.

The designers cranked up the clock speed for the 21164 and also reduced the shift, integer multiply, and floating-point multiply and divide cycle count latencies, as compared to the 21064. However, the simple approach of a fast but unaggressive decoder was retained, with multiple issue having to occur in order from a quadword-aligned instruction quartet; the decoder advanced only after everything in the current quartet had been issued.

The correct instruction mix for a four-issue cycle was two independent integer instructions, a floating-point multiply instruction, and an independent non-multiply floating-point instruction. However, these four instructions did not have ordering constraints within the quartet, since a *slotting stage* was included in the pipeline to route instructions from a two-quartet instruction-fetch buffer into the decoder. The two integer instructions could both be loads, and each load could be for either integer or floating-point values. To allow the compiler greater flexibility in branch target alignment and generating a correct instruction mix in each quartet, three flavors of nops were provided: integer unit, floating-point unit, and vanishing nops. Special provision was made for dual issue of a compare or logic instruction and a dependent conditional move or branch. Branches into the middle of a quartet were supported by having a valid bit on each instruction in the decoder. Exceptions on the 21164 were handled in the same manner as in the 21064: issue from the decoder stalled whenever a trap or exception barrier instruction was encountered.

In this second Alpha design, the branch prediction bits were removed from the instruction cache and instead were packaged in a 2048-entry BHT. The return address stack was also increased to 12 entries. A correctly predicted taken branch could result in a one-cycle bubble, but this bubble was often squashed by stalls of previous instructions within the issue stage.

A novel, but now common, approach to pipeline stall control was adopted in the 21164. The control logic checked for stall conditions in the early pipeline stages, but late-developing hazards such as cache miss and write buffer overflow were caught at the point of execution and the offending instruction and its successors were then *replayed*. This approach eliminated several critical paths in the design, and the handling of a load miss was specially designed so that no additional performance was lost due to the replay.

The on-chip cache memory consisted of an L1 instruction cache (8K bytes), a dual-ported L1 data cache (8K bytes), and a unified L2 cache (96K bytes). The split L1 caches provided the necessary bandwidth to the pipelines, but the size of the L1 data cache was limited because of the dual-port design. The L1 data cache provided two-cycle latency for loads and could accept two loads or one store per cycle; L2 access time was eight cycles. There was a six-entry miss address file (MAF) that sat between the L1 and L2 caches to provide nonblocking access to the L1 cache. The MAF merged nonsequential loads from the same L2 cache line, much the same way as large store buffers can merge stores to the same cache line; up to four destination registers could be remembered per missed address. There was also a two-entry bus address file (BAF) that sat between the L2 cache and the off-chip memory to provide nonblocking access for line-length refills of the L2 cache.

See Edmondson [1994], Edmondson et al. [1995a, b], and Bannon and Keller [1995] for details of the 21164 design. Circuit design is discussed by Benschneider et al. [1995] and Bowhill et al. [1995]. The Alpha 21164A is described by Gronowski et al. [1996].

8.3.1.3 Alpha 21264 (EV6) / 1997.
The 21264 was the first out-of-order implementation of the Alpha architecture. However, the in-order parts of the pipeline retain the efficiency of dealing with aligned instruction quartets, and instructions are preslotted into one of two sets of execution pipelines. Thus, it could be said that this design approach marries the efficiency of VLIW-like constraints on instruction alignment and slotting to the flexibility of an out-of-order superscalar. Jim Keller was the lead architect of the 21264.

A hybrid (or *tournament*) branch predictor is used in which a two-level adaptive local predictor is paired with a two-level adaptive global predictor. The local predictor contains 1024 ten-bit local history entries that index into a 1024-entry pattern history table, while the global predictor uses a 12-bit global history register that indexes into a separate 4096-entry pattern history table. A 4096-entry choice predictor is driven by the global history register and chooses between the local and global predictors. The instruction fetch is designed to speculate up through 20 branches.

Four instructions can be renamed per cycle, and these are then dispatched to either a 20-entry integer instruction queue or a 15-entry floating-point instruction queue. There are 80 physical integer registers (32 architectural, 8 privileged architecture library (PAL) shadow registers, and 40 renaming registers) and 72 physical floating-point registers (32 architectural and 40 renaming registers). The instruction quartets are retained in a 20-entry reorder buffer/active list, so that up to 80 instructions along with their renaming status can be tracked. A mispredicted branch requires one cycle to recover to the appropriate instruction quartet. Instructions can retire at the rate of two quartets per cycle, but the 21264 is unusual in that it can retire instructions whenever they and all previous instructions are past the point of possible exception and/or misprediction. This can allow retirement of instructions even before the execution results are calculated.

The integer instruction queue can issue up to four instructions per cycle, one to each of four integer function units. Each integer unit can execute add, subtract, and logic instructions. Additionally, one unit can execute branch, shift, and multimedia instructions; one unit can execute branch, shift, and multiply instructions; and the remaining two can each execute loads and stores. The integer register file is implemented as two identical copies so that enough register ports can be provided. Coherency between the two copies is maintained with a one-cycle latency between a write into one file and the corresponding update in the other. Gieseke et al. [1997] estimate that the performance penalty for the split integer clusters is 1%, whereas a unified integer cluster would have required a 22% increase in area, a 47% increase in data path width, and a 75% increase in operand bus length. A unified register file approach would also have limited the clock cycle time.

The floating-point instruction queue can issue up to two instructions per cycle, one to each of two floating-point function units. One floating-point unit can execute add, subtract, divide, and take the square root, while the other floating-point unit is dedicated to multiply. Floating-point add, subtract, and multiply are pipelined and have a four-cycle latency.

The 21264 instruction and data caches are each 64K bytes in size and are organized as two-way pseudo-set-associative. The data cache is cycled twice as fast as the processor clock, so that two loads, or a store and a victim extract, can be executed during each processor clock cycle. A 32-entry load reorder buffer and a 32-entry store reorder buffer are provided.

For more information on the 21264 microarchitecture, see Leibholz and Razdan [1997], Kessler et al. [1998], and Kessler [1999]. For some specifics on the logic design, see Gowan et al. [1998] and Matson et al. [1998].

8.3.1.4 Alpha 21364 (EV7) / 2001. The Alpha 21364 uses a 21264 (EV68) core and adds an on-chip L2 cache, two memory controllers, and a network interface. The L2 cache is seven-way set-associative and contains 1.75 Mbytes. The cache hierarchy also contains 16 victim buffers for L1 cast-outs, 16 victim buffers for L2 cast-outs, and 16 L1 miss buffers. The memory controllers support directory-based cache coherency and provide RAM bus interfaces. The network interface supports out-of-order transactions and adaptive routing over four links per processor, and it can provide a bandwidth of 6.4 Gbytes/s per link.

8.3.1.5 Alpha 21464 (EV8) / Canceled. The Alpha 21464 was an aggressive eight-wide superscalar design that included four-way simultaneous multithreading. The design was oriented toward high single-thread throughput, yet the chip area cost of adding simultaneous multithreading (SMT) control and replicated resources was minimal (reported to be 6%).

With up to two branch predictions performed each cycle, instruction fetch was designed to return two blocks, possibly noncontiguous, of eight instructions each. After fetch, the 16 instructions would be *collapsed* into a group of eight instructions, based on the branch predictions. Each group was then renamed and dispatched into a single 128-entry instruction queue. The queue was implemented with the dispatched instructions assigned age vectors, as opposed to the collapsing FIFO design of the instruction queues in the 21264.

Each cycle up to eight instructions would be issued to a set of 16 function units: eight integer ALUs, four floating-point ALUs, two load pipelines, and two store pipelines. The register file was designed to have 256 architected registers (64 each for the four threads) and an additional 256 registers available for renaming. Eight-way issue required the equivalent of 24 ports, but such a structure would be difficult to implement. Instead, two banks of 512 registers each were used, with each register being eight-ported. This structure required significantly more die area than the 64K-byte L1 data cache. Moreover, the integer execution pipeline, planned as requiring the equivalent of 18 stages, devoted three clock cycles to register file read. Several eight-entry register caches were included within the function units to provide forwarding (compare with the UltraSPARC-III working register file).

The chip design also included a system interconnect router for building a directory-based cache coherent NUMA system with up to 512 processors.

Alpha processor development, including the 21464, was canceled in June 2001 by Compaq in favor of switching to the Intel Itanium processors. Joel Emer gave an overview of the 21464 design in a keynote talk at PACT 2001, and his slides are available on the Internet. See also Preston et al. [2002]. Seznec et al. [2002] describe the branch predictor.

8.3.2 Hewlett-Packard PA-RISC Version 1.0

Hewlett-Packard's Precision Architecture (PA) was one of the first RISC architectures; it was designed between 1981 and 1983 by Bill Worley and Michael Mahon, prior to the introduction of MIPS and SPARC. It is a load/store architecture with many RISC-like qualities, and there is also a slight VLIW flavor to its instruction set architecture (ISA). In several cases, multiple operations can be specified in one instruction. Thus, while superscalar processors in the 32-bit PA line (PA-RISC version 1.0) were relatively unaggressive in superscalar *instruction* issue width, they were all capable of executing multiple *operations* per cycle. Indeed, the first dual-issue PA processor, the 7100, could issue up to four operations in a given cycle: an integer ALU operation, a condition test on the ALU result to determine if the next instruction would be nullified (i.e., predicated execution), a floating-point add, and an independent floating-point multiply. Moreover, these four operations could be issued while a previous floating-point divide was still in execution and while a cache miss was outstanding.

The 32-bit PA processors prior to the 7300LC were characterized by relatively low numbers of transistors per chip and instead emphasized the use of large off-chip caches. A simple five-stage pipeline was the starting point for each design, but careful attention was given to tailoring the pipelines to run as fast as the external cache SRAMs would allow. While small, specialized on-chip caches were introduced on the 7100LC and the 7200, the 7300LC featured large on-chip caches. The 64-bit, out-of-order 8000 reverted back to reliance on large, off-chip caches. Later versions of the 8x00 series have once again added large, on-chip caches as transistor budgets have allowed. These design choices resulted from the close attention HP system designers have paid to commercial workloads (e.g., transaction processing), which exhibit large working sets and thus poor locality for small on-chip caches.

The implementations listed next follow the HP tradition of team designs. That is, no one or two lead architects are identified. Perhaps more than other companies, HP has attempted to include compiler writers on these teams at an equal level with the hardware designers.

8.3.2.1 PA 7100 / 1992. The 7100 was the first superscalar implementation of the Precision Architecture series. It was a dual-issue design with one integer unit and three independent floating-point units. One integer unit instruction could be issued along with one floating-point unit instruction each cycle. The integer unit handled both integer and floating-point load/stores, while integer multiply was performed by the floating-point multiplier. Special pre-decode bits in the instruction cache were assigned on refill so that instruction issue was simplified. There were no ordering or alignment requirements for dual issue.

Branches on the 7100 were statically predicted based on the sign of the displacement. Precise exceptions were provided for a dual-issue instruction pair by delaying the writeback from the integer unit until the floating-point units had successfully written back.

A load could be issued each cycle, and returned data from a cache hit in two cycles. Special pairing allowed a dependent floating-point store to be issued in the same cycle as the result-producing floating-point operation. Loading to R0 provided for software-controlled data prefetching.

See Asprey et al. [1993] and DeLano et al. [1992] for more information on the PA 7100. The 7150 is a 125-MHz implementation of the 7100.

8.3.2.2 PA 7100LC and 7300LC / 1994 and 1996. The PA 7100LC was a low-cost, low-power extension of the 7100 that was oriented toward graphics and multimedia workstation use. It was available as a uniprocessor only, but it provided a second integer unit, a 1K-byte on-chip instruction cache, an integrated memory controller, and new instructions for multimedia support. Figure 8.6 illustrates the PA 7100 pipeline.

The integer units on the 7100LC were asymmetric, with only one having shift and bit-field circuitry. Given that there could be only one shift instruction per cycle, then either two integer instructions, or an integer instruction and a load/store, or an integer instruction and a floating-point instruction, or a load/store and a floating-point

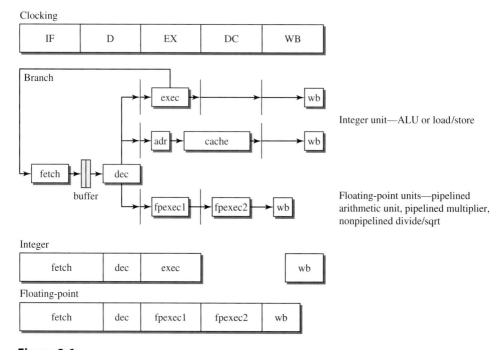

Figure 8.6
HP PA 7100 Pipeline Stages.

instruction could be issued in the same cycle. There was also a provision that two loads or two stores to the two words of a 64-bit aligned doubleword in memory could also be issued in the same cycle. This is a valuable technique for speeding up subroutine entry and exit.

Branches, and other integer instructions that can nullify the next sequential instruction, could be dual issued only with their predecessor instruction and not with their successor (e.g., a delayed branch cannot be issued with its branch delay slot). Instruction pairs that crossed cache line boundaries could be issued, except when the pair was an integer instruction and a load/store. The register scoreboard on the 7100LC also allowed write-after-write dependences to issue in the same cycle. However, to reduce control logic, the whole pipeline would stall on any operation longer in duration than two cycles; this included integer multiply, double-precision floating-point operations, and floating-point divide.

See Knebel et al. [1993], Undy et al. [1994], and the April 1995 special issue of the *Hewlett-Packard Journal* for more information on the PA 7100LC. The 7300LC is a derivative of the 7100LC with dual 64K-byte on-chip caches [Hollenbeck et al., 1996; Blanchard and Tobin, 1997; Johnson and Undy, 1997].

8.3.2.3 PA 7200 / 1994. The PA 7200 added a second integer unit and a 2K-byte on-chip assist cache for data. The instruction issue logic was similar to that of the 7100LC, but the pipeline did not stall on multiple-cycle operations. The 7200 provided multiple sequential prefetches for its instruction cache and also aggressively prefetched data. These data prefetches were internally generated according to the direction and stride of the address-register-update forms of the load/store instructions.

The decoding scheme on the 7200 was similar to that of the National Semiconductor Swordfish. The instruction cache expanded each doubleword with six pre-decode bits, some of which indicated data dependences between the two instructions and some of which were used to steer the instructions to the correct function units. These pre-decode bits were set upon cache refill.

The most interesting design twist to the 7200 was the use of an on-chip, fully associative assist cache of 64 entries, each being a 32-byte data cache line. All data cache misses and prefetches were directed to the assist cache, which had a FIFO replacement into the external data cache. A load/store hint was set in the instruction to indicate spatial locality only (e.g., block copy), so that the replacement of marked lines in the assist cache would bypass the external data cache. Thus cache pollution and unnecessary conflict misses in the direct-mapped external data cache were reduced.

See Kurpanek et al. [1994] and Chan et al. [1996] for more details on the PA 7200.

8.3.3 Hewlett-Packard PA-RISC Version 2.0

Michael Mahon and Jerry Hauck led the Hewlett-Packard efforts to extend Precision Architecture to 64 bits. PA-RISC 2.0 also includes multimedia extensions, called MAX [Lee and Huck, 1996]. The major change for superscalar implementations is the definition of eight floating-point condition bits rather than the original one. PA-RISC 2.0 adds a speculative cache line prefetch instruction that avoids invoking miss actions on a TLB miss, a weakly ordered memory model mode bit in the processor status word (PSW), hints in procedure calls and returns for maintaining a return-address prediction stack, and a fused multiply-add instruction. PA-RISC 2.0 further defines cache hint bits for loads and stores (e.g., for marking accesses as having spatial locality, as done for the HP PA 7200). PA-RISC 2.0 also uses a unique static branch prediction method: If register numbers are in ascending order in the compare and branch instruction, then the branch is predicted in one way; if they are in descending order, the branch is predicted in the opposite manner. The designers chose this covert manner of passing the static prediction information since there were no spare opcode bits available.

8.3.3.1 HP PA 8000 / 1996.
The PA 8000 was the first implementation of the 64-bit PA-RISC 2.0 architecture. This core is still used today in the various 8x00 chips. As shown in Figure 8.7, the PA 8000 has two 28-entry combined reservation station/reorder buffers and 10 function units: two integer ALUs, two shift/merge unit, two divide/square root unit, two multiply/accumulate units, and two load/store units. ALU instructions are dispatched into the ALU buffer, while memory instructions are dispatched into the memory buffer as well as a matching 28-entry address buffer. Some instructions, such as load-and-modify and branch, are dispatched to both buffers.

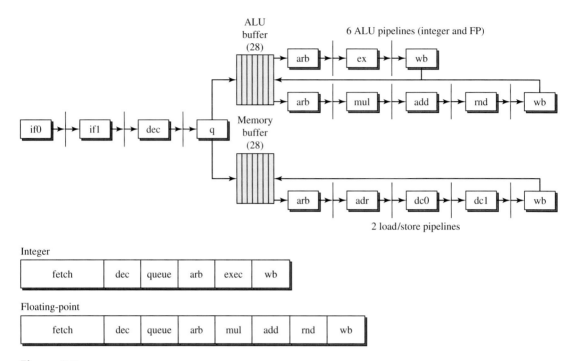

Figure 8.7
HP PA 8000 Pipeline Stages.

The combined reservation station/reorder buffers operate in an interesting divide-and-conquer manner [Gaddis and Lotz, 1996]. Instructions in the even-numbered slots in a buffer are issued to one integer ALU or one load/store unit, while instructions in the odd-numbered slots are issued to the other integer ALU or load/store unit. Additionally, arbitration for issue is done by subdividing a buffer's even half and odd half into four banks each (with sizes of 4, 4, 4, and 2). Thus there are four groups of four banks each. Within each group, the first ready instruction in the bank that contains the oldest instruction wins the issue arbitration. Thus, one instruction can be issued per group per cycle, leading to a maximum issue rate of two ALU instructions and two memory instructions per cycle. A large number of comparators is used to check register dependences when instructions are dispatched into the buffers, and an equally large number of comparators is used to match register updates to waiting instructions. Special propagate logic within the reservation station/reorder buffers handles carry-borrow dependences.

Because of the off-chip instruction cache, a taken branch on the 8000 can have a two-cycle penalty. However, a 32-entry, fully associative BTAC was used on the 8000 along with a 256-entry BHT. The BHT maintained a three-bit branch history register in each entry, and a prediction was made by majority vote of the history bits. A hit in the BTAC that leads to a correctly predicted taken branch has no penalty. Alternatively, prediction can be performed statically using a register number ordering scheme within the instruction format (see the introductory PA-RISC 2.0 paragraphs earlier). Static or dynamic prediction is selectable on a page basis. One suggestion made by HP is to profile dynamic library code, set the static prediction bits accordingly, and select static prediction for library pages. This preserves a program's dynamic history in the BHT across library calls.

See Hunt [1995] and Gaddis and Lotz [1996] for further description of the PA 8000.

8.3.3.2 PA 8200 / 1997. The PA 8200 is a follow-on chip that includes some improvements such as quadrupling the number of entries in the BHT to 1024 and allowing multiple BHT entries to be updated in a single cycle. The TLB entries are also increased from 96 to 120. See the special issue of the *Hewlett-Packard Journal* [1997] for more information on the PA 8000 and PA 8200.

8.3.3.3 PA 8500 / 1998. The PA 8500 is a shrink of the PA 8000 core and integrates 0.5 Mbyte of instruction cache and 1 Mbyte of data cache onto the chip. The 8500 changes the branch prediction method to use a 2-bit saturating agree counter in each BHT entry; the counter is decremented when a branch follows the static prediction and incremented when a branch mispredicts. The number of BHT entries is also increased to 2048. See Lesartre and Hunt [1997] for a brief description of the PA 8500.

8.3.3.4 Additional PA 8x00 Processors. The PA 8600 was introduced in 2000 and provides for lockstep operation between two chips for fault tolerant designs. The PA 8700, introduced in 2001, features increased cache sizes: a 1.5-Mbyte data cache and a 0.75-Mbyte instruction cache. See the Hewlett-Packard technical report [2000] for more details of the PA 8700.

Hewlett-Packard plans to introduce the PA 8800 and PA 8900 designs before switching its product line over to processors from the Intel Itanium processor family. The PA 8800 is slated to have dual PA 8700 cores, each with a 0.75-Mbyte data cache and a 0.75-Mbyte instruction cache, on-chip L2 tags, and a 32-Mbyte off-chip DRAM L2 cache.

8.3.4 IBM POWER

The design of the POWER architecture was led by Greg Grohoski and Rich Oehler and was based on the ideas of Cocke and Agerwala. Following the Cheetah and America designs, three separate function units were defined, each with its own register set. This approach reduced the design complexity for the initial implementations since instructions for different units do not need complex dependency checking to identify shared registers.

The architecture is oriented toward high-performance double-precision floating point. Each floating-point register is 64 bits wide, and all floating-point operations are done in double precision. Indeed, single-precision loads and stores require extra time in the early implementations because of converting to and from the internal double-precision format. A major factor in performance is the fused multiply-add instruction, a four-operand instruction that multiplies two operands, adds the product to a third, and stores the overall result in the fourth. This is exactly the operation needed for the inner product function found so frequently in inner loops of numerical codes. Brilliant logic design accomplished this operation in a two-pipe-stage design in the initial implementation. However, a side effect is that the addition must be done in greater than double precision, and this is visible in results that are slightly different from those obtained when normal floating-point rounding is performed after each operation.

Support for innermost loops in floating-point codes is seen in the use of the branch-and-count instruction, which can be fully executed in the branch unit, and in the renaming of floating-point registers for load instructions in the first implementation. Renaming the destination registers of floating-point loads is sufficient to allow multiple iterations of the innermost loop in floating-point codes to overlap execution, since the load in a subsequent iteration is not delayed by its reuse of an architectural register. Later implementations extend register renaming for all instructions.

Provision of precise arithmetic exceptions is obtained in POWER by the use of a mode bit. One setting serializes floating-point execution, while the other setting provides the faster alternative of imprecise exceptions.

Two major changes/extensions have been made to the POWER architecture. Apple, IBM, and Motorola joined forces in the early 1990s to define the PowerPC instruction set, which includes a subset of 32-bit instructions as well as 64-bit instructions. Also in the early 1980s, IBM Rochester defined the PowerPC-AS extensions to the 64-bit PowerPC architecture so that PowerPC processors could be used in the AS/400 computer systems.

The POWER family can be divided up into four major groups, with some of the more well-known members shown in the following table. (Note that there are many additional family members within the 32-bit PowerPC group that are not explicitly named, e.g., the 8xx embedded processor series.)

32-Bit POWER	32-Bit PowerPC	64-Bit PowerPC	64-Bit PowerPC-AS
RIOS	601	620	A30 (Muskie)
RSC	603	POWER3	A10 (Cobra)
POWER2	604	POWER4	A35 (Apache) / RS64
P2SC	740/750 (G3)	970 (G5)	A50 (Northstar) / RS64 II
	7400 (G4)		Pulsar / RS64 III
	7450 (G4+)		S-Star / RS64 IV

8.3.4.1 RIOS Pipelines / 1989.
Figure 8.8 depicts the pipelines in the initial implementation of POWER. These are essentially the same as the ones in the America processor designed by Greg Grohoski. The instruction cache and branch unit could fetch four instructions per cycle, even across cache line boundaries. During sequential execution these four instructions were placed in an eight-entry sequential instruction buffer. Although a predict-untaken policy was implemented with conditional issue/dispatch of sequential instructions, the branch logic inspected the first five entries of this buffer, and if a branch was found then a speculative fetch of four instructions at the branch target address was started. A special buffer held these target instructions. If the branch was not taken, the target buffer was flushed; however, if the branch was taken, the sequential buffer was flushed, the target instructions were moved to the sequential buffer, any conditionally dispatched sequential instructions were flushed, and the branch unit registers were restored from history registers as necessary. Sequential execution would then begin down the branch-taken path.

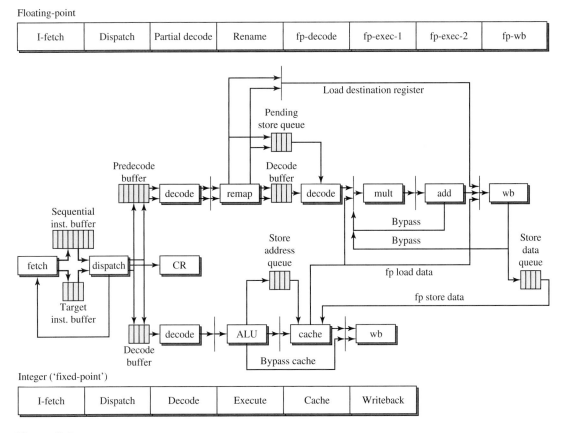

Figure 8.8
IBM POWER (RIOS) Pipeline Stages.

The instruction cache and branch unit dispatched up to four instructions each cycle. Two instructions could be issued to the branch unit itself, while two floating-point and integer (called *fixed-point* in the POWER) instructions could be dispatched to buffers. These latter two instructions could be both floating-point, both integer, or one of each. The fused multiply-add counted as one floating-point instruction. Floating-point loads and stores went to both pipelines, while other floating-point instructions were discarded by the integer unit and other integer instructions were discarded by the floating-point unit. If the instruction buffers were empty, then the first instruction of the appropriate type was allowed to issue into the unit. This dual dispatch into buffers obviated instruction pairing/ordering rules. Pre-decode tags were added to instructions on instruction cache refill to identify the required unit and thus speed up instruction dispatch.

The instruction cache and branch unit executed branches and condition code logic operations. There are three architected registers: a link register, which holds return addresses; a count register, which holds loop counts; and a condition register, which has eight separate condition code fields (CRi). CR0 is the default condition code for the integer unit, and CR1 is the default condition code for the floating-point unit. Explicit integer and floating-point compare instructions can specify any of the eight, but condition code updating as an optional side effect of execution occurs only to the default condition code. Multiple condition codes provide for reuse and also allow the compiler to substitute condition code logic operations in place of some conditional jumps in the evaluation of compound conditions. However, Hall and O'Brien [1991] indicated that the XL compilers at that point did not appear to benefit from the multiple condition codes.

The integer unit had four stages: decode, execute, cache access, and writeback. Rather than a traditional cache bypass for ALU results, the POWER integer unit passed ALU results directly to the writeback stage, which could write two integer results per cycle (this approach came from the 801 pipeline). Floating-point stores performed effective address generation and were then set aside into a store address queue until the floating-point data arrived later. There were four entries in this queue. Floating-point load data were sent to the floating-point writeback stage as well as to the first floating-point execution stage. This latter bypass allowed a floating-point load and a dependent floating-point instruction to be issued in the same cycle; the loaded value arrived for execution without a load penalty.

The floating-point unit accepted up to two instructions per cycle from its pre-decode buffer. Integer instructions were discarded by the pre-decode stage, and floating-point loads and stores were identified. The second stage in the unit renamed floating-point register references at a rate of two instructions per cycle; new physical registers were assigned as the targets of floating-point load instructions. (Thirty-eight physical registers were provided to map the 32 architected registers.)

At this point loads, stores, and ALU operations were separated; instructions for the latter two types were sent to their respective buffers. The store buffer thus allowed loads and ALU operations to bypass. The third pipe stage in the floating-point unit decoded one instruction per cycle and would read the necessary operands from the register file. The final three pipe stages were multiply, add, and writeback.

For more information on the POWER architecture and implementations, see the January 1990 special issue of *IBM Journal of Research and Development,* the IBM RISC System/6000 Technology book, Hester [1990], Oehler and Blasgen [1991], and Weiss and Smith [1994].

8.3.4.2 RSC / 1992. In 1992 the RSC was announced as a single-chip implementation of the POWER architecture. A restriction to one million transistors meant that the level of parallelism of the RIOS chip set could not be supported. The RSC was therefore designed with three function units (branch, integer, and floating-point) and the ability to issue/dispatch two instructions per cycle. An 8K-byte unified cache was included on chip; it was two-way set-associative, write-through, and had a line size of 64 bytes split into four sectors.

Up to four instructions (one sector) could be fetched from the cache in a cycle; these instructions were placed into a seven-entry instruction queue. The first three entries in the instruction queue were decoded each cycle, and either one of the first two instruction entries could be issued to the integer unit and the other dispatched to the floating-point unit. The integer unit was not buffered; it had a three-stage pipeline consisting of decode, execute, and writeback. Cache access could occur in either the execute or the writeback stage to help tolerate contention for access to the single cache. The floating-point unit had a two-entry buffer into which instructions were dispatched; this allowed the dispatch logic to reach subsequent integer and branch instructions more quickly. The floating-point unit did not rename registers.

The instruction prefetch direction after a branch instruction was encountered was predicted according to the sign of the displacement; however, an opcode bit could reverse the direction of this prediction. The branch unit was quite restricted and could independently handle only those branches that were dependent on the counter register and that had a target address in the same page as the branch instruction. All other branches had to be issued to the integer unit. There was no speculative execution beyond an unresolved branch.

Charles Moore was the lead designer for the RSC; see his paper [Moore et al., 1989] for a description of the RSC.

8.3.4.3 POWER2 / 1994. Greg Grohoski led the effort to extend the four-way POWER by increasing the instruction cache to 32K bytes and adding a second integer unit and a second floating-point unit; the result allowed six-way instruction issue and was called the POWER2. Additional goals were to process two branches per cycle and allow dependent integer instruction issue.

The POWER ISA was extended in POWER2 by the introduction of load/store quadword (128 bits), floating-point to integer conversions, and floating-point square root. The page table entry search and caching rule were also changed to reduce the expected number of cache misses during TLB miss handling.

Instruction fetch in POWER2 was increased to eight instructions per cycle, with cache line crossing permitted; and, the number of entries for the sequential and target instruction buffers was increased to 16 and 8, respectively. In sequential dispatch mode, the instruction cache and branch unit attempted to dispatch six instructions per cycle, and the branch unit inspected an additional two more instructions to look ahead for branches. In target dispatch mode, the instruction cache and branch unit prepared to dispatch up to four integer and floating-point instructions by placing them on the bus to the integer and floating-point units. This latter mode did not conditionally dispatch but did reduce the branch penalty for taken branches by up to two cycles. There were also two independent branch stations that could evaluate branch conditions and generate the necessary target addresses for the next target fetch. The major benefit of using two branch units was to calculate and prefetch the target address of a second branch that follows a *resolved-untaken first branch;* only the untaken-path instructions beyond one unresolved branch (either first or second) could be conditionally dispatched; a second unresolved branch stopped dispatch.

There were two integer units. Each had its own copy of the integer register file, and the hardware maintained consistency. Each unit could execute simple integer operations, including loads and stores. Cache control and privileged instructions were executed on the first unit, while the second unit executed multiplies and divides. The second unit provided integer multiplies in two cycles and could also execute two *dependent* add instructions in one cycle.

The integer units also handled load/stores. The data cache, including the directory, was fully dual-ported. In fact, the cache ran three times faster than the normal clock cycle time; each integer unit got a turn, sometimes in reverse order to allow a read to go first, and then the cache refill got a turn.

There were also two floating-point units, each of which could execute fused multiply-add in two cycles. A buffer in front of each floating-point ALU allowed one long-running instruction and one dependent instruction to be assigned to one of the ALUs, while other independent instructions subsequent to the dependent pair could be issued out of order to the second ALU. Each ALU had multiple bypasses to the other; however, only normalized floating-point numbers could be routed along these bypasses. Numbers that were denormalized or special-valued (e.g., not a number (NaN), infinity) had to be handled via the register file.

Arithmetic exceptions were imprecise on the POWER2, so the only precise-interrupt-generating instructions were load/stores and integer traps. The floating-point unit and the integer unit had to be synchronized whenever an interrupt-generating instruction was issued.

See Barreh et al. [1994], Shippy [1994], Weiss and Smith [1994], and White [1994] for more information on the POWER2. A single-chip POWER2 implementation is called the P2SC.

8.3.5 Intel i960

The i960 architecture was announced by Intel in 1988 as a RISC design for the embedded systems market. The basic architecture is integer-only and has 32 registers. The registers are divided into 16 global registers and 16 local registers, the latter of which are windowed in a nonoverlapping manner on procedure call/return. A numerics extension to the architecture provides for single-precision, double-precision, and extended-precision floating point; in this case, four 80-bit registers are added to the programming model. The i960 chief architect was Glen Myers.

The comparison operations in the i960 were carefully designed for pipelined implementations:

- A conditional compare instruction is available for use after a standard compare. The conditional compare does not execute if the first compare is true. This allows range checking to be implemented with only one conditional branch.

- A compare instruction with increment/decrement is provided for fast loop closing.

- A combined compare and branch instruction is provided for cases where an independent instruction cannot be scheduled by the compiler into the delay slot between a normal compare instruction and the subsequent conditional branch instruction.

The opcode name space was also carefully allocated so that the first 3 bits of an instruction easily distinguish between control instructions (*C-type*), register-to-register integer instructions (*R-type*), and load/store instructions (*M-type*). Thus dispatching to different function units can occur quickly, and pre-decode bits are unnecessary.

8.3.5.1 i960 CA / 1989. The i960 CA was introduced in 1989 and was the first superscalar microprocessor. It is unique among the early superscalar designs in that it is still available as a product today. Chief designers were Glenn Hinton and Frank Smith. The CA model includes on chip: an interrupt controller, a DMA controller, a bus controller, and 1.5K bytes of memory, which can be partially allocated for the register window stack and the remaining part used for low-latency memory. There are three units: instruction-fetch/branch (*instruction sequencer*), integer (*register side*), and address generation (*memory side*). The integer unit includes a single-cycle integer ALU and a pipelined multiplier/divider. The address generation unit controls access to memory and also handles accesses to the on-chip memory. The i960 CA pipeline is shown in Figure 8.9.

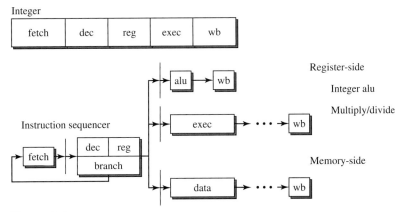

Figure 8.9
Intel i960 CA Pipeline Stages.

The decoder fetches four instructions as quickly as every two cycles and attempts to issue according to the following rules:

1. The first instruction in the four slots is issued if possible.
2. If the first instruction is a register-side instruction, that is, if it is neither a memory-side instruction nor a control instruction, then the second instruction is examined. If it is a memory-side instruction, then it is issued if possible.
3. If either one or two instructions have been issued and neither one was a control instruction, then all remaining instructions are examined. The first control instruction that is found is issued.

Thus, after a new instruction quadword has been fetched, there are nine possibilities for issue.

3-Issue	2-Issue, No R	2-Issue, No M	2-Issue, No C
R M C x	M C x x	R C x x	R M x x
R M x C	M x C x	R x C x	
	M x x C	R x x C	

(Here x represents an instruction that is not issued.)

Notice that in the lower rows of the table the control instruction is executed early, in an out-of-order manner. However, the instruction sequencer retains the current instruction quadword until all instructions have been issued. Thus, while the peak issue rate is three instructions in a given cycle, the maximum sustained issue rate is two per cycle. The instruction ordering constraint in the issue rules has been criticized as irregular and has been avoided by most other designers. However, the M-type load-effective-address (lda) instruction is general enough so that in many cases a pair of integer instructions can be *migrated* by an instruction scheduler or peephole optimizer into an equivalent pair of one integer instruction and one lda instruction. (See also the MM model, described in Section 8.3.5.2, which provides hardware-based instruction migration.)

The i960 CA includes a 1K-byte, two-way set-associative instruction cache. The set-associative design allows either one or both banks to be loaded (via a special instruction) with time-critical software routines and locked to prevent instruction cache misses. Moreover, a speculative memory fetch is started for each branch target address in anticipation of an instruction cache miss; this reduces the instruction cache miss penalty. Recovery from exceptions can be handled either by software or by use of an exception barrier instruction.

Please see Hinton [1989] and McGeady [1990a, b] for more details of the i960 CA. U.S. Statutory Invention Registration H1291 also describes the i960 CA.

8.3.5.2 Other Models of the i960. The i960 MM was introduced for military applications in 1991 and included both a 2K-byte instruction cache and a 2K-byte data cache on chip [McGeady et al., 1991]. The decoder automatically rewrote second integer instructions as equivalent lda instructions where possible. The MM model also included a floating-point unit to implement the numerics extension for the architecture. The CF model was announced in 1992 with a 4K-byte, two-way set-associative instruction cache and a 1K-byte, direct-mapped data cache.

The i960 Hx series of models provides a 16K-byte four-way set-associative instruction cache, an 8K-byte four-way set-associative data cache, and 2K-bytes of on-chip RAM.

8.3.6 Intel IA32—Native Approaches

The Intel IA32 is probably the most widely used architecture to be developed in the 50+ years of electronic computer history. Although its roots trace back to the 8080 8-bit microprocessor designed by Stanley Mazor, Federico Faggin, and Masatoshi Shima in 1973, Intel introduced the 32-bit computing model with the 386 in 1990 (the design was led by John Crawford and Patrick Gelsinger). The follow-on 486 (designed by John Crawford) integrated the FPU on chip and also used extensive pipelining to achieve single-cycle execution for many of the instructions. The next design, the Pentium, was the first superscalar implementation of IA32 brought to market.

Overall, the IA32 design efforts can be classified into whole-instruction (called *native*) approaches, like the Pentium, and decoupled microarchitecture approaches, like the P6 core and Pentium 4. Processors using the native approach are examined in this section.

8.3.6.1 Intel Pentium / 1993. Rather than being called the 586, the Pentium name was selected for trademark purposes. Actually there were a series of Pentium implementations, with different feature sizes, power management techniques, and clock speeds. The last implementation added the MMX multimedia instruction set extension [Peleg and Weiser, 1996; Lempel et al., 1997]. The Pentium chief architect was Don Alpert, who was assisted by Jack Mills, Bob Dreyer, Ed Grochowski, and Uri Weiser. Weiser led the initial design study in Israel that framed the P5 as a dual-issue processor, and Mills was instrumental in gaining final management approval of dual integer pipelines.

The Pentium was designed around two integer pipelines, U and V, that operated in lockstep manner. (The exception to this lockstep operation was that a paired instruction could stall in the execute stage of V without stalling the instruction in U.) The stages of these pipelines were very similar to the stages of the 486 pipeline. The first decode stage determined the instruction lengths and checked for dual issue. The second decode stage calculated the effective memory address so that the execute stage could access the data cache; this stage differed from its 486 counterpart in its ability to read both an index register and a base register in the same cycle and its ability to handle both a displacement and an immediate in the same cycle. The execute stage performed arithmetic and logical operations in one cycle if all operands were in registers, but it required multiple cycles for more complex instructions. For example, a common type of complex instruction, add register to memory, required three cycles in the execute stage: one to read the memory operand from cache, one to execute the add, and one to store the result back to cache. However, if two instructions of this form (read-modify-write) were paired, there were two additional stall cycles. Some integer instructions, such as shift, rotate, and add with carry, could only be performed in the U pipeline. Integer multiply was done by the floating-point pipeline, attached to the U pipeline, and stalled the pipelines for 10 cycles. The pipeline is illustrated in Figure 8.10.

Single issue down the U pipeline occurred for (1) complex instructions, including floating-point; (2) when the first instruction of a possible dual-issue pair was a branch; or (3) when the second instruction of a possible pair was dependent on the first (although WAR dependences were not checked and did not limit dual issue). Complex instructions generated a sequence of control words from a microcode sequencer in the D1 stage to control the pipelines for several cycles. There was special handling for pairing a flag-setting instruction with a dependent conditional branch, for pairing two push or two pop instructions in sequence (helpful for procedure entry/exit), and for pairing a floating-point stack register exchange (FXCH) and a floating-point arithmetic instruction.

The data cache on the Pentium was interleaved eight ways on 4-byte boundaries, with true dual porting of the TLB and tags. This allowed the U and V pipelines to access 32-bit doublewords from the data cache in parallel as long as there

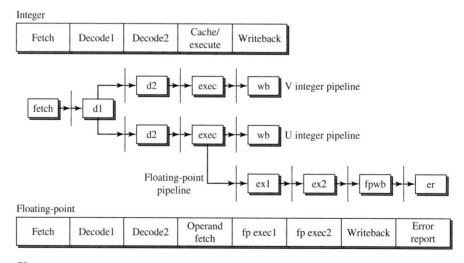

Figure 8.10

Intel Pentium Pipeline Stages.

was no bank conflict. The cache did not allocate lines on write misses, and thus dummy reads were sometimes inserted before a set of sequential writes as a compiler or hand-coded optimization.

The floating-point data paths were 80 bits wide, directly supporting extended precision operations, and the delayed exception model of the 486 was changed to predicting exceptions (called *safe instruction recognition*). The single-issue of floating-point instructions on the Pentium was not as restrictive a constraint as it would be on a RISC architecture, since floating-point instructions on the IA32 are allowed to have a memory operand. The cache design also supported double-precision floating-point loads and stores by using the U and V pipes in parallel to access the upper 32 bits and the lower 32 bits of the 64-bit double-precision value.

Branches were predicted in the fetch stage by use of a 256-entry BTB; each entry held the branch instruction address, target address, and two bits of history.

For more information on the original Pentium, see Alpert and Avnon [1993]. The Pentium MMX design included 57 multimedia instructions, larger caches, and a better branch prediction scheme (derived from the P6). Instruction-length decoding was also done in a separate pipeline stage [Eden and Kagan, 1997].

8.3.6.2 Cyrix 6x86 (M1) / 1994. Mark Bluhm and Ty Garibay led a design effort at Cyrix to improve on a Pentium-like dual-pipeline design. Their design, illustrated in Figure 8.11, was called the 6x86, and included register renaming (32 physical registers), forwarding paths across the dual pipelines (called X and Y in the 6x86), decoupled pipeline execution, the ability to swap instructions between the pipelines after the decode stages to support dynamic load balancing and stall avoidance, the ability to dual issue a larger variety of instruction pairs, the use of an eight-entry address stack to support branch prediction of return addresses, and

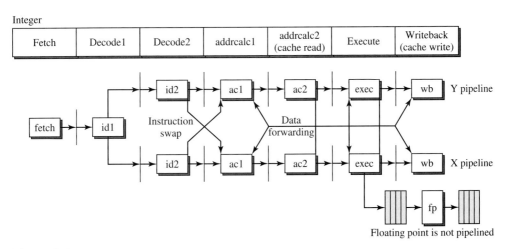

Figure 8.11
Cyrix 6x86 Pipeline Stages.

speculative execution past any combination of up to four branches and floating-point instructions. They also made several design decisions to better support older 16-bit IA32 programs as well as mixed 32-bit/16-bit programs.

The X and Y pipelines of the 6x86 had seven stages, compared with five stages of the U and V pipelines in the Pentium. The D1 and D2 stages of the Pentium were split into two stages each in the 6x86: two instruction decode stages and two address calculation stages. Instructions were obtained by using two 16-byte aligned fetches; even in the worst case of instruction placement, this provided at least 9 bytes per cycle for decoding. The first instruction decode stage identified the boundaries for up to two instructions, and the second decode stage identified the operands. The second decode stage of the 6x86 was optimized for processing instruction prefix bytes.

The first address calculation stage renamed the registers and flags and performed address calculation. To allow the address calculation to start as early as possible, it was overlapped with register renaming; thus, the address adder had to access values from a logical register file while the execute stage ALU had to access values from the renamed physical register file. A register scoreboard was used to track any pending updates to the logical registers used in address calculation and enforced address generation interlocks (AGIs). Coherence between the logical and physical copies of a register that was updated during address calculation, such as the stack pointer, was specially handled by the hardware. With this support, two pushes or two pops could be executed simultaneously. The segment registers were not renamed, but a segment register scoreboard was maintained by this stage and stalled dependent instructions until segment register updates were complete.

The second address calculation stage performed address translation and cache access for memory operands (the Pentium did this in the execute stage). Memory access exceptions were also handled by this stage, so certain instructions for which exceptions cannot be easily predicted had to be singly issued and executed serially from that point in the pipelines. An example of this is the return instruction in which a target address that is popped off the stack leads to a segmentation exception. For instructions with results going to memory, the cache access occurred in the writeback stage (the Pentium wrote memory results in the execute stage).

The 6x86 handled branches in the X pipeline and, as in the Pentium, special instruction pairing was provided for a compare and a dependent conditional jump. Unlike the Pentium, the 6x86 had the ability to dual issue a predicted-untaken branch in the X pipeline along with its fall-through instruction in the Y pipeline. The 6x86 also used a checkpoint-repair approach to allow speculative execution past predicted branches and floating-point instructions. Four levels of checkpoint storage were provided.

Memory-accessing instructions were typically routed to the Y pipeline, so that the X pipeline could continue to be used should a cache miss occur. The 6x86 provided special support to the repeat and string move instruction combination in which the resources of both pipelines were used to allow the move instruction to attain a speed of one cycle per iteration. Because of the forwarding paths between the X and Y pipelines (which were not present between the U and V pipelines on the Pentium) and because cache access occurred outside the execute stage, dependent instruction pairs could be dual-issued on the 6x86 when they had the form of a move memory-to-register instruction paired with an arithmetic instruction using that register, or the form of an arithmetic operation writing to a register paired with a move register-to-memory instruction using that register.

Floating-point instructions were handled by the X pipeline and placed into a four-entry floating-point instruction queue. At the point a floating-point instruction was known not to cause a memory-access fault, a checkpoint was made and instruction issue continued. This allowed the floating-point instruction to execute out of order. The 6x86 could also dual issue a floating-point instruction along with an integer instruction. However, in contrast to the Pentium's support of 80-bit extended precision floating point, the data path in the 6x86 floating-point unit was 64 bits wide and not pipelined. Also FXCH instructions could not be dual-issued with other floating-point instructions.

Cyrix chose to use a 256-byte fully associative instruction cache and a 16K-byte unified cache (four-way set-associative). The unified cache was 16-way interleaved (on 16-bit boundaries to provide better support for 16-bit code) and provided dual-ported access similar to the Pentium's data cache. Load bypass and load forwarding were also supported.

See Burkhardt [1994], Gwennap [1993], and Ryan [1994a] for overviews of the 6x86, which was then called the M1. McMahan et al. [1995] provide more details of the 6x86. A follow-on design, known as the M2 or 6x86MX, was designed by Doug Beard and Dan Green. It incorporated MMX instruction set extensions as well as increasing the unified cache to 64K bytes.

Cyrix continued to use the dual-pipeline native approach in several subsequent chip designs, and small improvements were made. For example, the Cayenne core allowed FXCH to dual issue in the FP/MMX unit. However, a decoupled design was started by Ty Garibay and Mike Shebanow in the mid-1990s and came to be called the Jalapeno core (also known as Mojave). Greg Grohoski took over as chief architect of this core in 1997. In 1999, Via bought both Cyrix and Centaur (designers of the WinChip series), and by mid-2000 the Cyrix design efforts were canceled.

8.3.7 Intel IA32—Decoupled Approaches

Decoupled efforts at building IA32 processors began at least in 1989, when NexGen publicly described its efforts for the F86 (later called the Nx586 and Nx686, and which became the AMD K6 product line). These efforts were influenced by the work of Yale Patt and his students on the high-performance substrate (HPS).

In this section, two outwardly scalar, but internally superscalar efforts are discussed first, and then the superscalar AMD and Intel designs are presented.

8.3.7.1 NexGen Nx586 (F86) / 1994.
The Nx586 appeared on the outside to be a scalar processor; however, internally it was a decoupled microarchitecture that operated in a superscalar manner. The Nx586 translated one IA32 instruction per cycle into one or more RISC86 instructions and then dispatched the RISC-like instructions to three function units: integer with multiply/divide, integer, and address generation.

Each function unit on the Nx586 had a 14-entry reservation station, where RISC86 instructions spent at least one cycle for renaming. Each reservation station also operated in a FIFO manner, but out-of-order issue of the RISC86 instructions could occur across function units. The major drawback to this arrangement is that if the first instruction in a reservation station must stall due to a data dependency, the complete reservation station is stalled.

The Nx586 required a separate FPU chip for floating-point, but it included two 16K-byte on-chip caches, dynamic branch prediction using an adaptive branch predictor, speculative execution, and register renaming using 22 physical registers. NexGen worked on this basic design for several years; three preliminary articles were presented at the 1989 Spring COMPCON on what was then called the F86. Later information can be found in Ryan [1994b]. Mack McFarland was the first NexGen architect, then Dave Stiles and Greg Favor worked on the Nx586, and later Korbin Van Dyke oversaw the actual implementation.

8.3.7.2 WinChip Series / 1997 to Present.
Glenn Henry has been working on an outwardly scalar, internally superscalar approach, similar to the NexGen effort, for the past decade. Although only one IA32 instruction can be decoded per cycle, dual issue of translated micro-operations is possible. One item of interest in the WinChip approach is that a load-ALU-store combination is represented as one micro-operation. See Diefendorff [1998] for a description of the 11-pipe-stage WinChip 4. Via is currently shipping the C3, which has a 16-pipe-stage core known as Nehemiah.

8.3.7.3 AMD K5 / 1995.
The lead architect of the K5 was Mike Johnson, whose 1989 Stanford Ph.D. dissertation was published as the first superscalar microprocessor design textbook [Johnson, 1991]. The K5 followed many of the design suggestions in his book, which was based on a superscalar AMD 29000 design effort.

In the K5, IA32 instructions were fetched from memory and placed into a 16K-byte instruction cache with additional pre-decode bits to assist in locating instruction fields and boundaries (see Figure 4.15). On each cycle up to 16 bytes were fetched from the instruction cache, based on the branch prediction scheme detailed in Johnson's book, and merged into a byte queue. According to this scheme, there could only be one branch predicted to be taken per cache line, so the cache lines were limited to 16 bytes to avoid conflicts among taken branches. Each cache line was initially marked as fall-through, and the marking was changed on each misprediction. The effect was about the same as using one history bit per cache line.

As part of filling a line in the instruction cache, each IA32 instruction was tagged with the number of micro-instructions (R-ops) that would be produced.

These tags acted as repetition numbers so that a corresponding number of decoders could be assigned to decode the instructions. In this manner, an IA32 instruction could be routed to one or more two-stage decoders without having to wait for control logic to propagate instruction alignment information across the decoders. The tradeoff is the increased instruction cache refill time required by the pre-decoding. There were four decoders in the K5, and each could produce one R-op per cycle. An interesting aspect of this process is that, depending on the sequential assignment of instruction + tag packets to decoders, the R-ops for one instruction might be split into different decoding cycles. Complex instructions overrode the normal decoding process and caused a stream of four R-ops per cycle to be fetched from a control store.

R-ops were renamed and then dispatched to six execution units: two integer units, two load/store units, a floating-point unit, and a branch unit. Each execution unit had a two-entry reservation station, with the exception that the floating-point reservation station had only one entry. Each reservation station could issue one R-op per cycle. With two entries in the branch reservation station, the K5 could speculatively execute past two unresolved branches.

R-ops completed and wrote their results into a 16-entry reorder buffer; up to four results could be retired per cycle. The reservation stations and reorder buffer entries handled mixed operand sizes (8, 16, and 32 bits) by treating each IA32 register as three separate items (low byte, high byte, and extended bytes). Each item had separate dependency-checking logic and an individual renaming tag.

The K5 had an 8K-byte dual-ported/four-bank data cache. As in most processors, stores were written upon R-op retirement. Unlike other processors, the refill for a load miss was not started until the load R-op became the oldest R-op in the reorder buffer. This choice was made to avoid incorrect accesses to memory-mapped I/O device registers. Starting the refills earlier would have required special case logic to handle the device registers.

The K5 was a performance disappointment, allegedly from design decisions made without proper workload information from Windows 3.x applications. An agreement with Compaq in 1995 to supply K5 chips fell through, and AMD bought NexGen (see Section 8.3.7.4). See Gwennap [1994], Halfhill [1994a], and Christie [1996] for more information on the design.

8.3.7.4 AMD K6 (NexGen Nx686) / 1996.

In 1995, AMD acquired NexGen and announced that the follow-on design to the Nx586 would be marketed as the AMD K6. That design, called the Nx686 and done by Greg Favor, extended the Nx586 design by integrating a floating-point unit as well as a multimedia operation unit onto the chip. The caches were enlarged, and the decode rate was doubled.

The K6 had three types of decoders operating in a mutually exclusive manner. There was a pair of *short decoders* that decoded one IA32 instruction each. These could produce one or two RISC86 operations each. There was an alternate *long decoder* that could handle a single, more complex IA32 instruction and produce up to four RISC86 operations. Finally, there was a *vector decoder* that provided an initial RISC86 operation group and then began streaming groups of RISC86 operations from a control store. The result was a maximum rate of four RISC86 operations per cycle from one of the three decoder types. Pre-decode bits assisted the K6 decoders, similar to the approach in the K5.

The K6 dispatched RISC86 operations into a 24-entry centralized reservation station, from which up to six RISC86 operations issued per cycle. The eight IA32 registers used in the instructions were renamed using 48 physical registers. Branch support included an 8192-entry BHT implementing adaptive branch prediction according to a global/adaptive/set (GAs) scheme. A 9-bit global branch history shift register and 4 bits from the instruction pointer were used to identify one out of the 8192 saturating 2-bit counters. There was also a 16-entry target instruction cache (16 bytes per line) and a 16-entry return address stack.

The data cache on the K6 ran twice per cycle to give the appearance of dual porting for one load and one store per cycle; this was chosen rather than banking in order to avoid dealing with bank conflicts.

See Halfhill [1996a] for a description of the K6. Shriver and Smith [1998] have written a book-length, in-depth case study of the K6-III.

8.3.7.5 AMD Athlon (K7) / 1999. Dirk Meyer and Fred Weber were the chief architects of the K7, later branded as the Athlon. The Athlon uses some of the same approaches as the K5 and K6; however, the most striking differences are in the deeper pipelining, the use of MacroOps, and special handling of floating-point/multimedia instructions as distinct from integer instructions. Stephan Meier led the floating-point part of the design.

The front-end, in-order pipeline for the Athlon consists of six stages (through dispatch). Branch prediction for the front end is handled by a 2048-entry BHT, a 2048-entry BTAC, and a 12-entry return stack. This scheme is simpler than the two-level adaptive scheme used in the K6. Decoding is performed by three *DirectPath* decoders that can produce one MacroOp each, or, for complex instructions, by a *VectorPath* decoder that sequences three MacroOps per cycle out of a control store. As in the K5 and K6, pre-decode bits assist in the decoding.

A *MacroOp* is a representation of an IA32 instruction of up to moderate complexity. A MacroOp is fixed length but can contain one or two *Ops*. For the integer pipeline, Ops can be of six types: load, store, combined load-store, address generation, ALU, and multiply. Thus, register-to-memory as well as memory-to-register IA32 instructions can be represented by a single MacroOp. For the floating-point pipeline, Ops can be of three types: multiply, add, or miscellaneous. The advantage of using MacroOps is the reduced number of buffer entries needed.

During the dispatch stage, MacroOps are placed in a 72-entry reorder buffer called the *instruction control unit* (ICU). This buffer is organized into 24 lines of three slots each, and the rest of the pipelines follow this three-slot organization. The integer pipelines are organized symmetrically with both an address generation unit and an integer function unit connected to each slot. Integer multiply is the only asymmetric integer instruction; it must be placed in the first slot since the integer multiply unit is attached to the first integer function unit. Floating-point and multimedia (MMX/3DNow! and later SSE) instructions have more restrictive slotting constraints.

From the ICU, MacroOps are placed either into the integer scheduler (18 entries, organized as six lines of three slots each) or the floating-point/multimedia scheduler (36 entries, organized as 12 lines of three slots each). The schedulers can schedule Ops individually and out of order, so that a MacroOp remains in the scheduler buffer until all its Ops are completed.

On the integer side, load and store Ops are sent to a 44-entry load/store queue for processing; the combined load-store Op remains in the load/store queue after the load is complete until the value to store is forwarded across a result bus; at that point it is ready to act as a store Op. The integer side also uses a 24-entry *integer future file and register file* (IFFRF). Integer operands or tags are read from this unit during dispatch, and integer results are written into this unit and the ICU upon completion. The ICU performs the update of the architected integer registers when the MacroOp retires.

Because of the IA32 floating-point stack model and the width of XMM registers, MacroOps sent to the floating-point/multimedia side are handled in a special manner and require additional pipeline stages. Rather than reading operands or tags at dispatch, floating-point/multimedia register references are later renamed using 88 physical registers. This occurs in three steps: first, in stage 7, stack register references are renamed into a linear map; second, in stage 8, these mapped references are renamed onto the physical registers; then, in stage 9, the renamed MacroOps are stored in the floating-point/multimedia scheduler. Because the operands are not read at dispatch on this side, an extra stage for reading the physical registers is needed. Thus, floating-point/multimedia execution does not start until stage 12 of the Athlon pipeline.

The Athlon has on-chip 64K-byte L1 caches and initially contained a controller for an off-chip L2 of up to 8 Mbytes with MOESI cache coherence. Later shrinks allowed for on-chip L2 caches. The L1 data cache is multibanked and supports two loads or stores per cycle. AMD licensed the Alpha 21264 bus, and the Athlon contains an on-chip bus controller.

See Diefendorff [1998] for a description of the Athlon.

8.3.7.6 Intel P6 Core (Pentium Pro / Pentium II / Pentium III) / 1996. The Intel P6 is discussed in depth in Chapter 7. The P6 core design team included Bob Colwell as chief architect, Glenn Hinton and Dave Papworth as senior architects, along with Michael Fetterman and Andy Glew. Figure 8.12 illustrates the P6 pipeline.

The P6 was Intel's first use of a decoupled microarchitecture that decomposed IA32 instructions. Intel calls the translated micro-instructions μops. An eight-stage fetch and translate pipeline allocates entries for the μops in a 40-entry reorder buffer and a 20-entry reservation station. Limitations of the IA32 floating-point stack model are removed by allowing FXCH (exchange) instructions to be inserted directly into the reorder buffer and tagged as complete after they are processed by the renaming hardware. These instructions never occupy reservation station slots.

Because of transistor count limitations on the instruction cache and the problem of branching to what a pre-decoder has marked as an interior byte of an instruction, extra pre-decode bits were rejected. Instead, fetch stages mark the

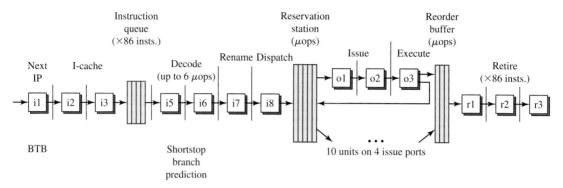

Figure 8.12
Intel P6 Pipeline Stages.

instruction boundaries for decoding. Up to three IA32 instructions can be decoded in parallel; but to obtain this maximum decoding effectiveness, the instructions must be arranged so that only the first one can generate multiple µops (up to four) while the other two instructions behind it must each generate one µop only. Instructions with operands in memory require multiple µops and therefore limit the decoding rate to one IA32 instruction per cycle. Extremely complex IA32 instructions (e.g., PUSHA) require that a long sequence of µops be fetched from a control store and dispatched into the processor over several cycles. Prefix bytes and a combination of both an immediate operand and a displacement addressing mode in the same instruction are also quite disruptive to decoding.

The reservation station is scanned in a FIFO-like manner each cycle in an attempt to issue up to four µops to five issue ports. *Issue ports* are a collection of two read ports and one write port to and from the reservation station. One issue port supports wide data paths and has six execution units of various types attached: integer, floating-point add, floating-point multiply, integer divide, floating-point divide, and integer shift. A second issue port handles µops for a second integer unit and a branch unit. The third port is dedicated to loads, while the fourth and fifth ports are dedicated to stores. In scanning of the reservation station, preference is given to back-to-back µops to increase the amount of operand forwarding among the execution units.

Branch handling uses a two-level adaptive branch predictor, and to assist branching when a branch is not found by a BTB lookup, a decoder *shortstop* in the middle of the front-end pipeline predicts the branch based on the sign of the displacement. A conditional move has been added to the IA32 ISA to help avoid some conditional branches, and new instructions to move the floating-point condition codes to the integer condition codes have also been added.

The Pentium Pro was introduced at 133 MHz, but the almost immediate availability of the 200-MHz version caught many in the industry by surprise, since its performance on the SPEC integer benchmarks exceeded even that of the contemporary 300-MHz DEC Alpha 21164. In the personal computer marketplace, the performance results were less dramatic. The designers traded off some 16-bit code performance (i.e., virtual 8086) to provide the best 32-bit performance possible; thus they chose to serialize the machine on such instructions as far calls and other segment register switches. While small-model 16-bit code runs well, large-model code (e.g., DOS and Windows programs) runs at only Pentium speed or worse.

See Gwennap [1995] and Halfhill [1995] for additional descriptions of the P6. Papworth [1996] discusses design tradeoffs made in the microarchitecture.

8.3.7.7 Intel Pentium 4 / 2001. The Pentium 4 design, led by Glenn Hinton, takes the same decoupled approach as in the P6 core, but the Pentium 4 looks even more like Yale Patt's HPS proposal by including a decoded instruction cache. This cache, called the *trace cache*, is organized to hold 2048 lines of six μops each in trace order, that is, with branch target μops placed immediately next to predict-taken branch μops. The fetch bandwidth of the trace cache is three μops per cycle.

The Pentium 4 is much more deeply pipelined than the P6 core, resulting in 30 or more pipeline stages. The number and actions of the back-end stages have not yet been disclosed, but the branch misprediction pipeline has been described. It has 20 stages from starting a trace cache access on a mispredicted path to restarting the trace cache on the correct path. The equivalent length in the P6 core is 10 stages.

There are two branch predictors in the Pentium 4, one for the front end of the pipeline and a smaller one for the trace cache itself. The front-end BTB has 4096 entries and reportedly uses a hybrid prediction scheme. If a branch misses in this structure, the front-end pipe stages will predict it based on the sign of the displacement, similar to the P6 shortstop.

Because the trace cache eliminates the need to re-decode recently executed IA32 instructions, the Pentium 4 uses a single decoder in its front end. Thus, compared to the P6 core, the Pentium 4 might take a few more cycles for decoding the first visit to a code segment but will be more efficient on subsequent visits. To maximize the hit rate in the trace cache, the Pentium 4 optimization manual advises against overuse of FXCH instructions (whereas P6 optimization encourages their use; see Section 8.3.7.6). Excessive loop unrolling should be avoided for the same reason.

Another difference between the two Intel designs is that the Pentium 4 does not store source and result values in the reservation stations and reorder buffer. Instead, it uses 128 physical registers for renaming the architected integer registers and a second set of 128 physical registers for renaming the floating-point stack and XMM registers. A front-end register alias table is used along with a retirement register alias table to keep track of the lookahead and retirement states.

μops are dispatched into two queues: one for memory operations and one for other operations. There are four issue ports, two of which handle load/stores and two of which handle the other operations. These latter two ports have multiple schedulers examining the μop queue and arbitrating for issue permission on the two ports. Some of these schedulers can issue one μop per cycle, but other *fast schedulers* can issue ALU μops twice per cycle. This double issue is because the integer ALUs are pipelined to operate in three half-cycles, with two half-cycle stages handling 16 bits of the operation each and the third half-cycle stage setting the flags. The overall effect is that the integer ALUs have one-half cycle effective latencies. Because of this staggered structure, dependent μops can be issued back-to-back in half cycles.

The Pentium 4 reorder buffer has 128 entries, and 126 μops can be in flight. The processor also has a 48-entry load queue and a 24-entry store queue, so that 72 of the 126 μops can be load/stores. The L1 data cache provides two-cycle latency for integer values and six-cycle latency for floating-point values. This cache also supports one load and one store per cycle. The schedulers speculatively issue μops that are dependent on loads so that the loaded values can be immediately forwarded to the dependent μops. However, if a load has a cache miss, the dependent μops must be replayed.

See Hinton et al. [2000] for more details of the Pentium 4.

8.3.7.8 Intel Pentium M / 2003. The Pentium M design was led by Simcha Gochman and is a low-power revision of the P6 core. The Intel team in Israel started their revision by adding streaming SIMD extensions (SSE2) and the Pentium 4 branch predictor to the basic P6 microarchitecture. They also extended branch prediction in two ways. The first is a loop detector that captures and stores loop counts in a set of hardware counters; this leads to perfect branch prediction of for-loops. The second extension is an adaptive indirect-branch prediction scheme that is designed for data-dependent indirect branches, such as are found in a bytecode interpreter. Mispredicted indirect branches are allocated new table entries in locations corresponding to the current global branch history shift register contents. Thus, the global history can be used to choose one predictor from among many possible instances of predictors for a data-dependent indirect branch.

The Pentium M team made two other changes to the P6 core. The first is that the IA32 instruction decoders have been redesigned to produce single, fused μops for load-and-operate and store instructions. In the P6 these instruction types can be decoded only by the complex decoder and result in two μops each. In the Pentium M they can be handled by any of the three decoders, and each type is now allocated a single reservation station entry and ROB entry. However, the μop scheduling logic recognizes and treats a fused-μop entry as two separate μops, so that the execution pipelines remain virtually the same. The retirement logic also recognizes a fused-μop entry as requiring two completions before retirement (compare with AMD Athlon MacroOps). A major benefit of this approach is a 10% reduction in the number of μops handled by the front-end and rear-end pipeline stages and consequent power savings. However, the team also reports a 5% increase in performance for integer code and 9% increase for floating-point code. This is due to the increased decoding bandwidth and to less contention for reservation station and ROB entries.

Another change made in the Pentium M is the addition of register tracking logic for the hardware stack pointer (ESP). The stack pointer updates that are required for push, pop, call, and return are done using dedicated logic and a dedicated adder in the front end, rather than sending a stack pointer adjustment μop through the execution pipelines for each update. Address offsets from the stack pointer are adjusted as needed for load and store μops that reference the stack, and a history buffer records the speculative stack pointer updates in case of a branch mispredict or exception (compare with the Cyrix 6x86 stack pointer tracking).

See Gochman et al. [2003] for more details.

8.3.8 x86-64

AMD has proposed a 64-bit extension to the x86 (Intel IA32) architecture. Chief architects of this extension were Kevin McGrath and Dave Christie. In the x86-64, compatibility with IA32 is paramount. The existing eight IA32 registers are extended to 64 bits in width, and eight more general registers are added. Also the SSE and SSE2 register set is doubled from 8 to 16 in size. See McGrath [2000] for a presentation of the x86-64 architecture.

8.3.8.1 AMD Opteron (K8) / 2003.
The first processor supporting the extended architecture is the AMD Opteron. An initial K8 project was led by Jim Keller but was canceled. The Opteron processor brought to market is an adaptation of the Athlon design, and this work was led by Fred Weber.

As compared to the Athlon (see Section 8.3.7.5), the Opteron retains the same three-slotted pipeline organization. The three regular decoders, now called *FastPath*, can handle more of the multimedia instructions without having to resort to the VectorPath decoder. The Opteron has two more front-end pipe stages than the Athlon (and fewer pre-decode bits in the instruction cache), so that integer instructions start execution in stage 10 rather than 8, and floating-point/multimedia instructions start in stage 14 rather than 12. Branch prediction is enhanced by enlarging the BHT to 16K entries. The integer scheduler and IFFRF sizes are increased to 24 and 40 entries, respectively, and the number of floating-point/multimedia physical registers is increased to 120.

The Opteron chip also contains three HyperTransport links for multiprocessor interconnection and an on-chip controller that integrates many of the normal Northbridge chip functions.

See Keltcher et al. [2003] for more information on the Opteron.

8.3.9 MIPS

The MIPS architecture is the quintessential RISC. It originated in research work on noninterlocked pipelines by John Hennessy of Stanford University, and the first design by the MIPS company included a noninterlocked load delay slot. The MIPS-I and -II architectures were defined in 1986 and 1990, respectively, by Craig Hansen. Earl Killian was the 64-bit MIPS-III architect in 1991. Peter Hsu started the MIPS-IV extensions at SGI prior to the SGI/MIPS merger; the R8000 and R10000/12000 implement MIPS-IV. MIPS-V was finalized by Earl Killian in 1995, with input from Bill Huffman and Peter Hsu, and includes the MIPS digital media extensions (MDMX).

MIPS is known for clean, fast pipeline design, and the R4000 designers (Peter Davies, Earl Killian, and Tom Riordan) chose to introduce a superpipelined (yet single-cycle ALU) processor in 1992 when most other companies were choosing a superscalar approach. Through simulation, the superpipeline design performed better on unrecompiled integer codes than a competing in-house superscalar design. This was because the superscalar required multiple integer units to issue two integer instructions per cycle but lacked the ability to issue dependent integer instruction pairs in the same cycle. In contrast, the superpipelined design ran the clock twice as fast and, by use of a single fast-cycle ALU, could issue the dependent integer instruction pair in only two of the fast cycles [Mirapuri et al., 1992]. It is interesting to compare this approach with the issue of dependent instructions using cascaded ALUs in the SuperSPARC, also a 1992 design. Also, the fast ALU idea is helpful to the Pentium 4 design.

8.3.9.1 MIPS R8000 (TFP) / 1994. The MIPS R8000 was superscalar but not superpipelined; this might seem an anomaly, and indeed, the R8000 was actually the final name for the tremendous floating-point (TFP) design that was started at Silicon Graphics by Peter Hsu. The R8000 was a 64-bit machine aimed at floating-point computation and seems in some ways a reaction to the IBM POWER. However, many of the main ideas in the R8000's design were inspired by the Cydrome Cydra-5. Peter Hsu was an alumnus of Cydrome, as were some of his design team: Ross Towle, John Brennan, and Jim Dehnert. Hsu also hired Paul Rodman and John Ruttenberg, who were formerly with Multiflow.

The R8000 is unique in separating floating-point data from integer data and addresses. The latter could be loaded into the on-chip 16K-byte cache, but floating-point data could not. This decision was made in an effort to prevent the poor temporal locality of floating-point data in many programs from rendering the on-chip cache ineffective. Instead the R8000 provided floating-point memory bandwidth using a large second-level cache that is two-way interleaved and has a five-stage access pipeline (two stages of which were included for chip crossings). Bank conflict was reduced by the help of a one-entry *address bellow;* this provided for reordering of cache accesses to increase the frequency of pairing odd and even bank requests.

A coherency problem could exist between the external cache and the on-chip cache when floating-point and integer data were mixed in the same structure or assigned to the same field (i.e., a union data structure). The on-chip cache prevented this by maintaining one valid bit per word (the MIPS architecture requires aligned accesses). Cache refill would set the valid bits, while integer and floating-point stores would set and reset the appropriate bits, respectively.

The R8000 issued up to four instructions per cycle to eight execution units: four integer, two floating-point, and two load/store. The integer pipelines inserted an empty stage after decode so that the ALU operation was in the same relative position as the cache access in the load/store pipelines. Thus there were no load/use delays, but address arithmetic stalled for one cycle when it depended on a loaded value.

A floating-point queue buffered floating-point instructions until they were ready to issue. This allowed the integer pipelines to proceed even when a floating-point load, with its five-cycle latency, was dispatched along with a dependent floating-point instruction. Imprecise exceptions were thus the rule for floating-point arithmetic, but there was a floating-point serialization mode bit to help in debugging, as in the IBM POWER.

A combined branch prediction and instruction alignment scheme similar to the one in the AMD K5 was used. There was a single branch prediction bit for each block of four instructions in the cache. A source bit mask in the prediction entry indicated how many valid instructions existed in the branch block, and another bit mask indicated where the branch target instruction started in the target block. Compiler support to eliminate the problem of two likely-taken branches being placed in the same block was helpful.

Hsu [1993, 1994] presents the R8000 in greater detail.

8.3.9.2 MIPS R10000 (T5) / 1996.
Whereas the R8000 was a multichip implementation, the R10000 (previously code-named the T5, and designed by Chris Rowen and Ken Yeager) is a single-chip implementation with a peak issue rate of five instructions per cycle. The sustained rate is limited to four per cycle. Figure 8.13 illustrates the MIPS R10000 pipeline.

Instructions on the R10000 are stored in a 32K-byte instruction cache with pre-decode bits and are fetched up to four per cycle from anywhere in a cache line. Decoding can run at a rate of four per cycle, and there is an eight-entry instruction buffer between the instruction cache and the decoder that allows fetch to continue even when decoding is stalled.

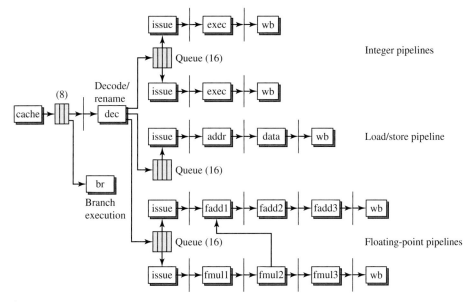

Figure 8.13
MIPS R10000 Pipeline Stages.

The decoder also renames registers. While the MIPS ISA defines 33 integer registers (31 plus two special registers for multiply and divide) and 31 floating-point registers, the R10000 has 64 physical registers for integers and 64 physical registers for floating-point. The current register mapping between logical registers and physical registers is maintained in two mapping tables, one for integer and one for floating-point.

Instructions are dispatched from the decoder into one of three instruction queues: integer, load/store, and floating-point. These queues serve the role of reservation stations, but they do not contain operand values, only physical register numbers. Operands are instead read from the physical register files during instruction issue out of a queue.

Each queue has 16 entries and supports out-of-order issue. Up to five instructions can be issued per cycle: two integer instructions can be issued to the two integer units, one of which can execute branches while the other contains integer multiply/divide circuitry; one load/store instruction can be issued to its unit; and one floating-point add instruction and one floating-point multiply instruction can be issued in parallel.

The implementation of the combined floating-point multiply-add instruction is unique in that an instruction of this type must first traverse the first two execution stages of the multiply pipeline and is then routed into the add pipeline, where it finishes normalization and writeback. Results from other floating-point operations can also be forwarded after two execution stages.

The processor keeps track of physical register assignments in a 32-entry active list of decoded instructions. The list is maintained in program order. An entry in this list is allocated for each instruction upon dispatch, and a *done flag* in each entry is initialized to zero. The indices of the active list entries are also used to tag the dispatched instructions as they are placed in the instruction queues. At completion, each instruction writes its result into its assigned physical register and sets its done flag to 1. In this manner the active list serves as a reorder buffer and supports in-order retirement (called *graduation*).

Entries in the active list contain the logical register number named as a destination in an instruction as well as the physical register previously assigned. This arrangement provides a type of history buffer for exception handling. Upon detecting an exception, instruction dispatching ceases; current instructions are allowed to complete; and then, the active list is traversed in reverse order, four instructions per cycle, unmapping physical registers by restoring the previous assignments to the mapping table. To provide precise exceptions, this process continues until the excepting instruction is unmapped. At that point, an exception handler can be called.

To make branch misprediction recovery faster, a checkpoint-repair scheme is used to make a copy of the register mapping tables and preserve the alternate path address at each branch. Up to four checkpoints can exist at one time, so the R10000 can speculatively execute past four branches. Recovery requires only one cycle to repair the mapping tables to the point of the branch and then restart instruction fetch at the correct address. Speculative instructions on the mispredicted path are flushed from the processor by use of a 4-bit branch mask added to each decoded instruction. The mask indicates if an instruction is speculative and on which of the four predicted branches it depends (multiple bits can be set). As branches are resolved, a correct prediction causes each instruction in the processor to reset the corresponding branch mask bit. Conversely, a misprediction causes each instruction with the corresponding bit set to be flushed.

Integer multiply and divide instructions have multiple destination registers (HI, LO) and thus disrupt normal instruction flow. The decoder in the R10000 stalls after encountering one of these instructions; also, the decoder will not dispatch a multiply or divide as the fourth instruction in a decode group. The reason for this is that special handling is required for the multiple destinations: two entries must be allocated in the active list for each multiply or divide.

Conditional branches are supported on the R10000 by a special condition file, in which the one bit per physical register is set to 1 whenever a result equal to zero is written into the physical register file. A conditional branch that compares against zero can immediately determine taken or not taken by checking the appropriate bit in the condition file, rather than read the value from the physical register file and check if it is zero. A 512-entry BHT is maintained for branch prediction, but there is no caching of branch target addresses. This results in a one-cycle penalty for correctly predicted taken branches.

Another interesting branch support feature of the R10000 is a branch link quadword, which holds up to four instructions past the most recent subroutine call. This acts as a return target instruction cache and supports fast returns from leaf subroutines. During initial design in 1994, a similar cache structure was proposed for holding up to four instructions on the fall-through path for the four most recent predicted-taken branches. Upon detecting a misprediction this *branch-resume cache* would allow immediate restart, and R10000 descriptions from 1994 and 1995 describe it as a unique feature. However, at best this mechanism only saves a single cycle over the simpler method of fetching the fall-through path instructions from the instruction cache, and it was left out of the actual R10000 chip.

To support strong memory consistency, the load/store instruction queue is maintained in program order. Two 16-by-16 matrices for address matching are used to determine load forwarding and also so that cache set conflicts can be detected and avoided.

See Halfhill [1994b] for an overview of the R10000. Yeager [1996] presents an in-depth description of the design, including details of the instruction queues and the active list. Vasseghi et al. [1996] presents circuit design details of the R10000.

The follow-on design, the R12000, increases the active list to 48 entries and the BHT to 2048 entries, and it adds a 32-entry two-way set-associative branch target cache. The recent R14000 and R16000 are similar to the R12000.

8.3.9.3 MIPS R5000 and QED RM7000 / 1996 and 1997.
The R5000 was designed by QED, a company started by Earl Killian and Tom Riordan, who also designed some of the R4x000 family members. The R5000 organization is very

simple and only provides one integer and one floating-point instruction issued per cycle (a level-1 design); however, the performance is as impressive as that of competing designs with extensive out-of-order capabilities. Riordan also extended this approach in the RM7000, which retains the R5000's dual-issue structure but integrates on one chip a 16K-byte four-way set-associative L1 instruction cache, a 16K-byte four-way set-associative L1 data cache, and a 256K-byte four-way set-associative L2 unified cache.

8.3.10 Motorola

Two Motorola designs have been superscalar, apart from processors in the PowerPC family.

8.3.10.1 Motorola 88110 / 1991.
The 88110 was a very aggressive design for its time (1991) and was introduced shortly after the IBM RS/6000 started gaining popularity. The 88110 was a dual-issue implementation of the Motorola 88K RISC architecture and extended the 88K architecture by introducing a separate extended-precision (80-bit) floating-point register file and by adding graphics instructions and nondelayed branches. The 88110 was notable for its 10 function units (see Figure 4.7) and its use of a history buffer to provide for precise exceptions and recovery from branch mispredictions. Keith Diefendorff was the chief architect; Willie Anderson designed the graphics and floating-point extensions; and Bill Moyer designed the memory system.

The 10 function units were the instruction-fetch/branch unit, load/store unit, bitfield unit, floating-point add unit, multiply unit, divide unit, two integer units, and two graphics units. Floating-point operations were performed using 80-bit extended precision. The integer and floating-point register files each had two dedicated history buffer ports to record the old values of two result registers per cycle. The history buffer provided 12 entries and could restore up to two registers per cycle.

On each cycle two instructions were fetched, unless the instruction pair crossed a cache line. The decoder was aggressive and tried to dual issue in each cycle. There was a one-entry reservation station for branches and a three-entry reservation station for stores; thus the processor performed in-order issue except for branches and stores. Instructions speculatively issued past a predicted branch were tagged as conditional and flushed if the branch was mispredicted; and any registers already written by mispredicted conditional instructions were restored using the history buffer. Conditional stores, however, were not allowed to update the data cache but remained in the reservation station until the branch was resolved.

Branches were statically predicted. A target instruction cache returned the pair of instructions at the branch's target address for the 32 most recently taken branches. The TIC was virtually addressed, and it had to be flushed on each context switch.

There was no register renaming, but instruction pairs with write-after-read dependences were allowed to dual issue, and dependent stores were allowed to dual issue with the result-producing instruction. The load/store unit had a four-entry load buffer and allowed loads to bypass stores.

There were two 80-bit writeback busses shared among the 10 function units. Because of the different latencies among the function units, instructions arbitrated for the busses. The arbitration priority was unusual in that it gave priority to lower-cycle-count operations and could thus further delay long-latency operations. This was apparently done in response to a customer demand for this type of priority.

Apple, Next, Data General, Encore, and Harris designed machines for the 88110 (with the latter three delivering 88110-based systems). However, Motorola had difficulty in manufacturing fully functional chips and canceled revisions and follow-on designs in favor of supporting the PowerPC. However, several of the cache, TLB, and bus design techniques for the 88110 were used in the IBM/Motorola PowerPC processors and in the Motorola 68060.

See Diefendorff and Allen [1992a,b] and Ullah and Holle [1993] for articles on the 88110.

8.3.10.2 68060 / 1993. The 68060 was the first superscalar implementation in the 68000 CISC architecture family to make it to market. Even though many of the earliest workstations used the 680x0 processors and Apple chose them for the Macintosh, the 680x0 family has been displaced by the more numerous IA32 and RISC designs. Indeed, Motorola had chosen in 1991 to target the PowerPC for the workstation market, and thus the 68060 was designed as a low-cost, low-power entrant in the embedded systems market. The architect was Joe Circello. The 68060 pipeline is illustrated in Figure 8.14.

The 68060 implementation has a decoupled microarchitecture that translates a variable-length 68000 instruction into a fixed-length format that completely identifies the resources required. The translated instructions are stored in a 16-entry FIFO buffer. Each entry has room for a 16-bit opcode, 32-bit extension words, and early decode information. Some of the complex instructions require more than one entry in the buffer. Moreover, some of the most complex 68040 instruction types

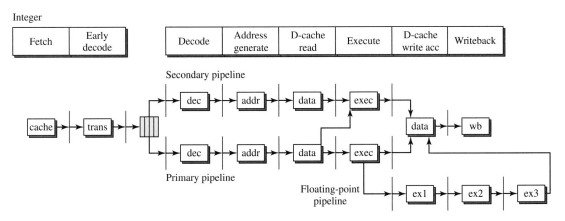

Figure 8.14
Motorola M68060 Pipeline Stages.

are not handled by the 68060 hardware but are instead implemented as traps to emulation software.

The 68060 contains a 256-entry branch cache with 2-bit predictors. The branch cache also uses branch folding, in which the branch condition and an address recovery increment are stored along with the target address in each branch cache entry. Each entry is tagged with the address of the instruction prior to the branch and thus allows the branch to be eliminated from the instruction stream sent to the FIFO buffer whenever the condition code bits satisfy the branch condition.

The issue logic attempts to in-order issue two instructions per cycle from the FIFO buffer to two four-stage operand-execution pipelines. The primary operand-execution pipeline can execute all instructions, including the initiation of floating-point instructions in a separate execution unit. The secondary operand-execution pipeline executes only integer instructions.

These dual pipelines must be operated in a lockstep manner, similar to the Pentium, but the design and the control logic are much more sophisticated. Each operand execution pipeline is composed of two pairs of fetch and execute stages; Motorola literature describes this as two RISC engines placed back to back. This is required for instructions with memory operands: the first pair fetches address components and uses an ALU to calculate the effective address (and starts the cache read), and the second pair fetches register operands and uses an ALU to calculate the operation result. Taking this further, by generalizing the effective address ALU, some operations can be executed by the first two stages in the primary pipeline and then have their results forwarded to the second pipeline in a cascaded manner. While some instructions are always executed in the first two stages, others are dynamically allocated according to the issue pair dependency; thus many times pairs of dependent instructions can be issued in the same cycle. Register renaming is also used to remove false dependences between issue pairs.

The data cache is four-way interleaved and allows one load and one nonconflicting store to execute simultaneously. See Circello and Goodrich [1993], Circello [1994], and Circello et al. [1995] for more detailed descriptions of the 68060.

8.3.11 PowerPC—32-bit Architecture

The PowerPC architecture is the result of cooperation begun in 1991 between IBM, Motorola, and Apple. IBM and Motorola set up the joint Somerset Design Center in Austin, Texas, and the POWER ISA and the 88110 bus interface were adopted as starting points for the joint effort. Single-precision floating-point, revised integer multiply and divide, load word and reserve and store word conditional, and support for both little-endian as well as big-endian were added to the ISA, along with the definition of a weakly ordered memory model and an I/O barrier instruction (the humorously named "eieio" instruction). Features removed include record locking, the multiplier-quotient (MQ) register and its associated instructions, and several bit-field and string instructions. Cache control instructions were also changed to provide greater flexibility. The lead architects were Rich Oehler (IBM), Keith Diefendorff (Motorola), Ron Hochsprung (Apple), and John Sell (Apple).

Diefendorff [1994], Diefendorff et al. [1994], and Diefendorff and Silha [1994] contain more information about the history of the PowerPC cooperation and the changes from POWER.

8.3.11.1 PowerPC 601 / 1993. The 601 was the first implementation of the PowerPC architecture and was designed by Charles Moore and John Muhich. An important design goal was time to market, so Moore's previous RSC design was used as a starting point. The bus and cache coherency schemes of the Motorola 88110 were also used to leverage Apple's previous 88110-based system designs. Compared to the RSC, the 601 unified cache was enlarged to 32K bytes and the TLB structure followed the 88110 approach of mapping pages and larger blocks.

Each cycle, the bottom four entries of an eight-entry instruction queue were decoded. Floating-point instructions and branches could be dispatched from any of the four entries, but integer instructions had to be issued from the bottom entry. A unique tagging scheme linked the instructions that were issued/dispatched in a given cycle into an *instruction packet*. The progress of this packet was monitored through the integer instruction that served as the anchor of the packet. If an integer instruction was not available to be issued in a given cycle, a nop was generated so that it could serve as the anchor for the packet. All instructions in a packet completed at the same time.

Following the RSC design, the 601 had a two-entry floating-point instruction queue into which instructions were dispatched, and it did not rename floating-point registers. The RSC floating-point pipeline stage design for multiply and add was reused. The integer unit also handled loads and stores, but there was a more sophisticated memory system than that in the RSC. Between the processor and the cache, the 601 added a two-entry load queue and a three-entry store queue. Between the cache and memory a five-entry memory queue was added to make the cache nonblocking. Branch instructions were predicted in the same manner as in the RSC, and conditional dispatch but not execution could occur past unresolved branches.

The designers added many multiprocessor capabilities to the 601. For example, the data cache implemented the MESI protocol, and the tags were double-pumped each cycle to allow for snooping. The writeback queue entries were also snooped so that refills could have priority without causing coherency problems.

See Becker et al. [1993], Diefendorff [1993], Moore [1993], Potter et al. [1994], and Weiss and Smith [1994] for more information on the 601.

8.3.11.2 PowerPC 603 / 1994. The 603 is a low-power implementation of the PowerPC that was designed by Brad Burgess, Russ Reininger, and Jim Kahle for small, single-processor systems, such as laptops. The 603 has separate 8K-byte instruction and data caches and five independent execution units: branch, integer, system, load/store, and floating-point. The system unit executes the condition code logic operations and instructions that move data to and from special system registers.

Two instructions are fetched each cycle from the instruction cache and sent to both the branch unit and a six-entry instruction queue. The branch unit can delete branches in the instruction queue when they do not change branch unit registers; otherwise, branches pass through the system unit. A decoder looks at the bottom two entries in the instruction queue and issues/dispatches up to two instructions per cycle. Dispatch includes reading register operands and assigning a rename register to destination registers. There are five integer rename registers and four floating-point rename registers.

There is a reservation station for each execution unit so that dispatch can occur even with data dependences. Dispatch also requires that an entry for each issued/dispatched instruction be allocated in the five-entry completion buffer. Instructions are retired from the completion buffer at a rate of two per cycle; retirement includes the updating of the register files by transferring the contents of the assigned rename registers. Because all instructions that change registers retire from the completion buffer in program order, all exceptions are precise.

Default branch prediction is based on the sign of the displacement, but a bit in the branch opcode can be used by the compilers to reverse the prediction. Speculative execution past one conditional branch is provided, with the speculative path able to follow an unconditional branch or a branch-on-count while conditional branches wait to be resolved. Branch misprediction is handled by flushing the predicted instructions and the completion buffer contents subsequent to the branch.

The load/store unit performs multiple accesses for unaligned operands and sequences multiple accesses for the load-multiple/store-multiple and string instructions. Loads are pipelined with a two-cycle latency; stores are not pipelined. Denormal floating-point numbers are supported by a special internal format, or a flush-to-zero mode can be enabled.

There are four power management modes: nap, doze, sleep, and dynamic. The dynamic mode allows idle execution units to reduce power consumption without impacting performance.

See Burgess et al. [1994a, 1994b] and the special issue of the *Communications of the ACM* on "The Making of the PowerPC" for more information. An excellent article describing the simulation studies of design tradeoffs for the 603 can be found in Poursepanj et al. [1994].

The 603e, done by Brad Burgess and Robert Golla, is a later implementation that doubles the sizes of the on-chip caches and provides the system unit with the ability to execute integer adds and compares. (Thus the 603e could be described as having a limited second integer unit.)

8.3.11.3 PowerPC 604 / 1994.

The 604 looks much like the standard processor design of Mike Johnson's textbook on superscalar design. As shown in Figure 8.15, there are six function units, each having a two-entry reservation station, and a 16-entry reorder buffer (*completion buffer*). Up to four instructions can be fetched per cycle into a four-entry decode buffer. These instructions are next placed into a four-entry dispatch buffer, which reads operands and performs register renaming. From this buffer, up to four instructions are dispatched per cycle to the six function units: a branch unit, two integer units, an integer multiply unit, a load/store unit, and a floating-point unit. Each of the integer units can issue an instruction

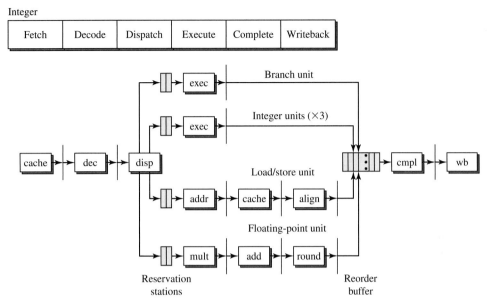

Figure 8.15
IBM/Motorola PowerPC 604 Pipeline Stages.

from either reservation station entry (i.e., out of order), whereas the reservation stations assigned to other units issue in order for the given instruction type but, of course, provide for interunit slip. There are two levels of speculative execution supported. The reorder buffer can retire up to four instructions per cycle.

Renaming is provided by a 12-entry rename buffer for integer registers, an eight-entry rename buffer for floating-point registers, and an eight-entry rename buffer for condition codes. Speculative execution is not allowed for stores, and the 604 also disallows speculative execution for logical operations on condition registers and integer arithmetic operations that use the carry bit.

Branch prediction on the 604 is supported by a 512-entry BHT, each entry having a 2-bit predictor, and a 64-entry fully associative BTAC. The decode stage recognizes and handles prediction for unconditional branches and branches that hit in the BHT but not in the BTAC. There is also special branch prediction for branches on the count register, which typically implement innermost loops. The dispatch logic stops collecting instructions for multiple dispatch when it encounters a branch, so only one branch per cycle is processed.

Peter Song was the chief architect of the 604. See Denman [1994] and Song et al. [1994] for more information on the 604. Denman et al. [1996] discuss a follow-on chip, the 604e, which is a lower-power, pin-compatible version. The 604e doubles the cache sizes and provides separate execution units for condition register operations and branches. Each of these two units has a two-entry reservation station, but dispatch is limited to one per cycle.

8.3.11.4 PowerPC 750 (G3) / 1997.
The chief architect of the 750 is Brad Burgess. The 750 is designed as a low-power chip with four pipeline stages, and it has less buffering than the 604. Burgess characterizes the design as "modest issue width, short pipeline, large caches, and an aggressive branch unit focused on resolving branches rather than predicting them." The 750 has six function units: two integer units, a system register unit, a load/store unit, a floating-point unit, and the branch unit. Unlike the 604, function units other than the branch unit and the load/store unit have only one entry each in their reservation stations.

Instructions are pre-decoded into a 36-bit format prior to storing in the L1 instruction cache. Four instructions can be fetched per cycle, and up to two instructions can be dispatched per cycle from the two bottom entries of a six-entry instruction buffer. Branches are processed as soon as they are recognized, and when predicted taken, they are deleted (*folded out*) from the instruction buffer and overlaid with instructions from the branch target path. Speculative execution is provided past one unresolved branch, and speculative fetching continues past two unresolved branches. An interesting approach to save space in the six-entry completion buffer is the "squashing" of nops and untaken branches from the instruction buffer prior to dispatch. These two types of instructions will not be allocated completion buffer entries, but an unresolved, predicted-untaken branch will be held in the branch unit until resolution so that any misprediction recovery can be performed.

The 750 includes a 64-entry four-way set-associative BTIC and a 512-entry BHT. The chip also includes a two-way set-associative level-two cache controller and the level-two tags; this supports 256K bytes, 512K bytes, or 1 Mbyte of off-chip SRAM. (The 740 version of the chip does not contain the L2 tags and controller.)

See Kennedy et al. [1997] for more details on the 750.

8.3.11.5 PowerPC 7400 (G4) and 7450 (G4+) / 1999 and 2001.
The 7400, a design led by Mike Snyder, is essentially the 750 with AltiVec added. A major redesign in the 74xx series occurred in 2001 with the seven-stage 7450. In this design, led by Brad Burgess, the issue and retire rates have been increased to three per cycle, and 10 function units are provided. The completion queue has been enlarged to 16 entries, and branch prediction has also been improved by quadrupling the BHT to 2048 entries and doubling the BTIC to 128 entries. A 256K-byte L2 cache is integrated onto the 7450 chip. See Diefendorff [1999] for a description of the 7450.

8.3.11.6 PowerPC e500 Core / 2001.
The e500 core implements the 32-bit embedded processor "Book E" instruction set. The e500 also implements the signal processing engine (SPE) extensions, which provide two-element vector operands, and the integer select extension, which provides for partial predication.

The e500 core is a two-way issue, seven-stage-pipeline design similar in some ways to the 7450. Branch prediction in the e500 core is provided by a single structure, a 512-entry BTB. Two instructions can be dispatched from the 12-entry instruction queue per cycle; and, both of these instructions can be moved into the four-entry general instruction queue, or one can be moved there and the other can be moved into the two-entry branch instruction queue. A single reservation station is placed between the branch instruction queue and the branch unit, and a similar arrangement occurs for each of the other functional units, which are fed instead by the general instruction queue. These units include two simple integer units, a multiple-cycle integer unit, and a load/store unit. The SPE instructions execute in the simple and multiple-cycle integer units along with the rest of the instructions. The completion queue has 14 entries and can retire up to two instructions per cycle.

8.3.12 PowerPC—64-bit Architecture

When the PowerPC architecture was defined in the early 1990s, a 64-bit mode of operation was also defined along with an 80-bit virtual address. See Peng et al. [1995] for a detailed description of the 64-bit PowerPC architecture.

8.3.12.1 PowerPC 620 / 1995. The 620 was the first 64-bit implementation of the PowerPC architecture and is detailed in Chapter 6. Its designers included Don Waldecker, Chin Ching Kau, and Dave Levitan. The four-way issue organization was similar to that of the 604 and used the same mix of function units. However, the 620 was more aggressive than the 604 in several ways. For example, the decode stage was removed from the instruction pipeline and instead replaced by pre-decoding during instruction cache refills (see Figure 6.2). The load/store unit reservation station was increased from two entries to three entries with out-of-order issue, and the branch unit reservation station was increased from two entries to four entries. To assist in speculating through the four branches, the 620 doubled the number of condition register fields rename buffers (to 16), and quadrupled the number of entries in the BTAC and the BHT (to 256 and 2048, respectively). Some implementation simplifications were (1) integer instructions that require two source registers could only be dispatched from the bottom two slots of the eight-entry instruction queue, (2) integer rename registers were cut from 12 in the 604 to 8 in the 620, and (3) reorder buffer entries were allocated and released in pairs.

The 620 implementation reportedly had bugs that initially restricted multiprocessor operation, and very few systems shipped with 620 chips. For more information on the design of the 620, see Thompson and Ryan [1994] and Levitan et al. [1995].

8.3.12.2 POWER3 (630) / 1998. Chapter 6 notes that the single floating-point unit in the 620 and the inability to issue more than one load plus one store per cycle are major bottlenecks in that design. These problems were addressed in the follow-on design, called at first the 630 but later known as the POWER3. Starting with the 620 core, the POWER3 doubled the number of floating-point and load/store units to two each. The data cache supported up to two loads, one store, and one refill each cycle; it also had four miss handling registers rather than the single register found in the 620. (See Table 6.11.)

Each of the eight function units in POWER3 could be issued an instruction each cycle (versus an issue limit of four in the 620). While branch and load/store instructions were issued in order from their respective reservation stations, the other five units could be issued instructions out of order. The completion buffer was doubled to 32 entries, and the number of rename registers was doubled and tripled for integer instructions and floating-point instructions, respectively. A four-stream hardware prefetch facility was also added.

A decision was made to not store operands in the reservation station entries; instead, operands were read from the physical registers in a separate pipe stage just prior to execution. Also, timing issues led to a separate finish stage prior to the commit stage. Thus the POWER3 pipeline has two additional stages as compared to the 620 and was able to reach a clock rate approximately three times faster than the 620.

See Song [1997b] and O'Connell and White [2000] for more information on the POWER3.

8.3.12.3 POWER4 / 2002. The IBM POWER4 is a high-performance multiprocessing system design. Jim Kahle and Chuck Moore were the chief architects. Each chip contains two processing cores, with each core having its own eight function units (including two floating-point units and two load/store units) and L1 caches but sharing a single unified L2 cache and L3 cache controller and directory. A single multichip module can package four chips, so the basic system building block is an eight-way SMP.

The eight-way issue core is equally as ambitious in design as the surrounding caches and memory access path logic. The traditional IBM brainiac style was explicitly discarded in POWER4 in favor of a deeply pipelined speed demon that even *cracks* some of the enhanced-RISC PowerPC instructions into separate, simpler internal operations. Up to 200 instructions can be in-flight.

Instructions are fetched based on a hybrid branch prediction scheme that is unusual in its use of 1-bit predictors rather than the more typical 2-bit predictors. A 16K-entry *selector* chooses between a 16K-entry local predictor and a gshare-like 16K-entry global predictor. Special handling of branch-to-link and branch-on-count instructions is also provided. POWER4 allows hint bits in the branch instructions to override the dynamic branch prediction.

In a scheme somewhat reminiscent of the PowerPC 601, *instruction groups* are formed to track instruction completion; however, in POWER4, the group is anchored by a branch instruction. Groups of five are formed sequentially, with the anchoring branch instruction in the fifth slot and nops used to pad out any unfilled slots. Condition register instructions must be specially handled, and they can only be assigned to the first or second slot of a group. The groups are tracked by use of a 20-entry *global completion table*.

Only one group can be dispatched into the issue queues per cycle, and only one group can complete per cycle. Instructions that require serialization form their own single-issue groups, and these groups cannot execute until they have no other uncompleted groups in front of them. Instructions that are cracked into two internal operations, like load-with-update, must have both internal operations in the same group. More complex instructions, like load-multiple, are cracked into several internal operations (called *millicoding*), and these operations must be placed into groups separated from other instructions. Upon an exception, instructions within the group from which the exception occurred are redispatched in separate, single-instruction groups.

Once in the issue queues, instructions and internal operations can issue out of order. There are 11 issue queues with a total of 78 entries among them. In a scheme somewhat reminiscent of the HP 8000, an even-odd distribution of the group slots to the issue queues and function units is used. An abundance of physical registers are provided, including 80 physical registers for the 32 architected general registers, 72 physical registers for the 32 architected floating-point registers, 16 physical registers for the architected link and count registers, and 32 physical registers for the eight condition register fields.

The POWER4 pipeline has nine stages prior to instruction issue (see Figure 6.6). Two of these stages are required for instruction fetch, six are used for instruction cracking and group formation, and one stage provides for resource mapping and dispatch. A simple integer instruction requires five stages during executing, including issue, reading operands, executing, transfer, and writing the result. Groups can complete in a final complete stage, making a 15-stage pipeline for integer instructions. Floating-point instructions require an extra five stages.

The on-chip caches include two 64K-byte instruction caches, two 32K-byte data caches, and a unified L2 cache of approximately 1.5 Mbytes. Each L1 data cache provides up to two reads and one store per cycle. Up to eight prefetch streams and an off-chip L3 of 32 Mbytes is supported. The L2 cache uses a seven-state, enhanced MESI coherency protocol, while the L3 uses a five-state protocol.

See Section 6.8 and Tendler et al. [2002] for more information on POWER4.

8.3.12.4 PowerPC 970 (G5) / 2003. The PowerPC 970 is a single-core version of the POWER4, and it includes the AltiVec extensions. The chief architect is Peter Sandon. An extra pipeline stage was added to the front end for timing purposes, so the 970 has a 16-stage pipeline for integer instructions. While two SIMD units have been added to make a total of 10 function units, the instruction group size remains at five and the issue limit remains at eight instructions per cycle. See Halfhill [2002] for details.

8.3.13 PowerPC-AS

Following a directive by IBM President Jack Kuehler in 1991, a corporate-wide effort was made to investigate standardizing on the PowerPC. Engineers from the AS/400 division in Rochester, Minnesota, had been working on a commercial RISC design (C-RISC) for the next generation of the single-level store AS/400 machines, but they were told to instead adapt the 64-bit PowerPC architecture. This extension, called Amazon and later PowerPC-AS, was designed by Andy Wottreng and Mike Corrigan at IBM Rochester, under the leadership of Frank Soltis.

Since the 64-bit PowerPC 620 was not ready, Rochester went on to develop the multichip A30 (Muskie), while Endicott developed the single-chip A10 (Cobra). These designs did not include the 32-bit PowerPC instructions, but the next Rochester design, the A35 (Apache), did. Apache was used in the RS/6000 series and called the RS64. See Soltis [2001] for details of the Rochester efforts.

Currently, PowerPC-AS processors, including the POWER4, implement the 228 64-bit PowerPC instruction set plus more than 150 AS-mode instructions.

8.3.13.1 PowerPC-AS A30 (Muskie) / 1995.
The A30 was a seven-chip, high-end, SMP-capable implementation. The design was based on a five-stage pipeline: fetch, dispatch, execute, commit, and writeback. Five function units were provided, and up to four instructions could be issued per cycle, in order. Hazard detection was done in the execute stage, rather than the dispatch stage, and floating-point registers were renamed to avoid hazards. The commit stage held results until they could be written back to the register files. Branches were handled using predict-untaken, but the branch unit could look up to six instructions back in the 16-entry current instruction queue and determine branch target addresses. An eight-entry branch target queue was used to prefetch taken-path instructions. Borkenhagen et al. [1994] describes the A30.

8.3.13.2 PowerPC-AS A10 (Cobra) and A35 (Apache, RS64) / 1995 and 1997.
The A10 was a single-chip, uniprocessor-only implementation with four pipeline stages and in-order issue of up to three instructions per cycle. No renaming was done. See Bishop et al. [1996] for more details. The A35 (Apache) was a follow-on design at Rochester that added the full PowerPC instruction set and multiprocessor support to the A10. It was a five-chip implementation and was introduced in 1997.

8.3.13.3 PowerPC-AS A50 (Star series) / 1998–2001.
In 1998, Rochester introduced the first of the multithreaded Star series of PowerPC-AS processors. This was the A50, also called Northstar and known as the RS64-II when used in RS/6000 systems. Process changes [specifically, copper interconnect and then silicon on insulator (SOI)] led to the A50 design being renamed as Pulsar / RS64-III and then i-Star. See Borkenhagen et al. [2000] for a description of the most recent member of the Star series, the s-Star or RS64-IV.

8.3.14 SPARC Version 8

The SPARC architecture is a RISC design derived from work by David Patterson at the University of California at Berkeley. One distinguishing feature of that early work was the use of register windows for reducing memory traffic on procedure calls, and this feature was adopted in SPARC by chief architect Robert Garner. The first SPARC processors implemented what was called version 7 of the architecture in 1986. It was highly pipeline oriented and defined a set of integer instructions, each of which could be implemented in one cycle of execution (integer multiply and divide were missing), and delayed branches. The architecture manual explicitly stated that "an untaken branch takes as much or more time than a taken branch." A floating-point queue was also explicitly defined in the architecture manual; it is a reorder buffer that can be directly accessed by exception handler software.

Although the version 7 architecture manual included suggested subroutines for integer multiply and divide, version 8 of the architecture in 1990 adopted integer multiply. The SuperSPARC and HyperSPARC processors implement version 8.

8.3.14.1 Texas Instruments SuperSPARC (Viking) / 1992.
The SuperSPARC was designed by Greg Blanck of Sun, with the implementation overseen by Steve Krueger of TI. The SuperSPARC issued up to three instructions per cycle in program order and was built around a control unit that handled branching, a floating-point unit, and a unique integer unit that contained three cascaded ALUs. These cascaded ALUs permitted the simultaneous issue of a dependent pair of integer instructions.

The SuperSPARC fetched an aligned group of four instructions each cycle. The decoder required one and one-half cycles and attempted to issue up to three instructions in the last half-cycle, in what Texas Instruments called a grouping stage. While some instructions were single-issue (e.g., register window save and restore, integer multiply), the grouping logic could combine up to two integer instructions, one load/store, and/or one floating-point instruction per group. The actual issue rules were quite complex and involved resource constraints such as a limit on the number of integer register write ports. An instruction group was said to be *finalized* after any control transfer instruction. In general, once issued, the group proceeded through the pipelines in lockstep manner. However, floating-point instructions would be placed into a four-entry instruction buffer to await floating-point unit availability and thereafter would execute independently. The SPARC floating-point queue was provided for dealing with any exceptions. As noted before, a dependent instruction (integer, store, or branch) could be included in a group with an operand-producing integer instruction due to the cascaded ALUs. This was not true for an operand-producing load; because of possible cache misses, any instruction dependent on a load had to be placed in the next group.

The SuperSPARC contained two four-instruction fetch queues. One was used for fetching along the sequential path, while the other was used to prefetch instructions at branch targets whenever a branch was encountered in the sequential path. Since a group finalized after a control transfer instruction, a delay slot instruction was placed in the next group. This group would be speculatively issued. (Thus the SuperSPARC was actually a predict-untaken design). If the branch was taken, the instructions in the speculative group, other than the delay slot instruction, would be squashed, and the prefetched target instructions would then be issued in the next group. Thus there was no branch penalty for a taken branch; rather there was a one-issue cycle between the branch group and the target group in which the delay slot instruction was executed by itself.

See Blanck and Krueger [1992] for an overview of SuperSPARC. The chip was somewhat of a performance disappointment, allegedly due to problems in the cache design rather than the core.

8.3.14.2 Ross HyperSPARC (Pinnacle) / 1993.
The HyperSPARC came to market a year or two after the SuperSPARC and was a less aggressive design in terms of multiple issue. However, its success in competing in performance against the SuperSPARC is another example, like Alpha versus POWER, of a speed demon versus a brainiac. The HyperSPARC specification was done by Raju Vegesna and the first simulator by Jim Monaco. A preliminary article on the HyperSPARC was published by Vegesna [1992].

The HyperSPARC had four execution units: integer, floating-point, load/store, and branch. Two instructions per cycle could be fetched from an 8K-byte on-chip instruction cache and placed into the decoder. The two-instruction-wide decoder was unaggressive and would not accept more instructions until both previously fetched instructions had been issued. The decoder also fetched register operand values.

Three special cases of dependent issue were supported: (1) sethi and dependent, (2) sethi and dependent load/store, and (3) an integer ALU instruction that sets the condition code and a dependent branch. Two floating-point instructions could also be dispatched into a four-entry floating-point prequeue in the same cycle, if the queue had room. There were several stall conditions, some of which involved register file port contention since there were only two read ports for the integer register file. Moreover, there were 53 single-issue instructions, including call, save, restore, multiply, divide, and floating-point compare.

The integer unit had a total of 136 registers, thus providing eight overlapping windows of 24 registers each and eight global registers. The integer pipeline, as well as the load/store and branch pipelines, consisted of four stages beyond the common fetch and decode: execute, cache read, cache write, and register update. The integer unit did not use the two cache-related stages, but they were included so that all non-floating-point pipelines would be of equal length. Integer multiply and divide were unusually long, 18 and 37 cycles, respectively; moreover, they stalled further instruction issue until they were completed.

The floating-point unit's four-entry prequeue and a three-entry postqueue together implemented the SPARC floating-point queue technique for out-of-order completions in the floating-point unit. The prequeue allowed the decoder to dispatch floating-point instructions as quickly as possible. Instructions in the floating-point prequeue were decoded in order and issued into the postqueue; each postqueue entry corresponded to an execution stage in the floating-point pipeline (execute-1, execute-2, round). A floating-point load and a dependent floating-point instruction could be issued/dispatched in the same cycle; however, the dependent instruction would spend two cycles in the prequeue before the loaded data were forwarded to the execute-1 stage. When a floating-point instruction and a dependent floating-point store were paired in the decoder, the store waited for at least two cycles in the decoder before the operation result entered the round stage and from there was forwarded to the load/store unit in the subsequent cycle.

8.3.14.3 Metaflow Lightning and Thunder / Canceled.
The Lightning and Thunder were out-of-order execution SPARC designs by Bruce Lightner and

Val Popescu. These designs used a centralized reservation station approach called deferred-scheduling register-renaming instruction shelf (DRIS). Thunder was described at the 1994 Hot Chips and was an improved three-chip version of the four-chip Lightning, which was designed in 1991. Thunder issued up to four instructions per cycle to eight execution units: three integer units, two floating-point units, two load/store units, and one branch unit. Branch prediction was dynamic and included return address prediction. See Lightner and Hill [1991] and Popescu et al. [1991] for articles on Lightning, and see Lightner [1994] for a presentation on Thunder. Neither design was delivered, and Hyundai was assigned the patents.

8.3.15 SPARC Version 9

The 64-bit SPARC instruction set is known as version 9. The revisions were decided by a large committee with more than 100 meetings. Major contributors were Dave Ditzel (chairman), Joel Boney, Steve Chessin, Bill Joy, Steve Kleiman, Steve Kruger, Dave Weaver, Winfried Wilcke, and Robert Yung. The goals of the version 9 architecture also included avoiding serialization points. Thus, there are now four separate floating-point condition codes as well as a new type of integer branch that conditionally branches on the basis of integer register contents, giving the effect of multiple integer condition codes. Version 9 also added support for nonfaulting speculative loads, branch prediction bits in the branch instruction formats, conditional moves, and a memory-barrier instruction for a weakly ordered memory model.

8.3.15.1 HaL SPARC64 / 1995.

The SPARC64 was the first of several implementations of the SPARC version 9 architecture that were planned by HaL, including a multiprocessor version with directory-based cache coherence. The HaL designs use a unique three-level memory management scheme (with regions, views, and then pages) to reduce the amount of storage required for mapping tables for its 64-bit address space. The SPARC64 designers were Hisashige Ando, Winfried Wilcke, and Mike Shebanow.

The windowed register file contained 116 integer registers, 78 of which were bound at any given time to form four SPARC register windows. This left 38 free integer registers to be used for renaming. There were also 112 floating-point registers, 32 of which were bound at any given time to single-precision and another 32 of which were bound to double-precision. This left 48 free floating-point registers to be used in renaming. The integer register file had 10 read ports and 4 write ports, while the floating-point register file had 6 read ports and 3 write ports.

The SPARC64 had four 64K-byte, virtually addressed, four-way set-associative caches (two were used for instructions, and two were used for data; this allowed two nonconflicting load/stores per cycle). A real address table was provided for inverse mapping of the data caches, and nonblocking access to the data caches (with load merging) was also provided using eight reload buffers. For speeding up instruction access, a level-0 4K-byte direct-mapped instruction cache was provided in which SPARC instructions were stored in a partially decoded internal format; this format included room for partially calculated branch target addresses. A 2-bit branch history was also provided for each instruction in the level-0 instruction cache.

Up to four instructions were dispatched per cycle, with some limits according to instruction type, into four reservation stations. There was an 8-entry reservation station for four integer units (two integer ALUs, an integer multiply unit, and an integer divide unit); an 8-entry reservation station for two address generation units; an 8-entry reservation station for two floating-point units (a floating-point multiplier-adder unit and a floating-point divider); and a 12-entry reservation station for two load/store units. Register renaming was performed during dispatch. A load or store instruction was dispatched to both the address generation unit reservation station and the load/store unit reservation station. The effective address was sent from the address generation unit to a value cache associated with the load/store reservation station.

While some designs provide for an equal number of instructions to be dispatched, issued, completed, and retired during a given cycle, the SPARC64 had a wide variance. In a given cycle, up to four instructions could dispatch, up to seven instructions could issue, up to ten could execute, up to nine instructions could complete, up to eight instructions could commit, and up to four instructions could retire. A maximum of 64 instructions could be active at any point, and the hardware kept track of these in the *A ring* via individually assigned 6-bit serial numbers. The A ring operated in a checkpoint-repair manner to provide branch misprediction recovery, and there was room for 16 checkpoints (at branches or instructions that modified unrenamed control registers). Four pointers were used to update the A ring: last issued serial number (ISN), last committed serial number (CSN), resource recovery pointer (RRP), and noncommitted memory serial number pointer (NCSNP), which allowed aggressive scheduling of loads and stores. A pointer to the last checkpoint was appended to each instruction to allow for a one-cycle recovery to the checkpoint. For trapping instructions that were not aligned on a checkpoint, the processor could undo four instructions per cycle.

The integer instruction pipeline had seven stages: fetch, dispatch, execute, write, complete, commit, and retire. A decode stage was missing since the decoding was primarily accomplished as instructions were loaded into the level-0 instruction cache. The complete stage checked for errors/exceptions; the commit stage performed the in-order update of results into the architectural state; and the retire stage deallocated any resources. Two extra execution stages were required for load/stores. Using the trap definitions in version 9, the SPARC64 could rename trap levels, and this allowed the processor to speculatively enter traps that were detected during dispatch.

See Chen et al. [1995], Patkar et al. [1995], Simone et al. [1995], Wilcke [1995], and Williams et al. [1995] for more details of SPARC64. The Simone paper details several interesting design tradeoffs, including special priority logic for issuing condition-code-modifying instructions.

HaL was bought by Fujitsu, which produced various revisions of the basic design, called the SPARC64-II, -III, GP, and -IV (e.g., increased level-0 instruction cache and BHT sizes). A two-level branch predictor and an additional pipeline stage for dispatch were introduced in the SPARC64-III [Song, 1997a]. An ambitious new core, known as the SPARC64 V, was an eight-way issue design using a trace cache and value prediction. Mike Shebanow, the chief architect, described this design at the 1999 Microprocessor Forum [Diefendorff, 1999b] and at a seminar presentation at Stanford University in 1999 [Shebanow, 1999]. Fujitsu canceled this project and instead introduced another revision of the original core under the name SPARC64-V in 2003 [Krewell, 2002].

8.3.15.2 UltraSPARC-I / 1995.
The UltraSPARC-I was designed by Les Kohn, Marc Tremblay, Guillermo Maturana, and Robert Yung. It provided four-way issue to nine function units (two integer ALUs, load/store, branch, floating-point add, floating-point multiply, floating-point divide/square root, graphics add, and graphics multiply). A set of 30 or so graphics instructions was introduced for the UltraSPARC and is called the visual instruction set (VIS). Block load/store instructions and additional register windows were also provided in the UltraSPARC-I. Figure 8.16 illustrates the UltraSPARC-I pipeline.

The UltraSPARC-I was not an ambitious out-of-order design as were many of its contemporaries. The design team extensively simulated many designs, including various forms of out-of-order processing. They reported that an out-of-order approach would have cost a 20% penalty in clock cycle time and would have likely increased the time to market by three to six months. Instead, high performance was sought by including features such as speculative, nonfaulting loads, which the UltraSPARC compilers can use to perform aggressive global code motion.

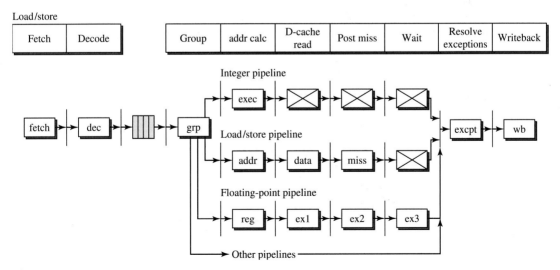

Figure 8.16
Sun UltraSPARC-I Pipeline Stages.

Building on the concepts of grouping and fixed-length pipeline segments as found in the SuperSPARC and HyperSPARC, the UltraSPARC-I performed in-order issue of groups of up to four instructions each. The design provided precise exceptions by discarding the traditional SPARC floating-point queue in favor of padding out all function unit pipelines to four stages each. Exceptions in longer-running operations (e.g., divide, square root) were predicted.

Speculative issue was provided using a branch prediction mechanism similar to Johnson's proposal for an extended instruction cache. An instruction cache line in UltraSPARC-I contained eight instructions. Each instruction pair had a 2-bit history, and each instruction quad had a 12-bit next-cache-line field. The history and next-line field were used to fill the instruction buffer, and this allocation of history bits was claimed to improve prediction accuracy by removing interference between multiple branches that map to the same entry in a traditional BHT. Branches were resolved after the first execute stage in the integer and floating-point pipelines.

The UltraSPARC-I was relatively aggressive in its memory interface. The instruction cache used a set prediction method that provided the access speed of a direct-mapped cache while retaining the reduced conflict behavior of a two-way set-associative cache. There was a nine-entry load buffer and an eight-entry store buffer. Load bypass was provided as well as write merging of the last two store buffer entries.

See Wayner [1994], Greenley et al. [1995], Lev et al. [1995], and Tremblay and O'Connor [1996] for descriptions of the UltraSPARC-I processor. Tremblay et al. [1995] discusses some of the tradeoff decisions made during the design of this processor and its memory system. Goldman and Tirumalai [1996] discuss the UltraSPARC-II, which adds memory system enhancements, such as prefetching, to the UltraSPARC-I core.

8.3.15.3 UltraSPARC-III / 2000.

Gary Lauterbach is the chief designer for the UltraSPARC-III, which retains the in-order issue approach of its predecessor. The UltraSPARC-III pipeline, however, is extended to 14 stages with careful attention to memory bandwidth. There are two integer units, a memory/special-instruction unit, a branch unit, and two floating-point units. Instructions are combined into groups of up to four instructions, and each group proceeds through the pipelines in lockstep manner. Grouping rules are reminiscent of the SuperSPARC and UltraSPARC-I. However, as in the Alpha 21164, the UltraSPARC-III rejects a global stall signaling scheme and instead adopts a replay approach.

Branches are predicted using a form of gshare with 16K predictors. Pipeline timing considerations led to a design with the pattern history table being held in eight banks. The xor result of 11 bits from the program counter and 11 bits from the global branch history shift register is used to read out one predictor per bank, and then an additional three low-order bits from the program counter are used in the next stage to select among the eight predictors. Simulations indicated that this approach has similar accuracy to the normal gshare scheme.

A four-entry miss queue for holding fall-through instructions is used along with a 16-entry instruction queue (although sometimes described as having

20 entries) to reduce the branch misprediction penalty for untaken branches. Conditional moves are available in the instruction set for partial predication, but the code optimization section of the manual advises that code performance is better with conditional branches than with conditional moves if the branches are fairly predictable.

To reduce the number of integer data forwarding paths, a variant of a future file, called the working register file, is used. Results are written to this structure in an out-of-order manner and are thus available to dependent instructions as early as possible. Registers are not renamed or tagged. Instead, age bits are included in the decoded instruction fields along with destination register IDs and are used to eliminate WAW hazards. WAR hazards are prevented by reading operands in the issue ("dispatch") stage. Precise exceptions are supported by not updating the architectural register file until the last stage, after all possible exceptions have been checked. If recovery is necessary, the working register file can be reloaded from the architectural register file in a single cycle.

The UltraSPARC-III has a 32K-byte L1 instruction cache and 64K-byte L1 data cache. The data cache latency is two cycles, which derives from a sum-addressed memory technique. A 2K-byte write cache allows the data cache to appear as write-through but defers the actual L2 update until a line has to be evicted from the write cache itself. Individual byte valid bits allow for storing only the changed bytes in the write cache and also support write-merging. A 2K-byte triple-ported prefetch cache is provided, which on each clock cycle can provide two independent 8-byte reads and receive 16 bytes from the main memory. In addition to the available software prefetch instructions, a hardware prefetch engine can detect the stride of a load instruction within a loop and automatically generates prefetch requests. Also included on-chip is a memory controller and cache tags for an 8-Mbyte L2 cache.

See Horel and Lauterbach [1999] and Lauterbach [1999] for more information on the UltraSPARC-III. The working register file is described in more detail in U.S. Patent 5,964,862. The UltraSPARC-IV is planned to be a chip multiprocessor with two UltraSPARC-III cores.

8.4 Verification of Superscalar Processors

Charles Moore (RSC, PPC 601, and POWER4) recently started a series of articles in *IEEE Micro* about the challenges of complexity faced by processor design teams [2003]. He suggested that a design team was in trouble when there were any less than two verification people assigned to clean up after each architect. While this simple and humorous rule of thumb may not hold in every case, it is true that superscalar processors are some of the most complex types of logical designs. It is not unusual to have over 100 in-flight instructions that may interact with each other in various ways and interact with corner cases, such as exceptions and faults. The combinatorial explosion of possible states is overwhelming. Indeed, several architects have chosen simpler in-order design strategies explicitly to reduce complexity and thereby improve time-to-market.

A study of verification techniques is beyond the scope of this chapter. However, since verification plays such an important role in the design process, a sampling of references to verification efforts for commercial superscalar processors follows. Articles on design and verification techniques used for Alpha processors includes Kantrowitz and Noack [1996], Grundmann et al. [1997], Reilly [1997], Dohm et al. [1998], Taylor et al. [1998], and Lee and Tsien [2001]. An article by Monaco et al. [1996] is a study of functional verification for the PPC 604, while an article by Ludden et al. [2002] is a more recent study of the same topic for POWER4. As a further sample of the approaches taken by industry design teams, Turumella et al. [1995] review design verification for the HaL SPARC64, Mangelsdorf et al. [1997] discuss the verification of the HP PA-8000, and Bentley and Gray [2001] present the verification techniques used for the Intel Pentium 4.

8.5 Acknowledgments to the Author (Mark Smotherman)

Several people were very helpful in providing information on the superscalar processors covered in this chapter: Tilak Agerwala, Fran Allen, Gene Amdahl, Erich Bloch, Pradip Bose, Fred Brooks, Brad Burgess, John Cocke, Lynn Conway, Marvin Denman, Mike Flynn, Greg Grohoski, Marty Hopkins, Peter Song, and Ed Sussenguth, all associated with IBM efforts; Don Alpert, Gideon Intrater, and Ran Talmudi, who worked on the NS Swordfish; Mitch Alsup, Joe Circello, and Keith Diefendorff, associated with Motorola efforts; Pete Bannon, John Edmondson, and Norm Jouppi, who worked for DEC; Mark Bluhm, who worked for Cyrix; Joel Boney, who worked on the SPARC64; Bob Colwell, Andy Glew, Mike Haertel, and Uri Weiser, who worked on Intel designs; Josh Fisher and Bill Worley of HP; Robert Garner, Sharad Mehrotra, Kevin Normoyle, and Marc Tremblay of Sun; Earl Killian, Kevin Kissell, John Ruttenberg, and Ross Towle, associated with SGI/MIPS; Steve Krueger of TI; Woody Lichtenstein and Dave Probert, both of whom worked on the Culler 7; Tim Olson; Yale Patt of the University of Texas and inventor of HPS; Jim Smith of the University of Wisconsin, designer of the ZS-1, and co-enthusiast for a processor pipeline version of *Jane's Fighting Ships;* and John Yates, who worked on the Apollo DN10000. Peter Capek of IBM was also instrumental in helping me obtain information on the ACS. I also want to thank numerous students at Clemson, including Michael Laird, Stan Cox, and T. J. Tumlin.

REFERENCES

Processor manuals are available from the individual manufacturers and are not included in the references.

> Acosta, R., J. Kjelstrup, and H. Torng: "An instruction issuing approach to enhancing performance in multiple functional unit processors," *IEEE Trans. on Computers,* C-35, 9, September 1986, pp. 815–828.
>
> Allen, F.: "The history of language processor technology in IBM," *IBM Journal of Research and Development,* 25, 5, September 1981, pp. 535–548.

Alpert, D., and D. Avnon: "Architecture of the Pentium microprocessor," *IEEE Micro,* 11, 3, June 1993, pp. 11–21.

Asprey, T., G. Averill, E. DeLano, R. Mason, B. Weiner, and J. Yetter: "Performance features of the PA7100 microprocessor," *IEEE Micro,* 13, 3, June 1993, pp. 22–35.

Bailey, D.: "High-performance Alpha microprocessor design," *Proc. Int. Symp. VLSI Tech., Systems and Appls.,* Taipei, Taiwan, June 1999, pp. 96–99.

Bannon, P., and J. Keller: "The internal architecture of Alpha 21164 microprocessor," *Proc. COMPCON,* San Francisco, CA, March 1995, pp. 79–87.

Barreh, J., S. Dhawan, T. Hicks, and D. Shippy: "The POWER2 processor," *Proc. COMPCON,* San Francisco, CA, Feb.–March 1994, pp. 389–398.

Bashe, C., L. Johnson, J. Palmer, and E. Pugh: *IBM's Early Computers.* Cambridge, MA: M.I.T. Press, 1986.

Becker, M., M. Allen, C. Moore, J. Muhich, and D. Tuttle: "The PowerPC 601 microprocessor," *IEEE Micro,* 13, 5, October 1993, pp. 54–67.

Benschneider, B., et al.: "A 300-MHz 64-b quad issue CMOS RISC microprocessor," *IEEE Journal of Solid-State Circuits,* 30, 11, November 1995, pp. 1203–1214. [21164]

Bentley, B., and R. Gray, "Validating the Intel Pentium 4 Processor," *Intel Technical Journal,* quarter 1, 2001, pp. 1–8.

Bishop, J., M. Campion, T. Jeremiah, S. Mercier, E. Mohring, K. Pfarr, B. Rudolph, G. Still, and T. White: "PowerPC AS A10 64-bit RISC microprocessor," *IBM Journal of Research and Development,* 40, 4, July 1996, pp. 495–505.

Blanchard, T., and P. Tobin: "The PA 7300LC microprocessor: A highly integrated system on a chip," *Hewlett-Packard Journal,* 48, 3, June 1997, pp. 43–47.

Blanck, G., and S. Krueger: "The SuperSPARC microprocessor," *Proc. COMPCON,* San Francisco, CA, February 1992, pp. 136–141.

Borkenhagen, J., R. Eickemeyer, R. Kalla, and S. Kunkel: "A multithreaded PowerPC processor for commercial servers," *IBM Journal of Research and Development,* 44, 6, November 2000, pp. 885–898. [SStar/RS64-IV]

Borkenhagen, J., G. Handlogten, J. Irish, and S. Levenstein: "AS/400 64-bit PowerPC-compatible processor implementation," *Proc. ICCD,* Cambridge, MA, October 1994, pp. 192–196.

Bose, P.: "Optimal code generation for expressions on super scalar machines," *Proc. AFIPS Fall Joint Computer Conf.,* Dallas, TX, November 1986, pp. 372–379.

Bowhill, W., et al.: "Circuit implementation of a 300-MHz 64-bit second-generation CMOS Alpha CPU," *Digital Technical Journal,* 7, 1, 1995, pp. 100–118. [21164]

Buchholz, W., *Planning a Computer System.* New York: McGraw-Hill, 1962. [IBM Stretch]

Burgess, B., M. Alexander, Y.-W. Ho, S. Plummer Litch, S. Mallick, D. Ogden, S.-H. Park, and J. Slaton: "The PowerPC 603 microprocessor: A high performance, low power, superscalar RISC microprocessor," *Proc. COMPCON,* San Francisco, CA, Feb.–March 1994a, pp. 300–306.

Burgess, B., N. Ullah, P. Van Overen, and D. Ogden: "The PowerPC 603 microprocessor," *Communications of the ACM,* 37, 6, June 1994b, pp. 34–42.

Burkhardt, B.: "Delivering next-generation performance on today's installed computer base," *Proc. COMPCON,* San Francisco, CA, Feb.–March 1994, pp. 11–16. [Cyrix 6x86]

Chan, K., et al.: "Design of the HP PA 7200 CPU," *Hewlett-Packard Journal*, 47, 1, February 1996, pp. 25–33.

Chen, C., Y. Lu, and A. Wong: "Microarchitecture of HaL's cache subsystem," *Proc. COMPCON*, San Francisco, CA, March 1995, pp. 267–271. [SPARC64]

Christie, D.: "Developing the AMD-K5 architecture," *IEEE Micro*, 16, 2, April 1996, pp. 16–26.

Circello, J., and F. Goodrich: "The Motorola 68060 microprocessor," *Proc. COMPCON*, San Francisco, CA, February 1993, pp. 73–78.

Circello, J., "The superscalar hardware architecture of the MC68060," Hot Chips VI, videotaped lecture, August 1994, **http://murl.microsoft.com/LectureDetails.asp?490.**

Circello, J., et al.: "The superscalar architecture of the MC68060," *IEEE Micro*, 15, 2, April 1995, pp. 10–21.

Cocke, J.: "The search for performance in scientific processors," *Communications of the ACM*, 31, 3, March 1988, pp. 250–253.

Cohler, E., and J. Storer: "Functionally parallel architecture for array processors," *IEEE Computer*, 14, 9, Sept. 1981, pp. 28–36. [MAP 200]

DeLano, E., W. Walker, J. Yetter, and M. Forsyth: "A high speed superscalar PA-RISC processor," *Proc. COMPCON*, San Francisco, CA, February 1992, pp. 116–121. [PA 7100]

Denman, M., "PowerPC 604 RISC microprocessor," Hot Chips VI, videotaped lecture, August 1994, **http://murl.microsoft.com/LectureDetails.asp?492.**

Denman, M., P. Anderson, and M. Snyder: "Design of the PowerPC 604e microprocessor," *Proc. COMPCON*, Santa Clara, CA, February 1996, pp. 126–131.

Diefendorff, K., "PowerPC 601 microprocessor," Hot Chips V, videotaped lecture, August 1993, **http://murl.microsoft.com/LectureDetails.asp?483.**

Diefendorff, K.: "History of the PowerPC architecture," *Communications of the ACM*, 37, 6, June 1994, pp. 28–33.

Diefendorff, K., "K7 challenges Intel," *Microprocessor Report*, 12, 14, October 26, 1998a, pp. 1, 6–11.

Diefendorff, K., "WinChip 4 thumbs nose at ILP," *Microprocessor Report*, 12, 16, December 7, 1998b, p. 1.

Diefendorff, K., "PowerPC G4 gains velocity," *Microprocessor Report*, 13, 14, October 25, 1999a, p. 1.

Diefendorff, K., "Hal makes Sparcs fly," *Microprocessor Report*, 13, 15, November 15, 1999b, pp. 1, 6–12.

Diefendorff, K., and M. Allen: "The Motorola 88110 superscalar RISC microprocessor," *Proc. COMPCON*, San Francisco, CA, February 1992a, pp. 157–162.

Diefendorff, K., and M. Allen: "Organization of the Motorola 88110 superscalar RISC microprocessor," *IEEE Micro*, 12, 2, April 1992b, pp. 40–63.

Diefendorff, K., R. Oehler, and R. Hochsprung: "Evolution of the PowerPC architecture," *IEEE Micro*, 14, 2, April 1994, pp. 34–49.

Diefendorff, K., and E. Silha: "The PowerPC user instruction set architecture," *IEEE Micro*, 14, 5, December 1994, pp. 30–41.

Ditzel, D., and H. McLellan: "Branch folding in the CRISP microprocessor: Reducing branch delay to zero," *Proc. ISCA*, Philadelphia, PA, June 1987, pp. 2–9.

Ditzel, D., H. McLellan, and A. Berenbaum: "The hardware architecture of the CRISP microprocessor," *Proc. ISCA,* Philadelphia, PA, June 1987, pp. 309–319.

Dohm, N., C. Ramey, D. Brown, S. Hildebrandt, J. Huggins, M. Quinn, and S. Taylor, "Zen and the art of Alpha verification," *Proc. ICCD,* Austin, TX, October 1998, pp. 111–117.

Eden, M., and M. Kagan: "The Pentium processor with MMX technology," *Proc. COMPCON,* San Jose, CA, February 1997, pp. 260–262.

Edmondson, J., "An overview of the Alpha AXP 21164 microarchitecture," Hot Chips VI, videotaped lecture, August 1994, **http://murl.microsoft.com/LectureDetails.asp?493.**

Edmondson, J., et al.: "Internal organization of the Alpha 21164, a 300-MHz 64-bit quad-issue CMOS RISC microprocessor," *Digital Technical Journal,* 7, 1, 1995a, pp. 119–135.

Edmondson, J., P. Rubinfeld, R. Preston, and V. Rajagopalan: "Superscalar instruction execution in the 21164 Alpha microprocessor," *IEEE Micro,* 15, 2, April 1995b, pp 33–43.

Fisher, J.: "Very long instruction word architectures and the ELI-512," *Proc. ISCA,* Stockholm, Sweden, June 1983, pp. 140–150.

Flynn, M.: "Very high-speed computing systems," *Proc. IEEE,* 54, 12, December 1966, pp. 1901–1909.

Gaddis, N., and J. Lotz: "A 64-b quad-issue CMOS RISC microprocessor," *IEEE Journal of Solid-State Circuits,* 31, 11, November 1996, pp. 1697–1702. [PA 8000]

Gieseke, B., et al.: "A 600MHz superscalar RISC microprocessor with out-of-order execution," *Proc. IEEE Int. Solid-State Circuits Conference,* February 1997. pp. 176–177. [21264]

Gochman, S., et al., "The Pentium M processor: Microarchitecture and performance," *Intel Tech. Journal,* 7, 2, May 2003, pp. 21–36.

Goldman, G., and P. Tirumalai: "UltraSPARC-II: The advancement of ultracomputing," *Proc. COMPCON,* Santa Clara, CA, February 1996, pp. 417–423.

Gowan, M., L. Brio, and D. Jackson, "Power considerations in the design of the Alpha 21264 microprocessor," *Proc. Design Automation Conf.,* San Francisco, CA, June 1998, pp. 726–731.

Greenley, D., et al.: "UltraSPARC: The next generation superscalar 64-bit SPARC," *Proc. COMPCON,* San Francisco, CA, March 1995, pp. 442–451.

Gronowski, P., et al.: "A 433-MHz 64-b quad-issue RISC microprocessor," *IEEE Journal of Solid-State Circuits,* 31, 11, November 1996, pp. 1687–1696. [21164A]

Grundmann, W., D. Dobberpuhl, R. Almond, and N. Rethman, "Designing high performance CMOS microprocessors using full custom techniques," *Proc. Design Automation Conf.,* Anaheim, CA, June 1997, pp. 722–727.

Gwennap, L.: "Cyrix describes Pentium competitor," *Microprocessor Report,* 7, 14, October 25, 1993, pp. 1–6. [M1/6x86]

Gwennap, L.: "AMD's K5 designed to outrun Pentium," *Microprocessor Report,* 8, 14, October 14, 1994, pp. 1–7.

Gwennap, L.: "Intel's P6 uses decoupled superscalar design," *Microprocessor Report,* 9, 2, February 16, 1995, pp. 9–15.

Halfhill, T.: "AMD vs. Superman," *Byte,* 19, 11, November 1994a, pp. 95–104. [AMD K5]

Halfhill, T.: "T5: Brute force," *Byte,* 19, 11, November 1994b, pp. 123–128. [MIPS R10000]

Halfhill, T.: "Intel's P6," *Byte,* 20, 4, April 1995, p. 435. [Pentium Pro]

Halfhill, T.: "AMD K6 takes on Intel P6," *Byte,* 21, 1, January 1996a, pp. 67–72.

Halfhill, T.: "PowerPC speed demon," *Byte,* 21, 12, December 1996b, pp. 88NA1–88NA8.

Halfhill, T., "IBM trims Power4, adds AltiVec," *Microprocessor Report,* October 28, 2002.

Hall, C., and K. O'Brien: "Performance characteristics of architectural features of the IBM RISC System/6000," *Proc. ASPLOS-IV,* Santa Clara, CA, April 1991, pp. 303–309.

Hester, P., "Superscalar RISC concepts and design of the IBM RISC System/6000," videotaped lecture, August 1990, **http://murl.microsoft.com/LectureDetails.asp?315.**

Hewlett-Packard: "PA-RISC 8x00 family of microprocessors with focus on PA-8700," Hewlett Packard Corporation, Technical White Paper, April 2000.

Hinton, G.: "80960—Next generation," *Proc. COMPCON,* San Francisco, CA, March 1989, pp. 13–17.

Hinton, G., D. Sager, M. Upton, D. Boggs, D. Carmean, A. Kyker, and P. Roussel: "The microarchitecture of the Pentium 4 processor," *Intel Technology Journal,* Quarter 1, 2001, pp. 1–12.

Hollenbeck, D., A. Undy, L. Johnson, D. Weiss, P. Tobin, and R. Carlson: "PA7300LC integrates cache for cost/performance," *Proc. COMPCON,* Santa Clara, CA, February 1996, pp. 167–174.

Horel, T., and G. Lauterbach: "UltraSPARC-III: Designing third generation 64-bit performance," *IEEE Micro,* 19, 3, May–June, 1999, p. 85.

Horst, R., R. Harris, and R. Jardine: "Multiple instruction issue in the NonStop Cyclone processor," *Proc. ISCA,* Seattle, WA, May 1990, pp. 216–226.

Hsu, P., "Silicon GraphicsTFP micro-supercomputer chipset," Hot Chips V, videotaped lecture, August 1993, **http://murl.microsoft.com/LectureDetails.asp?484.** [R8000]

Hsu, P.: "Designing the TFP microprocessor," *IEEE Micro,* 14, 2, April 1994, pp. 23–33. [MIPS R8000]

Hunt, J.: "Advanced performance features of the 64-bit PA-8000," *Proc. COMPCON,* San Francisco, CA, March 1995, pp. 123–128.

IBM: *IBM RISC System/6000 Technology.* Austin, TX: IBM Corporation, 1990, p. 421.

Johnson, L., and S. Undy: "Functional design of the PA 7300LC," *Hewlett-Packard Journal,* 48, 3, June 1997, pp. 48–63.

Jouppi, N., and D. Wall: "Available instruction-level parallelism for superscalar and superpipelined machines," *Proc. ASPLOS-III,* Boston, MA, April 1989, pp. 272–282.

Kantrowitz, M., and L. Noack, "I'm done simulating: Now what? Verification coverage analysis and correctness checking of the DECchip 21164 Alpha microprocessor," *Proc. Design Automation Conf.,* Las Vegas, NV, June 1996, pp. 325–330.

Keltcher, C., K. McGrath, A. Ahmed, and P. Conway, "The AMD Opteron processor for multiprocessor servers," *IEEE Micro,* 23, 2, March–April 2003, pp. 66–76.

Kennedy, A., et al.: "A G3 PowerPC superscalar low-power microprocessor," *Proc. COMPCON,* San Jose, CA, February 1997, pp. 315–324. [PPC 740/750, but one diagram lists this chip as the 613.]

Kessler, R.: "The Alpha 21264 microprocessor," *IEEE Micro,* 19, 2, March–April 1999, pp. 24–36.

Kessler, R., E. McLellan, and D. Webb: "The Alpha 21264 microprocessor architecture," *Proc. ICCD,* Austin, TX, October 1998, pp. 90–95.

Knebel, P., B. Arnold, M. Bass, W. Kever, J. Lamb, R. Lee, P. Perez, S. Undy, and W. Walker: "HP's PA7100LC: A low-cost superscalar PA-RISC processor," *Proc. COMPCON,* San Francisco, CA, February 1993, pp. 441–447.

Krewell, K., "Fujitsu's SPARC64 V is real deal," *Microprocessor Report,* 16, 10, October 21, 2002, pp. 1–4.

Kurpanek, G., K. Chan, J. Zheng, E. DeLano, and W. Bryg: "PA7200: A PA-RISC processor with integrated high performance MP bus interface," *Proc. COMPCON,* San Francisco, CA, Feb.–March 1994, pp. 375–382.

Lauterbach, G.: "Vying for the lead in high-performance processors," *IEEE Computer,* 32, 6, June 1999, pp. 38-41. [UltraSPARC III]

Lee, R., and J. Huck, "64-bit and multimedia extensions in the PA-RISC 2.0 architecture," *Proc. COMPCON,* Santa Clara, CA, February 1996, pp. 152–160.

Lee, R., and B. Tsien, "Pre-silicon verification of the Alpha 21364 microprocessor error handling system," *Proc. Design Automation Conf.,* Las Vegas, NV, 2001, pp. 822–827.

Leibholz, D., and R. Razdan: "The Alpha 21264: A 500 MHz out-of-order-execution microprocessor," *Proc. COMPCON,* San Jose, CA, February 1997, pp. 28–36.

Lempel, O., A. Peleg, and U. Weiser: "Intel's MMX technology—A new instruction set extension," *Proc. COMPCON,* San Jose, CA, February 1997, pp. 255–259.

Lesartre, G., and D. Hunt: "PA-8500: The continuing evolution of the PA-8000 family," *Proc. COMPCON,* San Jose, CA, February 1997.

Lev, L., et al.: "A 64-b microprocessor with multimedia support," *IEEE Journal of Solid-State Circuits,* 30, 11, November 1995, pp. 1227–1238. [UltraSPARC]

Levitan, D., T. Thomas, and P. Tu: "The PowerPC 620 microprocessor: A high performance superscalar RISC microprocessor," *Proc. COMPCON,* San Francisco, CA, March 1995, pp. 285–291.

Lichtenstein, W.: "The architecture of the Culler 7," *Proc. COMPCON,* San Francisco, CA, March 1986, pp. 467–470.

Lightner, B., "Thunder SPARC processor," Hot Chips VI, videotaped lecture, August 1994, **http://murl.microsoft.com/LectureDetails.asp?494.**

Lightner, B., and G. Hill: "The Metaflow Lightning chipset," *Proc. COMPCON,* San Francisco, CA, February 1991, pp. 13–18.

Liptay, J. S.: "Design of the IBM Enterprise System/9000 high-end processor," *IBM Journal of Research and Development,* 36, 4, July 1992, pp. 713–731.

Ludden, J., et al.: "Functional verification of the POWER4 microprocessor and POWER4 multiprocessor systems," *IBM Journal of Research and Development,* 46, 1, 2002, pp. 53–76.

Mangelsdorf, et al.: "Functional verification of the HP PA 8000 processor," *HP Journal,* 48, 4, August 1997, pp. 22–31.

Matson, M., et al., "Circuit implementation of a 600 MHz superscalar RISC microprocessor," *Proc. ICCD,* Austin, TX, October 1998, pp. 104–110. [Alpha 21264]

May, D., R. Shepherd, and P. Thompson, "The T9000 Transputer," *Proc. ICCD,* Cambridge, MA, October 1992, pp. 209–212.

McGeady, S.: "The i960CA superscalar implementation of the 80960 architecture," *Proc. COMPCON,* San Francisco, CA, February 1990a, pp. 232–240.

McGeady, S.: "Inside Intel's i960CA superscalar processor," *Microprocessors and Microsystems,* 14, 6, July/August 1990b, pp. 385–396.

McGeady, S., R. Steck, G. Hinton, and A. Bajwa: "Performance enhancements in the superscalar i960MM embedded microprocessor," *Proc. COMPCON,* San Francisco, CA, February 1991, pp. 4–7.

McGrath, K., "x86-64: Extending the x86 architecture to 64 bits," videotaped lecture, September 2000, **http://murl.microsoft.com/LectureDetails.asp?690.**

McLellan, E.: "The Alpha AXP architecture and 21064 processor," *IEEE Micro,* 11, 3, June 1993, pp. 36–47.

McMahan, S., M. Bluhm, and R. Garibay, Jr.: "6x86: The Cyrix solution to executing x86 binaries on a high performance microprocessor," *Proc. IEEE,* 83, 12, December 1995, pp. 1664–1672.

Mirapuri, S., M. Woodacre, and N. Vasseghi: "The Mips R4000 processor," *IEEE Micro,* 12, 2, April 1992, pp. 10–22.

Monaco, J., D. Holloway, and R. Raina: "Functional verification methodology for the PowerPC 604 microprocessor," *Proc. Design Automation Conf.,* Las Vegas, NV, June 1996, pp. 319–324.

Montanaro, J.: "The design of the Alpha 21064 CPU chip," videotaped lecture, April 1992, **http://murl.microsoft.com/LectureDetails.asp?373.**

Moore, C.: "The PowerPC 601 microprocessor," *Proc. COMPCON,* San Francisco, CA, February 1993, pp. 109–116.

Moore, C.: "Managing the transition from complexity to elegance: Knowing when you have a problem," *IEEE Micro,* 23, 5, Sept.–Oct. 2003, pp. 86–88.

Moore, C., D. Balser, J. Muhich, and R. East: "IBM single chip RISC processor (RSC)," *Proc. ICCD,* Cambridge, MA, October 1989, pp. 200–204.

O'Connell, F., and S. White: "POWER3: The next generation of PowerPC processors," *IBM Journal of Research and Development,* 44, 6, November 2000, pp. 873–884.

Oehler, R., and M. Blasgen: "IBM RISC System/6000: Architecture and performance," *IEEE Micro,* 11, 3, June 1991, pp. 54–62.

Papworth, D.: "Tuning the Pentium Pro microarchitecture," *IEEE Micro,* 16, 2, April 1996, pp. 8–15.

Patkar, N., A. Katsuno, S. Li, T. Maruyama, S. Savkar, M. Simone, G. Shen, R. Swami, and D. Tovey: "Microarchitecture of HaL's CPU," *Proc. COMPCON,* San Francisco, CA, March 1995, pp. 259–266. [SPARC64]

Patt, Y., S. Melvin, W-M. Hwu, M. Shebanow, C. Chen, and J. Wei: "Run-time generation of HPS microinstructions from a VAX instruction stream," *Proc. MICRO-19,* New York, December 1986, pp. 75–81.

Peleg, A., and U. Weiser: "MMX technology extension to the Intel architecture," *IEEE Micro,* 16, 4, August 1996, pp. 42–50.

Peng, C. R., T. Petersen, and R. Clark: "The PowerPC architecture: 64-bit power with 32-bit compatibility," *Proc. COMPCON,* San Francisco, CA, March 1995, pp. 300–307.

Popescu, V., M. Schultz, J. Spracklen, G. Gibson, and B. Lightner: "The Metaflow architecture," *IEEE Micro,* 11, 3, June 1991, pp. 10–23.

Potter, T., M. Vaden, J. Young, and N. Ullah: "Resolution of data and control-flow dependencies in the PowerPC 601," *IEEE Micro,* 14, 5, October 1994, pp. 18–29.

Poursepanj, A., D. Ogden, B. Burgess, S. Gary, C. Dietz, D. Lee, S. Surya, and M. Peters: "The PowerPC 603 Microprocessor: Performance analysis and design tradeoffs," *Proc. COMPCON,* San Francisco, CA, Feb.–March 1994, pp. 316–323.

Preston, R., et al., "Design of an 8-wide superscalar RISC microprocessor with simultaneous multithreading," *Proc. ISSCC,* San Francisco, CA, February 2002, p. 334. [Alpha 21464]

Pugh, E., L. Johnson, and J. Palmer: *IBM's 360 and Early 370 Systems.* Cambridge, MA: MIT Press, 1991.

Rau, B., C. Glaeser, and R. Picard: "Efficient code generation for horizontal architectures: Compiler techniques and architectural support," *Proc. ISCA,* Austin, TX, April 1982, pp. 131–139. [ESL machine, later Cydrome Cydra-5]

Reilly, M.: "Designing an Alpha processor," *IEEE Computer,* 32, 7, July 1999, pp. 27–34.

Riseman, E., and C. Foster: "The inhibition of potential parallelism by conditional jumps," *IEEE Trans. on Computers,* C-21, 12, December 1972, pp. 1405–1411.

Ryan, B.: "M1 challenges Pentium," *Byte,* 19, 1, January 1994a, pp. 83–87. [Cyrix 6x86]

Ryan, B.: "NexGen Nx586 straddles the RISC/CISC divide," *Byte,* 19, 6, June 1994b, p. 76.

Schorr, H.: "Design principles for a high-performance system," *Proc. Symposium on Computers and Automata,* New York, April 1971, pp. 165–192. [IBM ACS]

Seznec, A., S. Felix, V. Krishnan, and Y. Sazeides: "Design tradeoffs for the Alpha EV8 conditional branch predictor," *Proc. ISCA,* Anchorage, AK, May 2002, pp. 295–306.

Shebanow, M., "SPARC64 V: A high performance and high reliability 64-bit SPARC processor," videotaped lecture, December 1999, **http://murl.microsoft.com/LectureDetails.asp?455.**

Shen, J. P., and A. Wolfe: "Superscalar processor design," Tutorial, ISCA, San Diego, CA, May 1993.

Shippy, D.: "POWER2+ processor," Hot Chips VI, videotaped lecture, August 1994, **http://murl.microsoft.com/LectureDetails.asp?495.**

Shriver, B., and B. Smith: *The Anatomy of a High-Performance Microprocessor: A Systems Perspective.* Los Alamitos, CA: IEEE Computer Society Press, 1998. [AMD K6-III]

Simone, M., A. Essen, A. Ike, A. Krishnamoorthy, T. Maruyama, N. Patkar, M. Ramaswami, M. Shebanow, V. Thirumalaiswamy, and D. Tovey: "Implementation tradeoffs in using a restricted data flow architecture in a high performance RISC microprocessor," *Proc. ISCA,* Santa Margherita Ligure, Italy, May 1995, pp. 151–162. [HaL SPARC64]

Sites, R.: "Alpha AXP architecture," *Communications of the ACM,* 36, 2, February 1993, pp. 33–44.

Smith, J. E.: "Decoupled access/execute computer architectures," *Proc. ISCA,* Austin, TX, April 1982, pp. 112–119.

Smith, J. E.: "Decoupled access/execute computer architectures," *ACM Trans. on Computer Systems,* 2, 4, November 1984, pp. 289–308.

Smith, J. E., G. Dermer, B. Vanderwarn, S. Klinger, C. Rozewski, D. Fowler, K. Scidmore, and J. Laudon: "The ZS-1 central processor," *Proc. ASPLOS-II,* Palo Alto, CA, October 1987, pp. 199–204.

Smith, J. E., and T. Kaminski: "Varieties of decoupled access/execute computer architectures," *Proc. 20th Annual Allerton Conf. on Communication, Control, and Computing,* Monticello, IL, October 1982, pp. 577–586.

Smith, J. E., and S. Weiss: "PowerPC 601 and Alpha 21064: A tale of two RISCs," *IEEE Computer,* 27, 6, June 1994, pp. 46–58.

Smith, J. E., S. Weiss, and N. Pang: "A simulation study of decoupled architecture computers," *IEEE Trans. on Computers,* C-35, 8, August 1986, pp. 692–702.

Smith, M., M. Johnson, and M. Horowitz: "Limits on multiple instruction issue," *Proc. ASPLOS-III,* Boston, MA, April 1989, pp. 290–302.

Soltis, F. *Fortress Rochester: The Inside Story of the IBM iSeries.* Loveland, CO: 29th Street Press, 2001.

Song, P., "HAL packs SPARC64 onto single chip," *Microprocessor Report,* 11, 16, December 8, 1997a, p. 1.

Song, P., "IBM's Power3 to replace P2SC," *Microprocessor Report,* 11, 15, November 17, 1997b, pp. 23–27.

Song, S., M. Denman, and J. Chang: "The PowerPC 604 RISC microprocessor," *IEEE Micro,* 14, 5, October 1994, pp. 8–17.

Special issue: "The IBM RISC System/6000 processor," *IBM Journal of Research and Development,* 34, 1, January 1990.

Special issue: "Alpha AXP architecture and systems," *Digital Technical Journal,* 4, 4, 1992.

Special issue: "Digital's Alpha chip project," *Communications of the ACM,* 36, 2, February 1993.

Special issue: "The making of the PowerPC," *Communications of the ACM,* 37, 6, June 1994.

Special issue: "POWER2 and PowerPC architecture and implementation," *IBM Journal of Research and Development,* 38, 5, September 1994.

Special issue: *Hewlett-Packard Journal,* 46, 2, April 1995. [HP PA 7100LC]

Special issue: *Hewlett-Packard Journal,* 48, 4, August 1997. [HP PA 8000 and PA 8200]

Special issue: *IBM Journal of Research and Development,* 46, 1, 2002. [POWER4]

Sporer, M., F. Moss, and C. Mathias: "An introduction to the architecture of the Stellar graphics supercomputer," *Proc. COMPCON,* San Francisco, CA, 1988, pp. 464–467. [GS-1000]

Sussenguth, E.: "Advanced Computing Systems," video-taped talk, Symposium in Honor of John Cocke, IBM T. J. Watson Research Center, Yorktown Heights, NY, June 18, 1990.

Taylor, S., et al.: "Functional verification of a multiple-issue, out-of-order, superscalar Alpha microprocessor," *Proc. Design Automation Conf.,* San Francisco, CA, 1998, pp. 638–643.

Tendler, J., J. Dodson, J. Fields, H. Le, and B. Sinharoy, "POWER4 system microarchitecture," *IBM Journal of Research and Development,* 46, 1, 2002, pp. 5–26.

Thompson, T., and B. Ryan: "PowerPC 620 soars," *Byte,* 19, 11, November 1994, pp. 113–120.

Tjaden, G., and M. Flynn: "Detection of parallel execution of independent instructions," *IEEE Trans. on Computers,* C-19, 10, October 1970, pp. 889–895.

Tremblay, M., D. Greenly, and K. Normoyle: "The design of the microarchitecture of the UltraSPARC I," *Proc. IEEE,* 83, 12, December 1995, pp. 1653–1663.

Tremblay, M., and J. M. O'Connor: "UltraSPARC I: A four-issue processor supporting multimedia," *IEEE Micro,* 16, 2, April 1996, pp. 42–50.

Turumella, B., et al.: "Design verification of a super-scalar RISC processor," *Proc. Fault Tolerant Computing Symposium,* Pasadena, CA, June 1995, pp. 472–477. [HaL SPARC64]

Ullah, N., and M. Holle: "The MC88110 implementation of precise exceptions in a superscalar architecture," *ACM Computer Architecture News,* 21, 1, March 1993, pp. 15–25.

Undy, S., M. Bass, D. Hollenbeck, W. Kever, and L. Thayer: "A low-cost graphics and multimedia workstation chip set," *IEEE Micro,* 14, 2, April 1994, pp. 10–22. [HP 7100LC]

Vasseghi, N., K. Yeager, E. Sarto, and M. Seddighnezhad: "200-MHz superscalar RISC microprocessor," *IEEE Journal of Solid-State Circuits,* 31, 11, November 1996, pp. 1675–1686. [MIPS R10000]

Vegesna, R.: "Pinnacle-1: The next generation SPARC processor," *Proc. COMPCON,* San Francisco, CA, February 1992, pp. 152–156. [HyperSPARC]

Wayner, P.: "SPARC strikes back," *Byte,* 19, 11, November 1994, pp. 105–112. [UltraSPARC]

Weiss, S., and J. E. Smith: *POWER and PowerPC.* San Francisco, CA: Morgan Kaufmann, 1994.

White, S.: "POWER2: Architecture and performance," *Proc. COMPCON,* San Francisco, CA, Feb.–March 1994, pp. 384–388.

Wilcke, W.: "Architectural overview of HaL systems," *Proc. COMPCON,* San Francisco, CA, March 1995, pp. 251–258. [SPARC64]

Williams, T., N. Patkar, and G. Shen: "SPARC64: A 64-b 64-active-instruction out-of-order-execution MCM processor," *IEEE Journal of Solid-State Circuits,* 30, 11, November 1995, pp. 1215–1226.

Wilson, J., S. Melvin, M. Shebanow, W.-M. Hwu, and Y. Patt: "On tuning the microarchitecture of an HPS implementation of the VAX," *Proc. Micro-20,* Colorado Springs, CO, December 1987, pp. 162–167. [This proceeding is hard to obtain, but the paper also appears in reduced size in *SIGMICRO Newsletter,* 19, 3, September 1988, pp. 56–58.]

Yeager, K.: "The MIPS R10000 superscalar microprocessor," *IEEE Micro,* 16, 2, April 1996, pp. 28–40.

HOMEWORK PROBLEMS

P8.1 Although logic design techniques and microarchitectural tradeoffs can be treated as independent design decisions, explain the typical pairing of synthesized logic and a brainiac design style versus full custom logic and a speed-demon design style.

P8.2 In the late 1950s, the Stretch designers placed a limit of 23 gate levels on any logic path. As recently as 1995, the UltraSPARC-I was designed with 20 gate levels per pipe stage. Yet many designers have tried to drastically reduce this number. For example, the ACS had a target of five gate levels of logic per stage, and the Ultra-SPARC-III uses the equivalent of eight gate levels per stage. Explain the rationale for desiring low gate-level counts. (You may also want to examine the lower level-count trend in recent Intel processors, as discussed by Hinton et al. [2001].)

P8.3 Prepare a table comparing the approaches to floating-point arithmetic exception handling found on these IBM designs: Stretch, ACS, RIOS, PowerPC 601, PowerPC 620, POWER4.

P8.4 Consider Table 8-2. Can you identify any trends? If so, suggest a rationale for each trend you identify.

P8.5 Explain the market forces that led to the demise of the Compaq/DEC Alpha. Are there any known blemishes in the Alpha instruction set that make high-performance implementations particularly difficult or inefficient? Is the Alpha tradition of full custom logic design too labor or resource intensive?

P8.6 Compare how the Compaq/DEC Alpha and IBM RIOS eliminated the type of complex instruction pairing rules that are found in the Intel i960 CA.

P8.7 Explain the importance of caches in HP processor designs. How was the assist cache used in the HP 7200 both a surprise and a natural development in the HP design philosophy?

P8.8 Find a description of load/store locality hints in the Itanium Processor Family. Compare the Itanium approach with the approaches used in the HP 7200 and MIPS R8000.

P8.9 Consider the IBM RIOS.

(a) The integer unit pipeline design is the same as the pipeline used in the IBM 801. Explain the benefit of routing the cache bypass path directly from the ALU stage to the writeback stage as opposed to this bypass being contained within the cache stage (and thus having ALU results required to flow through the ALU/cache and cache/writeback latches as done in the simple five- and six-stage scalar pipeline designs of Chapter 2). What is the cost of this approach in terms of the integer register file design?

(b) New physical registers are assigned only for floating-point loads. For what types of code segments is this sufficient?

(c) Draw a pipeline timing diagram showing that a floating-point load and a dependent floating-point instruction can be fetched, dispatched, and issued together without any stalls resulting.

P8.10 Why did the IBM RIOS provide three separate logic units, each with a separate register set? This legacy has been carried into the PowerPC instruction set. Is this legacy a help, hindrance, or inconsequential to high-issue-rate PowerPC implementations?

P8.11 Identify market and/or design factors that have led to the long life span of the Intel i960 CA.

P8.12 Is the Intel P6 a speed demon, a brainiac, or both? Explain your answer.

P8.13 Consider the completion/retirement logic of the PowerPC designs. How are the 601, 620, and POWER4 related?

P8.14 Draw a pipeline timing diagram illustrating how the SuperSPARC processor deals with a delayed branch and its delay slot instruction.

P8.15 The UltraSPARC-I provides in-order completion at the cost of empty stages and additional forwarding paths.

  (a) Give a list of the pipe stage destinations for the forwarding paths that must accompany an empty integer pipeline stage.

  (b) List the possible sources of inputs to the multiplexer that fronts one leg of the ALU in the integer execution pipe stage. (*Note:* This is more than the empty integer stages.)

  (c) Describe how the number of forwarding paths was reduced in the UltraSPARC-III, which had even more pipeline stages.

CHAPTER 9

Gabriel H. Loh

Advanced Instruction Flow Techniques

> **CHAPTER OUTLINE**
>
> 9.1 Introduction
> 9.2 Static Branch Prediction Techniques
> 9.3 Dynamic Branch Prediction Techniques
> 9.4 Hybrid Branch Predictors
> 9.5 Other Instruction Flow Issues and Techniques
> 9.6 Summary
>
> References
> Homework Problems

9.1 Introduction

In Chapter 5, it was stated that the instruction flow, or the processing of branches, provides an upper bound on the throughput of all subsequent stages. In particular, conditional branches in programs are a serious bottleneck to improving the rate of instruction flow and, hence, the performance of the processor. Before a conditional branch is resolved in a pipelined processor, it is unknown which instructions should follow the branch. To increase the number of instructions that execute in parallel, modern processors make a *branch prediction* and speculatively execute the instructions in the predicted path of program control flow. If the branch is discovered later on to have been mispredicted, actions are taken to recover the state of the processor to the point before the mispredicted branch, and execution is resumed along the correct path.

The penalty associated with mispredicted branches in modern pipelined processors has a great impact on performance. The performance penalty is increased as the pipelines deepen and the number of outstanding instructions increases. For example, the AMD Athlon processor has 10 stages in the integer pipeline [Meyer, 1998], while the Intel NetBurst microarchitecture used in the Pentium 4 processor is "hyper-pipelined" with a 20-stage branch misprediction penalty [Hinton et al., 2001]. Several studies have suggested that the processor pipeline depth may continue to grow to 30 to 50 stages [Hartstein and Puzak, 2002; Hrishikesh et al., 2002]. Wide-issue superscalar processors further exacerbate the problem by creating a greater demand for instructions to execute. Despite the huge body of existing research in branch predictor design, these microarchitecture trends toward deeper and wider designs will continue to create a demand for more accurate branch prediction algorithms.

Processing conditional branches has two major components: predicting the branch direction and predicting the branch target. Sections 9.2 through 9.4, the bulk of this chapter, focus on the former problem of predicting whether a conditional branch is taken or not taken. Section 9.5 discusses the problem of branch target prediction and other issues related to effective instruction delivery.

Over the past two to three decades, there has been an incredible body of published research on the problem of predicting conditional branches and fetching instructions. The goal of this chapter is to take all these papers and distill the information down to the key ideas and concepts. Absolute comparisons such as whether one branch prediction algorithm is more accurate than another are difficult to make since such comparisons depend on a large number of assumptions such as the instruction set architecture, die area and clock frequency limitations, and choice of applications. The text makes note of techniques that have been implemented in commercial processors, but this does not necessarily imply that these algorithms are inherently better than some of the alternatives covered in this chapter. This chapter surveys a wide breadth of techniques with the aim of making the reader aware of the design issues and known methods in dealing with instruction flow.

The predictors described in this chapter are organized by how they make their predictions. Section 9.2 covers static branch predictors, that is, predictors that do not make use of any run-time information about branch behavior. Section 9.3 explains a wide variety of dynamic branch prediction algorithms, that is, predictors that can monitor branch behavior while the program is running and make future predictions based on these observations. Section 9.4 describes hybrid branch predictors that combine the strengths of multiple simpler predictors to form a better overall predictor.

9.2 Static Branch Prediction Techniques

Static branch prediction algorithms tend to be very simple and by definition do not incorporate any feedback from the run-time environment. This characteristic is both the strength and weakness of static prediction algorithms. By not paying any attention to the dynamic run-time behavior of a program, the branch prediction is incapable of adapting to changes in branch outcome patterns. These patterns may vary based on the input set for the program or different phases of a program's execution.

The advantage of static branch prediction techniques is that they are very simple to implement, and they require very little hardware resources. Static branch prediction algorithms are of less interest in the context of future-generation, large transistor budget, very large-scale integration (VLSI) processors because the additional area for more effective dynamic branch predictors can be afforded. Nevertheless, static branch predictors may still be used as components in more complex hybrid branch predictors or as a simpler fallback predictor when no other prediction information is available.

Profile-based static prediction can achieve better performance than simpler *rule-based* algorithms. The key assumption underlying profile-based approaches is that the actual run-time behavior of a program can be approximated by different runs of the program on different data sets. In addition to the branch outcome statistics of sample executions, profile-based algorithms may also take advantage of information that is available at compile time such as the high-level structure of the program. The main disadvantage with profile-based techniques is that profiling must be part of the compilation phase of the program, and existing programs cannot take advantage of the benefits without being recompiled. If the branch behavior statistics collected from the training runs are not representative of the branch behavior in the actual run, then the profile-based predictions may not provide much benefit.

This section continues with a brief survey of some of the rule-based static branch prediction algorithms and then presents an overview of profile-based static branch prediction.

9.2.1 Single-Direction Prediction

The simplest branch prediction strategy is to predict that the direction of all branches will always go in the same direction (always taken or always not taken). Older pipelined processors, such as the Intel i486 [Intel Corporation, 1997], used the always-not-taken prediction algorithm. This trivial strategy simplifies the task of fetching instructions because the next instruction to fetch after a branch is always the next sequential instruction in the static order of the program. Apart from cache misses and branch mispredictions, the instructions will be fetched in an uninterrupted stream. Unfortunately, branches are more often taken than not taken. For integer benchmarks, branches are taken approximately 60% of the time [Uht, 1997].

The opposite strategy is to always predict that a branch will be taken. Although this usually achieves a higher prediction accuracy rate than an always-not-taken strategy, the hardware is more complex. The problem is that the branch target address is generally unavailable at the time the branch prediction is made. One solution is to simply stall the front end of the pipeline until the branch target has been computed. This wastes processing slots in the pipeline (i.e., this causes pipeline bubbles) and leads to reduced performance. If the branch instruction specifies its target in a PC-relative fashion, the destination address may be computed in as little as an extra cycle of delay. Such was the case for the early MIPS R-series pipelines [Kane and Heinrich, 1992]. In an attempt to recover some of the lost processing cycles due to the pipeline bubbles, a *branch delay slot* after the branch instruction was architected into the ISA. That is, the instruction immediately following a branch instruction is always executed regardless of the outcome of the branch. In theory, the branch delay slots can then be filled with useful instructions, although studies have shown that compilers cannot effectively make use of all the available delay slots [McFarling and Hennessy, 1986]. Faster cycle times may introduce more pipeline stages before the branch target calculation has completed, thus increasing the number of wasted cycles.

9.2.2 Backwards Taken/Forwards Not-Taken

A variation of the single-direction static prediction approach is the backwards taken/forwards not-taken (BTFNT) strategy. A *backwards branch* is a branch instruction that has a target with a lower address (i.e., one that comes earlier in the program). The rationale behind this heuristic is that the majority of backwards branches are loop branches, and since loops usually iterate many times before exiting, these branches are most likely to be taken. This approach does not require any modifications to the ISA since the sign of the target displacement is already encoded in the branch instruction. Many processors have used this prediction strategy; for example, the Intel Pentium 4 processor uses the BTFNT approach as a backup strategy when its dynamic predictor is unable to provide a prediction [Intel Corporation, 2003].

9.2.3 Ball/Larus Heuristics

Some instruction set architectures provide the compiler an interface through which *branch hints* can be made. These hints are encoded in the branch instructions, and an implementation of an ISA may choose to use these hints or not. The compiler can make use of these branch hints by inserting what it believes are the most likely outcomes of the branches based on high-level information about the structure of the program. This kind of static prediction is called *program-based* prediction.

There are branches in programs that almost always go in the same direction, but knowing the direction may require some high-level understanding of the programming language or the application itself. For example, consider the following code:

```
void * p = malloc (numBytes);
if (p == NULL)
    errorHandlingFunction( );
```

Except in very exceptional conditions, the call to `malloc` will return a valid pointer, and the following if-statement's condition will be false. Predicting the conditional branch that corresponds to this if-statement with a static prediction will result in perfect prediction rates (for all practical purposes).

Ball and Larus [1993] introduced a set of heuristics based on program structure to statically predict conditional branches. These rules are listed in Table 9.1. The heuristics make use of branch opcodes, the operands to branch instructions, and attributes of the instruction blocks that succeed the branch instructions in an attempt to make predictions based on the knowledge of common programming

Table 9.1
Ball and Larus's static branch prediction rules

Heuristic Name	Description
Loop branch	If the branch target is back to the head of a loop, predict taken.
Pointer	If a branch compares a pointer with NULL, or if two pointers are compared, predict in the direction that corresponds to the pointer being not NULL, or the two pointers not being equal.
Opcode	If a branch is testing that an integer is less than zero, less than or equal to zero, or equal to a constant, predict in the direction that corresponds to the test evaluating to false.
Guard	If the operand of the branch instruction is a register that gets used before being redefined in the successor block, predict that the branch goes to the successor block.
Loop exit	If a branch occurs inside a loop, and neither of the targets is the loop head, then predict that the branch does not go to the successor that is the loop exit.
Loop header	Predict that the successor block of a branch that is a loop header or a loop preheader is taken.
Call	If a successor block contains a subroutine call, predict that the branch goes to that successor block.
Store	If a successor block contains a store instruction, predict that the branch does not go to that successor block.
Return	If a successor block contains a return from subroutine instruction, predict that the branch does not go to that successor block.

idioms. In some situations, more than one heuristic may be applicable. For these situations, there is an ordering of the heuristics, and the first rule that is applicable is used. Ball and Larus evaluated all permutations of their rules to decide on the best ordering. Some of the rules capture the intuition that tests for exceptional conditions are rarely true (e.g., pointer and opcode rules), and some other rules are based on assumptions of common control flow patterns (the loop rules and the call/return rules).

9.2.4 Profiling

Profile-based static branch prediction involves executing an instrumented version of a program on sample input data, collecting statistics, and then feeding back the collected information to the compiler. The compiler makes use of the profile information to make static branch predictions that are inserted into the final program binary as branch hints.

One simple approach is to run the instrumented binary on one or more sample data sets and determine the frequency of taken branches for each static branch instruction in the program. If more than one data set is used, then the measured frequencies can be weighted by the number of times each static branch was executed. The compiler inserts branch hints corresponding to the more frequently observed branch directions during the sample executions. If during the profiling run, a branch was observed to be taken more than 50% of the time, then the compiler would set the branch hint bit to predict-taken. In Fisher and Freudenberger [1992], such an experiment was performed, and it was found that for some benchmarks, different runs of a program were successful at predicting future runs on different data sets. In other cases, the success varied depending on how representative the sample data sets were.

The advantage of profile-based prediction techniques and the other static branch prediction algorithms is that they are very simple to implement in hardware. One disadvantage of profile-based prediction is that once the predictions are made, they are forever "set in stone" in the program binary. If an input set causes branching behaviors that are different from the training sets, performance will suffer. Additionally, the instruction set architecture must provide some interface to the programmer or compiler to insert branch hints.

Except for the always-taken and always-not-taken approaches, rule-based and profile-based branch prediction have the shortcoming that the branch instruction must be fetched from the instruction cache to be able to read the prediction embedded in the branch hint. Modern processors use multicycle pipelined instruction caches, and therefore the prediction for the next instruction must be available several cycles before the current instruction is fetched. In the following section, the dynamic branch prediction algorithms only make use of the address of the current branch and other information that is immediately available.

9.3 Dynamic Branch Prediction Techniques

Although static branch prediction techniques can achieve conditional branch prediction rates in the 70% to 80% range [Calder et al., 1997], if the profiling information is not representative of the actual run-time behavior, prediction accuracy can suffer greatly. Dynamic branch prediction algorithms take advantage of the run-time information available in the processor, and can react to changing branch patterns. Dynamic branch predictors typically achieve branch prediction rates in the range of 80% to 95% (for example, see McFarling [1993] and Yeh and Patt [1992]).

There are some branches that static prediction approaches cannot handle, but the branch behavior is still fundamentally very predictable. Consider a branch that is always taken during the first half of the program, and then is always not taken in the second half of the program. Profiling will reveal that the branch is taken 50% of the time, and any static prediction will result in a 50% prediction accuracy. On the other hand, if we simply predict that the branch will go in the same direction as the last time we encountered the branch, we can achieve nearly perfect prediction, with only a single misprediction at the halfway point of the program when the branch changes directions. Another situation where very predictable branches cannot be determined at compile time is where a branch's direction depends on the program's input. As an example, a program that performs matrix computations may have different algorithms optimized for different sized matrices. Throughout the program, there may be branches that check the size of the matrix and then branch to the appropriate optimized code. For a given execution of this program, the matrix size is constant, and so these branches will have the same direction for the entire execution. By observing the run-time behavior, a dynamic branch predictor could easily predict all these branches. On the other hand, the compiler does not have any idea what size the matrices will be and is incapable of making much more than a blind guess.

Dynamic branch predictors may require a significant amount of chip area to implement, especially when more complex algorithms are used. For small processors, such as older-generation CPUs or processors targeted for embedded systems, the additional area for these prediction structures may simply be too expensive. For larger, future-generation, wide-issue superscalar processors, accurate conditional branch prediction is critical. Furthermore, these processors have much larger chip areas, and so considerable resources may be dedicated to the implementation of more sophisticated dynamic branch predictors. An additional benefit of dynamic branch prediction is that performance enhancements can be realized without profiling all the applications that one wishes to run, and recompilation is not needed so existing binary executables can benefit.

This section describes many of the dynamic branch prediction algorithms that have been published. Many of these prediction algorithms are important on their own, and some have even been implemented in commercial processors. In Section 9.4, we will also explore ways of composing more than one of these predictors into more powerful hybrid branch predictors.

This section has been divided into three parts based on the characteristics of the prediction algorithms. Section 9.3.1 covers several fundamental prediction schemes that are the basis for many of the more sophisticated algorithms. Section 9.3.2 describes predictors that address the branch aliasing problem. Section 9.3.3 covers prediction schemes that make use of a wider variety of information in making predictions.

9.3.1 Basic Algorithms

Most dynamic branch predictors have their roots in one or more of the basic algorithms described here.

9.3.1.1 Smith's Algorithm. The main idea behind the majority of dynamic branch predictors is that each time the processor discovers the true outcome of a branch (whether it is taken or not taken), it makes note of some form of context so that the next time it encounters the same situation, it will make the same prediction. An analogy for branch prediction is the problem of navigating in a car to get from one place to another where there are forks in the road. The driver just wants to keep driving as fast as she can, and so each time she encounters a fork, she can just guess a direction and keep on going. At the same time, her "copilot" (who happens to be slow at map reading) is trying to keep up. When he realizes that they made a wrong turn, he notifies the driver and then she will have to backtrack and then resume along the correct route.

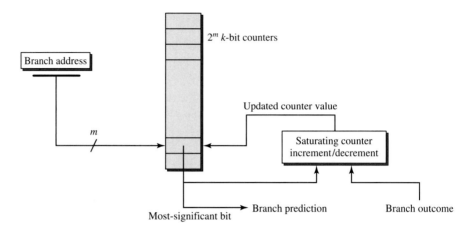

Figure 9.1
Smith Predictor with a 2^m-entry Table of Saturating k-bit Counters.

If these two friends frequently drive in this area, the driver can do better than blindly guessing at each fork in the road. She might notice that they always end up making a right turn at the intersection with the pizza shop and always make a left at the supermarket. These landmarks form the context for the driver's predictions. In a similar fashion, dynamic branch predictors make note of context (in the form of branch history), and then make their predictions based on this information.

Smith's algorithm [1981] is one of the earliest proposed dynamic branch direction prediction algorithms, and one of the simplest. The predictor consists of a table that records for each branch whether or not the previous instances were taken or not taken. This is analogous to our driver keeping track in her head of the cross streets for each intersection and remembering if they went left or right. The cross streets correspond to the branch addresses, and the left/right decisions correspond to taken/not-taken branch outcomes. Because the predictor tracks whether a branch is in a mostly taken mode or a mostly not-taken mode, the name *bimodal* predictor is also commonly used for the Smith predictor.

The Smith predictor consists of a table of 2^m counters, where each counter tracks the past branch directions. Since there are only 2^m entries, the branch address [program counter (PC)] is hashed down to m bits.[1] Each counter in the table has a width of k bits. The most-significant bit of the counter is used for the branch direction prediction. If the most-significant bit is a one, then the branch is predicted to be taken; if the most significant bit is a zero, the branch is predicted to be not-taken. Figure 9.1 illustrates the hardware for Smith's algorithm. The notation Smith$_\kappa$ means Smith's algorithm with $k = \kappa$.

---

[1] In his 1981 paper, Smith proposed an exclusive-OR hashing function, although most modern implementations use a simple (PC mod 2^m) hashing function which requires no logic to implement. Typically, a few of the least-significant bits are ignored due to the fact that for an ISA with instruction word sizes that are powers of two, the lower bits will always be zero.

After a branch has resolved and its true direction is known, the counter is updated depending on the branch outcome. If the branch was taken, then the counter is incremented only if the current value is less than the maximum possible. For instance, a k-bit counter will *saturate* at $2^k - 1$. If the branch was not taken, then the counter is decremented if the current value is greater than zero.[2] This simple finite state machine is also called a *saturating k-bit counter,* or an *up-down counter*. The counter will have a higher value if the corresponding branch was often taken in the last several encounters of this branch. The counter will tend toward lower values when the recent branches have mostly been not taken. The case of Smith's algorithm when $k = 1$ simply keeps track of the last outcome of a branch that mapped to the counter.

Some branches are predominantly biased toward one direction. A branch at the end of a for loop is usually taken, except for the case of the loop exit. This one exceptional case is called an *anomalous decision*. The outcomes of several of the most recent branches to map to the same counter can be used if $k > 1$. By using the histories of several recent branches, the counter will not be thrown off by a single anomalous decision. The additional bits add some *hysteresis* to the predictor's state. Smith also calls this *inertia*.

Returning to the analogy of the driver, it may be the case that she almost always makes a left turn at a particular intersection, but most recently she ended up having to make a right turn instead because her and her friend had to go to the hospital due to an emergency. If our driver only remembered her most recent trip, then she would predict to make a right turn again the next time she was at this intersection. On the other hand, if she remembered the last several trips, she would realize that more often than not she ended up making a left turn. Using additional bits in the counter allows the predictor to effectively remember more history.

The 2-bit saturating counter (2bC) is used in many branch prediction algorithms. There are four possible states: 00, 01, 10, 11. States 00 and 01, called strongly not-taken (SN) and weakly not-taken (WN), respectively, provide predictions of not-taken. States 10 and 11, called weakly taken (WT) and strongly taken (ST), respectively, provide a taken-branch prediction. The reason states 00 and 11 are called "strong" is that the same outcome must have occurred multiple times to reach that state.

Figure 9.2 illustrates a short sequence of branches and the predictions made by Smith's algorithm for $k = 1$ ($Smith_1$) and $k = 2$ ($Smith_2$). Prior to the anomalous decision, both versions of Smith's algorithm predict the branches accurately. On the anomalous decision (branch C), both predictors mispredict. On the following branch D, $Smith_1$ mispredicts again because it only remembers the most recent branch and predicts in the same direction. This occurs despite the fact that the vast majority of prior branches were taken. On the other hand, $Smith_2$ makes the correct decision because its prediction is influenced by several of the most recent branches instead of the single most recent branch. For such anomalous decisions, $Smith_1$ makes two mispredictions while $Smith_2$ only errs once.

---

[2]The original paper presented the counter as using values from -2^{k-1} up to $2^{k-1} - 1$ in two's complement notation. The complement of the most-significant bit is then used as the branch direction prediction. The formulation presented here is used in the more recent literature.

Branch	Branch Direction	Smith₁ State	Smith₁ Prediction	Smith₂ State	Smith₂ Prediction
A	1	1	1	11	1
B	1	1	1	11	1
C	0	1	1 (misprediction)	11	1 (misprediction)
D	1	0	0 (misprediction)	10	1
E	1	1	1	11	1
F	1	1	1	11	1

Figure 9.2
A Comparison of a Smith₁ and a Smith₂ Predictor on a Sequence of Branches with a Single Anomalous Decision.

Practically every dynamic branch prediction algorithm published since Smith's seminal paper uses saturating counters. For tracking branch directions, 2-bit counters provide better prediction accuracies than 1-bit counters due to the additional hysteresis. Adding a third bit only improves performance by a small increment. In many branch predictor designs, this incremental improvement is not worth the 50% increase in area of adding an additional bit to every 2-bit counter.

9.3.1.2 Two-Level Prediction Tables. Yeh and Patt [1991; 1992; 1993] and Pan et al. [1992] proposed variations of the same branch prediction algorithms called two-level adaptive branch prediction and correlation branch prediction, respectively. The two-level predictor employs two separate levels of branch history information to make the branch prediction.

Using the car navigation analogy, the Smith predictor parallel for driving was to remember what decision was made at each intersection. The car-navigation equivalent to the two-level predictor is for our driver to remember the exact sequence of turns made before arriving at the current intersection. For example, to drive from her apartment to the bank, our driver makes three turns: a left turn, another left, and then a right. To drive from the mall to the bank, she also makes three turns, but they are a right, a left, and then another right. If she finds herself at the bank and remembers that she most recently went right, left, and right, then she could guess that she just came from the mall and is heading home and make her next routing decision accordingly.

The *global-history* two-level predictor uses a history of the most recent branch outcomes. These outcomes are stored in the *branch history register* (BHR). The BHR is a shift register where the outcome of each branch is shifted into one end, and the oldest outcome is shifted out of the other end and discarded. The branch outcomes are represented by zeros and ones, which correspond to not-taken and taken, respectively. Therefore, an *h*-bit branch history register records the *h* most recent branch outcomes. The branch history is the first level of the global-history two-level predictor.

The second level of the global-history two-level predictor is a table of saturating 2-bit counters (2bCs). This table is called the *pattern history table* (PHT). The PHT is indexed by a concatenation of a hash of the branch address with the contents of the BHR. This is analogous to our driver using a combination of the intersection as well as the most recent turn decisions in making her prediction. The counter in the indexed PHT entry provides the branch prediction in the same fashion as the Smith predictor (prediction is determined by the most-significant bit of the counter). Updates to the counter are also the same as for the Smith predictor counters: saturating increment on a taken branch, and saturating decrement on a not-taken branch.

Figure 9.3 shows the hardware organization of a sample global-history two-level predictor. This predictor uses the outcomes of the four most recent branch instructions and 2 bits from the branch address to form an index into a 64-entry PHT. With *h* bits of branch history and *m* bits of branch address, the PHT has 2^{h+m} entries. When using only *m* bits of branch address (where *m* is less than the total width of the PC), the branch address must be hashed down to *m* bits, similar to the Smith predictor. Note that in the example in Figure 9.3, this means that *any* branch address that ends in 01 will share the same entries as the branch depicted in the figure. Using the car navigation analogy again, this is similar to our driver remembering Elm Street as simply "Elm," which may cause confusion when she encounters an Elm Road, Elm Lane, or Elm Boulevard. Note that this problem is not unique to the two-level predictor, and that it can affect the Smith predictor as well. Since the

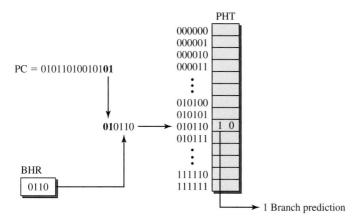

Figure 9.3
A Global-History Two-Level Predictor with a 4-bit Branch History Register.

Smith predictor does not use branch history, all its index bits are from the branch address, which reduces this branch conflict problem.

The size of the global-history two-level predictor depends on the total available hardware budget. For a 4K-byte budget, the PHT would have 16,384 entries (4,096 bytes times 4 two-bit counters per byte). In general, for an X K-byte budget, the PHT will contain $4X$ counters. There is a tradeoff between the number of branch address bits used and the length of the BHR, since the sum of their lengths must be equal to the number of index bits. Using more branch address bits reduces the branch conflict problem, whereas using more branch history allows the predictor to correlate against more complex branch history patterns. The optimal balance depends on many factors such as how the compiler arranges the code, the program being run, the instruction set architecture, and the input set to the program.

The intuition behind using the global branch history is that the behavior of a branch may be linked or *correlated* with a different earlier branch. For example, the branches may test conditions that involve the same variable. Another more common situation is that one branch may guard an instruction that modifies a variable that the second branch tests. Figure 9.4 shows a code segment where an if-statement (branch A) determines whether or not the variable x gets assigned a different value. Later in the program, another if-statement (branch C) tests the value of x. If A's condition was true, then x gets assigned the value of 3, and then C will evaluate $x \leq 0$ as false. On the other hand, if A's condition was false, then x retains its original value of 0, and C will evaluate $x \leq 0$ as true. The behavior (outcome) of the branch corresponding to if-statement C is strongly correlated to the outcome of if-statement A. A branch predictor that tracks the outcome of if-statement A could potentially achieve perfect prediction for the branch of if-statement C. Notice that there could be an intervening branch B that does not affect the outcome of C (that is, C is not correlated to B). Such irrelevant branches

```
x = 0;
if (someCondition) {           /* branch A */
    x = 3;
}
if (someOtherCondition) {      /* branch B */
    y += 19;
}
if (x <= 0) {                  /* branch C */
    doSomething( );
}
```

Figure 9.4

A Sample Code Segment with Correlated Branches.

increase the training time of global history predictors because the predictor must learn to ignore these irrelevant history bits.

Another variation of the two-level predictor is the *local-history* two-level predictor. Whereas global history tracks the outcomes of the last several branches encountered, local history tracks the outcomes of the last several encounters of only the current branch. Using the car navigation analogy again, our driver might make a right turn at a particular intersection during the week on her way to work, but on the weekends she makes left turns at the exact same intersection to go downtown. The turns she made to get to that intersection (i.e., the global history) might be the same regardless of the day of week. On the other hand, if she remembers that over the last several days she went R, R, R, L, then today is probably Sunday and she would predict that making a left turn is the correct decision. To remember a driving decision history for each intersection requires our driver to remember much more information than a simple global history, but there are some patterns that are easier to predict using such a local history.

To implement a local-history branch predictor, the single global BHR is replaced by one BHR per branch. The collection of BHRs form a *branch history table* (BHT). A global BHR is really a degenerate case where the BHT has only a single entry. The branch address is used to select one of the entries in the BHT, which provides the local history. The contents of the selected BHR are then combined with the PC in the same fashion as the global-history two-level predictor to index into the PHT. The most-significant bit of the counter provides the branch prediction, and the update of the counter is also the same as the Smith predictor. To update the history, the most recent branch outcome is shifted into the selected entry from the BHT.

Figure 9.5 shows the hardware organization for an example local-history two-level predictor. The BHT has eight entries and is indexed by the three least-significant bits of the branch address. The PHT in this example has 128 entries, which uses a

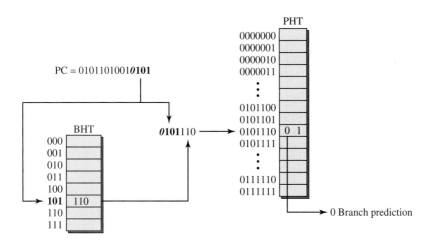

Figure 9.5
A Local-History Two-Level Predictor with an Eight-Entry BHT and a 3-bit History Length.

7-bit index ($\log_2 128 = 7$). Since the history length is 3 bits long, the other 4 index bits come from the branch address. These bits are concatenated together to select a counter from the PHT which provides the final prediction.

The tradeoffs in sizing a local-history two-level predictor are more complex than the case of the global-history predictor. In addition to balancing the number of history and address bits for the PHT index, there is also a tradeoff between the number of bits dedicated to the BHT and the number of bits dedicated to the PHT. In the BHT, there is also a balance between the number of entries and the width of each entry (i.e., the history length). A local-history two-level predictor with an L-entry BHT and an h-bit history and that uses m bits of the branch address for the PHT index requires a total size of $Lh + 2^{h+m+1}$ bits. The Lh bits are for the BHT, and the PHT has 2^{h+m} entries, each 2 bits wide (the +1 in the exponent).

Figure 9.6(a) shows an example predictor with an 8-entry BHT, a 4-bit history length, and a 64-entry PHT (only the first 16 entries are shown). The last four outcomes for the branch at address 0xC084 have been T, N, T, N. To select a BHR, we hash the branch address down to three bits by using the 3 least-significant bits of the address (100 in binary). Note that the selected branch history register's contents are 1010, which corresponds to a history of TNTN. With a 64-entry PHT, the size of the PHT index is 6 bits, of which 4 will be from the branch history. This

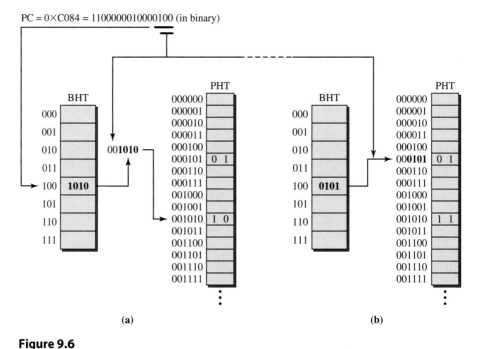

Figure 9.6
An Example Lookup on a Two-Level Branch Predictor: (a) Making the Prediction, and (b) After the Predictor Update.

leaves only 2 bits from the branch address. The concatenation of the branch address bits and the branch history selects one of the counters, whose most significant bit indicates a taken-branch prediction.

After the actual branch outcome has been computed during the execute stage of the pipeline, both the BHT and PHT need to be updated. Assuming that the actual outcome was a taken branch, a 1 is shifted into the BHR as shown in Figure 9.6(b). The PHT is updated as per the Smith predictor algorithm, and the 2-bit counter gets incremented from 10 to 11 (in binary). Note that for the next time the branch at 0xC084 is encountered, the BHR now contains the pattern 0101 which selects a different entry in the PHT. The Intel P6 microarchitecture uses a local-history two-level predictor with a 4-bit history length (see Chapter 7).

By tracking the behavior of each branch individually, a predictor can detect patterns that are *local* to a particular branch, like the alternating pattern shown in Figure 9.6. As a second example, consider a loop-closing branch with a short iteration count that exhibits the pattern 1110111011101..., where again a 1 denotes a taken branch, and a 0 denotes a not-taken branch. By tracking the last several outcomes of only this particular branch, the PHT will quickly learn this pattern. Figure 9.7 shows a 16-entry PHT with the entries corresponding to predicting this pattern (no branch address bits are used for this PHT index). When the last four outcomes of this branch are 1101, the loop has not yet terminated and the next time the branch will again be taken. Every time the processor encounters the pattern 1101, the following branch is always taken. This results in incrementing the corresponding saturating counter at index 1101 every time this pattern occurs. After the pattern has occurred a few times, the PHT will predict taken because the counter indexed by 1101 will now remember (by having the state ST) that the

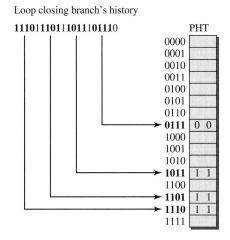

Figure 9.7
Local History Predictor Example.

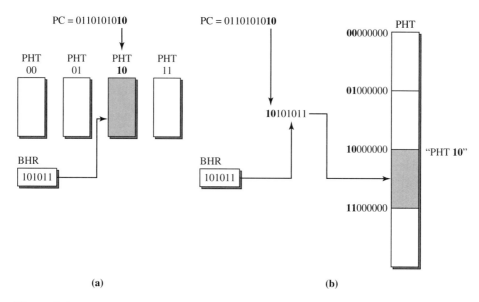

Figure 9.8
Alternative Views of the PHT Organization: (a) A Collection of PHTs, and (b) A Single Monolithic PHT.

following instance of this branch will be taken. When the last four outcomes are 0111, the entry indexed by 0111 has a not-taken prediction stored (SN) state. The PHT basically learns a mapping of the form "when I see a history pattern of X, the outcome of the next branch is usually Y."

Some texts and research papers view the two-level predictors as having multiple PHTs. The branch address hash selects one of the PHTs, and then the branch history acts as a subindex into the chosen PHT. This view of the two-level predictor is shown in Figure 9.8(a). The monolithic PHT shown in Figure 9.8(b) is actually equivalent. This chapter uses the monolithic view of the PHT because it reduces the PHT design tradeoffs down to a conceptually simpler problem of deciding how to allocate the bits of the index (i.e., how many index bits come from the PC and how long is the history length?).

Yeh and Patt [1993] also introduced a third variation that utilizes a BHT that uses an arbitrary hashing function to divide the branches into different *sets*. Each group shares a single BHR. Instead of using the least-significant bits of the branch address to select a BHR from the BHT, other example set-partitioning functions use only the higher-order bits of the PC, or divide based on opcode. This type of history is called *per-set* branch history, and the table is called a *per-set branch history table* (SBHT). Yeh and Patt use the letters G (for global), P (for per-address) and S (for per-set) to denote the different variations of the two-level branch prediction algorithm.

The choice of nonhistory bits used in the PHT index provide for several additional variations. The first option is to simply ignore the PC and use only the BHR to index into the PHT. All branches thus share the entries of the PHT, and this is called a *global pattern history table* (gPHT), which is used in the example of Figure 9.7. The second alternative, already illustrated in Figure 9.6, is to use the lower bits of the PC to create a *per-address pattern history table* (pPHT). The last variation is to apply some other hashing function (analogous to the hashing function for the per-set BHT) to provide the nonhistory index bits for a *per-set pattern history table* (sPHT).

Yeh and Patt use the letters g, p, and s to indicate these three indexing variations. Combined with the three branch history options (G, P, and S), there are a total of nine variations of two-level predictors using this taxonomy. The notation presented by Yeh and Patt is of the form xAy, where x is G, P, or S, and y is g, p, or s. Therefore, the nine two-level predictors are GAg, GAp, GAs, PAg, PAp, PAs, SAg, SAp, and SAs. In general, the two-level predictors identify patterns of branch outcomes and associate a prediction with each pattern. This captures correlations with complex branch patterns that the simpler Smith predictors cannot track.

9.3.1.3 Index-Sharing Predictors.
The two-level algorithm requires the branch predictor designer to make a tradeoff between the width of the BHR (the number of history bits to use) and the number of branch address bits used to index the PHT. For a fixed PHT size, employing a larger number of history bits reveals more opportunities to correlate with more distant branches, but this comes at the cost of using fewer branch address bits. For example, consider again our car navigation analogy. Assume that our driver has a limited memory and can only remember a sequence of at most six letters. She could choose to remember the first five letters of the street name and the one most recent turn decision. This allows her to distinguish between many street names, but has very little decision history information to correlate against. Alternatively, she could choose to only remember the first two letters of the street name, while recording the four most recent turn decisions. This provides more decision history, but she may get confused between *Br*oad Street and *Br*idge Street.

Note that if the history length is long, the frequently occurring history patterns will map into the PHT in a very sparse distribution. For example, consider the local history pattern used in Figure 9.7. Since the history length is 4 bits long, there are 16 entries in the PHT. But for the particular pattern used in the example, only 4 of the 16 entries are ever accessed. This indicates that the index formation for the two-level predictor introduces inefficiencies.

McFarling [1993] proposed a variation of a global-history two-level predictor called *gshare*. The gshare algorithm attempts to make better use of the index bits by hashing the BHR and the PC *together* to select an entry from the PHT. The hashing function used is a bit-wise exclusive-OR operation. The combination of the BHR and PC tends to contain more information due to the nonuniform distribution of PC values and branch histories. This is called *index sharing*.

Figure 9.9 illustrates a set of PC and branch history pairs and the resulting PHT indices used by the GAp and gshare algorithms. Because the GAp algorithm is forced to trade off the number of bits used between the BHR width and the PC bits used, some information from one of these two sources must be left out. In the

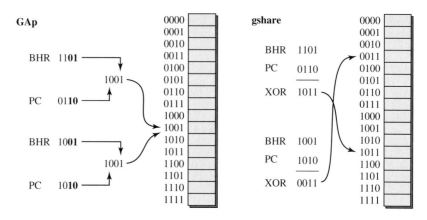

Figure 9.9
Indexing Example with a Global-History Two-Level Predictor and the gshare Predictor.

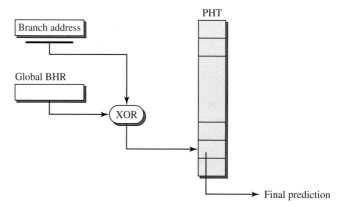

Figure 9.10
The gshare Predictor.

example, the GAp algorithm uses 2 bits from the PC and 2 bits from the global history. Notice that even though the overall PC and history bits are different, using only 2 bits from each causes the two to both map to entry 1001. On the other hand, the exclusive-OR of the 4 bits of the branch address with the full 4 bits of the global history yields different distinct PHT indices.

The hardware for the gshare predictor is shown in Figure 9.10. The circuit is very similar to the global history two-level predictor, except that the concatenation operator for the PHT index has been replaced with an XOR operator. If the number of global history bits used h is less than the number of branch address bits used m, then the global history is XORed with the *upper* h bits of the m branch address bits. The reason for this is that the upper bits of the PC tend to be sparser than the lower-order bits.

Evers et al. [1996] proposed a variation of the gshare predictor that uses a per-address branch history table to store local branch history. The *pshare* algorithm is the local-history analog of the gshare algorithm. The low-order bits of the branch address are used to index into the first-level BHT in the same fashion as the PAg/PAs/PAp two-level predictors. Then the contents of the indexed BHR are XORed with the branch address to form the PHT index.

Index sharing predictors are commonly used in modern branch predictors. For example, the IBM Power4 microprocessor's global-history predictor uses an 11-bit global history (BHR) and a 16,384-entry PHT [Tendler et al., 2002]. The Alpha 21264 also makes use of a global history predictor with a 12-bit global history and a 4096-entry PHT [Kessler, 1999]. The amount of available storage, commonly measured in bytes of state, is often the deciding factor in how many PHT entries to use. More recently with the steady increase in clock frequencies, the latency of the PHT access is also becoming a limiting factor in the size of the PHT.

The Power4's 16,384-entry PHT requires 2K bytes of storage since each entry is a 1-bit (1/8 byte) counter. The number of bits of history to use is limited by the PHT size and may also depend on the target set of applications. If the most frequently executed programs exhibit behavior that requires a longer branch history to capture, then it is likely that a longer BHR will be employed. The exact behavior of branches will depend on the compiler, the instruction set, and the input to the program, thus making it difficult to choose an optimal history length that performs well across a wide range of applications.

9.3.1.4 Reasons for Mispredictions.
Branch mispredictions can occur for a variety of reasons. Some branches are simply hard to predict. Other mispredictions are due to the fact that any realistic branch predictor is limited in size and complexity.

There are several cases where a branch is fundamentally unpredictable. The first time the predictor encounters a branch, it has no past information about how the branch behaves, and so at best the predictor could make a random choice and expect a 50% prediction rate. With predictors that use branch histories, a similar situation occurs any time the predictor encounters a new branch history pattern. A predictor needs to see a particular branch (or branch history) a few times before it learns the proper prediction that corresponds to the branch (or branch history). During this *training* period, it is unlikely that the predictor will perform very well. For a branch history of length n, there are 2^n possible branch history patterns, and so the training time for a predictor increases with the history length. If the program enters a new phase of execution (for example, a compiler going from parsing to type-checking), branch behaviors may change and the predictor must relearn the new patterns.

Another case where branches are unpredictable is when the data involved in the program are intrinsically random. For example, a program that processes compressed data may have many hard-to-predict branches because well-compressed input data will appear to be random. Other application areas that may have hard-to-predict branches include cryptography and randomized algorithms.

The physical constraints on the size of branch predictors introduces additional sources of branch mispredictions. For example, if a branch predictor has a 128-entry table of counters, and there are 129 distinct branches in a program, then there will be at least one entry that has two different branches mapped to it. If one of these branches is always taken and the other is always not taken, then they will *interfere* with each other and cause branch mispredictions. Such interference is called *negative interference*. If both of these branches are always taken (or both are always not taken), they would still interfere, but no additional mispredictions would be generated; this is called *neutral interference*. Interference is also called *aliasing* because both branches are aliases for the same predictor entry.

Aliasing can occur even if there are more predictor entries than branches. With a 128-entry table, let us assume that the hashing function is the remainder of the branch address when divided by 128 (i.e., index := address mod 128). There may only be two branches in the entire program, but if their addresses are 131 and 259, then both branches will still map to predictor entry 3. This is called *conflict aliasing*. This is similar to the case in our car driving analogy where our driver gets confused by Broad St. and Bridge St. because she is only remembering the first two letters and they happen to be the same.

Some branches are predictable, but a particular predictor may still mispredict the branch because the predictor does not have the right information. For example, consider a branch that is strongly correlated with the ninth-most recent branch. If a predictor only uses an 8-bit branch history, then the predictor will not be able to accurately make this prediction. Similarly, if a branch is strongly correlated to a previous local history bit, then it will be difficult for a global history predictor to make the right prediction.

More sophisticated prediction algorithms can deal with some classes or types of mispredictions. For capacity problems, the only solution is to increase the size of the predictor structures. This is not always possible due to die area, latency, and/or power constraints. For conflict aliasing, a wide variety of algorithms have been developed to address this problem, and many of these are described in Section 9.3.2. Furthermore, many algorithms have been developed to make use of different types of information (such as global vs. local branch histories or short vs. long histories), and these are covered in Section 9.3.3.

9.3.2 Interference-Reducing Predictors

The PHT used in the two-level and gshare predictors is a direct-mapped, tagless structure. Aliasing occurs between different address-history pairs in the PHT. The PHT can be viewed as a cache-like structure, and the three-C's model of cache misses [Hill, 1987; Sugumar and Abraham, 1993] gives rise to an analogous model for PHT aliasing [Michaud et al., 1997]. A particular address-history pair can "miss" in the PHT for the following reasons:

1. *Compulsory aliasing* occurs the first time the address-history pair is ever used to index the PHT. The only recourse for compulsory aliasing is to initialize the PHT counters in such a way that the majority of such lookups

still yield accurate predictions. Fortunately, Michaud et al. show that compulsory aliasing accounts for a very small fraction of all branch prediction lookups (much less than 1% on the IBS benchmarks [Richard Uhlig et al., 1995]).

2. *Capacity aliasing* occurs because the size of the current working set of address-history pairs is greater than the capacity of the PHT. This aliasing can be mitigated by increasing the PHT size.

3. *Conflict aliasing* occurs when two different address-history pairs map to the same PHT entry. Increasing the PHT size often has little effect on reducing conflict aliasing. For caches, the associativity can be increased or a better replacement policy can be used to reduce the effects of conflicts.

For caches, the standard solution for conflict aliasing is to increase the associativity of the cache. Even for a direct-mapped cache, address tags are necessary to determine whether the cached item belongs to the requested address. Branch predictors are different because tags are not required for proper operation. In many cases, there are other ways to use the available transistor budget to deal with conflict aliasing than the use of associativity. For example, instead of adding a 2-bit tag to every saturating 2-bit counter, the size of the predictor could instead be doubled. Sections 9.3.2.1 to 9.3.2.6 describe a variety of ways to deal with the problem of interference in branch predictors. These predictors are all global-history predictors because global-history predictors are usually more accurate than local-history predictors, but the ideas are equally applicable to local-history predictors as well. Note that many of these algorithms are often referred to as *two-level* branch predictors, since they all use a first level of branch history and a second level of counters or other state that provides the final prediction.

9.3.2.1 The Bi-Mode Predictor. The Bi-Mode predictor uses multiple PHTs to reduce the effects of aliasing [Lee et al., 1997]. The Bi-Mode predictor consists of two PHTs (PHT_0 and PHT_1), both indexed in a gshare fashion. The indices used on the PHTs are identical. A separate *choice predictor* is indexed with the lower-order bits of the branch address only. The choice predictor is a table of 2-bit counters (identical to a $Smith_2$ predictor), where the most-significant bit indicates which of the two PHTs to use. In this manner, the branches that have a strong taken bias are placed in one PHT and the branches that have a not-taken bias are separated into the other PHT, thus reducing the amount of destructive interference. The two PHTs have identical sizes, although the choice predictor may have a different number of entries.

Figure 9.11 illustrates the hardware for the Bi-Mode predictor. The branch address and global branch history are hashed together to form an index into the PHTs. The same index is used on both PHTs, and the corresponding predictions are read. Simultaneously, the low-order bits of the branch address are used to index the choice predictor table. The prediction from the choice predictor drives the select line of a multiplexer to choose one of the two PHT banks.

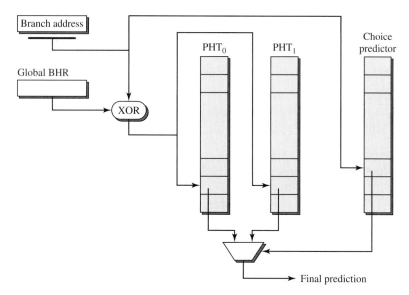

Figure 9.11
The Bi-Mode Predictor.

The rationale behind the Bi-Mode predictor is that most branches are biased toward one direction or the other. The choice predictor effectively remembers what the bias of each branch is. Branches that are more strongly biased toward one direction all use the same PHT. The result is that even if two branches map to the same entry in this PHT, they are more likely to go in the same direction. The result is that an opportunity for negative interference has been converted to neutral interference.

The PHT bank selected by the choice predictor is always updated when the final branch outcome has been determined. The other PHT bank is not updated. The choice predictor is always updated with the branch outcome, except in the case where the choice predictor's direction is the opposite of the branch outcome, but the overall prediction of the selected PHT bank was correct. These update rules implement a *partial update* policy.

9.3.2.2 The gskewed Predictor. The gskewed algorithm divides the PHT into three (or more) banks. Each bank is indexed by a different hash of the address-history pair. The results of these three lookups are combined by a majority vote to determine the overall prediction. The intuition is that if the hashing functions are different, even if two address-history pairs destructively alias to the same PHT entry in one bank, they are unlikely to conflict in the other two banks. The hashing functions f_0, f_1, and f_2 presented in Michaud et al. [1997] have the property that if $f_0(x_1) = f_0(x_2)$, then $f_1(x_1) \neq f_1(x_2)$ and $f_2(x_1) \neq f_2(x_2)$ if $x_1 \neq x_2$. That is, if two addresses conflict in one PHT, they are guaranteed to not conflict with each other in the other two PHTs. For three banks of 2^n-entry PHTs, the definitions of the three hashing functions are

$$f_0(x, y) = H(y) \oplus H^{-1}(x) \oplus x \qquad (9.1)$$

$$f_1(x, y) = H(y) \oplus H^{-1}(x) \oplus y \qquad (9.2)$$

$$f_2(x, y) = H^{-1}(y) \oplus H(x) \oplus x \qquad (9.3)$$

where $H(b_n, b_{n-1}, \ldots, b_3, b_2, b_1) = (b_n \oplus b_1, b_n, b_{n-1}, \ldots, b_3, b_2)$, H^{-1} is the inverse of H, and x and y are each n bits long. For the gskewed algorithm, the arguments x and y of the hashing functions are the n low-order bits of the branch address, and the n most recent global branch outcomes, respectively.

The amount of conflict aliasing is a result of the hashing function used to map the PC-history pair into a PHT index. Although the gshare exclusive-OR hash can remove certain types of interference, it can also introduce interference as well. Two different PC values and two different histories can still result in the same index. For example, the PC-history pairs (PC = 0110) $\oplus$ (history = 1100) = 1010, and 1101 $\oplus$ 0111 = 1010 map to the same index.

The hardware for the gskewed predictor is illustrated in Figure 9.12. The branch address and the global branch history are hashed separately with the three hashing functions (9.1) through (9.3). Each of the three resulting indices is used to address a different PHT bank. The direction bits from the 2-bit counters in the PHTs are combined with a majority function to make the final prediction.

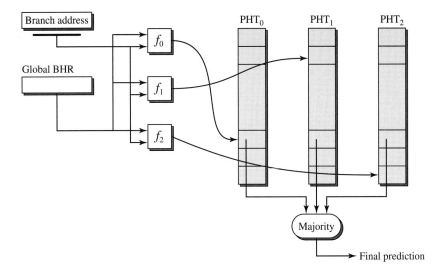

Figure 9.12
The gskewed Predictor.

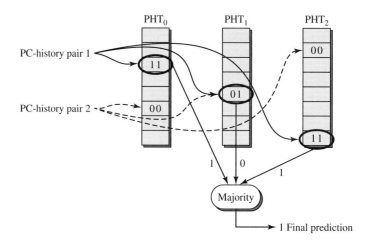

Figure 9.13

A gskewed Example Showing How the Majority Function Can Tolerate Interference in One of the PHT Banks.

Figure 9.13 shows a gskewed predictor example with two sets of PC-history pairs corresponding to two different branches. In this example, one PC-history pair corresponds to a strongly taken branch, whereas the other PC-history pair corresponds to a strongly not-taken branch. The two branches map to the same entry in PHT_1 which causes destructive interference. The hashing functions (9.1) through (9.3) guarantee that a conflict in one PHT means there are no conflicts between these two branches in both of the other two PHTs. As a result, the majority function can effectively mask the one disagreeing vote and still provide the correct prediction.

Two different update policies for the gskewed algorithm are *total update* and *partial update*. The total update policy treats each of the PHT banks identically and updates all banks with the branch outcome. The partial update policy does not update a bank if that particular bank mispredicted, but the overall prediction was correct. The partial update policy improves the overall prediction rate of the gskewed algorithm. When only one of the three banks mispredicts, it is not updated, thus allowing it to contribute to the correct prediction of another address-history pair.

The choice of the branch history length involves a tradeoff between capacity and aliasing conflicts. Shorter branch histories tend to reduce the amount of possible aliasing because there are fewer possible address-branch history pairs. On the other hand, longer histories tend to provide better branch prediction accuracy because there is more correlation information available. A modification to the gskewed predictor is the *enhanced* gskewed predictor. In this variation, PHT banks 1 and 2 are indexed in the usual fashion using the branch address, global history, and the hashing functions f_1 and f_2, while PHT bank 0 is indexed only by the lower bits of the program counter. The rationale behind this approach is as follows. When the history length becomes larger, the number of branches between one instance of a branch address-branch history pair, and another identical instance tends to increase. This increases the probability that aliasing will occur in the meantime and corrupt one of the banks. Since the first bank of the enhanced gskewed predictor is addressed by the branch address only, the distance between successive accesses will be shorter, and so the likelihood that an unrelated branch aliases to the same entry in PHT_0 is decreased.

A variant of the enhanced gskewed algorithm was selected to be used in the Alpha EV8 microprocessor [Seznec et al., 2002], although the EV8 project was eventually cancelled in a late phase of development.

9.3.2.3 The Agree Predictor.
The gskewed algorithm attempts to reduce the effects of conflict aliasing by storing the branch prediction in multiple locations. The agree predictor reduces destructive aliasing interference by reinterpreting the PHT counters as a *direction agreement* bit [Spangle et al., 1997].

When two address-history pairs map into the same PHT entry, there are two types of interference that can result. The first is *destructive* or *negative* interference. Destructive interference occurs when the counter updates of one address-history pair corrupt the stored state of a different address-history pair, thus causing more mispredictions. The address-history pairs that result in destructive interference are each trying to update the counter in opposite directions; that is, one address-history pair is consistently incrementing the counter, and the other pair attempts to decrement the counter. The other type of interference is *neutral* interference where the PHT entry correctly predicts the branch outcomes for both address-history pairs.

Regardless of the actual direction of the history-address pairs, branches tend to be heavily biased in one direction or the other. In other words, in an infinite-entry PHT where there is no interference, the majority of counters will be either in the strongly taken (ST) or strongly not-taken (SN) states.

The agree predictor stores the most likely predicted direction in a separate *biasing bit*. This biasing bit may be stored with the branch target buffer (see Section 9.5.1.1) line of the corresponding branch, or in some other separate hardware structure. The biasing bit may be initialized to the outcome of the first instance of the branch, or it may be a branch hint inserted by the compiler. Instead of predicting the branch direction, the PHT counter now predicts whether or not the branch will go in the same direction as the corresponding biasing bit. Another interpretation is that the PHT counter predicts whether the branch outcome will *agree* with the biasing bit.

Figure 9.14 illustrates the hardware for the agree predictor. Like the gshare algorithm, the branch address and global branch history are combined to index into the PHT. At the same time, the branch address is also used to look up the biasing bit. If the most-significant bit of the indexed PHT counter is a one (predict agreement with the biasing bit), then the final branch prediction is equal to the biasing bit. If the most significant bit is a zero (predict disagreement with the biasing bit), then the complement of the biasing bit is used for the final prediction. The number of biasing bits stored is generally different than the number of PHT entries.

After a branch instruction has resolved, the corresponding PHT counter is updated based on whether or not the actual branch outcome agreed with the biasing bit. In this fashion, two different address-history pairs may conflict and map to the same PHT entry, but if their corresponding biasing bits are set accurately, the predictions will not

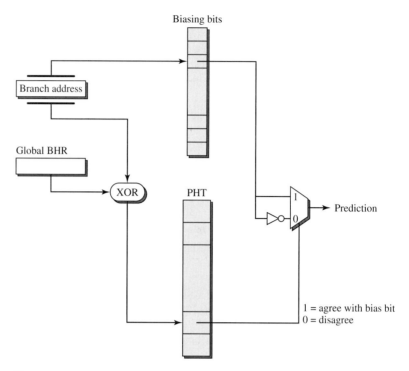

Figure 9.14
The Agree Predictor.

be affected. The agree prediction mechanism is used in the HP PA-RISC 8700 processor [Hewlett Packard Corporation, 2000]. Their biasing bits are determined by a combination of compiler analysis of the source code and profile-based optimization.

9.3.2.4 The YAGS Predictor. The Bi-Mode predictor study demonstrated that the separation of branches into two separate mostly taken and mostly not-taken substreams is beneficial. The yet another global scheme (YAGS) approach is similar to the Bi-Mode predictor, except that the two PHTs record only the instances that do not agree with the direction bias [Eden and Mudge, 1998]. The PHTs are replaced with a T-cache and an NT-cache. Each cache entry contains a 2-bit counter and a small tag (6 to 8 bits) to record the branch instances that do not agree with their overall bias. If a branch does not have an entry in the cache, then the selection counter is used to make the prediction. The hardware is illustrated in Figure 9.15.

To make a branch prediction with the YAGS predictor, the branch address indexes a *choice PHT* (analogous to the choice predictor of the Bi-Mode predictor). The 2-bit counter from the choice PHT indicates the bias of the branch *and* is used to select one of the two caches. If the choice PHT counter indicates taken, then the NT-Cache is consulted. The NT-Cache is indexed with a hash of the branch address and the global history, and the stored tag is compared to the least-significant bits of the

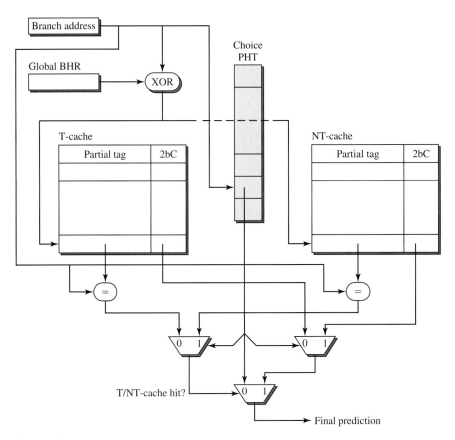

Figure 9.15
The YAGS Predictor.

branch address. If a tag match occurs, then the prediction is made by the counter from the NT-cache, otherwise the prediction is made from the choice PHT (predict taken). The actions taken for a choice PHT prediction of not-taken are analogous.

At a conceptual level, the idea behind the YAGS predictor is that the choice PHT provides the prediction "rule," and then the T/NT-caches record only the "exceptions to the rule," if any exist. Most of the components in Figure 9.15 are simply for detecting if there is an exception (i.e., hit in the T/NT-caches) and then for selecting the appropriate prediction.

After the branch outcome is known, the choice PHT is updated with the same partial update policy used by the Bi-Mode choice predictor. The NT-cache is updated if it was used, or if the choice predictor indicated that the branch was taken, but the actual outcome is not-taken. Symmetric rules apply for the T-cache.

In the Bi-Mode scheme, the second-level PHTs must store the directions for all branches, even though most of these branches agree with the choice predictor. The Bi-Mode predictor only reduces aliasing by dividing the branches into two substreams. The insight for the YAGS predictor is that the PHT counter values in the second-level PHTs of the Bi-Mode predictor are mostly redundant with the information conveyed by the choice predictor, and so it only allocates hardware resources to make note of the cases where the prediction does not match the bias.

In the YAGS study, two-way associativity was also added to the T-cache and NT-cache, which only required the addition of 1 bit to maintain the LRU state. The tags that are already stored are reused for the purposes of associativity, and only an extra comparator and simple logic need to be added. The replacement policy is LRU, with the exception that if the counter of an entry in the T-cache indicates not-taken, it is evicted first because this information is already captured by the choice PHT. The reverse rule applies for entries in the NT-cache. The addition of two-way associativity slightly increases prediction accuracy, although it adds some additional hardware complexity as well.

9.3.2.5 Branch Filtering.
Branches tend to be highly biased toward one direction or the other, and the Bi-Mode algorithm works well because it sorts the branches based on their bias which reduces negative interference. A different approach called *branch filtering* attempts to remove the highly biased branches from the PHT, thus reducing the total number of branches stored in the PHT which helps to alleviate capacity and conflict aliasing [Change et al., 1996]. The idea is to keep track of how many times each branch has gone in the same direction. If a branch has taken the same direction more than a certain number of times, then it is "filtered" in that it will no longer make updates to the PHT.

Figure 9.16 shows the organization of the branch counting table and the PHT, along with the logic for detecting whether a branch should be filtered. Although this figure shows branch filtering with a gshare predictor, the branch filtering technique could be applied to other prediction algorithms as well. An entry in the branch counting table tracks the branch direction, and how many consecutive times the branch has taken that direction. If the direction changes, the new direction is stored and the count is reset. If the counter has been incremented to its maximum value, then the corresponding branch is deemed to be very highly biased. At this point, this branch will no longer update the PHT, and the branch count table provides the prediction. If at any point the direction changes for this branch, then the counter is reset and the PHT once again takes over making the predictions.

Branch filtering effectively removes branches corresponding to error-checking code, such as the almost-never-taken malloc checking branch in the example from Section 9.2.3, and other dynamically constant branches. Although the branch counting table has been described here as a separate entity, the counter and direction bit would actually be part of the branch target buffer, which is described in Section 9.5.1.1.

9.3.2.6 Selective Branch Inversion.
The previous several branch prediction schemes all aim to provide better branch prediction rates by reducing the amount of interference in the PHT (interference avoidance). Another approach, *selective branch inversion* (SBI), attacks the interference problem differently by using interference

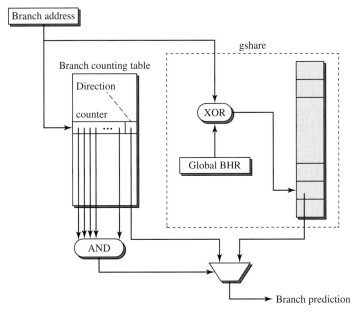

Figure 9.16
Branch Filtering Applied to a gshare Predictor.

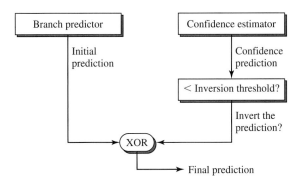

Figure 9.17
Selective Branch Inversion Applied to a Generic Branch Predictor.

correction [Argon et al., 2001; Manne et al., 1999]. The idea is to estimate the *confidence* of each branch prediction; if the confidence is lower than some threshold, then the direction of the branch prediction is inverted. See Section 9.5.2 for an explanation of predicting branch confidence. A generic SBI predictor is shown in Figure 9.17. Note that the SBI technique can be applied to any existing branch prediction scheme. An SBI gskewed or SBI Bi-Mode predictor achieves better prediction rates by performing both interference avoidance and interference correction.

9.3.3 Predicting with Alternative Contexts

Correlating branch predictors are basically simple pattern-recognition mechanisms. The predictors learn mappings from a *context* to a branch outcome. That is, every time a branch outcome becomes known, the predictor makes note of the current context. In the future, should the same context arise, the predictor will make a prediction that corresponds to the previous time(s) it encountered that context. So far, the predictors described in this chapter have used some combination of the branch address and the branch outcome history as the context for making predictions.

There are many design decisions that go into choosing the context for a branch predictor. Should the predictor use global history or local history? How many of the most recent branch outcomes should be used? How many bits of the branch address should be included? How is all of this information combined to form the final context?

In general, the more context a predictor uses, the more opportunities it has for detecting correlations. Using the same example given in Section 9.3.1.4, a branch correlated to a branch outcome nine branches ago will not be accurately predicted by a predictor that makes use of a history that is only eight branches deep. That is, an eight-deep branch history does not provide the proper context for making this prediction.

The predictors described here all improve prediction accuracies by making use of better context. Some use a greater amount of context, some use different contexts for different branches, and some use additional types of information beyond the branch address and the branch history.

9.3.3.1 Alloyed History Predictors.

The GA* predictors are able to make predictions based on correlations with the global branch history. The PA* predictors use correlations with local, or per-address, branch history. Programs may contain some branches whose outcomes are well predicted by global-history predictors *and* other branches that are well predicted by local-history predictors. On the other hand, some branches require *both* global branch history and per-address branch history to be correctly predicted. Mispredictions due to using the wrong type of history or only one type when more than one are needed are called *wrong-history mispredictions*.

An alloyed branch predictor removes some of these wrong-history mispredictions by using *both* global and local branch history [Skadron et al., 2003]. A per-address BHT is maintained as well as a global branch history register. Bits from the branch address, the global branch history, and the local branch history are all concatenated together to form an index into the PHT. The combined global/local branch history is called *alloyed* branch history. This approach allows both global and local correlations to be distinguished by the same structure. Alloyed branch history also enables the branch predictor to detect correlations that simultaneously depend on both types of history; this class of predictions is one that could not be successfully predicted by either a global-history predictor or a local-history predictor alone.

Alloyed predictors can also be classified as MAg/MAs/MAp predictors (M for "merged" history), where the second-level table can be indexed in the same way as the two-level predictors. Therefore, the three basic alloyed predictors are MAg, MAp, and

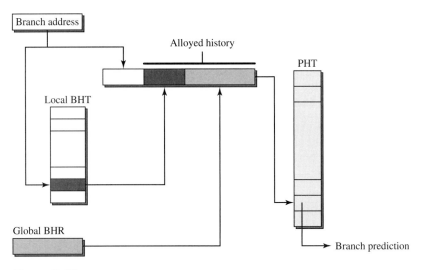

Figure 9.18
The Alloyed History Predictor

MAs. Alloyed history versions of other branch prediction algorithms are also possible, such as mshare (alloyed history gshare), or mskewed (alloyed history gskewed).

Figure 9.18 illustrates the hardware organization for the alloyed predictor. Like the PAg/PAs/PAp two-level predictors, the low-order bits of the branch address are used to index into the local history BHT. The corresponding local history is then concatenated with the contents of the global BHR and the bits from the branch address. This index is used to perform a lookup in the PHT, and the corresponding counter is used to make the final branch prediction. The branch predictor designer must make a tradeoff between the width of the global BHR and the width of the per-address BHT entries.

9.3.3.2 Path History Predictors. With the outcome history-based approaches to branch prediction, it may be the case that two very different paths of the program execution may have overlapping branch address and branch history pairs. For example, in Figure 9.19, the program may reach branch X in block D by going through blocks A, C, and D, or going through B, C, and D. When attempting to predict branch X in block D, the branch address and the branch histories for the last two global branches are identical for either ACD or BCD. Depending on the *path* by which the program arrived at block D, branch X is primarily not-taken (for path ACD), or primarily taken (for path BCD). When using only the branch outcome history, the different branch outcome patterns will cause a great deal of interference in the corresponding PHT counter.

Path-based branch correlation has been proposed to make better branch predictions when dealing with situations like the example in Figure 9.19. Instead of storing the last n branch outcomes, k bits from each of the last n branch addresses

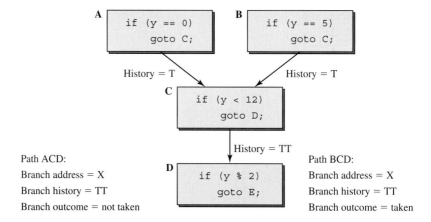

Figure 9.19
Path History Example.

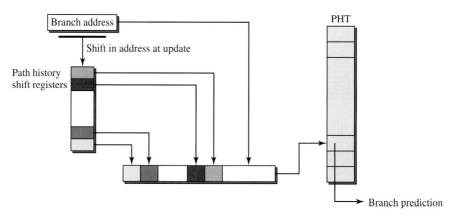

Figure 9.20
A Path History Predictor.

are stored [Nair, 1995; Reches and Weiss, 1997]. The concatenation of these nk bits encodes the branch path of the last n branches, also called the *path history*, thus potentially allowing the predictor to differentiate between the two very different branch behaviors in the example of Figure 9.19. Combined with a subset of the branch address bits of the current branch, this forms an index into a PHT. The prediction is then made in the same way as a normal two-level predictor.

Figure 9.20 illustrates the hardware for the path history branch predictor. The bits from the last n branches are concatenated together to form a path history. The path history is then concatenated with the low-order bits of the current branch address. This index is used to perform a lookup in the PHT, and the final prediction is made. After the branch is processed, bits from the current branch address are added to the path history, and the oldest bits are discarded. The path history register can be implemented with shift registers. The number of bits per branch address to be stored k, the number of branches in the path history n, and the number of bits from the current branch address m all must be carefully chosen. The PHT has 2^{nk+m} entries, and therefore the area requirements can become prohibitive for even moderate values of n, k, and m. Instead of concatenating the n branch addresses, combinations of shifting, rotating, and hashing (typically using XORs) can be used to compress the $nk + m$ bits down to a more manageable size [Stark et al., 1998].

9.3.3.3 Variable Path Length Predictors. Some branches are correlated to branch outcomes or branch addresses that occurred very recently. Incorporating a longer history introduces additional bits that do not provide any additional information to the predictor. In fact, this useless context can degrade the performance of the predictor because the predictor must figure out what parts of the context are irrelevant, which in turn increases the training time. On the other hand, some branches are strongly correlated to older branches, which requires the predictor to make use of a longer history if these branches are to be correctly predicted.

One approach to dealing with the varying history length requirements of branches is to use different history lengths for each branch [Stark et al., 1998]. The following description uses path history, but the idea of using different history lengths can be applied to branch outcome histories as well. Using n different hashing functions $f_1, f_2, \ldots, f_n$, hash function f_i creates a hash of the last i branch addresses in the path history. The hash function used may be different between different branches, thus allowing for variable-length path histories. The selection of which hash function to use can be determined statically by the compiler, chosen with the aid of program profiling, or dynamically selected with additional hardware for tracking how well each of the hash functions is performing.

The elastic history buffer (EHB) uses a variable outcome history length [Tarlescu et al., 1996]. A profiling phase statically chooses a branch history length for each static branch. The compiler communicates the chosen length by using branch hints.

9.3.3.4 Dynamic History Length Fitting Predictors. The optimal history length to use in a predictor varies between applications. Some applications may have program behaviors that change frequently and are better predicted by more adaptive short-history predictors because short-history predictors require less time to train. Other programs may have distantly correlated branches, which require long histories to detect the patterns. By fixing the branch history length to some constant, some applications may be better predicted at the cost of reduced performance for others. Furthermore, the optimal history length for a program may change during the execution of the program itself. Any multiphased computation such as a compiler may exhibit very different branch patterns in the different phases of execution. A short history may be optimal in one phase, and a longer history may provide better prediction accuracy in the next phase.

Dynamic history length fitting (DHLF) addresses the problem of varying optimal history lengths. Instead of fixing the history length to some constant, the predictor uses different history lengths and attempts to find the length that minimizes branch mispredictions [Juan et al., 1998]. For applications that require shorter histories, a DHLF predictor will tune itself to consider fewer branch outcomes; for benchmarks that require longer histories, a DHLF predictor will adjust for that situation as well. The DHLF technique can be applied to all kinds of correlating predictors (gshare, Bi-Mode, gskewed, etc.).

9.3.3.5 Loop Counting Predictors.

In general, the termination of a for-loop is difficult to predict using any of the algorithms already presented in this section. Each time a for-loop is encountered, the number of iterations executed is often the same as the previous time the loop was encountered. A simple example of this is the inner loop of a matrix multiply algorithm where the number of iterations is equal to the matrix block size. Because of the consistent number of iterations, the loop exit branch should be very easy to predict. Unfortunately, a branch history register–based approach would require BHR sizes greater than the number of iterations of the loop. Beyond a small number of iterations, the storage requirements for such a predictor become prohibitive, because the PHT size is exponential in the history length.

The Pentium-M processor uses a loop predictor in conjunction with a branch history–based predictor [Gochman et al., 2003]. The loop predictor consists of a table where each entry contains fields to record the current iteration count, the iteration limit, and the direction of the branch. This is illustrated in Figure 9.21. A *loop branch* is one that always goes the same direction (either taken or not-taken) followed by a single instance where the branch direction is the opposite, and then this pattern repeats. A traditional loop-closing branch has a pattern of 111...1110111...1110111..., but the Pentium-M loop predictor can also handle the

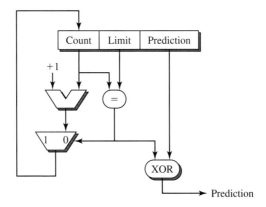

Figure 9.21
A Single Entry of the Loop Predictor Table Used in the Pentium-M Processor.

opposite pattern of 000 . . . 0001000 . . . 0001000 . . . . The limit field stores a count of the number of iterations that were observed for the previous invocation of the loop. When a loop exit is detected, the counter value is copied to the limit field and then the counter is reset for the next run of the loop. The prediction field records the predominant direction of the branch. As long as the counter is less than the limit, the loop predictor will use the prediction field. When the counter reaches the limit, this indicates that the predictor has reached the end of the loop, and so it predicts in the opposite direction as that stored in the prediction field.

While loop-counting predictors are useful in hybrid predictors (see Section 9.4), they provide very poor performance when used by themselves because they cannot capture nonloop behaviors.

9.3.3.6 The Perceptron Predictor. By maintaining larger branch history registers, the additional history stored provides more opportunities for correlating the branch predictions. There are two major drawbacks with this approach. The first is that the size of the PHT is exponential in the width of the BHR. The second is that many of the history bits may not actually be relevant, and thus act as training "noise." Two-level predictors with large BHR widths take longer to train.

One solution to this problem is the Perceptron predictor [Jimenez and Lin, 2003]. Each branch address (*not* address-history pair) is mapped to a single entry in a Perceptron table. Each entry in the table consists of the state of a single Perceptron. A Perceptron is the simplest form of a neural network [Rosenblatt, 1962]. A Perceptron can be trained to learn certain boolean functions.

In the case of the Perceptron branch predictor, each bit x_i of the input x is equal to 1 if the branch was taken ($BHR_i = 1$) and x_i is equal to -1 if the branch was not taken ($BHR_i = 0$). There is one special *bias* input x_0 which is always 1. The Perceptron has one weight w_i for each input x_i, including one weight w_0 for the bias input. The Perceptron's output y is computed as

$$y = w_0 + \sum_{i=1}^{n}(w_i \cdot x_i)$$

If y is negative, the branch is predicted to be not taken. Otherwise the branch is predicted to be taken.

After the branch outcome is available, the weights of the Perceptron are updated. Let $t = -1$ if the branch was not taken, and $t = 1$ if the branch was taken. In addition, let $\theta > 0$ be a *training threshold*. The variable y_{out} is computed as

$$y_{out} = \begin{cases} 1 & \text{if } y > \theta \\ 0 & \text{if } -\theta \leq y \leq \theta \\ -1 & \text{if } y < -\theta \end{cases}$$

Then if y_{out} is not equal to t, all the weights are updated as $w_i = w_i + tx_i$, $i \in \{0, 1, 2, \ldots, n\}$. Intuitively, $-\theta \leq y \leq \theta$ indicates that the Perceptron has not been trained to a state where the predictions are made with high confidence. By setting y_{out} to zero, the condition $y_{out} \neq t$ will always be true, and the Perceptron's weights will be updated (training continues). When the correlation is large, the magnitude of the weight will tend to become large.

One limitation of using the Perceptron learning algorithm is that only *linearly separable* functions can be learned. Linearly separable boolean functions are those where all instances of outputs that are 1 can be separated in hyperspace from all instances whose outputs are 0 by a hyperplane. In Jimenez and Lin [2003], it is shown that for half of the SPEC2000 integer benchmarks, over 50% of the branches are linearly inseparable. The Perceptron predictor generally performs better than gshare on benchmarks that have more linearly separable branches, whereas gshare outperforms the Perceptron predictor on benchmarks that have a greater number of linearly inseparable branches.

The Perceptron predictor can adjust the weights corresponding to each bit of the history, since the algorithm can effectively "tune out" any history bits that are not relevant (low correlation). Because of this ability to selectively filter the branches, the Perceptron often attains much faster training times than conventional PHT-based approaches.

Figure 9.22 illustrates the hardware organization of the Perceptron predictor. The lower-order bits of the branch address are used to index into the table of

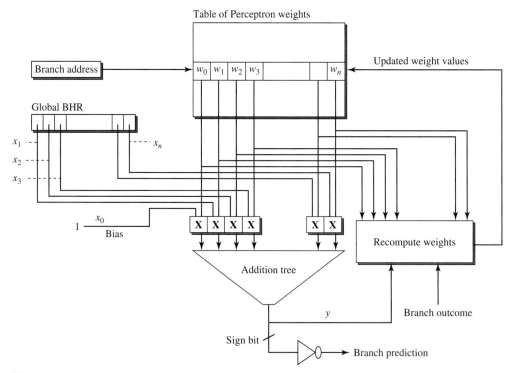

Figure 9.22
The Perceptron Predictor.

Perceptrons in a per-address fashion. The weights of the selected Perceptron and the BHR are forwarded to a block of combinatorial logic that computes y. The prediction is made based on the complement of the sign bit (most-significant bit) of y. The value of y is also forwarded to an additional block of logic and combined with the actual branch outcome to compute the updated values of the weights of the Perceptron.

The design space for the Perceptron branch predictor appears to be much larger than that of the gshare and Bi-Mode predictors, for example. The Perceptron predictor has four parameters: the number of Perceptrons, the number of bits of history to use, the width of the weights, and the learning threshold. There is an empirically derived relation for the optimal threshold value as a function of the history length. The threshold θ should be equal to $\lfloor 1.93h +14 \rfloor$, where h is the history length. The number of history bits that can potentially be used is still much larger than in the gshare predictors (and similar schemes).

Similar to the alloyed history two-level branch predictors, alloyed history Perceptron predictors have also been proposed. For n bits of global history and m bits of local history, each Perceptron uses $n + m + 1$ weights (+1 for the bias) to make a branch prediction.

9.3.3.7 The Data Flow Predictor. The Perceptron predictor makes use of a long-history register and effectively finds the highly correlated branches by assigning them higher weights. The majority of these branches are correlated for two reasons. The first is that a branch may guard instructions that affect the test condition of the later branch, such as branch A from the example in Figure 9.4. These branches are called *affector* branches. The second is that the two branches operate on similar data. The Perceptron attempts to find the highly correlated branches in a fuzzy fashion by assigning larger weights to the more correlated branches. Another approach to find the highly correlated branches from a long-branch-history register is the *data flow* branch predictor that explicitly tracks register dependences [Thomas et al., 2003].

The main idea behind the data flow branch predictor is to explicitly track which previous branches are affector branches for the current branch. The *affector register file* (ARF) stores one bitmask per architected register, where the entries of the bitmask correspond to past branches. If the ith most recent branch is an affector for register R, then the ith most recent bit in entry R of the ARF will be set. For register updating instructions of the form Ra = Rb op Rc, the ARF entry for Ra is set equal to the bitwise-OR of the ARF entries for Rb and Rc, with the least-significant bit (most recent branch) set to 1. This is illustrated in Figure 9.23. Setting the least-significant bit to one indicates that the most recent branch (b0) guards an instruction that modifies Ra. Note that the entries for Rb and Rc also have their corresponding affector bits set. The OR of the ARF entries for the operands makes the current register inherit the affectors of its operands. In this fashion, an ARF entry records a bitmask that specifies all the affector branches that can potentially affect the register's value. On a branch instruction, the ARF is updated by shifting all entries to the left by one and filling in the least-significant bit with zero.

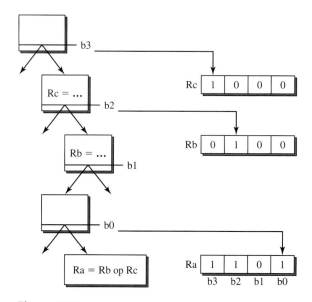

Figure 9.23
Affector Register File Update for a Register Writing Instruction.

The ARF specifies a set of potentially important branches, because it is these branches that can affect the values of the condition registers. A conditional branch compares one or more register values and evaluates some condition on these values (e.g., equal to zero or greater than). To make a prediction, the data flow predictor uses the ARF entries corresponding to the operand(s) of the current branch; if there are two operands, then a final bitmask is formed by the exclusive-OR of the respective ARF entries. The affector bitmask is ANDed with the global history register, which isolates the global history outcomes for the affector branches only. The branch history register is likely to be larger than the index into the PHT, and so the masked version of the history register still needs to be hashed down to an appropriate size. This final hashed version of the masked history register indexes into a PHT to provide the final prediction. The overall organization of the data flow predictor is illustrated in Figure 9.24.

In the original data flow branch predictor study, the predictor was presented as a *corrector predictor,* which is basically a secondary predictor that backs up some other prediction mechanism. The idea is basically the same as the overriding predictor organization explained in Section 9.5.4.2. A primary predictor provides most of the predictions, and the data flow predictor attempts to learn and provide corrections for the branches that the primary predictor does not properly handle. For this reason, the PHT entries of the data flow predictor may be augmented with partial tags (see partial resolution in Section 9.5.1.1) and set associativity. This allows the data flow predictor to carefully identify and correct only the branches it knows about. Because the data flow predictor only attempts to correct a select set of branches, its total size may be smaller than other conventional stand-alone branch predictors.

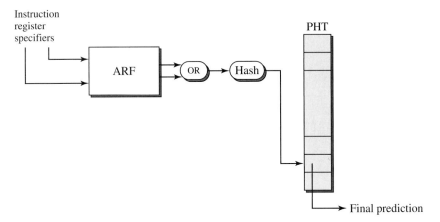

Figure 9.24
The Data Flow Branch Predictor.

9.4 Hybrid Branch Predictors

Different branches in a program may be strongly correlated with different types of history. Because of this, some branches may be accurately predicted with global history–based predictors, while others are more strongly correlated with local history. Programs typically contain a mix of such branch types, and for example, choosing to implement a global history–based predictor may yield poor prediction accuracies for the branches that are more strongly correlated with their own local history.

To a certain degree, the alloyed branch predictors address this issue, but a tradeoff must be made between the number of global history bits used and the number of local history bits used. Furthermore, the alloyed branch predictors cannot effectively take advantage of predictors that use other forms of information, such as the loop predictor.

This section describes algorithms that employ two or more single-scheme branch prediction algorithms and combine these multiple predictions together to make one final prediction.

9.4.1 The Tournament Predictor

The simplest and earliest proposed multischeme branch predictor is the *tournament* algorithm [McFarling, 1993]. The predictor consists of two component predictors P_0 and P_1 and a *meta-predictor* M. The component predictors can be any of the single-scheme predictors described in Section 9.3, or even one of the hybrid predictors described in this section.

The meta-predictor M is a table of 2-bit counters indexed by the low-order bits of the branch address. This is identical to the lookup phase of $Smith_2$, except that a (meta-)prediction of zero indicates that P_0 should be used, and a (meta-)prediction of one indicates that P_1 should be used (the meta-prediction is made from the most-significant bit of the counter). The meta-predictor makes a prediction of which predictor will be correct.

After the branch outcome is available, P_0 and P_1 are updated according to their respective update rules. Although the meta-predictor M is structurally identical to $Smith_2$, the update rules (i.e., state transitions) are different. Recall that the 2-bit counters used in the predictors are finite state machines (FSMs), where the inputs are typically the branch outcome and the previous state of the FSM. For the meta-predictor M, the inputs are now c_0, c_1, and the previous FSM state, where c_i is one if P_i predicted correctly. Table 9.2 lists the state transitions. When P_1's prediction was correct and P_0 mispredicted, the corresponding counter in M is incremented, saturating at a maximum value of 3. Conversely, when P_1 mispredicts and P_0 predicts correctly, the counter is decremented, saturating at zero. If both P_0 and P_1 are correct, or both mispredict, the counter in M is unmodified.

Figure 9.25a illustrates the hardware for the tournament selection mechanism with two generic component predictors P_0 and P_1. The prediction lookups on P_0, P_1, and M are all performed in parallel. When all three predictions have been made, the meta-prediction is used to drive the select line of a multiplexer to choose between the predictions of P_0 and P_1. Figure 9.25b illustrates an example tournament

Table 9.2
Tournament meta-predictor update rules

c_0 (P_0 Correct?)	c_1 (P_1 Correct?)	Modification to M
0	0	Do nothing
0	1	Saturating increment
1	0	Saturating decrement
1	1	Do nothing

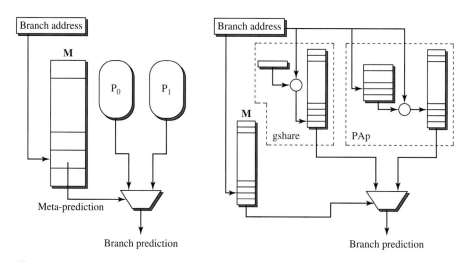

Figure 9.25
(a) The Tournament Selection Mechanism. (b) Tournament Hybrid with gshare and PAp.

selection predictor with gshare and PAp component predictors. A hybrid predictor similar to the one depicted in Figure 9.25b was implemented in the Compaq Alpha 21264 microprocessor [Kessler, 1999]. The local history component used a 1024-entry BHT with 10-bit per-branch histories. This 10-bit history is then used to index into a single 1024-entry PHT. The global history component uses a 12-bit history that indexes into a 4096-entry PHT of 2-bit counters. The meta-predictor also uses a 4096-entry table of counters.

Like the two-level branch predictors, the tournament's meta-predictor can also make use of branch history. It has been shown that a global branch outcome history hashed with the PC (similar to gshare) provides better overall prediction accuracies [Chang et al., 1995].

Either or both of the two components of a tournament hybrid predictor may themselves be hybrid predictors. By recursively arranging multiple tournament meta-predictors into a tree, any number of predictors may be combined [Evers, 2000].

9.4.2 Static Predictor Selection

Through profiling and program-based analysis, reasonable branch prediction rates can be achieved for many programs with static branch prediction. The downside of static branch prediction is that there is no way to adapt to unexpected branch behavior, thus leaving the possibility for undesirable worst-case behaviors. Grunwald et al. [1998] proposed using profiling techniques, but limited only to the meta-predictor. The entire multischeme branch predictor supports two or more component predictors, all which may be dynamic. The selection of which component to use is determined statically and encoded in the branch instruction as branch hints. The meta-predictor requires no additional hardware except for a single multiplexer to select between the component predictors' predictions.

The proposed process of determining the static meta-predictions is a lot more involved than traditional profiling techniques. Training sets are used to execute the programs to be profiled, but the programs are not executed on native hardware. Instead, a processor simulator is used to fully simulate the branch prediction structures in addition to the functional behavior of the program. The component predictor that is correct with the highest frequency is selected for each static branch. This may not be practical since the simulator may be very slow and full knowledge of the component branch predictors' implementations may not be available.

There are several advantages to the static selection mechanism. The first is that the hardware cost is negligible (a single additional n-to-1 multiplexer for n component predictors). The second advantage is that each static branch is assigned to one and only one component branch predictor. This means that the average number of static branches per component is reduced, which alleviates some of the problems of conflict and capacity aliasing. Although meta-predictions are performed statically, the underlying branch predictions still incorporate dynamic information, thus reducing the potential effects of worst-case branch patterns. The disadvantages include the overhead associated with simulating branch prediction structures during the profiling phase, the fact that the branch hints are not available until after the instruction fetch has been completed, and the fact that the number of component predictors is limited by the number of hint bits available in a single branch instruction.

9.4.3 Branch Classification

The branch classification meta-prediction algorithm is similar to the static selection algorithm and may even be viewed as a special case of static selection [Chang et al., 1994]. A profiling phase is first performed, but, in contrast to static selection, only the branch taken rates are collected (similar to the profile-based static branch prediction techniques described in Section 9.2.4). Each static branch is placed in one of six branch classes depending on its taken rate. Those which are heavily biased in one direction, defined as having a taken rate or not-taken rate of less than 5%, are statically predicted. The remaining branches are predicted using a tournament hybrid method.

The overall predictor has the structure of a static selection multischeme predictor with three components (P_0, P_1, and P_2). P_0 is a static not-taken branch predictor. P_1 is a static taken branch predictor. P_2 is itself another multischeme branch predictor, consisting of a tournament meta-predictor M and two component predictors, $P_{2,0}$ and $P_{2,1}$. The two component predictors of P_2 can be chosen to be any dynamic or static branch prediction algorithms, but are typically a global history predictor and a local history predictor. The branch classification algorithm has the advantage that easily predicted branches are removed from the dynamic branch prediction structures, thus reducing the number of potential sources for aliasing conflicts. This is similar to the benefits provided by branch filtering.

Figure 9.26 illustrates the hardware for a branch classification meta-predictor with static taken and non-taken predictors, as well as two unspecified generic components $P_{2,0}$ and $P_{2,1}$, and a tournament selection meta-predictor to choose between the two dynamic components. Similar to the static hybrid selection

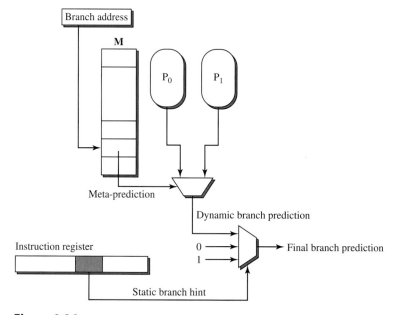

Figure 9.26
The Branch Classification Mechanism.

mechanism, the branch classification hint is not available to the predictor until after instruction fetch has completed.

9.4.4 The Multihybrid Predictor

Up to this point, none of the multischeme meta-predictors presented are capable of *dynamically* selecting from more than two component predictors (except for recursively using the tournament meta-predictor). By definition, a single tournament meta-predictor can only choose between two components. The static selection approach cannot dynamically choose any of its components. The branch classification algorithm can statically choose one of three components, but the dynamic selector used only chooses between two components.

The multihybrid branch predictor does allow the dynamic selection between an arbitrary number of component predictors [Evers et al., 1996]. The lower bits of the branch address are used to index into a table of prediction selection counters. Each entry in the table consists of n 2-bit saturating counters, $c_1, c_2, \ldots c_n$, where c_i is the counter corresponding to component predictor P_i. The components that have been predicting well have higher counter values. The meta-prediction is made by selecting the component whose counter value is 3 (the maximum) and a predetermined priority ordering is used to break ties. All counters are initialized to 3, and the update rules guarantee that at least one counter will have the value of 3. To update the counters, if at least one component with a counter value of 3 was correct, then the counter values corresponding to components that mispredicted are decremented (saturating at zero). Otherwise, the counters corresponding to components that predicted correctly are incremented (saturating at 3).

Figure 9.27 illustrates the hardware organization for the multihybrid meta-predictor with n component predictors. The branch address is used to look up an entry in the table of prediction selection counters, and each of the n counters is

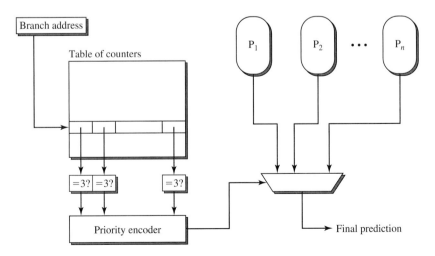

Figure 9.27
The Multihybrid Predictor.

checked for a value of 3. A priority encoder generates the index for the component with a counter value of 3 and the highest priority in the case of a tie. The index signal is then forwarded to the final multiplexer that selects the final prediction.

Unlike the static selection or even the branch classification meta-prediction algorithms, the multihybrid meta-predictor is capable of dynamically handling any number of component branch predictors.

9.4.5 Prediction Fusion

All the hybrid predictors described so far use a *selection* mechanism to choose one out of n predictions. By singling out a single predictor, selection-based hybrids throw out any useful information conveyed by the other predictors. Another approach, called *prediction fusion,* attempts to combine the predictions from all n individual predictors in making the final prediction [Loh and Henry, 2002]. This allows the hybrid predictor to leverage the information available from all component predictors, potentially making use of both global- and local-history components, or short- and long-history components, or some combination of these.

Prediction fusion covers a wide variety of predictors. Selection-based hybrid predictors are special cases of fusion predictors where the fusion mechanism ignores $n - 1$ of the inputs. The gskewed predictor can be thought of as a prediction fusion predictor that uses three gshare predictors with different hashing functions as inputs, and a majority function as the fusion mechanism.

One fusion-based hybrid predictor is the *fusion table*. Like the multihybrid predictor, the fusion table can take the predictions from an arbitrary number of subpredictors. For n predictors, the fusion table concatenates the corresponding n predictions together into an index. This index, combined with bits from the PC and possibly the global branch history form a final index into a table of saturating counters. The most significant bit of the indexed saturating counter provides the final prediction. This is illustrated in Figure 9.28.

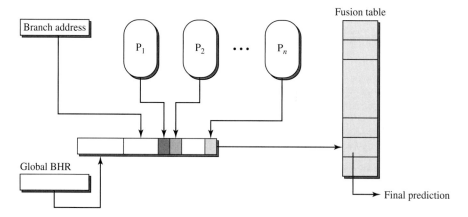

Figure 9.28
The Fusion Table Hybrid Predictor.

The fusion table provides a way to correlate branch outcomes to multiple branch predictions. The fusion table can remember any arbitrary mapping of predictions to branch outcome. For example, in the case where a branch is always taken if exactly one of two predictors is taken, the entries in the fusion table that correspond to this situation will be trained to predict taken, while the other entries that correspond to the predictor both predicting taken or both predicting not-taken will train to predict not-taken.

The fusion table hybrid predictor is very effective because it is very flexible. With a combination of global- and local-history components, and short- and long-history components, the fusion table can accurately capture a wide array of branch behaviors.

9.5 Other Instruction Flow Issues and Techniques

Predicting the direction of conditional branches is only one of several issues in providing a high rate of instruction fetch. This section covers these additional problems such as taken-branch target prediction, branch confidence prediction, predictor-cache organizations and interactions, fetching multiple instructions in parallel, and coping with faster clock speeds.

9.5.1 Target Prediction

For conditional branches, predicting whether the branch is taken or not-taken is only half of the problem. After the direction of a branch is known, the actual target address of the next instruction along the predicted path must also be determined. If the branch is predicted to be not-taken, then the target address is simply the current branch's address plus the size of an instruction word. If the branch is predicted to be taken, then the target will depend on the type of branch. Target prediction must also cover unconditional branches (branches that are always taken).

There are two common types of branch targets. Branch targets may be PC-relative, which means that the taken target is always at the current branch's address plus a constant (the constant may be negative). A branch target can also be *indirect,* which means that the target is computed at run time. An indirect branch target is read from a register, sometimes with a constant offset added to the contents of the register. Indirect branches are frequently used in object-oriented programs (such as the C++ vtable that determines the correct method to invoke for classes using inheritance), dynamically linked libraries, subroutine returns, and sometimes multitarget control constructs (i.e., C `switch` statements).

9.5.1.1 Branch Target Buffers.
The target of a branch is usually predicted by a branch target buffer (BTB), sometimes also called a branch target address cache (BTAC) [Lee and Smith, 1984]. The BTB is a cache-like structure that stores the last seen target address for a branch instruction. When making a branch prediction, the traditional branch predictor provides a predicted direction. In parallel, the processor uses the current branch's PC to index into the BTB. The BTB is typically a tagged structure, often implemented with some degree of set associativity.

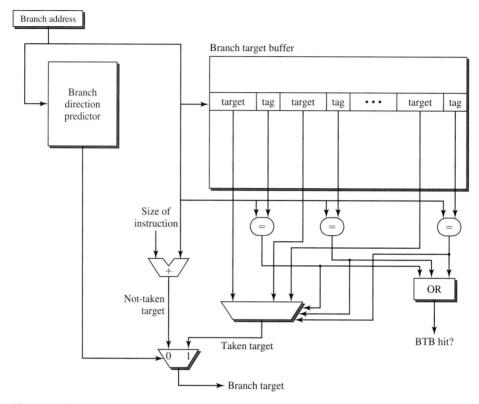

Figure 9.29
The Branch Target Buffer, a Generic Branch Predictor, and the Target Selection Logic.

Figure 9.29 shows the organization of the branch predictor and the BTB. If the branch predictor predicts not-taken, the target is simply the next sequential instruction. If the branch predictor predicts taken and there is a hit in the BTB, then the BTB's prediction is used as the next instruction's address. It is also possible that there is a taken-branch prediction, but there is a miss in the BTB. In this situation, the processor may stall fetching until the target is known. If the branch has a PC-relative target, then the fetch only stalls for a few cycles to wait for the completion of the instruction fetch from the instruction cache, the target offset extraction from the instruction word, and the addition of the offset to the current PC to generate the actual target. Another approach is to fall back to the not-taken target on a BTB miss.

Different strategies may be used for maintaining the information stored in the BTB. A simple approach is to store the targets of all branches encountered. A slightly better use of the BTB is to only store the targets of taken branches. This is because if a branch is predicted to be not taken, the next address is easily computed. By filtering out the not-taken targets, the prediction rate of the BTB may be improved by a decrease in interference.

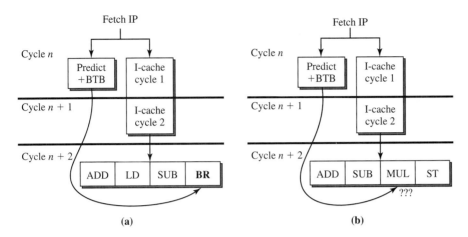

Figure 9.30
Timing Diagram Illustrating That a Branch Prediction Occurs before the Instruction Fetch Completes: (a) When a Branch Is Present in the Fetch Block, and (b) When There Is No Branch Present in the Fetch Block.

In a pipelined processor, the instruction cache access may require multiple cycles to fetch an instruction. After a branch target has been predicted, the processor can immediately proceed with the fetch of the next instruction. There is a potential problem in that until the instruction has been fetched and decoded, how does the processor know if the next instruction is a branch or not? Figure 9.30 illustrates a branch predictor with a two-cycle instruction cache. In cycle n, the current branch address (IP) is fed to the branch predictor and BTB to predict the target of the next fetch block. At the same time, the instruction cache access for the current block is started. By the end of cycle n, the branch predictor and BTB have provided a direction and target prediction for the branch highlighted in bold in Figure 9.30(a).

Note that during cycle n when the branch prediction is made, it is not known that there will be a branch in the corresponding block. Figure 9.30(b) shows an example where there are no branches present in the fetch block. Since there are no branches, the next block to fetch is simply the next sequential block. But during cycle n when the branch prediction is made, the predictor does not know that there are no branches, and may even provide a target address that corresponds to a taken branch! A predicted taken branch that has no corresponding branch instruction is sometimes called a *phantom* branch or a *bogus* branch. In cycle $n + 2$, the decode logic can detect that there are no branches present and, if there was a taken-branch prediction, the predictor and instruction cache accesses can be redirected to the correct next-block address. Phantom branches incur a slight performance penalty because the delay between branch prediction and phantom branch detection causes bubbles in the fetch pipeline.

When there are no branches present in a fetch block, the correct next-fetch address is the next sequential instruction block. This is equivalent to a not-taken branch prediction, which is why only a *taken* branch prediction without a corresponding branch introduces a phantom branch. If the BTB is only ever updated with the targets of taken branches, and the next block of instructions does not contain *any* branches, then there will always be a BTB miss. If the processor uses a fallback to the not-taken strategy, then this will result in correct next-instruction address prediction when no branches are present, thus removing the phantom branches.

Address tags are typically fairly large, and so BTBs often use *partial resolution* [Fagin and Russell, 1995]. With partial resolution, only a subset of the tags are stored in the BTB entry. This allows for a decrease in the storage requirements, but opens up the opportunity for false hits. Two instructions with different addresses may both hit in the same BTB entry because the subset of bits used in the tag are identical, but there are differences in the address bits somewhere else. A BTB typically has fewer entries than a direction predictor because it must store an entire target address per entry (typically over 30 bits per entry), whereas the direction predictor only stores a small 2-bit counter per entry. The slight increase in mispredictions due to false hits is usually worth the decrease in structure size provided by partial resolution. Note that false hits can enable phantom branches to occur again, but if the false hit rate is low, then this will not be a serious problem.

9.5.1.2 Return Address Stack.
Function calls frequently occur in programs. Both the jump into the function and the jump back out (the return) are usually unconditional branches. The target of a jump into a function is typically easy to predict. A branch instruction that jumps to `printf` will likely jump to the same place every time it is encountered. On the other hand, the return from the `printf` function may be difficult to predict because `printf` could be called from many different places in a program.

Most instruction set architectures support subroutine calls by providing a means of storing the subroutine return address. When executing a jump to a subroutine, the address of the instruction that sequentially follows the jump is stored into a register. This address is then typically stored on the stack and used as a jump address at the end of the function when the return is called.

The return address stack (RAS) is a special branch target predictor that only provides predictions for subroutine returns [Kaeli and Emma, 1991]. When a jump into a function happens, the return address is pushed onto the RAS, as shown in Figure 9.31(a). During this initial jump, the RAS does not provide a prediction and the target must be predicted from the regular BTB. At some later point in the program when the program returns from the subroutine, the top entry of the RAS is popped and provides the correct target prediction as shown in Figure 9.31(b). The stack can store multiple return addresses, and so returns from nested functions will also be properly predicted.

The return address stack does not guarantee perfect prediction of return target addresses. The stack has limited capacity, and therefore functions that are too deeply nested will cause a stack overflow. The RAS is often implemented as a circular buffer, and so an overflow will cause the most recent return address to overwrite the oldest return address. When the stack unwinds to the return that was

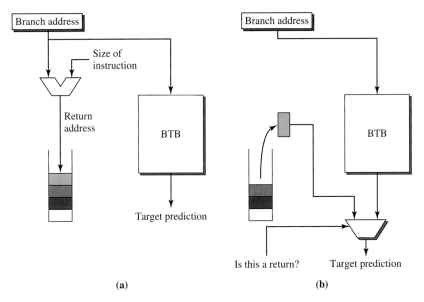

Figure 9.31
(a) Return Address Push on Jump to Subroutine. (b) Return Address Pop on Subroutine Return.

overwritten, a target misprediction will occur. Another source of RAS misprediction is irregular code that does not have matched subroutine calls and returns. Usage of the C library functions `setjmp` and `longjmp` could result in the RAS containing many incorrect targets.

Usage of the RAS requires knowing whether a branch is a function call or a return. This information is typically not available until after the instruction has been fetched. For a subroutine call, the target is predicted by the BTB, and so this will not introduce any bubbles into the fetch pipeline. For a subroutine return, the BTB may provide an initial target prediction. After the instruction has actually been fetched, it will be known that it is a return. At this point, the instruction flow may be corrected by squashing any instructions incorrectly fetched (or in the process of being fetched) and then resuming fetch from the return target provided by the RAS. Without the RAS, the target misprediction would not be detected until the return address has been loaded from the program stack into a register and the return instruction has been executed.

Return address stacks are implemented in almost all current mircroprocessors. An example is the Pentium 4, which uses a 16-entry return address stack [Hinton et al., 2001]. The RAS is also sometimes referred to as a *return stack buffer* (RSB).

9.5.2 Branch Confidence Prediction

Some branches are easy to predict, while others cause great trouble for the branch predictor. Branch confidence prediction does not make any attempt to predict the outcome of a branch, but instead makes a prediction about a branch prediction.

The purpose of branch confidence prediction is to guess or estimate how certain the processor is about a particular branch prediction. For example, the selective branch inversion technique (see Section 9.3.2.6) switches the direction of the initial branch prediction when the confidence is predicted to be very low. The confidence prediction detects cases where the branch direction predictor is consistently doing the wrong thing, and then selective branch inversion (SBI) uses this information to rectify the situation. There are many other applications of branch confidence information. This section first discusses techniques for predicting branch confidence, and then surveys some of the applications of branch confidence prediction.

9.5.2.1 Prediction Mechanisms. With branch confidence prediction, the information used is whether branch predictions are correct or not, as opposed to whether the prediction is taken or not-taken [Jacobson et al., 1996]. Figure 9.32 shows a branch confidence predictor that uses a global branch outcome history as context in a fashion similar to a gshare predictor, but the PHT has been replaced by an array of *correct/incorrect registers* (CIRs). A CIR is a shift register similar to a BHR in conventional branch predictors, but instead of storing the history of branch directions, the CIR stores the history of whether or not the branch was correctly predicted. Assuming that a 0 indicates a correct prediction, and a 1 indicates a misprediction, four correct predictions followed by two mispredictions followed by three more correct predictions would have a CIR pattern of 000011000.

To generate a final confidence prediction of high confidence or low confidence, the CIR must be processed by a *reduction function* to produce a single bit. The ones-counting approach counts the number of 1s in the CIR (that is, the number of mispredictions). The confidence predictor assumes that a large number of recent mispredictions indicates that future predictions will also likely be incorrect. Therefore, a higher ones-count indicates lower confidence. A more efficient implementation replaces the CIR shift register with a saturating counter. Each time

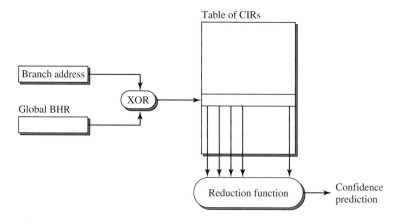

Figure 9.32
A Branch Confidence Predictor.

there is a correct prediction, the counter is incremented. The counter is decremented for a misprediction. If the counter has a large value, then it means that the branch predictions have been mostly correct, and therefore a large CIR counter value indicates high confidence. To detect n consecutive correct predictions, a shift register CIR needs to be n bits wide. On the other hand, a counter-based CIR only requires $\lceil \log_2 n \rceil$ bits. An alternative implementation uses resetting counters where each misprediction causes the counter to reset to zero instead of decrementing the counter. The counter value is now equal to the number of branches since the last misprediction seen by the CIR. Because the underlying branch prediction algorithms are already very accurate, the patterns observed in the shift-based CIRs are dominated by all zeros (no recent mispredictions) or a single one (only one recent misprediction). Since the resetting counter tracks the distance since the last misprediction, it approximately represents the same information.

The structure of the branch confidence predictor is very similar to branch direction predictors. Some of the more advanced techniques used for branch direction predictors could also be applied to confidence predictors.

9.5.2.2 Applications. Besides the already discussed selective branch inversion technique, branch confidence prediction has many other potential applications.

An alternative approach to predicting conditional branches and speculatively going down one path or the other is to fetch and execute from both the taken and not-taken paths at the same time. This technique, called *eager execution,* guarantees that the processor will perform some useful work, but it also guarantees that some of the instructions fetched and executed will be discarded. Allowing the processor to "fork" every conditional branch into two paths rapidly becomes very expensive since the processor must be able to track all the different paths and flush out different sets of instructions as different branches are resolved. Furthermore, performing eager execution for a highly predictable branch wastes resources. The wrong path will use up fetch bandwidth, issue slots, functional units, and cache bandwidth that could have otherwise been used for the correct path instructions. *Selective eager execution* limits the harmful effects of uncontrolled eager execution by limiting dual-path execution to only those branches that are deemed to be difficult, i.e., low-confidence branches [Klauser et al., 1998]. A variation of eager execution called *disjoint eager execution* is discussed in Chapter 11 [Uht, 1997].

Branch mispredictions are a big reason for performance degradations, but they also represent a large source of wasted power and energy. All the instructions on a mispredicted path are eventually discarded, and so all the power spent on fetching, scheduling, and executing these instructions is energy spent for nothing. Branch confidence can be used to decrease the power consumption of the processor. When a low-confidence branch is encountered, instead of making a branch prediction that has a poor chance of being correct, the processor can simply stall the front-end and wait until the actual branch outcome has been computed. This reduces instruction-level parallelism by covering up any parallelism blocked by this control dependency, but greatly reduces the power wasted by branch mispredictions.

Branch confidence can also be used to modify the fetch policies of simultaneously multithreaded (SMT) processors (see Chapter 11). An SMT processor fetches and executes instructions from multiple threads using the same hardware. The fetch engine in an SMT processor will fetch instructions for one thread in a cycle, and then depending on its fetch policy, it may choose to fetch instructions from another thread on the following cycle. If the current thread encounters a low-confidence branch, then the fetch engine could stop fetching branches from the current thread and start fetching instructions from another thread that are more likely to be useful. In this fashion, the execution resources that would have been wasted on a likely branch misprediction are usefully employed by another thread.

9.5.3 High-Bandwidth Fetch Mechanisms

For a superscalar processor to execute multiple instructions per cycle, the fetch engine must also be able to fetch multiple instructions per cycle. Fetching instructions from an instruction cache is typically limited to accessing a single cache line per cycle. Taken branches disrupt the instruction delivery stream from the instruction cache because the next instruction to fetch is likely to be in a different cache line. For example, consider a cache line that stores four instruction words where one is a taken branch. If the taken branch is the first instruction in the cache line, then the instruction cache will only provide one useful instruction (i.e., the branch). If the taken branch is in the last position, then the instruction cache can provide four useful instructions.

The maximum number of instructions fetched per cycle is bounded by the number of words in the instruction cache line (assuming a limit of one instruction cache access per cycle). Unfortunately, increasing the size of the cache line has only very limited effectiveness in increasing the fetch bandwidth. For typical integer applications, the number of instructions between the target of a branch and the next branch (i.e., a basic block) is only about five to six instructions. Assuming six instructions per basic block, and that 60% of branches are taken, the expected number of instructions between taken branches is 15. Unfortunately, there are many situations where the rate of taken branches is close to 100% (for example, a loop). In these cases, the instruction fetch engine will only be able to provide at most one iteration of the loop per cycle.

A related problem is the situation where a block of instructions is split across cache lines. If the first of four sequential instructions to be fetched is located at the end of a cache line, then two cache accesses are necessary to fetch all four instructions.

This section describes two mechanisms for providing instruction fetch bandwidth that can handle multiple basic blocks per cycle.

9.5.3.1 The Collapsing Buffer. The *collapsing buffer* scheme uses a combination of an interleaved BTB to provide multiple target predictions, a banked instruction cache to provide more than one line of instructions in parallel, and masking and alignment (collapsing) circuitry to compact the statically nonsequential instructions [Conte et al., 1995].

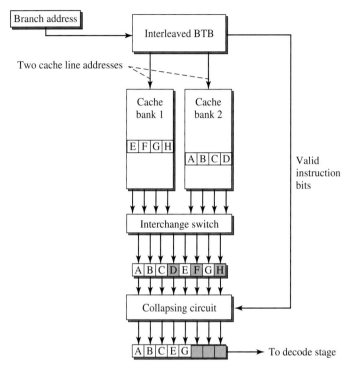

Figure 9.33
The Collapsing Buffer Fetch Organization.

Figure 9.33 shows the organization of the collapsing buffer. In this figure, the cache lines are four instruction words wide, and the cache has been broken into two banks. The instructions to be fetched are A, B, C, E, and G. In a conventional instruction cache, only instructions A, B, and C would be fetched in a single cycle. The branch target buffer provides the predictions that instruction C is a taken branch (with target address E) and that instruction E is also a taken branch with target address G. One cache bank fetches the cache line containing instructions A, B, C, and D, and the other cache bank provides instructions E, F, G, and H. If both cache lines are in the same bank, then the collapsing buffer fetch mechanism will only provide at most one cache line's worth of instructions.

After the two cache lines have been fetched, the cache lines go through an interchange switch that swaps the two lines if the second cache line contains the earlier instructions. The interleaved BTB provides valid instruction bits that specify which entries in the cache lines are part of the predicted path. A collapsing circuit, implemented with shifting logic, collapses the disparate instructions into one contiguous sequence to be forwarded to the decoder. The instructions to be removed are shaded in Figure 9.33. Note that in this example, the branch C crosses cache lines. The branch E is an *intrablock* branch where the branch target resides in the same cache line as the branch. In this example, the collapsing buffer has provided five instructions whereas a traditional instruction cache would only fetch three.

A shift-logic based collapsing circuit suffers from a long latency. Another way to implement the collapsing buffer is with a crossbar. A crossbar allows an arbitrary permutation of its inputs, and so even the interchange network is not needed. Because of this, the overall latency for interchange and collapse may be reduced, despite the relatively complex nature of crossbar networks.

The collapsing buffer adds some complex circuitry to the fetch path. The interchange switch and collapsing circuit add considerable latency to the front-end pipeline. This extra latency would take the form of additional pipeline stages, which increases the branch misprediction penalty. The organization of the collapsing buffer is difficult to scale to support fetching from more than two cache lines per cycle.

9.5.3.2 Trace Cache. The collapsing buffer fetch mechanism highlights the fact that many dynamically sequential instructions are not physically located in contiguous locations. Taken branches and cache line alignment problems frequently disrupt the fetch engine's attempt to provide a continuous high-bandwidth stream of instructions. The *trace cache* attempts to alleviate this problem by storing logically sequential instructions in the same consecutive physical locations [Friendly et al., 1997; Rotenberg et al., 1996; 1997].

A *trace* is a dynamic sequence of instructions. Figure 9.34(a) shows a sequence of instructions to be fetched and their locations in a conventional instruction cache. Instruction B is a predicted taken branch to C. Instructions C and D are split across two separate cache lines. Instruction D is another predicted taken branch to E. Instructions E through J are split across two cache lines, but this is to be expected since there are more instructions in this group than the width of a cache line. With a conventional fetch architecture, it will take at least five cycles to fetch these 10 instructions because the instructions are scattered over five different

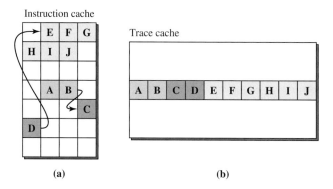

Figure 9.34
(a) Ten Instructions in an Instruction Cache. (b) The Same 10 Instructions in a Trace Cache.

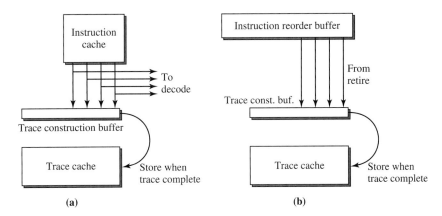

Figure 9.35
(a) Fetch-Time Trace Construction. (b) Completion-Time Trace Construction.

cache lines. Even with a collapsing buffer, it would still take three cycles (maximum fetch rate of two lines per cycle).

The trace cache takes a different approach. Instead of attempting to fetch from multiple locations and stitch the instructions back together like the collapsing buffer, the trace cache stores the entire trace in one physically contiguous location as shown in Figure 9.34(b). The trace cache can deliver this entire 10-instruction trace in a single lookup without any complicated reshuffling or realignment of instructions.

Central to the trace cache fetch mechanism is the task of trace construction. Trace construction primarily occurs at one of two locations. The first possibility is to perform trace construction at fetch time, as shown in Figure 9.35(a). As instructions are fetched from the conventional instruction cache, a trace construction buffer stores the dynamic sequence of instructions. When the trace is complete, which may be determined by various constraints such as the width of the trace cache or a limit on the number of branches per trace, this newly constructed trace is stored into the trace cache. In the future, when this same path is encountered, the trace cache can provide all of the instructions in a single access.

The other point for trace construction is at the back end of the processor when instructions retire. Figure 9.35(b) shows how as the processor back end retires instructions in order, these instructions are placed into a trace construction buffer. When the trace is complete, the trace is stored into the trace cache and a new trace is started. One advantage of back-end trace construction is that the circuitry is not in the branch misprediction pipeline, and the trace constructor may take more cycles to construct traces.

A trace entry consists of the instructions in the trace, and the entry also contains tags corresponding to the starting points of each basic block included in the trace. To perform a lookup in the trace cache, the fetch engine must provide the trace cache with the addresses of all basic block starting addresses on the predicted path. If all addresses match, then there is a trace cache hit. If some prefix of the addresses match (e.g., the first two addresses match but the third does not), then it is possible to provide only the subset of the trace that corresponds to the predicted path. For a high rate of fetch, the trace cache requires the front end to perform multiple branch predictions per cycle. Adapting conventional branch predictors to perform multiple predictions while maintaining reasonable access latencies is a challenging design task.

An alternative to making multiple branch predictions per cycle is to treat a trace as the fundamental basic unit and perform *trace prediction*. Each trace has a unique identifier defined by the starting PC and the outcomes of all conditional branches in the trace. The trace predictor's output is one of these trace identifiers. This approach provides trace-level sequencing of instructions.

Even with trace-level sequencing, some level of instruction-level sequencing (i.e., conventional fetch) must still be provided. At the start of a program, or when a program enters new regions of code, the trace cache will not have constructed the appropriate traces and the trace predictor has not learned the trace-to-trace transitions. In this situation, a conventional instruction cache and branch predictor provide the instructions at a slower rate until the new traces have been constructed and the trace predictor has been properly trained.

The Intel Pentium 4 microarchitecture employs a trace cache, but no first-level instruction cache [Hinton et al., 2001]. Figure 9.36 shows the block-level organization of the Pentium 4 fetch and decode engine. When the trace cache is in use, the trace cache BTB provides the fetch addresses and next-trace predictions. If the predicted trace is not in the trace cache, then instruction-level sequencing occurs, but the instructions are fetched from the level-2 instruction/data cache. This increases the number of cycles to fetch an instruction when there is a trace cache miss. These instructions are then decoded, and these *decoded instructions* are stored in the trace cache. Storing the decoded instructions allows instructions fetched from the trace cache to skip over the decode stage of the pipeline.

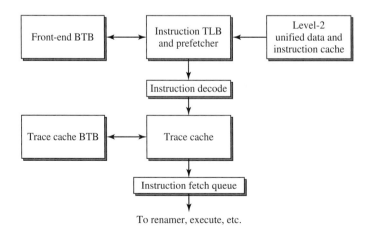

Figure 9.36
Intel Pentium 4 Trace Cache Organization.

9.5.4 High-Frequency Fetch Mechanisms

Processor pipeline depths and processor frequencies are rapidly increasing. This has a twofold impact on the design of branch predictors and fetch mechanisms. With deeper pipelines, the need for more accurate branch prediction increases due to the increased misprediction penalty. With faster clock speeds, the prediction and cache structures must have faster access times. To achieve a faster access time, the predictor and cache sizes must be reduced, which in turn increases the number of mispredictions and cache misses. This section describes a technique to provide faster single-cycle instruction cache lookups and a second technique for combining multiple branch predictors with different latencies.

9.5.4.1 Line and Way Prediction.

To provide one instruction cache access per cycle, the instruction cache must have a lookup latency of a single cycle, and the processor must compute or predict the next cache access by the start of the next cycle. Typically, the program counter provides the index into the instruction cache for fetch, but this is actually more information than is strictly necessary. The processor only needs to know the specific location in the instruction cache where the next instruction should be fetched from. Instead of predicting the next instruction address, *line prediction* predicts the cache line number where the next instruction is located [Calder and Grunwald, 1995].

In a line-predicted instruction cache, each cache line stores a next-line prediction in addition to the instructions and address tag. Figure 9.37 illustrates a line-predicted instruction cache. In the first cycle shown, the instruction cache has been

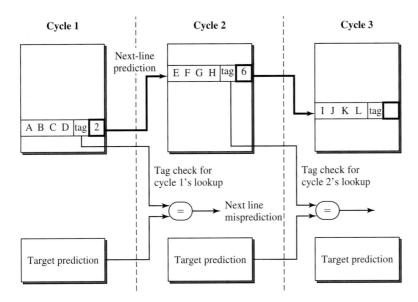

Figure 9.37
Direct-Mapped Cache with Next-Line Prediction.

accessed and provides the instructions and the stored next-line prediction. On the second cycle, the next-line prediction (highlighted in bold) is used as an index into the instruction cache providing the next line of instructions. This allows the instruction cache access to start before the branch predictor has completed its target prediction. After the target prediction has been computed, the already fetched cache line's tag is compared to the target prediction. If there is a match, then the line prediction was correct and the fetched cache line contained the correct instructions. If there is a mismatch, then the line prediction was wrong, and a new instruction cache access is initiated with the predicted target address. When the line predictor is correct, then a single-cycle instruction fetch is achieved. A next-line misprediction causes the injection of one pipeline bubble for the cycle spent fetching the wrong cache line.

Line prediction allows the front end to continually fetch instructions from the instruction cache, but it does not directly address the latency of a cache access. Direct-mapped caches have faster access times than set-associative caches, but suffer from higher miss rates due to conflicts. Set-associative caches have lower miss rates than direct-mapped caches, but the additional logic for checking multiple tags and performing way-selection greatly increases the lookup time. The technique of *way prediction* allows the instruction cache to be accessed with the latencies of direct-mapped caches while still retaining the miss rates of set-associative caches. With way prediction, a cache lookup only accesses a single way of the cache structure. Accessing only a single way appears much like an access to a direct-mapped cache because all the logic for supporting set associativity has been removed. Similar to line prediction, a verification of the way prediction must be performed, but this occurs off the critical path. If a way misprediction is detected, then another cache access is needed to provide the correct instructions, which results in a pipeline bubble. By combining both line prediction and way prediction, an instruction cache can fetch instructions every cycle at an aggressive clock speed. Way prediction can also be applied to the data cache to decrease access times.

9.5.4.2 Overriding Predictors. Deeper processor pipelines enable greater increases to the processor clock frequency. Although high clock speeds are generally associated with high throughput, the fast clock and deep pipeline have a compounding effect on branch predictors and the front end in general. A faster clock speed means that there is less time to perform a branch prediction. To achieve a single-cycle branch prediction, the sizes of the branch predictor tables, such as the PHT, must be reduced. Smaller branch prediction structures lead to more capacity and conflict aliasing and, therefore, to more branch mispredictions. The branch misprediction penalty has also increased because the number of pipe stages has increased. Therefore, the aggressive pipelining and clock speed have increased the number of branch mispredictions as well as the performance penalty for a misprediction. Trying to increase the branch prediction rate may require larger structures which will impact the clock speed. There is a tradeoff between the fetch efficiency and the clock speed and pipeline depth.

An *overriding predictor* organization attempts to rectify this situation by using two different branch predictors [Jimenez, 2002; Jimenez et al., 2000]. The

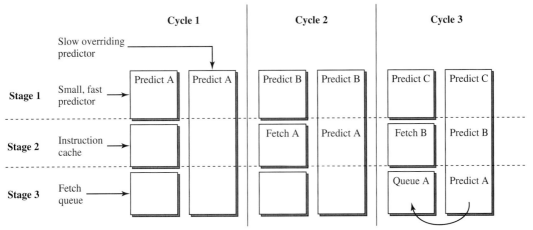

Figure 9.38
Organization of a Fast Predictor and a Slower Overriding Predictor.

first branch predictor is a small but fast, single-cycle predictor. This predictor will generally have only mediocre prediction rates due to its limited size but will still manage to provide accurate predictions for a reasonable number of branches. The second predictor is much larger and requires multiple cycles to access, but it is much more accurate.

The operation of the overriding predictor organization proceeds as follows and is illustrated in Figure 9.38. The first predictor makes an initial target prediction (A), and instruction fetch uses this prediction to start the fetch of the instructions. At the same time, the second predictor also starts its prediction lookup, but this prediction will not be available for several cycles. In cycle 2, while waiting for the second predictor's prediction, the first predictor provides another prediction so that the instruction cache can continue fetching more instructions. A lookup in the second predictor for this branch is also started, and therefore the second predictor must be pipelined. The predictions and fetches continue in a pipelined fashion until the second predictor has finished its prediction of the original instruction A in cycle 3. At this point, this more accurate prediction is compared to the original "quick and dirty" prediction. If the predictions match, then the first predictor was correct (with respect to the second predictor) and fetch can continue. If the predictions do not match, then the second predictor *overrides* the first prediction. Any further fetches that have been initiated in the meantime are flushed from the pipeline (i.e., A, B, and C are converted to bubbles), and the first predictor and instruction cache are reset to the target of the overridden branch.

There are four possible outcomes between the two branch predictors. If both predictors have made the correct prediction, then there are no bubbles injected.

If the first predictor is wrong, but the overriding predictor is correct, then a few bubbles are injected (equal to the difference in access latencies of the two predictors, sometimes called the *override latency*), but this is still a much smaller price to pay than a full pipeline flush that would occur if there was no overriding predictor. If both predictors mispredict, then the penalty is just a regular pipeline flush that would have occurred even in the idealistic case where the better second predictor has a single-cycle latency. If the first predictor was actually correct, and the second predictor caused an erroneous override, then the mispredict penalty is equal to a pipeline flush plus the override latency. The overriding predictor organization provides an overall benefit because the frequency of correct overrides is much greater than that of erroneous overrides.

The Alpha 21264 uses a technique that is similar to an overriding predictor configuration. The fast, but not as accurate, first-level predictor is the combination of the instruction cache next-line and next-way predictor. This implicitly provides a branch prediction with a target that is the address of the cache line in the predicted line and way. The more accurate hybrid predictor (described in Section 9.4.1) with a two-cycle latency provides a second prediction in the following cycle. If this prediction results in a target that is different than that chosen by the next-line/next-way predictors, then that instruction is flushed and the fetch restarts at the newly predicted target [Kessler, 1999].

9.6 Summary

This chapter has provided an overview of many of the ideas and concepts proposed to address the problems associated with providing an effective instruction fetch bandwidth. The problem of predicting the direction of conditional branches has received a large amount of attention, and much research effort has produced a myriad of prediction algorithms. These techniques target different challenges associated with predicting conditional branches with high accuracy. Research in history-based correlating branch predictors has been very influential, and such predictors are used in almost all modern superscalar processors. Other ideas such as hybrid branch predictors, various branch target predictor strategies, and other instruction delivery techniques have also been adopted by commercial processors.

Besides the conditional branch prediction problem, this chapter has also surveyed some of the other design issues and problems related to a processor's front-end microarchitecture. Predicting branch targets, fetching instructions from the cache hierarchy, and delivering the instructions to the rest of the processor are all important, and an effective fetch engine cannot be designed without paying close attention to all these components.

Although some techniques have had more influence than others (as measured by whether they were ever implemented in a real processor), there is no single method that is the absolute best way to predict branches or fetch instructions. As with any real project, a processor design involves many engineering tradeoffs and the best techniques for one processor may be completely inappropriate for another. There is a lot of engineering and a certain degree of art to designing a well-balanced and effective fetch engine. This chapter has been written to broaden your knowledge and understanding of advanced instruction flow techniques, and hopefully it may inspire you to someday help advance the state of the art as well!

REFERENCES

Aragon, J. L., Jose Gonzalez, Jose M. Garcia, and Antonio Gonzalez: "Confidence estimation for branch prediction reversal," *Lecture Notes in Computer Science,* 2228, 2001, pp. 214–223.

Ball, Thomas, and James R. Larus: "Branch prediction for free," *ACM SIGPLAN Symposium on Principles and Practice of Parallel Programming,* May 1993, pp. 300–313.

Calder, Brad, and Dirk Grunwald: "Next cache line and set prediction," *Int. Symposium on Computer Architecture,* June 1995, pp. 287–296.

Calder, Brad, Dirk Grunwald, Michael Jones, Donald Lindsay, James Martin, Michael Mozer, and Benjamin Zorn: "Evidence-based static branch prediction using machine learning," *ACM Trans. on Programming Languages and Systems,* 19, 1, January 1997, pp. 188–222.

Chang, Po-Yung, Marius Evers, and Yale N. Patt: "Improving branch prediction accuracy by reducing pattern history table interference," *Int. Conference on Parallel Architectures and Compilation Techniques,* October 1996, pp. 48–57.

Chang, Po-Yung, Eric Hao, and Yale N. Patt: "Alternative implementations of hybrid branch predictors," *Int. Symposium on Microarchitecture,* November 1995, pp. 252–257.

Chang, Po-Yung, Eric Hao, Tse-Yu Yeh, and Yale N. Patt: "Branch classification: A new mechanism for improving branch predictor performance," *Int. Symposium on Microarchitecture,* November 1994, pp. 22–31.

Conte, Thomas M., Kishore N. Menezes, Patrick M. Mills, and Burzin A. Patel: "Optimization of instruction fetch mechanisms for high issue rates," *Int. Symposium on Computer Architecture,* June 1995, pp. 333–344.

Eden, N. Avinoam, and Trevor N. Mudge: "The YAGS branch prediction scheme," *Int. Symposium on Microarchitecture,* December 1998, pp. 69–77.

Evers, Marius: "Improving branch prediction by understanding branch behavior," PhD Thesis, University of Michigan, 2000.

Evers, Marius, Po-Yung Chang, and Yale N. Patt: "Using hybrid branch predictors to improve branch prediction accuracy in the presence of context switches," *Int. Symposium on Computer Architecture,* May 1996, pp. 3–11.

Evers, Marius, Sanjay J. Patel, Robert S. Chappell, and Yale N. Patt: "An analysis of correlation and predictability: What makes two-level branch predictors work," *Int. Symposium on Computer Architecture,* June 1998, pp. 52–61.

Fagin, B., and K. Russell: "Partial resolution in branch target buffers," *Int. Symposium on Microarchitecture,* December 1995, pp. 193–198.

Fisher, Joseph A., and Stephan M. Freudenberger: "Predicting conditional branch directions from previous runs of a program," *Symposium on Architectural Support for Programming Languages and Operating Systems,* October 1992, pp. 85–95.

Friendly, Daniel H., Sanjay J. Patel, and Yale N. Patt: Alternative fetch and issue techniques for the trace cache mechanism," *Int. Symposium on Microarchitecture,* December 1997, pp. 24–33.

Gochman, Simcha, Ronny Ronen, Ittai Anati, Ariel Berkovitz, Tsvika Kurts, Alon Naveh, Ali Saeed, Zeev Sperber, and Robert C. Valentine: "The Intel Pentium M processor: Microarchitecture and performance," *Intel Technology Journal,* 7, 2, May 2003, pp. 21–36.

Grunwald, Dirk, Donald Lindsay, and Benjamin Zorn: "Static methods in hybrid branch prediction," *Int. Conference on Parallel Architectures and Compilation Techniques,* October 1998, pp. 222–229.

Hartstein, A., and Thomas R. Puzak: "The optimum pipeline depth for a microprocessor," *Int. Symposium on Computer Architecture,* May 2002, pp. 7–13.

Hewlett Packard Corporation: *PA-RISC 2.0 Architecture and Instruction Set Manual,* 1994.

Hewlett Packard Corporation: "PA-RISC 8x00 Family of Microprocessors with Focus on PA-8700," *Technical White Paper,* April 2000.

Hill, Mark D.: "Aspects of cache memory and instruction buffer performance," PhD Thesis, University of California, Berkeley, November 1987.

Hinton, Glenn, Dave Sager, Mike Upton, Darrell Boggs, Doug Karmean, Alan Kyler, and Patrice Roussel: "The microarchitecture of the Pentium 4 processor," *Intel Technology Journal,* Q1, 2001.

Hrishikesh, M. S., Norman P. Jouppi, Keith I. Farkas, Doug Burger, Stephen W. Keckler, and Primakishore Shivakumar: "The optimal useful logic depth per pipeline stage is 6–8 FO4," *Int. Symposium on Computer Architecture,* May 2002, pp. 14–24.

Intel Corporation: *Embedded Intel 486 Processor Hardware Reference Manual.* Order Number: 273025-001, July 1997.

Intel Corporation: *IA-32 Intel Architecture Optimization Reference Manual.* Order Number 248966-009, 2003.

Jacobson, Erik, Eric Rotenberg, and James E. Smith: "Assigning confidence to conditional branch predictions," *Int. Symposium on Microarchitecture,* December 1996, pp. 142–152.

Jimenez, Daniel A.: "Delay-sensitive branch predictors for future technologies," PhD Thesis, University of Texas at Austin, January 2002.

Jimenez, Daniel A., Stephen W. Keckler, and Calvin Lin: "The impact of delay on the design of branch predictors," *Int. Symposium on Microarchitecture,* December 2000, pp. 4–13.

Jimenez, Daniel A., and Calvin Lin: "Neural methods for dynamic branch prediction," *ACM Trans. on Computer Systems,* 20, 4, February 2003, pp. 369–397.

Juan, Toni, Sanji Sanjeevan, and Juan J. Navarro: "Dynamic history-length fitting: A third level of adaptivity for branch prediction," *Int. Symposium on Computer Architecture,* June 1998, pp. 156–166.

Kaeli, David R., and P. G. Emma: "Branch history table prediction of moving target branches due to subroutine returns," *Int. Symposium on Computer Architecture,* May 1991, pp. 34–41.

Kane, G., and J. Heinrich: *MIPS RISC Architecture.* Englewood Cliffs, NJ: Prentice-Hall, 1992.

Kessler, R. E.: "The Alpha 21264 Microprocessor," *IEEE Micro Magazine,* 19, 2, March–April 1999, pp. 24–26.

Klauser, Artur, Abhijit Paithankar, and Dirk Grunwald: "Selective eager execution on the polypath architecture," *Int. Symposium on Computer Architecture,* June 1998, pp. 250–259.

Lee, Chih-Chieh, I-Cheng K. Chan, and Trevor N. Mudge: "The Bi-Mode branch predictor," *Int. Symposium on Microarchitecture,* December 1997, pp. 4–13.

Lee, Johnny K. F., and Alan Jay Smith: "Branch prediction strategies and branch target buffer design," *IEEE Computer,* 17, 1, January 1984, pp. 6–22.

Loh, Gabriel H., and Dana S. Henry: "Predicting conditional branches with fusion-based hybrid predictors," *Int. Conference on Parallel Architectures and Compilation Techniques,* September 2002, pp. 165–176.

Manne, Srilatha, Artur Klauser, and Dirk Grunwald: "Branch prediction using selective branch inversion," *Int. Conference on Parallel Architectures and Compilation Techniques,* October 1999, pp. 48–56.

McFarling, Scott: "Combining branch predictors," TN-36, Compaq Computer Corporation Western Research Laboratory, June 1993.

McFarling, Scott, and John L. Hennessy: "Reducing the cost of branches," *Int. Symposium on Computer Architecture,* June 1986, pp. 396–404.

Meyer, Dirk: "AMD-K7 technology presentation," *Microprocessor Forum,* October 1998.

Michaud, Pierre, Andre Seznec, and Richard Uhlig: "Trading conflict and capacity aliasing in conditional branch predictors," *Int. Symposium on Computer Architecture,* June 1997, pp. 292–303.

Nair, Ravi: "Dynamic path-based branch correlation," *Int. Symposium on Microarchitecture,* December 1995, pp. 15–23.

Pan, S. T., K. So, and J. T. Rahmeh: "Improving the accuracy of dynamic branch prediction using branch correlation," *Symposium on Architectural Support for Programming Languages and Operating Systems,* October 1992, pp. 12–15.

Reches, S., and S. Weiss: "Implementation and analysis of path history in dynamic branch prediction schemes," *Int. Conference on Supercomputing,* July 1997, pp. 285–292.

Rosenblatt, F.: *Principles of Neurodynamics: Perceptrons and the Theory of Brain Mechanisms.* Spartan Books, 1962.

Rotenberg, Eric, S. Bennett, and James E. Smith: "Trace cache: A low latency approach to high bandwidth instruction fetching," *Int. Symposium on Microarchitecture,* December 1996, pp. 24–35.

Rotenberg, Eric, Quinn Jacobson, Yiannakis Sazeides, and Jim Smith: "Trace processors," *Int. Symposium on Microarchitecture,* December 1997, pp. 138–148.

Seznec, Andre, Stephen Felix, Venkata Krishnan, and Yiannakis Sazeides: "Design tradeoffs for the Alpha EV8 conditional branch predictor," *Int. Symposium on Computer Architecture,* May 2002, pp. 25–29.

Skadron, Kevin, Margaret Martonosi, and Douglas W. Clark: "A Taxonomy of Branch Mispredictions, and Alloyed Prediction as a Robust Solution to Wrong-History Mispredictions," *Int'l Conference on Parallel Architectures and Compilation Techniques,* September 2001, pp. 199–206.

Smith, Jim E.: "A study of branch prediction strategies," *Int. Symposium on Computer Architecture,* May 1981, pp. 135–148.

Sprangle, Eric, Robert S. Chappell, Mitch Alsup, and Yale N. Patt: "The agree predictor: A mechanism for reducing negative branch history interference," *Int. Symposium on Computer Architecture,* June 1997, pp. 284–291.

Stark, Jared, Marius Evers, and Yale N. Patt: "Variable path branch prediction," *ACM SIGPLAN Notices,* 33, 11, 1998, pp. 170–179.

Sugumar, Rabin A., and Santosh G. Abraham: "Efficient simulation of caches under optimal replacement with applications to miss characterization," *ACM Sigmetrics,* May 1993, pp. 284–291.

Tarlescu, Maria-Dana, Kevin B. Theobald, and Guang R. Gao: "Elastic history buffer: A low-cost method to improve branch prediction accuracy," *Int. Conference on Computer Design,* October 1996, pp. 82–87.

Tendler, Joel M., J. Steve Dodson, J. S. Fields, Jr., Hung Le, and Balaram Sinharoy: "POWER4 system microarchitecture," *IBM Journal of Research and Development,* 46, 1, January 2002, pp. 5–25.

Thomas, Renju, Manoj Franklin, Chris Wilkerson, and Jared Stark: "Improving branch prediction by dynamic dataflow-based identification of correlated branches from a large global history," *Int. Symposium on Computer Architecture,* June 2003, pp. 314–323.

Uhlig Richard, David Nagle, Trevor Mudge, Stuart Sechrest, Joel Emer: "Instruction fetching: coping with code bloat" *The 22nd Int. Symposium on Computer Architecture,* June 1995, pp. 345–356.

Uht, Augustus K., Vijay Sindagi, and Kelley Hall: "Disjoint eager execution: An optimal form of speculative execution," *Int. Symposium on Microarchitecture,* November 1995, pp. 313–325.

Uht, Augustus K.: "Branch effect reduction techniques," *IEEE Computer,* 30, 5, May 1997, pp. 71–81.

Yeh, Tse-Yu, and Yale N. Patt: "Two-level adaptive branch prediction," *Int. Symposium on Microarchitecture,* November 1991, pp. 51–61.

Yeh, Tse-Yu, and Yale N. Patt: "Alternative implementations of two-level adaptive branch prediction," *Int. Symposium on Computer Architecture,* May 1992, pp. 124–134.

Yeh, Tse-Yu, and Yale N. Patt: "A comparison of dynamic branch predictors that use two levels of branch history," *Int. Symposium on Computer Architecture,* 1993, pp. 257–266.

HOMEWORK PROBLEMS

P9.1 Profiling a program has indicated that a particular branch is taken 53% of the time. How effective are the following at predicting this branch and why? (a) Always-taken static prediction. (b) Bimodal/Smith predictor. (c) Local-history predictor. (d) Eager execution. State your assumptions.

P9.2 Assume that a branch has the following sequence of taken (T) and not-taken (N) outcomes:

T, T, T, N, N, T, T, T, N, N, T, T, T, N, N

What is the prediction accuracy for a 2-bit counter (Smith predictor) for this sequence assuming an initial state of strongly taken?

P9.3 What is the minimum local history length needed to achieve perfect branch prediction for the branch outcome sequence used in Problem 9.2?

Draw the corresponding PHT and fill in each entry with one of T (predict taken), N (predict not-taken), or X (doesn't matter).

P9.4 Suppose that most of the branches in a program only need a 6-bit global history predictor to be accurately predicted. What are the advantages and disadvantages to using a longer history length?

P9.5 Conflict aliasing occurs in conventional caches when two addresses map to the same line of the cache. Adding tags and associativity is one of the common ways to reduce the miss rate of caches in the presence of conflict aliasing. What are the advantages and disadvantages of adding set associativity to a branch prediction data structure (e.g., PHT)?

P9.6 In some sense, there is no way to make a "broken" branch predictor. For example, a predictor that always predicted the wrong branch direction (0% accuracy) would still result in correct program execution, because the correct branch direction will be computed later in the pipeline and the misprediction will be corrected. This behavior makes branch predictors difficult to debug.

Suppose you just invented a new branch prediction algorithm and implemented it in a processor simulator. For a particular program, this algorithm should achieve a 93% prediction accuracy. Unbeknownst to you, a programming error on your part has caused the simulated predictor to report a 95% accuracy. How would you go about verifying the correctness of your branch predictor implementation (beyond just double-checking your code)?

P9.7 The path history example from Figure 9.19 showed a situation where the global *branch outcome* history was identical for two different program paths. Does the global path history provide a superset of the information contained in the global branch outcome history? If not, describe a situation where the same global path can result in two different global branch histories.

P9.8 Most proposed hybrid predictors involve the combination of a global-history predictor with a local-history predictor. Explain the benefits, if any, of combining two global-history predictors (possibly of different types like Bi-Mode and gskewed, for example) in a hybrid configuration. If there is no advantage to a global-global hybrid, explain why.

P9.9 For branches with a PC-relative target address, the address of the next instruction on a taken branch is always the same (not including self-modifying code). On the other hand, indirect jumps may have different targets on each execution. A BTB only records the most recent branch target and, therefore, may be ineffective at predicting frequently changing targets of an indirect jump. How could the BTB be modified to improve its prediction accuracy for this scenario?

P9.10 Branch predictors are usually assumed to provide a single branch prediction on every cycle. An alternative is to build a predictor with a two-cycle latency that attempts to predict the outcome of not only the current branch, but the next branch as well (i.e., it provides two predictions, but only on every other cycle). This approach still provides an average prediction rate of one branch prediction per cycle. Explain the benefits and shortcomings of this approach as compared to a conventional single-cycle branch predictor.

P9.11 A trace cache's next-trace predictor relies on the program to repeatedly execute the same sequences of code. Subroutine returns have very predictable targets, but the targets frequently change from one invocation of the subroutine to the next. How do frequently changing return addresses impact the performance of a trace cache in terms of hit rates and next-trace prediction?

P9.12 Traces can be constructed in either the processor's front end during fetch, or in the back end at instruction commit. Compare and contrast front-end and back-end trace construction with respect to the amount of time between the start of trace construction and when the trace can be used, branch misprediction delays, branch/next-trace prediction, performance, and interactions with the rest of the microarchitecture.

P9.13 Overriding predictors use two different predictors to provide a quick and dirty prediction and a slower but better prediction. This scheme could be generalized to a hierarchy of predictors with an arbitrary depth. For example, a three-level overriding hierarchy would have a quick and inaccurate first predictor, a second predictor that provides somewhat better prediction accuracy with a moderate delay, and then finally a very accurate but much slower third predictor. What are the difficulties involved in implementing, for example, a 10-level hierarchy of overriding branch predictors?

P9.14 Implement one of the dynamic branch predictors described in this chapter in a processor simulator. Compare its branch prediction accuracy to that of the default predictors.

P9.15 Devise your own original branch prediction algorithm and implement it in a processor simulator. Compare its branch prediction accuracy to other known techniques. Consider the *latency* of a prediction lookup when designing the predictor.

P9.16 A processor's branch predictor only provides mediocre prediction accuracy. Does it make sense to implement a large instruction window for this processor? State as many reasons as you can for *and* against implementing a larger instruction window in this situation.

CHAPTER 10

Advanced Register Data Flow Techniques

CHAPTER OUTLINE

10.1 Introduction
10.2 Value Locality and Redundant Execution
10.3 Exploiting Value Locality without Speculation
10.4 Exploiting Value Locality with Speculation
10.5 Summary

References
Homework Problems

10.1 Introduction

As we have learned, modern processors are fundamentally limited in performance by two program characteristics: control flow and data flow. The former was examined at length in our study of advanced instruction fetch techniques such as branch prediction, trace caches, and other high-bandwidth solutions to Flynn's bottleneck [Tjaden and Flynn, 1970]. Historically, these techniques have proved to be quite effective and many have been widely adopted in today's advanced processor designs. Nevertheless, resolving the limitations that control flow places on processor performance continues to be an extremely important area of research and advanced development. In Chapter 11, we will revisit this issue and focus on an active area of research that attempts to exploit multiple simultaneous flows of control to overcome bottlenecks caused by inaccuracies in branch prediction and inefficiencies in branch resolution. Before we do so, however, we will take a closer look at the performance limits that are caused by a program's data flow.

Earlier sections have already focused on resolving performance limitations caused by false or name dependences in a program. As the reader may recall, false dependences are caused by reuse of storage locations during program execution. Such reuse is induced by the fact that programmers and compilers must specify temporary operands with a finite number of unique register identifiers and are forced to reuse register identifiers once all available identifiers have been allocated. Furthermore, even if the instruction set provided the luxury of an unlimited number of registers and register identifiers, program loops induce reuse of storage identifiers, since multiple instances of a single static loop body can be in flight at the same time. Hence, false or name dependences are unavoidable. As we learned in Chapter 5, the underlying technique employed to resolve false dependences is to dynamically rename each destination operand to a unique storage location, and hence avoid unnecessary serialization of multiple writes to a shared location. This process of register renaming, first introduced as Tomasulo's algorithm in the IBM S/360-91 [Tomasulo, 1967] in the late 1960s, and detailed in Chapter 5, effectively removes false dependences and allows instructions to execute subject only to their true dependences. As has been the case with branch prediction, this technique has proved very effective, and various forms of register renaming have been implemented in numerous high-performance processor designs over the past four decades.

In this chapter, we turn our attention to techniques that attempt to elevate performance beyond what is achievable simply by eliminating false data dependences. A processor that executes instructions at a rate limited only by true data dependences is said to be operating at the *data flow limit*. Informally, a processor has achieved the data flow limit when each instruction in a program's dynamic data flow graph executes as soon as its source operands become available. Hence, an instruction's scheduled execution time is determined solely by its position in the data flow graph, where its position is defined as the longest path that leads to it in the data flow graph. For example, in Figure 10.1, instruction C is executed in cycle 2 because its true data dependences position it after instructions A and B, which execute in cycle 1. Recall that in a data flow graph

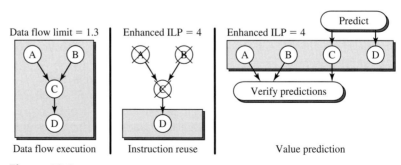

Figure 10.1

Exceeding the Instruction-Level Parallelism (ILP) Dictated by the Data Flow Limit.

the nodes represent instructions, the edges represent data dependences between instructions, and the edges are weighted with the result latency of the producing instruction.

Given a data flow graph, we can compute a lower bound for a program's execution time by computing the height (i.e., the length of the longest existing path) of the data flow graph. The *data flow limit* represents this lower bound and, in turn, determines the maximum achievable rate of instruction execution (or ILP), which is defined as the number of instructions in the program divided by the height of the data flow graph. Just as an example, refer to the simple data flow graph shown on the left-hand side of Figure 10.1, where the maximum achievable ILP as determined by the data flow limit can be computed as 4 instructions

3 cycles of latency on the longest path through the graph = 1.3

In this chapter, we focus on two techniques—value prediction and instruction reuse—that exploit a program characteristic termed *value locality* to accelerate processing of instructions beyond the classic data flow limit. In this context, *value locality* describes the likelihood that a program instruction's computed result—or a similar, predictable result—will recur later during the program's continued execution. More broadly, the value locality of programs captures the empirical observation that a limited set of unique values constitute the majority of values produced and consumed by real programs. This property is analogous to the temporal and spatial locality that caches and memory hierarchies rely on, except that it describes the values themselves, rather than their storage locations.

The two techniques we consider exploit value locality by either nonspeculatively reusing the results of prior computation (in instruction reuse) or by speculatively predicting the results of future computation based on the results of prior executions (in value prediction). Both approaches allow a processor to obtain the results of an instruction earlier in time than its position in the data flow graph might indicate, and both are able to reduce the effective height of the graph, thereby increasing the rate of instruction execution beyond the data flow limit. For example, as shown in the middle of Figure 10.1, an instruction reuse scheme might recognize that instructions A, B, and C are repeating an earlier computation and could reuse the results of that earlier computation and allow instruction D to execute immediately, rather than having to wait for the results of A, B, and C. This would result in an effective throughput of four instructions per cycle. Similarly, the right side of Figure 10.1 shows how a data value prediction scheme could be used to enhance available instruction-level parallelism from a meager 1.3 instructions per cycle to an ideal 4 instructions per cycle by correctly predicting the results of instructions A, B, and C. Since A and B are predicted correctly, C need not wait for them to execute. Similarly, since C is correctly predicted, D need not wait for C to execute. Hence, all four instructions execute in parallel.

Figure 10.1 also illustrates a key distinction between instruction reuse and value prediction. In the middle case, invoking reuse completely avoids execution of instructions A, B, and C. In contrast, on the right, value prediction avoids the serializing effect of these instructions, but is not able to prevent their execution.

This distinction arises from a fundamental difference between the two techniques: instruction reuse guarantees value locality, while value prediction only predicts it. In the latter case, the processor must still verify the prediction by executing the predicted instructions and comparing their results to the predicted results. This is similar to branch prediction, where the outcome of the branch is predicted, almost always correctly, but the branch must still be executed to verify the correctness of the prediction. Of course, verification consumes execution bandwidth and requires a comparison mechanism for validating the results. Conversely, instruction reuse provides an a priori guarantee of correctness, so no verification code is needed. However, as we will find out in Section 10.3.2, this guarantee of correctness, while seemingly attractive, carries with it some baggage that can increase implementation cost and reduce the effectiveness of instruction reuse.

Neither value prediction nor instruction reuse, only relatively recently introduced in the literature, has yet been implemented in a real design. However, both demonstrate substantial potential for improving the performance of real programs, particularly programs where true data dependences—as opposed to structural or control dependences—place limits on achievable instruction-level parallelism. As with any new idea, there are substantial challenges involved in realizing that performance potential and reducing it to practice. We will explore some of these challenges and identify which have known realizable solutions and which require further investigation.

First, we will examine instruction reuse, since it has its roots in a historical and well-known program optimization called *memoization*. Memoization, which can be performed manually by the programmer, or automatically by the compiler, is a technique for short-circuiting complex computations by dynamically recording the outcomes of such computations. Subsequent instances of such computations then perform table lookups and reuse the results of prior computations whenever a new instance matches the same preconditions as an earlier instance. As may be evident to the reader, memoization is a nonspeculative technique, since it requires precisely correct preconditions to be satisfied before computation reuse is invoked. Similarly, instruction reuse is also nonspeculative and can be viewed as a hardware implementation of memoization at the instruction level.

Next, we will examine value prediction, which is fundamentally different due to its speculative nature. Rather than reusing prior executions of instructions, value prediction instead seeks to predict the outcome of a future instance of an instruction, based on prior outcomes. In this respect it is very similar to widely used history-based dynamic branch predictors (see Chapter 5), with one significant difference. While branch predictors collect outcome histories that can be quite deep (up to several dozen prior instances of branches can contribute their outcome history to the prediction of a future instance), the information content of the property they are predicting is very small, corresponding only to a single state bit that determines whether the branch is taken. In contrast, value predictors attempt to forecast full 32- or 64-bit values computed by register-writing instructions. Naturally, the challenges of accurately generating such predictions require much wider (full operand width) histories and additional mechanisms for avoiding mispredictions. Furthermore, generating predictions is only a small part of the implementation challenges required to realize value prediction's performance potential. Just as with branch prediction, mechanisms for speculative execution based on predicted values as well as prediction verification and misprediction recovery, are all required for correct operation.

We begin with a discussion of value locality and its causes, and then consider many aspects of both nonspeculative techniques (e.g., instruction reuse) and speculative techniques (e.g., value prediction) for exploiting value locality. We examine all aspects of such techniques in detail; show how these techniques, though seemingly different, are actually closely related; and also describe how the two can be hybridized by combining elements of instruction reuse with an aggressive implementation of value prediction to reduce the cost of prediction verification.

10.2 Value Locality and Redundant Execution

In this section, we further explore the concept of *value locality,* which we define as the likelihood of a previously seen value recurring repeatedly within a storage location [Lipasti et al., 1996; Lipasti and Shen, 1996]. Although the concept is general and can be applied to any storage location within a computer system, here we consider the value locality of general-purpose or floating-point registers immediately following instructions that write those registers. A plethora of previous work on dynamic branch prediction has focused on an even more restricted application of value locality, namely, the prediction of a single condition bit based on its past behavior. Many of the ideas in this chapter can be viewed as a logical continuation of that body of work, extending the prediction of a single bit to the prediction of an entire 32- or 64-bit register.

10.2.1 Causes of Value Locality

Intuitively, it seems that it would be a very difficult task to discover any useful amount of value locality in a register. After all, a 32-bit register can contain any one of over four billion values—how could one possibly predict which of those is even somewhat likely to occur next? As it turns out, if we narrow the scope of our prediction mechanism by considering each static instruction individually, the task becomes much easier, and we are able to accurately predict a significant fraction of values being written to the register file.

What is it that makes these values predictable? After examining a number of real-world programs, we have found that value locality exists primarily because real-world programs, run-time environments, and operating systems are *general by design.* That is, not only are they implemented to handle contingencies, exceptional conditions, and erroneous inputs, all of which occur relatively rarely in real life, but they are also often designed with future expansion and code reuse in mind. Even code that is aggressively optimized by modern, state-of-the-art compilers exhibits these tendencies. The following empirical observations result from our examination of many real programs, and they should help the reader understand why value locality exists:

- *Data redundancy*. Frequently, the input sets for real-world programs contain data that have little variation. Examples of this are sparse matrices that contain many zeros, text files with white space, and empty cells in spreadsheets.

- *Error checking.* Checks for infrequently occurring conditions often compile into loads of what are effectively run-time constants.
- *Program constants.* It is often more efficient to generate code to load program constants from memory than code to construct them with immediate operands.
- *Computed branches.* To compute a branch destination, say for a switch statement, the compiler must generate code to load a register with the base address for the branch jump table, which is often a run-time constant.
- *Virtual function calls.* To call a virtual function, the compiler must generate code to load a function pointer, which can often be a run-time constant.
- *Glue code.* Because of addressability concerns and linkage conventions, the compiler must often generate glue code for calling from one compilation unit to another. This code frequently contains loads of instruction and data addresses that remain constant throughout the execution of a program.
- *Addressability.* To gain addressability to nonautomatic storage, the compiler must load pointers from a table that is not initialized until the program is loaded, and thereafter remains constant.
- *Call-subgraph identities.* Functions or procedures tend to be called by a fixed, often small, set of functions, and likewise tend to call a fixed, often small, set of functions. Hence, the calls that occur dynamically often form identities in the call graph for the program. As a result, loads that restore the link register as well as other callee-saved registers can have high value locality.
- *Memory alias resolution.* The compiler must be conservative about stores that may alias with loads, and will frequently generate what appear to be redundant loads to resolve those aliases. These loads are likely to exhibit high degrees of value locality.
- *Register spill code.* When a compiler runs out of registers, variables that may remain constant are spilled to memory and reloaded repeatedly.
- *Convergent algorithms.* Often, value locality is caused by algorithms that the programmer chose to implement. One common example is convergent algorithms, which iterate over a data set until global convergence is reached; quite often, local convergence will occur before global convergence, resulting in redundant computation in the converged areas.
- *Polling algorithms.* Another example of how algorithmic choices can induce value locality is the use of polling algorithms instead of more efficient event-driven algorithms. In a polling algorithm, the most likely outcome is that the event being polled for has not yet occurred, resulting in redundant computation to repeatedly check for the event.

Naturally, many of these observations are subject to the particulars of the instruction set, compiler, and run-time environment being employed, and one could argue that some could be eliminated with changes to the ISA, compiler, or run-time environment, or by applying aggressive link-time or run-time code optimizations. However, such changes and improvements have been slow to appear; the aggregate effect of the listed (and other) factors on value locality is measurable and significant today on the two modern RISC instruction sets that we examined, both of which provide state-of-the-art compilers and run-time systems. It is worth pointing out, however, that the value locality of particular static loads in a program can be significantly affected by compiler optimizations such as loop unrolling, loop peeling, and tail replication, since these types of transformations tend to create multiple instances of a load that may now exclusively target memory locations with high or low value locality.

10.2.2 Quantifying Value Locality

Figure 10.2 shows the value locality for load instructions in a variety of benchmark programs. The value locality for each benchmark is measured by counting the number of times each static load instruction retrieves a value from memory

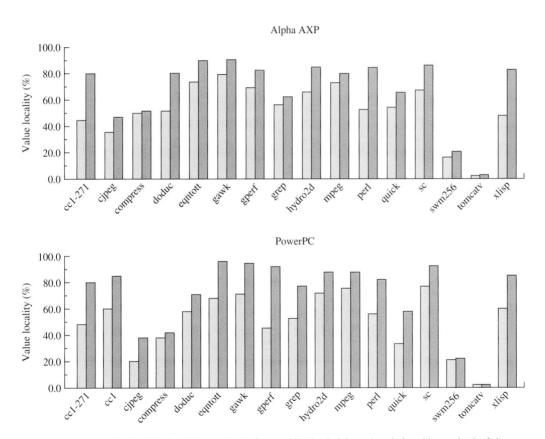

The light bars show value locality for a history depth of one, while the dark bars show it for a history depth of sixteen.

Figure 10.2

Load Value Locality.

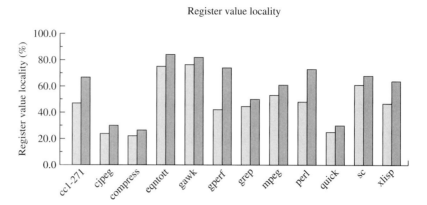

The light bars show value locality for a history depth of one, while the dark bars show it for a history depth of four.

Figure 10.3

Register Value Locality.

that matches a previously seen value for that static load, and dividing by the total number of dynamic loads in the benchmark. Two sets of numbers are shown, one (light bars) for a history depth of 1 (i.e., check for matches against only the most recently retrieved value), while the second set (dark bars) has a history depth of 16 (i.e., check against the last 16 unique values). We see that even with a history depth of 1, most of the integer programs exhibit load value locality in the 50% range, while extending the history depth to 16 (along with a hypothetical perfect mechanism for choosing the right one of the 16 values) can improve that to better than 80%. What this means is that the vast majority of static loads exhibit very little variation in the values that they load during the course of a program's execution. Unfortunately, three of these benchmarks (*cjpeg, swm256,* and *tomcatv*) demonstrate poor load value locality.

Figure 10.3 shows the *average value locality* for all instructions that write an integer or floating-point register in each of the benchmarks. The value locality of each static instruction is measured by counting the number of times that instruction writes a value that matches a previously seen value for that static instruction and dividing by the total number of dynamic occurrences of that instruction. The average value locality of a benchmark is the dynamically weighted average of the value localities of all the static instructions in that benchmark. Two sets of numbers are shown, one (light bars) for a history depth of one (i.e., we check for matches against only the most recently written value), while the second set (dark bars) has a history depth of four (i.e., we check against the last four unique values). We see that even with a history depth of one, most of the programs exhibit value locality in the 40% to 50% range (average 51%), while extending the history depth to four (along with a perfect mechanism for choosing the right one of the four values) can improve that to the 60% to 70% range (average 66%). What that means is that a majority of static instructions exhibit very little variation in the values that they write into registers during the course of a program's execution. Once again, three of these benchmarks—*cjpeg, compress,* and *quick*—demonstrate poor register value locality.

In summary, all the programs studied here, and many others studied exhaustively elsewhere, demonstrate significant amounts of value locality, for both load instructions and all register-writing instructions [Lipasti et al., 1996; Lipasti and Shen, 1996; 1997; Mendelson and Gabbay, 1997; Gabbay and Mendelson, 1997; 1998a; 1998b; Sazeides and Smith, 1997; Calder et al., 1997; 1999; Wang and Franklin, 1997; Burtscher and Zorn, 1999; Sazeides, 1999]. This property has been independently verified for at least a half-dozen different instruction sets and compilers and a large number of workloads including both user-state and kernel-state execution.

10.3 Exploiting Value Locality without Speculation

The widespread occurrence of value locality in real programs creates opportunities for increasing processor performance. As we have already outlined, both speculative and nonspeculative techniques are possible. We will first describe nonspeculative techniques for exploiting value locality, since related techniques have been known for a long time. A recent proposal has reinvigorated interest in such techniques by advocating *instruction reuse* [Sodani and Sohi, 1997; 1998; Sodani, 2000], which is a pure hardware technique for reusing the result of a prior execution of an instruction. In its simplest form, an instruction reuse mechanism avoids the structural and data hazards caused by execution of an instruction whenever it discovers an identical instruction execution within its history mechanism. In such cases, it simply reuses the historical outcome saved in the instruction reuse buffer and discards the fetched instruction without executing it. Dependent instructions are able to issue and execute immediately, since the result is available right away. Because of value locality, such reuse is often possible since many static instructions repeatedly compute the same result.

10.3.1 Memoization

Instruction reuse has its roots in a historical and well-known program optimization called *memoization*. Memoization, which can be performed manually by the programmer or automatically by the compiler, is a technique for short-circuiting complex computations by dynamically recording the outcomes of such computations and reusing those outcomes whenever possible. For example, each <operand, result> pair resulting from calls to the function *fibonacci(x)* shown in Figure 10.4 can be recorded in a memoization table. Subsequent instances of such computations then perform table lookups and reuse the results of prior computations whenever a new instance matches the same preconditions as an earlier instance. Continuing our example, a memoized version of the *fibonacci(x)* function checks to see if the current call matches an earlier call, and then returns the value of the earlier call immediately, rather than executing the full routine to recompute the Fibonacci series sum.

```
/* fibonacci series computation */          /* linked list example */
int fibonacci(x) {                          int ordered_linked_list_insert(record *x) {
  int result = 0;                             int position=0;
  if (x==0)                                   record *c,*p;
    result = 0;                               c=head;
  else if (x<3)                               while (c && (c->data < x->data)) {
    result = 1;                                 ++position;
  else {                                        p = c;
    result = fibonacci(x-2);                    c = c->next;
    result += fibonacci(x-1);                 }
  }                                           if (p) {
  return result;                                x->next = p->next;
}                                               p->next = x;
/* memoized version */                        } else
int memoized_fibonacci(x) {                     head = x;
  if (seen_before(x))                         return position;
    return memoized_result(x);              }
  else {
    int result = fibonacci(x);
    memoize(x,result);
    return result;
  }
}
```

The call to *fibonacci(x)*, shown on the left, can easily be memoized, as shown in the *memoized_fibonacci(x)* function. The call to *ordered_linked_list(record *x)* would be very difficult to memoize due to its reliance on global variables and side effect updates to those global variables.

Figure 10.4

Memoization Example.

Besides the overhead of recording and checking for memoized results, the main shortcoming of memoization is that any computation that is memoized must be guaranteed to be free of side effects. That is, the computation must not itself modify any global state, nor can it rely on external modifications to the global state. Rather, all its inputs must be clearly specified so the memoization table lookup can verify that they match the earlier instance; and all its outputs, or effects on the rest of the program, must also be clearly specified so the reuse mechanism can perform them correctly. Again, in our simple *fibonacci(x)* example, the only input is the operand *x*, and the only output is the Fibonacci series sum corresponding to *x*, making this an excellent candidate for memoization. On the other hand, a procedure such as *ordered_linked_list_insert(record *x)*, also shown in Figure 10.4, would be a poor candidate for memoization, since it both depends on the global state (a global *head* pointer for the linked list as well as the nodes in the linked list) and modifies the global state by updating the *next* pointer of a linked list element. Correct memoization of this type of function would require checking that the head pointer and none of the elements of the list had changed since the previous invocation.

Nevertheless, memoization is a powerful programming technique that is widely deployed and can be very effective. Clearly, memoization is a nonspeculative technique, since it requires precisely correct preconditions to be satisfied before reuse is invoked.

10.3.2 Instruction Reuse

Conceptually, instruction reuse is nothing more than a hardware implementation of memoization at the instruction level. It exposes additional instruction-level parallelism by decoupling the execution of a consumer instruction from its producers whenever it finds that the producers need not be executed. This is possible whenever the reuse mechanism finds that a producer instruction matches an earlier instance in the reuse history and is able to safely reuse the results of that prior instance. Sodani and Sohi's initial proposal for instruction reuse advocated reuse of an individual machine instruction whenever the operands to that instruction were shown to be invariant with respect to a prior instance of that instruction [Sodani and Sohi, 1997]. A more advanced mechanism for recording and reusing sequences of data-dependent instructions was also described. This mechanism stored the data dependence relationships between instructions in the reuse history table and could automatically reuse a data flow region of instructions (i.e., a subgraph of the dynamic data flow graph) whenever all the inputs to that region were shown to be invariant. Subsequent proposals have also considered expanding the reuse scope to include basic blocks as well as instruction traces fetched from a trace cache (refer to Chapter 5 for more details on how trace caches operate).

All these proposals for reuse share the same basic approach: the execution of an individual instruction or set of instructions is recorded in a history structure that stores the result of the computation for later reuse. The set of instructions can be defined by either control flow (as in basic block reuse and trace reuse) or data flow (as in data flow region reuse). The history structure must have a mechanism that guarantees that its contents remain coherent with subsequent program execution. Finally, the history structure has a lookup mechanism that allows subsequent instances to be checked against the stored instances. A hit or match during this lookup triggers the reuse mechanism, which allows the processor to skip execution of the reuse candidates. As a result, the processor eliminates the structural and data dependences caused by the reuse candidates and is able to fast-forward to subsequent program instructions. This process is summarized in Figure 10.5.

10.3.2.1 The Reuse History Mechanism.
Any implementation of reuse must have a mechanism for remembering, or memoizing, prior executions of instructions or sequences of instructions. This history mechanism must associate a set of preconditions with a previously computed result. These preconditions must exactly specify both the computation to be performed as well as all the *live inputs,* or operands that can affect the outcome of the computation. For instruction reuse, the computation to be performed is specified by a program counter (PC) tag that uniquely identifies a static instruction in the processor's address space, while the live inputs are both register and memory operands to that static instruction. For block reuse, the computation is specified by the address range of the instructions in the basic block, while the live inputs are all the source register and memory operands that are live on entry to the basic block. For trace reuse, the computation corresponds to a trace cache entry, which is uniquely identified by the fetch address and a set of conditional branch

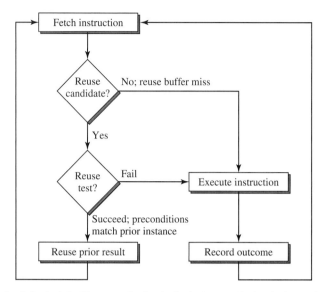

After an instruction is fetched, the history mechanism is checked to see whether the instruction is a candidate for reuse. If so, and if the instructions preconditions match the historical instance, the historical instance is reused and the fetched instruction is discarded. Otherwise, the instruction is executed as always, and its outcome is recorded in the history mechanism.

Figure 10.5
Instruction Reuse.

outcomes that specify the control flow path of the trace. By extension, all operands that are live on entry to the trace must also be specified.

The key attribute of the preconditions stored in the reuse buffer is that they uniquely specify the set of events that led to the computation of the memoized result. Hence, if that precise set of events ever occurs again, the computation need not be performed again. Instead, the memoized result can be substituted for the result of the repeated computation. However, just as with the memoization example in Figure 10.4, care must be taken that the preconditions in fact fully specify all the events that might affect the outcome of the computation. Otherwise, the reuse mechanism may introduce errors into program execution.

Indexing and Updating the Reuse Buffer. The history mechanism, or *reuse buffer,* is illustrated in Figure 10.6. It is usually indexed by low-order bits of the PC, and it can be organized as a direct-mapped, set-associative, or fully associative structure. Additional index information can be provided by including input operand value bits in the index and/or the tag; such an approach enables multiple instances of the same static instruction, but with varying input operands, to coexist in the reuse buffer. The reuse buffer is updated dynamically, as instructions or groups of instructions retire from the execution window; this may require a multi-ported or heavily banked structure to accommodate high throughput. There are

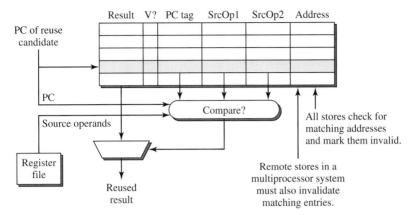

The instruction reuse buffer stores all the preconditions required to guarantee correct reuse of prior instances of instructions. For ALU and branch instructions, this includes a PC tag and source operand values. For loads and stores, the memory address must also be stored, so that intervening writes to that address will invalidate matching reuse entries.

Figure 10.6
Instruction Reuse Buffer.

also the usual design space issues regarding replacement policy and writeback policy (for multilevel history structures), similar to design issues for caches and cache hierarchies.

Reuse Buffer Organization. The reuse buffer can be organized to store history for individual instructions (i.e., each entry corresponds to a single instruction), for basic blocks, for traces (effectively integrating reuse history in the trace cache), or for data flow regions. There are scalability issues related to tracking live inputs for large numbers of instructions per reuse entry. For example, a basic block history mechanism may have to store up to a dozen or more live inputs and half as many results, given a basic block size of six or more instructions, each with two source operands and one destination. Similar scalability problems exist for proposed trace reuse mechanisms, which attempt to reuse entire traces of up to 16 instructions. Imagine increasing the width of the one-instruction-wide structure shown in Figure 10.6 to accommodate 16 instances of all the columns. Clearly, building such wide structures and wide comparators for checking reuse preconditions presents a challenging task.

Specifying Live Inputs. Live register inputs to a reuse entry can be specified either by *name* or by *value*. Specifying by name means recording either the architected register number for a register operand or the address for a memory operand. Specifying by value means recording the actual value of the operand instead of its name. Either way, all live inputs must be specified to maintain correctness, since failure to specify a live input can lead to incorrect reuse, where a computation is reused even though a subtle change to an unrecorded live input could cause a different result to occur. Sodani and Sohi investigated mechanisms that specified register operands both by name and by value, but only considered specifying memory operands by name. The example reuse buffer in Figure 10.6 specifies register source operands by value and memory locations by name.

Validating Live Inputs. To validate the live inputs of a reuse candidate, one must verify that the inputs stored in the reuse entry match the current architected values of those operands; this process is called the *reuse test*. Unless all live inputs are validated, reuse must not occur, since the reused result may not be correct. For named operands, this property is guaranteed by a coherence mechanism (explained next) that checks all program writes against the reuse buffer. For operands specified by value, the reuse mechanism must compare the current architected values against those in the reuse entry to check for a match. For register operands, this involves reading the current values from the architected register file and comparing them to the values stored in the reuse entry. Note that this creates considerable additional demand for read ports into the physical register file, since all operands for all reuse candidates must be read simultaneously. For memory operands specified by value, performing the reuse test would involve fetching the operand values from memory in order to compare them. Clearly, there is little to be gained here, since fetching the operands from memory in order to compare them is no less work than performing the memory operation itself. Hence, all reuse proposals to date specify memory operands by name, rather than by value. In Figure 10.6, each reuse candidate must fetch its source operands from the register file and compare them with the values stored in the reuse buffer.

Reuse Buffer Coherence Mechanism. To guarantee correctness, the reuse buffer must remain coherent with program execution that occurs between insertion of an entry into the reuse buffer and any subsequent reuse of that entry. To remain coherent, any intervening writes to either registers or memory that conflict with named live inputs must be properly reflected in the reuse buffer. The coherence mechanism is responsible for tracking all writes performed by the program (or other programs running on other processors in a multiprocessor system) and making sure that any named live inputs that correspond to those writes are marked invalid in the reuse structure. This prevents invalid reuse from occurring in cases where a named live input has changed. If live inputs are specified by value, rather than by name, intervening writes need not be detected, since the live input validation will compare the resulting architected and historic values and will trigger reuse only when the values match. Note that for named inputs, the coherence mechanism must perform an associative lookup over all the live inputs in the reuse buffer for every program write. For long names (say, 32- or 64-bit memory addresses), this associative lookup can be prohibitively expensive even for modest history table sizes. In Figure 10.6, all stores executed by the processor must check for matching entries in the reuse buffer and must invalidate the entry if its address matches the store. Similarly, in a multiprocessor system, all remote writes must invalidate matching entries in the reuse buffer.

As a final note, in systems that allow self-modifying code, the coherence mechanism must also track writes to the instruction addresses that are stored in the reuse buffer and must invalidate any matching reuse entries. Failure to do so could result in the reuse of an entry that no longer corresponds to the current program image. Similarly, the semantics of instructions that are used to invalidate instruction cache entries (e.g., *icbi* in the PowerPC architecture) must be extended to also invalidate reuse buffer entries with matching tags.

10.3.2.2 Reuse Mechanism. Finally, to gain performance benefit from reuse, the processor must be able to eliminate or reduce data and structural dependences for reused instructions by omitting the execution of these instructions and skipping ahead to subsequent work. This seems straightforward, but may require nontrivial modifications to the processor's data and control paths. First, reuse candidates (whether individual instructions or groups of instructions) must inject their results into the processor's architected state; since the data paths for doing so in real processors often only allow functional units to write results into the register file, this will probably involve adding write ports to an already heavily multiported physical register file. Second, instruction wakeup and scheduling logic will have to be modified to accommodate reused instructions with effectively zero cycles of result latency. Third, the reuse candidates must enter the processor's reorder buffer in order to maintain support for precise exceptions, but must simultaneously bypass the issue queues or reservation stations; this nonstandard behavior will introduce additional control path complexity. Finally, reused memory instructions must still be tracked in the processor's load/store queue (LSQ) to maintain correct memory reference ordering. Since LSQ entries are typically updated after instruction issue based on addresses generated during execution, this may also entail additional data paths and LSQ write ports that allow updates to occur from an earlier (prior to issue or execute) pipeline stage.

In summary, implementing instruction reuse will require substantial redesign or modification of existing control and data paths in a modern microprocessor design. This requirement may be the reason that reuse has not yet appeared in any real designs; the changes are substantial enough that they are likely to be incorporated only into a brand-new, clean-slate design.

10.3.3 Basic Block and Trace Reuse

Subsequent proposals have extended Sodani and Sohi's original proposal for instruction reuse to encompass sets of instructions defined by control flow [Huang and Lilja, 1999; Gonzalez et al., 1999]. In these proposals, similar mechanisms for storing and looking up reuse history are employed, but at the granularity of basic blocks or instruction traces. In both cases, the control flow unit (either basic block or trace) is treated as an atomically reusable computation. In other words, partial reuse due to partial matching of input operands is disallowed. Expanding the scope of instruction reuse to basic blocks and traces increases the potential benefit per reuse instance, since a substantial chunk of instructions can be directly bypassed. However, it also decreases the likelihood of finding a matching reuse entry, since the likelihood that a set of a half-dozen or dozen live inputs are identical to a previous computation is much lower than the likelihood of finding individual instructions within those groups that can be reused. Also, as discussed earlier, there are scalability issues related to conducting a reuse test for the large numbers of live inputs that basic blocks and traces can have. Only time will tell if reuse at a coarser control-flow granularity will prove to be more effective than instruction-level reuse.

10.3.4 Data Flow Region Reuse

In contrast to subsequent approaches that attempt to reuse groups of instructions based on control flow, Sodani also proposed an approach for storing and reusing data flow regions of instructions (the S_{n+d} and S_{v+d} schemes). This approach requires a bookkeeping scheme that embeds pointers in the reuse buffer to connect data-dependent instructions. These pointers can then be traversed to reuse entire subgraphs of the data flow graph; this is possible since the reuse property is transitive with respect to the data flow graph. More formally, any instruction whose data flow antecedents are all reuse candidates (i.e., they all satisfy the reuse test) is also a reuse candidate. By applying this principle inductively, a reusable data flow region can be constructed, resulting in a set of connected instructions that are all reusable. The reusable region is constructed dynamically by following the data dependence pointers embedded in the reuse table. Dependent instructions are connected by these edges, and any successful reuse test results are propagated along these edges to dependent instructions. The reuse test for the dependent instructions simply involves checking that all live input operands originate in instructions that were just reused or otherwise pass the reuse test. If this condition is satisfied, meaning that all operands are found to be invariant or to originate from reused antecedents, the dependent instructions themselves can be reused. The reuse test can be performed either by name (in the S_{n+d} scheme) or by value (in the S_{v+d} scheme).

Maintaining the integrity of the data dependence pointers presents a difficult challenge in a dynamically managed structure: Whenever an entry in the reuse buffer is replaced, all pointers to that entry become stale. All these stale pointers must be found and removed to prevent subsequent accesses to the reuse buffer from resulting in incorrect transitive propagation of reusability. Sodani proposed an associative lookup mechanism that automatically invalidates all such pointers on every replacement. Clearly, the expense and complexity of associative lookup coupled with frequent replacement prevent this from being a scalable solution. Alternative schemes that store dependence pointers in a separate, smaller structure which can feasibly support associative lookup are also possible, though unexplored in the current literature.

Subsequent work by Connors and Hwu [1999] proposes implementing region-level reuse strictly in software by modifying the compiler to generate code that performs the reuse test for data flow regions constructed by the compiler. This approach checks the live input operands and invokes region reuse by omitting execution of the region and immediately writing its results to the architected state whenever a matching history entry is found. In fact, this work takes us full circle back to software-based memoization techniques and establishes that automated, profile-driven techniques for memoization are indeed feasible and desirable.

10.3.5 Concluding Remarks

In summary, various schemes for reuse of prior computation have been proposed. These proposals are conceptually similar to the well-understood technique of memoization and vary primarily in the granularity of reuse and details of implementation. They all rely on the program characteristic of value locality, since without it, the likelihood of identifying reuse candidates would be very low. Reuse techniques have not been adopted in any real designs to date; yet they show significant performance potential if all the implementation challenges can be successfully overcome.

10.4 Exploiting Value Locality with Speculation

Having considered nonspeculative techniques for exploiting value locality and enhancing instruction-level parallelism, we now address speculative techniques for doing the same. Before delving into the details of value prediction, we step back to consider a theoretical basis for speculative execution—the *weak dependence model* [Lipasti and Shen, 1997; Lipasti, 1997].

10.4.1 The Weak Dependence Model

As we have learned in our study of techniques for removing false dependences, the implied inter-instruction precedences of a sequential program are an overspecification and need not be rigorously enforced to meet the requirements of the sequential execution model. The actual program semantics and inter-instruction dependences are specified by the control flow graph (CFG) and the data flow graph (DFG). As long as the serialization constraints imposed by the CFG and the DFG are not violated, the execution of instructions can be overlapped and reordered to achieve better performance by avoiding the enforcement of implied but unnecessary precedences. This can be achieved by Tomasulo's algorithm or more recent, modern reorder-buffer-based implementations. However, true inter-instruction dependences must still be enforced. To date, all machines enforce such dependences in a rigorous fashion that involves the following two requirements:

- Dependences are determined in an absolute and exact way; that is, two instructions are identified as either dependent or independent, and when in doubt, dependences are pessimistically assumed to exist.

- Dependences are enforced throughout instruction execution; that is, the dependences are never allowed to be violated, and are enforced continuously while the instructions are in flight.

Such a traditional and conservative approach for program execution can be described as adhering to the *strong dependence model*. The traditional strong dependence model is overly rigorous and unnecessarily restricts available parallelism. An alternative model that enables aggressive techniques such as value prediction is the *weak dependence model,* which specifies that:

- Dependences need not be determined exactly or assumed pessimistically, but instead can be optimistically approximated or even temporarily ignored.

- Dependences can be temporarily violated during instruction execution as long as recovery can be performed prior to affecting the permanent machine state.

The advantage of adopting the weak dependence model is that the program semantics as specified by the CFG and DFG need not be completely determined before the machine can process instructions. Furthermore, the machine can now speculate aggressively and temporarily violate the dependences as long as corrective measures are in place to recover from misspeculation. If a significant percentage of the speculations are correct, the machine can effectively exceed the performance limit imposed by the traditional strong dependence model.

Conceptually speaking, a machine that exploits the weak dependence model has two interacting engines. The front-end engine assumes the weak dependence model and is highly speculative. It tries to make predictions about instructions in order to aggressively process instructions. When the predictions are correct, these speculative instructions effectively will have skipped over or *folded out* certain pipeline stages. The back-end engine still uses the strong dependence model to validate the speculations, to recover from misspeculation, and to provide history and guidance information to the speculative engine. In combining these two interacting engines, an unprecedented level of instruction-level parallelism can be harvested without violating the program semantics. The edges in the DFG that represent inter-instruction dependences are now enforced in the critical path only when misspeculations occur. Essentially, these dependence edges have become probabilistic and the serialization penalties incurred due to enforcing these dependences are eliminated or masked whenever correct speculations occur. Hence, the traditional *data flow limit* based on the length of the critical path in the DFG is no longer a hard limit that cannot be exceeded.

10.4.2 Value Prediction

We learned in Section 10.2.2 that the register writes in many programs demonstrate a significant degree of value locality. This discovery opens up exciting new possibilities for the microarchitect. Since the results of many instructions can be accurately predicted before they are issued or executed, dependent instructions are no longer bound by the serialization constraints imposed by operand data flow. Instructions can now be scheduled speculatively with additional degrees of freedom to better utilize existing functional units and hardware buffers and are frequently able to complete execution sooner since the critical paths through dependence graphs have been collapsed. However, in order to exploit value locality and reap all these benefits, a variety of hardware mechanisms must be implemented: one for accurately predicting values (the *value prediction unit*); microarchitectural support for executing with speculative values; a mechanism for verifying value predictions; and finally a recovery mechanism for restoring correctness in cases where incorrectly predicted values were introduced into the program's execution.

10.4.3 The Value Prediction Unit

The value prediction unit is responsible for generating accurate predictions for speculative consumption by the processor core. The two competing factors that determine the efficacy of the value prediction unit are *accuracy* and *coverage;* a third factor related to coverage is the predictor's *scope*. Accuracy measures the predictor's ability to avoid mispredictions, while coverage measures the predictor's ability to predict as many instruction outcomes as possible. A predictor's scope describes the set of instructions that the predictor targets. Achieving high accuracy (e.g., few mispredictions) generally implies trading off some coverage, since any scheme that eliminates mispredictions will likely also eliminate some correct predictions. Conversely, achieving high coverage will likely reduce accuracy for the same reason: Aggressively pursuing every prediction opportunity is likely to result in a larger number of mispredictions.

Grasping the tradeoff between accuracy and coverage is easy if you consider the two extreme cases. At one extreme, a predictor can achieve 100% coverage by indiscriminately predicting all instructions; this will result in poor accuracy, since many instructions are inherently unpredictable and will be mispredicted. At the other extreme, a predictor can achieve 100% accuracy by not predicting any instructions and eliminating all mispredictions; of course, this will result in 0% coverage since none of the predictable instructions will be predicted either. The designer's challenge is to find a point between these two extremes that provides both high accuracy and high coverage.

Limiting the *scope* of the value predictor to focus on a particular class of instructions (e.g., load instructions) or some other dynamically or statically determined subset can make it easier to improve accuracy and/or coverage for that subset, particularly with a fixed implementation cost budget.

Building a value prediction unit that achieves the right balance of accuracy and coverage requires careful tradeoff analysis that must consider the performance effects of variations in coverage (i.e., proportional variation in freedom for scheduling of instructions for execution and changes in the height of the dynamic data flow graph) and variations in accuracy (i.e., fewer or more frequent mispredictions). This analysis will vary depending on minute structural and timing details of the microarchitecture being considered and requires detailed register-transfer-level simulation for correct tradeoff analysis. The analysis is further complicated by the fact that greater coverage does not always result in better performance, since only a relatively small subset of predictions are actually critical for performance. Similarly, improved accuracy may not improve performance either, since the mispredictions that were eliminated may also not have been critical for performance. A recent study by Fields, Rubin, and Bodik [2001] quantitatively demonstrates this by directly measuring the critical path of a program's execution and showing that relatively few correct value predictions actually remove edges along the critical path. They suggest limiting the value predictor's scope to only those instructions that are on the critical (i.e., longest) path in the program's data flow graph.

10.4.3.1 Prediction Accuracy. A naive value prediction scheme would simply endorse all possible predictions generated by the prediction scheme and supply them as speculative operands to the execution core. However, as published reports have shown, value predictors vary dramatically in their accuracy, at times providing as few as 18% correct predictions. Clearly, naive consumption of incorrect predictions is not only intellectually unsatisfying; it can lead to performance problems due to misprediction penalties. While it is theoretically possible to implement misprediction recovery schemes that have no direct performance penalty, practical difficulties will likely preclude such schemes (we discuss one possible approach in Section 10.4.4.5 under the heading Data Flow Eager Execution). Hence, beginning with the initial proposal for value prediction, researchers have described confidence estimation techniques for improving predictor accuracy.

Confidence Estimation. Confidence estimation techniques associate a confidence level with each value prediction, and they are used to filter incorrect predictions to improve predictor accuracy. If a prediction exceeds some confidence threshold, the processor core will actually consume the predicted value. If it does not, the predicted value is ignored and execution proceeds nonspeculatively, forcing the dependent operations to wait for the producer to finish computing its result. Typically, confidence levels are established with a history mechanism that increments a counter for every correct prediction and decrements or resets the counter for every incorrect prediction. Usually, there is a counter associated with every entry in the value prediction unit, although multiple counters per entry and multiple entries per counter have also been studied. The classification table shown in Figure 10.7 is a simple example of a confidence estimation mechanism. The design space for confidence estimators has been explored quite extensively in the literature to date and is quite similar to the design space for dynamic branch predictors (as discussed in Chapter 5). Design parameters include the choice of single or multiple levels of history; indexing with prediction outcome history, PC value, or some hashed combination; the number of states and transition functions in the predictor entry state machines; and so on. Even a relatively simple confidence estimation scheme, such as the one described in Figure 10.7, can provide prediction accuracy that eliminates more than 90% of all mispredictions while sacrificing less than 10% of coverage.

10.4.3.2 Prediction Coverage. The second factor that measures the efficacy of a value prediction unit is prediction coverage. The simple value predictors that were initially proposed simply remembered the previous value produced by a particular static instruction. An example of such a *last value predictor* is shown in Figure 10.7. Every time an instruction executes, the *value prediction table* (VPT) is updated with its result. As part of the update, the confidence level in the classification table is incremented if the prior value matched the actual outcome, and decremented otherwise. The next time the same static instruction is fetched, the previous value is

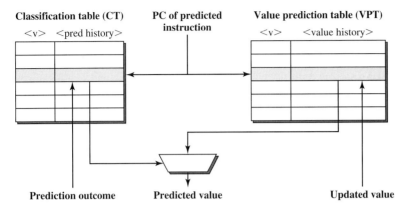

The internal structure of a simple value prediction unit (VPU). The VPU consists of two tables: the *classification table* (CT) and the *value prediction table* (VPT), both of which are direct-mapped and indexed by the instruction address (PC) of the instruction being predicted. Entries in the CT contain two fields: the *valid* field, which consists of either a single bit that indicates a valid entry or a partial or complete tag field that is matched against the upper bits of the PC to indicate a valid field; and the *prediction history*, which is a saturating counter of 1 or more bits. The prediction history is incremented or decremented whenever a prediction is correct or incorrect, respectively, and is used to classify instructions as either predictable or unpredictable. This classification is used to decide whether or not the result of a particular instruction should be predicted. Increasing the number of bits in the saturating counter adds hysteresis to the classification process and can help avoid erroneous classifications by ignoring anomalous values and/or destructive interference.

Figure 10.7
Value Prediction Unit.

retrieved along with the current confidence level. If the confidence level exceeds a fixed threshold, the predicted value is used; otherwise, it is discarded.

Simple last value predictors provide roughly 40% coverage over a set of general-purpose programs. Better coverage can be obtained with more sophisticated predictors that either provide additional context to allow the predictor to choose from multiple prior values (*history-based predictors*) or are able to detect predictable sequences and compute future, previously unseen, values (*computational predictors*).

History-Based Predictors. The simplest history-based predictors remember the most recent value written by a particular static instruction and predict that the same value will be computed by the next dynamic instance of that instruction. More sophisticated predictors provide a means for storing multiple different values for each static instruction, and then use some scheme to choose one of those values as the predicted one. For example, the *last-n value predictor* proposed by Burtscher and Zorn [1999] uses a scheme of prediction outcome histories to choose one of n values stored in the value prediction table. Alternatively, the *finite-context-method* (FCM) predictor proposed by Sazeides and Smith [1997] also stores multiple values, but chooses one based on a finite context of recent values observed during program execution, rather than strictly by PC value. This value context is analogous to the branch outcome context captured by a branch history register that is used successfully to implement two-level branch predictors. The FCM scheme is able to capture periodic sequences of values, such as the set of pointer addresses loaded by the traversal of a linked list. The FCM predictor has been shown to reach prediction coverage in excess of 90% for certain workloads, albeit with considerable implementation cost for storing multiple values and their contexts.

Computational Predictors. Computational predictors attempt to capture a predictable pattern in the sequence of values generated by a static instruction and then compute the next instance in the sequence. They are fundamentally different from history-based predictors since they are able to generate predicted values that have not occurred in prior program execution. Gabbay and Mendelson [1997] first proposed a *stride predictor* that detects a fixed stride in the value sequence and is able to compute the next value by adding the observed stride to the prior value. A stride predictor requires additional hardware: to detect strides it must use a 32- or 64-bit subtraction unit to extract the stride and a comparator to check the extracted stride against the previous stride instance; it needs additional space in the value prediction table to store the stride value and some additional confidence estimation bits to indicate a valid stride; and, finally, it needs an adder to add the prior value to the stride to compute each new prediction. Stride prediction can be quite effective for certain workloads; however, it is not clear if the additional storage, arithmetic hardware, and complexity are justified.

More advanced computational predictors have been discussed, but none have been formally proposed to date. Clearly, there is a continuum in the design space for computational predictors between the two extremes of history-based prediction with no computational ability and full-blown preexecution, where all the architected state is made available as *context* to the predictor, and which simply anticipates the semantics of the actual program to precompute its results. While the latter extreme is obviously neither practical nor useful, since it simply replicates the functionality of the processor's execution core, the interesting question that remains is whether there is a useful middle ground where at least a subset of program computation can be abstracted to the point that a computational predictor of reasonable cost is able to replicate it with high accuracy. Clearly, sophisticated branch predictors are able to abstract 95% or more of many programs' control flow behavior; whether sophisticated computational value predictors can ever reach the same goal for a program's data flow remains an open question.

Hybrid Predictors. Finally, analogous to the hybrid or combining branch predictors described in Chapter 9, various schemes that combine multiple heterogeneous predictors into a single whole have been proposed. Such a hybrid prediction scheme might combine a last value predictor, a stride predictor, and a finite-context predictor in an attempt to reap the benefits of each. Hybrid predictors can enable not only better overall coverage, but can also allow more efficient and smaller implementations of advanced prediction schemes, since they can be targeted only to the subset of static instructions that require them. A very effective hybrid predictor was proposed by Wang and Franklin [1997].

Implementation Issues. Several studies have examined various implementation issues for value prediction units. These issues encompass the size, organization, accessibility, and sensitivity to update latency of value prediction structures, and they can be difficult to solve, particularly for complex computational and hybrid predictors. In general, solutions such as clever hash functions for indexing the tables and banking the structure to enable multiple simultaneous accesses have been shown to work well. A recent proposal that shifts complex value predictor access to completion time, and stores the results of that access in a simple, direct-mapped table or directly in a trace cache entry, is able to shift much of the access complexity away from the timing-critical front end of the processor pipeline [Lee and Yew, 2001]. Another intriguing proposal refrains from storing values in a separate history structure by instead predicting that the needed value is already in the register file, and storing a pointer to the appropriate register [Tullsen and Seng, 1999]. Surprisingly, this approach works reasonably well, especially if the compiler allocates register names with some knowledge of the values stored in the registers.

10.4.3.3 Prediction Scope.
The final factor determining the efficacy of a value prediction unit is its intended prediction scope. The initial proposal for value prediction focused strictly on load instructions, limiting its scope to a subset of instructions generally perceived to be critical for performance. Reducing load latency by predicting and speculatively consuming the values returned by those loads has been shown to improve performance and reduce the effect of structural hazards for highly contended cache ports, and should increase memory-level parallelism by allowing loads that would normally be blocked by a data flow–antecedent cache miss to execute in parallel with the miss.

The majority of proposed prediction schemes target all register-writing instructions. However, there are some interesting exceptions. Sodani and Sohi [1998] point out that register contents that are directly used to resolve conditional branches should probably not be predicted, since such value predictions are usually less accurate than the tailored predictions made by today's sophisticated branch predictors. This issue was sidestepped in the initial value prediction work, which used the PowerPC instruction set architecture, in which all conditional branches are resolved using dedicated condition registers. Since only general-purpose registers were predicted, the detrimental effect of value mispredictions misguidedly overriding correct branch predictions was kept to a minimum. In instruction sets similar to MIPS or PISA (used in Sodani's work), there are no condition registers, so a scheme that value predicts all general-purpose registers will also predict branch source operands and can directly and adversely affect branch resolution.

Several researchers have proposed focusing value predictions on only those data dependences that are deemed critical for performance [Calder et al., 1999]. This has several benefits: The extra work of useless predictions can be avoided; predictors with better accuracy and coverage and lower implementation cost can be devised; and mispredictions that occur for useless predictions can be reduced or eliminated. Fields, Rubin, and Bodik [2001] demonstrate many of these benefits in their recent proposal for deriving data dependence criticality by a novel approach to monitoring out-of-order instruction execution.

10.4.4 Speculative Execution Using Predicted Values

Just as with instruction reuse, value prediction requires microarchitectural support for taking advantage of the early availability of instruction results. However, there is a fundamental difference in the required support due to the speculative nature of value prediction. Since instruction reuse is preceded by a reuse test that guarantees its correctness, the microarchitectural changes outlined in Section 10.3.2.2 consist primarily of additional bandwidth into the bookkeeping structures within an out-of-order superscalar processor. In contrast, value prediction—an inherently speculative technique—requires more pervasive support in the microarchitecture to handle detection of and recovery from misspeculation. Hence, value prediction implies microarchitectural support for *value-speculative execution,* for *verifying predictions,* and for *misprediction recovery.* We will first describe a minimal approach for supporting value-speculative execution; then we will discuss more advanced verification and recovery strategies.

10.4.4.1 Straightforward Value Speculation.
At first glance, it seems that speculative execution using predicted values maps quite naturally onto the structures that a modern out-of-order superscalar processor already provides. First of all, to support value speculation, we need a mechanism for storing and forwarding predictions from the value prediction unit to the dependent instructions: the existing rename buffers or rename registers serve this purpose quite well. Second, we need a mechanism to issue dependent instructions speculatively; the standard out-of-order issue logic, with minor modifications, will work for this purpose as well. Third, we need a mechanism for detecting mispredicted values. The obvious solution is to augment the reservation stations to hold the predicted output values for each instruction, and provide additional data paths from the reservation station and the functional unit output to a comparator that checks these values for equality and signals a misprediction when the comparison fails. Finally, we need a way to recover from mispredictions. If we treat value mispredictions the same way we treat branch mispredictions, we can simply recycle the branch misprediction recovery mechanism that flushes out speculative instructions and refetches all instructions following the mispredicted one. Surprisingly, these minimal modifications are sufficient for correctness in a uniprocessor system,[1] and can even provide nontrivial speedup as long as the predictor is highly accurate and mispredictions are relatively rare. However, more sophisticated verification and recovery techniques can lead to higher-performance designs, but require additional complexity. We discuss such techniques in the following.

---

[1] A recent publication discusses why they are not sufficient in a cache-coherent multiprocessor: essentially, value prediction removes the natural reference ordering between data-dependent loads by allowing a dependent load to execute before a preceding load that computes its address; multiprocessor programs that rely on such dependence ordering for correctness can fail with the naive value prediction scheme described here. The interested reader is referred to Martin et al. [2001] for further details.

10.4.4.2 Prediction Verification. Prediction verification is analogous to the reuse test that guarantees correctness for instruction reuse. In other words, it must guarantee that the predicted outcome of a value-predicted instruction matches the actual outcome, as determined by the architected state and the semantics of the instruction. The most straightforward approach to verification is to execute the predicted instruction and then compare the outcome of the execution with the value prediction. Naively, this implies appending an ALU-width comparator to each functional unit to verify predictions. Since the latency through a comparator is equivalent to the delay through an ALU, most proposals have assumed an extra cycle of latency to determine whether or not a misprediction occurred.

Prediction verification serves two purposes. The first is to trigger a recovery action whenever a misprediction occurs; possible recovery actions are discussed in Section 10.4.4.4. The second purpose is more subtle and occurs when there is no misprediction: The fact that a correct prediction was verified may now need to be communicated to dependent instructions that have executed speculatively using the prediction. Depending on the recovery model, such speculatively executed instructions may continue to occupy resources within the processor window until they are found to be nonspeculative. For example, in a conventional out-of-order microprocessor, instructions can only enter the issue queues or reservation stations in program order. Once they have issued and executed, there is no data or control path that enables placing them back in the issue queue to reissue. In such a microarchitecture, an instruction that consumed a predicted source operand and issued speculatively would need to remain in the issue queue or reservation station in case it needed to reissue with a future corrected operand. Since issue queue slots are an important and performance-critical hardware resource, timely notification of the fact that an instruction's input operands were not mispredicted can be important for reducing structural hazards.

As mentioned, the most straightforward approach for misprediction detection is to wait until a predicted instruction's operands are available before executing the instruction and comparing its result with its predicted result. The problem with this approach is that the instruction's operands themselves may be speculative (that is, the producer instructions may have been value predicted, or, more subtly, some data flow antecedent of the producer instructions may have been value predicted). Since speculative input operands beget speculative outputs, a single predicted value can propagate transitively through a data flow graph for a distance limited only by the size of the processor's instruction window, creating a wavefront of speculative operand values (see Figure 10.8). If a speculative operand turns out to be incorrect, verifying an instruction's own prediction with that incorrect operand may cause the verification to succeed when it should not or to fail when it should succeed. Neither of these is a correctness issue; the former case will be caught since the incorrect input operand will eventually be detected when the misprediction that caused it is verified, while the latter case will only cause unnecessary invocations of the recovery mechanism. However, for this very reason, the latter can cause a performance problem, since correctly executed instructions are reexecuted unnecessarily.

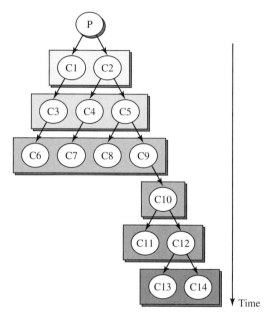

The speculative operand wavefront traverses the dynamic data flow graph as a result of the predicted outcome of instruction P. Its consumers C1 and C2 propagate the speculative property to their consumers C3, C4, and C5, and so on. Serial propagation of prediction verification status propagates through the data flow graph in a similar manner. Parallel propagation, which requires a tag broadcast mechanism, allows all speculatively executed dependent instructions to be notified of verification status in a single cycle.

Figure 10.8
The Speculative Operand Wavefront.

Speculative Verification. A similar problem arises when speculative operands are used to resolve branch instructions. In this scenario, a correctly predicted branch can be resolved incorrectly due to an incorrect value prediction, resulting in a branch misprediction redirect. The straightforward solution to these two problems is to disallow prediction verification (whether value or branch) with speculative inputs. The shortcoming of this solution is that performance opportunity is lost whenever a correct speculative input would have appropriately resolved a mispredicted branch or corrected a value misprediction. There is no definitive answer as to the importance of this performance effect; however, the recent trend toward deep execution pipelines that are very performance-sensitive to branch mispredictions would lead one to believe that any implementation decision that delays the resolution of incorrectly predicted branches is the wrong one.

Propagating Verification Results. As an additional complication, in order to delay verification until all input operands are nonspeculative, there must be a mechanism in place that informs the instruction whether its input operands have been verified. In its simplest form, such a mechanism is simply the reorder buffer (ROB); once an instruction becomes the oldest in the ROB, it can infer that all its data flow antecedents are verified, so it can now also be verified. However, delaying verification until an instruction is next to commit has negative performance implications, particularly for mispredicted conditional branch instructions. Hence, a mechanism that propagates verification status of operands through the data flow graph is desirable. Two fundamental design alternatives exist: The verification status can be propagated serially, along the data dependence edges, as instructions are verified; or it can be broadcast in parallel. Serial propagation can be piggybacked on the existing broadcast result used to wake up dependent instructions in out-of-order execution. Parallel broadcast is more expensive, and it implies tagging operand values with all speculative data flow antecedents, and then broadcasting these tags as the predictions are verified. Parallel broadcast has a significant latency benefit, since entire dependence chains can become nonspeculative in the cycle following verification of some long-latency instruction (e.g., cache miss) at the head of the chain. As discussed, this instantaneous commit can reduce structural hazards by freeing up issue queue or reservation station slots right away, instead of waiting for serial propagation through the data flow graph.

10.4.4.3 Data Flow Region Verification. One interesting opportunity for improving the efficiency of value prediction verification arises from the concept of data flow regions. Recall that data flow regions are subgraphs of the data flow graph that are defined by the set of instructions that are reachable from a set of live inputs. As proposed by Sodani and Sohi [1997], a data flow region can be reused en masse if the set of live inputs to the region meets the reuse test. The same property can also be exploited to verify the correctness of all the value predictions that occur in a data flow region. A mechanism similar to the one described in Section 10.3.4 can be integrated into the value prediction unit to construct data flow regions by storing data dependence pointers in the value prediction table. Subsequent invocation of value predictions from a self-consistent data flow region then leads to a reduction in verification scope. Namely, as long as the data flow region mechanism guarantees that all the predictions within the region are consistent with each other, only the initial predictions that correspond to the live inputs to the data flow region need to be verified. Once these initial predictions are verified via conventional means, the entire data flow region is known to be verified, and the remaining instructions in the region need not ever be executed or verified.

This approach is strikingly similar to data flow region reuse, and it requires quite similar mechanisms in the value prediction table to construct data flow region information and guarantee its consistency (these issues are discussed in greater detail in Section 10.3.4). However, there is one fundamental difference: data flow region reuse requires the live inputs to the data flow region to be either unperturbed (if the reuse test is performed by name) or unchanged and available in the register file (if the reuse test is performed by value). Integrating data flow regions with value prediction, however, avoids these limitations by deferring the reuse test indefinitely, until the live inputs are available within the processor's execution window. Once the live inputs have all been verified, the entire data flow region can be notified of its nonspeculative status and can retire without ever executing. This should significantly reduce structural dependences and contention for functional units for programs where reusable data flow regions make up a significant portion of the instructions executed.

10.4.4.4 Misprediction Recovery via Refetch. There are two approaches to recovering from value mispredictions: *refetch* and *selective reissue*. As already mentioned, refetch-based recovery builds on the branch misprediction recovery mechanism which is present in almost every modern superscalar processor. In this approach, value mispredictions are treated exactly as branch mispredictions: All instructions that follow the mispredicted instruction in program order are flushed out of the processor, and instruction fetch is redirected to refetch these instructions. The architected state is restored to the instruction boundary following the mispredicted instruction, and the refetched instructions are guaranteed to not be polluted by any mispredicted values, since such mispredicted values do not survive the refetch.

The most attractive feature of refetch-based misprediction recovery is that it requires very few changes to the processor, assuming the mechanism is already in place for redirecting mispredicted branches. On the other hand, it has the obvious drawback that the misprediction penalty is quite severe. Studies have shown that in a processor with a refetch policy for recovering from value mispredictions, highly accurate value prediction is a requirement for gaining performance benefit. Without highly accurate value prediction—usually brought about by a high-threshold confidence mechanism—performance can in fact degrade due to the excessive refetches. Unfortunately, a high-threshold confidence mechanism also inevitably reduces prediction coverage, resulting in a processor design that fails to capture all the potential performance benefit of value prediction.

10.4.4.5 Misprediction Recovery via Selective Reissue. Selective reissue provides a potential solution to the performance limitations of refetch-based recovery. With selective reissue, only those instructions that are data dependent on a mispredicted value are required to reissue. Implementing selective reissue requires a mechanism for propagating misprediction information through the data flow graph to all dependent instructions. Just as was the case for propagating verification information, reissue information can also be propagated serially or in parallel. A serial mechanism can easily piggyback on the existing result bus that is used to wake up dependent instructions in an out-of-order processor. With serial propagation, the delay for communicating a reissue condition is proportional to the data flow distance from the misprediction to the instruction that must reissue. It is conceivable, although unlikely, that the reissue message will never catch up with the speculative operand wavefront illustrated in Figure 10.8, since both propagate through the data flow graph at the same rate of one level per cycle. Furthermore, even if the reissue message does eventually reach the speculative operand wavefront, the serial propagation delay of the reissue message can cause excessive wasted execution along the speculative operand wavefront. Hence, researchers have also proposed broadcast-based mechanisms that communicate reissue commands in parallel to all dependent instructions.

In such a parallel mechanism, speculative value-predicted operands are provided with a unique tag, and all dependent instructions that execute with such operands must propagate those tags to their dependent instructions. On a misprediction, the tag corresponding to the mispredicted operand is broadcast so that all data flow descendants realize they must reissue and reexecute with a new operand. Figure 10.9 illustrates a possible implementation of value prediction with parallel-broadcast selective reissue.

Misprediction Penalty with Selective Reissue. With refetch-based misprediction recovery, the misprediction penalty is comparable to a branch misprediction penalty and can run to a dozen or more cycles in recent processors with very deep

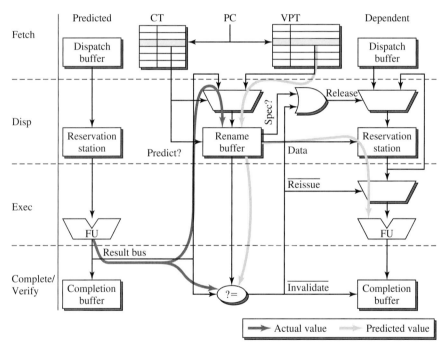

The dependent instruction shown on the right uses the predicted result of the instruction on the left, and is able to issue and execute in the same cycle. The VP Unit predicts the values during fetch and dispatch, then forwards them speculatively to subsequent dependent instructions via a rename buffer. The dependent instruction is able to issue and execute immediately, but is prevented from completing architecturally and retains possession of its reservation station until its inputs are no longer speculative. Speculatively forwarded values are tagged with the uncommitted register writes they depend on, and these tags are propagated to the results of any subsequent dependent instructions. Meanwhile, the predicted instruction executes on the right, and the predicted value is verified by a comparison against the actual value. Once a prediction is verified, its tag is broadcast to all active instructions, and all the dependent instructions can either release their reservation stations and proceed into the completion unit (in the case of a correct prediction), or restart execution with the correct register values (if the prediction was incorrect).

Figure 10.9
Example of Value Prediction with Selective Reissue.

pipelines. The goal of selective reissue is to mitigate this penalty by reducing the number of cycles that elapse between determining that a misprediction occurred and correctly re-executing data-dependent instructions. Assuming a single additional cycle for prediction verification, the apparent best case would be a single cycle of misprediction penalty. That is to say, the dependent instruction executes one cycle later than it would have had there been no value prediction.

The penalty occurs only when a dependent instruction has already executed speculatively but is waiting in its reservation station for one of its predicted inputs to be verified. Since the value comparison takes an extra cycle beyond the pipeline result latency, the dependent instruction will reissue and execute with the correct value one cycle later than it would have had there been no prediction. In addition, the earlier incorrect speculative issue may have caused a structural hazard that prevented other useful instructions from dispatching or executing. In those cases where the dependent instruction has not yet executed (due to structural or other unresolved data dependences), there is no penalty, since the dependent instruction can issue as soon as the actual computed value is available, in parallel with the value comparison that verifies the prediction.

Data Flow Eager Execution. It is possible to reduce the misprediction penalty to zero cycles by employing *data flow eager execution*. In such a scheme, the dependent instruction is speculatively re-executed as soon as the nonspeculative operand becomes available. In other words, there is a second shadow issue of the dependent instruction as if there had been no earlier speculative one. In parallel with this second issue, the prediction is verified, and in case of correct prediction, the shadow issue is squashed. Otherwise, the shadow issue is allowed to continue, and execution continues as if the value prediction had never occurred, with effectively zero cycles of misprediction penalty. Of course, the data flow eager shadow issue of all instructions that depend on value predictions consumes significant additional execution resources, potentially overwhelming the available functional units and slowing down computation. However, given a wide machine with sufficient execution resources, this may be a viable alternative for reducing the misprediction penalty. Prediction confidence could also be used to gate data flow eager execution. In cases where prediction confidence is high, eager execution is disabled; in cases where confidence is low, eager execution can be used to mitigate the misprediction penalty.

The Effect of Scheduling Latency. In a canonical out-of-order processor that implements the modern equivalent of Tomasulo's algorithm, instruction scheduling decisions are made in a single cycle immediately preceding the actual execution of the instructions that are selected for execution. Such a scheme allows the scheduler to react immediately to dynamic events, such as detection of store-to-load aliases or cache misses, and issue alternative, independent instructions in subsequent cycles. However, cycle time constraints have led recent designs to abandon this property, resulting in instruction schedulers that create an execution schedule several cycles in advance of the actual execution. This effect, called the *scheduling latency,* inhibits the scheduler's ability to react to dynamic events. Of course, value misprediction detection is a dynamic event, and the fact that several modern processor designs (e.g., Alpha 21264 and Intel Pentium 4) have multicycle scheduling latency will necessarily increase the value misprediction penalty on such machines. In short, the value misprediction penalty is in fact the sum of the scheduling latency and the verification latency. Hence, a processor with three-cycle scheduling latency and a one-cycle verification latency would have a value misprediction latency of four cycles. However, even in such designs it is possible to reduce the misprediction penalty via *data flow eager execution*. Of course, the likelihood that execution resources will be overwhelmed by this approach increases with scheduling latency, since the number of eagerly executed and squashed instructions is proportional to this latency.

10.4.4.6 Further Implications of Selective Reissue

Memory Data Dependences. Selective reissue requires all data-dependent instructions to reissue following a value misprediction. While it is fairly straightforward to identify register data dependences and reissue dependent instructions, memory data dependences can cause a subtle problem. Namely, memory data dependences are defined by register values themselves; if the register values prove to be incorrect due to a value misprediction, memory data dependence information may need to be reconstructed to guarantee correctness and to determine which additional instructions are in fact dependent on the value misprediction. For example, if the mispredicted value was used either directly or indirectly to compute an address for a load or store instruction, the load/store queue or any other alias resolution mechanism within the processor may have incorrectly concluded that the load or store is or is not aliased with some other store or load within the processor's instruction window. In such cases, care must be taken to ensure that memory dependence information is recomputed for the load or store whose address was polluted by the value misprediction. Alternatively, the processor can disallow the use of value-predicted operands in address generation for loads or stores. Of course, doing so will severely limit the ability of value prediction to improve memory-level parallelism. Note that this problem does not occur with a refetch recovery policy, since memory dependence information is explicitly recomputed for all instructions following the value misprediction.

Changes to Scheduling Logic. Reissuing instructions requires nontrivial changes to the scheduling logic of a conventional processor. In normal operation, instructions issue only one time, once their input operands become available and a functional unit is available. However, with selective reissue, an instruction may have to issue multiple times, once with speculative operands and again with corrected operands. All practical out-of-order implementations partition the active instruction window into two disjoint sets: instructions waiting to issue (these are the instructions still in reservation stations or issue queues), and instructions that have issued but are waiting to retire. This partitioning is driven by cycle-time demands that limit the total number of instructions that can be considered for issue in a single cycle. Since instructions that have already issued need not be considered for reissue, they are moved out of reservation stations or issue queues into the second partition (instructions waiting to retire).

Unfortunately, with selective reissue, a clean partition is no longer possible, since instructions that issued with speculative operands may need to reissue, and hence should not leave the issue queue or reservation station. There are two solutions to this problem: either remove speculatively issued instructions from the reservation stations, but provide an additional mechanism to reinsert them if they need to reissue; or keep them in the reservation stations until their input operands are no longer speculative. The former solution introduces significant additional complexity into the front-end control and data paths and must also deal with a possible deadlock scenario. One such scenario occurs when all reservation station entries are full of newer instructions that are data dependent on an older instruction that needs to be reinserted into a reservation station so that it can reissue. Since there are no reservation stations available, and none ever become available since all the newer instructions are waiting for the older instruction, the older instruction cannot make forward progress, and can never retire, leading to a deadlocked system. Note that refetch-based recovery does not have this problem, since all newer data-dependent instructions are flushed out of the reservation stations upon misprediction recovery.

Hence, the latter solution of forcing speculatively issued instructions to retain their reservation station entries is proposed most often. Of course, this approach requires a mechanism for promoting speculative operands to nonspeculative status. A parallel or serial mechanism like the ones described in Section 10.4.4.2 will suffice for this purpose. In addition to the complexity introduced by having to track the verification status of operands, this solution has the additional slight problem that it increases the occupancy of the reservation station entries. Without value prediction, a dependent instruction releases its reservation station in the same cycle that it issues, which is the cycle following computation of its last input operand. With the proposed scheme, even though the instruction may have issued much earlier with a value-predicted operand, the reservation station itself is occupied for one additional cycle beyond operand availability, since the entry is not released until after the predicted operand is verified, one cycle later than it is computed.

Existing Support for Data Speculation. Note that existing processors that do not implement value prediction, but do support other forms of data speculation (for example, speculating that a load is not aliased to a prior store), may already support a limited form of selective reissue. The Intel Pentium 4 is one such processor and implements selective reissue to recover from cache hit speculation; here, data from a cache access are forwarded to dependent instructions before the tag match that validates a cache hit has completed. If there is a tag mismatch, the dependent instructions are selectively reissued. If this kind of selective reissue scheme already exists, it can also be used to support value misprediction recovery. However, the likelihood of being able to reuse an existing mechanism is reduced by the fact that existing mechanisms for selective reissue are often tailored for speculative conditions that are resolved within a small number of cycles (e.g., tag mismatch or alias resolution). The fact that the speculation window extends only for a few cycles allows the speculative operand wavefront (see Figure 10.8) to propagate through only a few levels in the data flow graph, which in turn limits the total number of instructions that can issue speculatively. If the selective reissue mechanism exploits this property and is somehow restricted to handling only a small number of dependent operations, it is not useful for value prediction, since the speculation window for value prediction can extend to tens or hundreds of cycles (e.g., when a load that misses the cache is value predicted) and can encompass the processor's entire instruction window. However, the converse does hold: If a processor implements selective reissue to support value prediction, the same mechanism can be reused to support recovery for other forms of data speculation.

10.4.5 Performance of Value Prediction

Numerous published studies have examined the performance potential of value prediction. The results have varied widely, with reported performance effects ranging from minor slowdowns to speedups of 100% or more. Achievable performance depends heavily on many of the factors already mentioned, including particular details of the machine model and pipeline structure, as well as workload choice. Some of the factors affecting performance are

- The degree of value locality present in the programs or workloads.

- The dynamic dependence distance between correctly predicted instructions and the instructions that consume their results. If the compiler has already scheduled dependent instructions to be far apart, reducing result latency with value prediction may not provide much benefit.

- The instruction fetch rate achieved by the machine model. If the fetch rate is fast relative to the pipeline's execution rate, value prediction can significantly improve execution throughput. However, if the pipeline is fetch-limited, value prediction will not help much.

- The coverage achieved by the value prediction unit. Clearly, the more instructions are predicted, the more performance benefit is possible. Conversely, poor coverage results in limited opportunity.

- The accuracy of the value prediction unit. Achieving a high ratio between correct and incorrect predictions is critical for reaping significant performance benefit, since mispredictions can slow the processor down.

- The misprediction penalty of the pipeline implementation. As discussed, both the recovery policy (refetch versus reissue) and the efficiency of the recovery policy can severely affect the performance impact of mispredictions. Generally speaking, deeper pipelines that require speculative scheduling will have greater misprediction penalties and will be more sensitive to this effect.

- The degree to which a program is limited by data flow dependences. If a program is primarily performance-limited by something other than data dependences, eliminating data dependences via value prediction will not result in much benefit. For example, if a program is limited by instruction fetch, branch mispredictions, structural hazards, or memory bandwidth, it is unlikely that value prediction will help performance.

In summary, the performance effects of value prediction are not yet fully understood. What is clear is that under a large variety of instruction sets, benchmark programs, machine models, and misprediction recovery schemes, nontrivial speedup is achievable and has been reported in the literature.

As an indication of the performance potential of value prediction, some performance results for an idealized machine model are shown in Figure 10.10. This idealized machine model measures one possible data flow limit, since, for all practical purposes, parallel issue in this model is restricted only by the following three factors:

- Branch prediction accuracy, with a minimum redirect penalty of three cycles
- Fetch bandwidth (single taken branch per cycle)
- Data flow dependences
- A value misprediction penalty of one cycle

This machine model reflects idealized performance in most respects, since the misprediction penalties are very low and there are no structural hazards. However, we consider history-based value predictors only, so later studies that employed

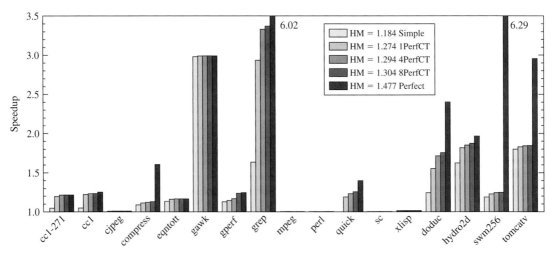

The Simple configuration employs a straightforward last-value predictor. The 1PerfCT, 4PerfCT, and 8PerfCT configurations use perfect confidence, eliminating all mispredictions while maximizing coverage, and choosing from a value history of 1, 4, or 8 last values, respectively. The Perfect configuration eliminates all true data dependences and indicates the overall performance potential.

Figure 10.10

Value Prediction Speedup for an Idealized Machine Model.

Table 10.1
Value prediction unit configurations

Configuration	Value Prediction Table		Confidence Table	
	Direct-Mapped Entries	History Depth	Direct-Mapped Entries	Bits/Entry
Simple	4096	1	1024	2-bit up-down saturating counter
1PerfCT	4096	1	∞	Perfect
4PerfCT	4096	4/Perfect selector	∞	Perfect
8PerfCT	4096	8/Perfect selector	∞	Perfect
Perfect	∞	Perfect	∞	Perfect

computational or hybrid predictors have shown dramatically higher potential speedup. Figure 10.10 shows speedup for five different value prediction unit configurations, which are summarized in Table 10.1. Attributes that are marked *perfect* in Table 10.1 indicate behavior that is analogous to *perfect caches;* that is, a mechanism that always produces the right result is assumed. More specifically, in the 1PerfCT, 4PerfCT, and 8PerfCT configurations, we assume an *oracle confidence table (CT)* that is able to correctly identify all predictable and unpredictable register writes. Furthermore, in the 4PerfCT and 8PerfCT configurations, we assume a perfect mechanism for choosing which of the four (or eight) values stored in the value history is the correct one. Note that this is an idealized version of the last-n predictor proposed by Burtscher and Zorn [1999]. Moreover, we assume that the perfect configuration can always correctly predict a value for every register write, effectively removing all data dependences from execution. Of these configurations, the only value prediction unit configuration that we know how to build is the simple one, while the other four are merely included to measure the potential contribution of improvements to both value prediction table (VPT) and CT prediction accuracy.

The results in Figure 10.10 clearly demonstrate that even simple predictors are capable of achieving significant speedup. The difference between the simple and 1PerfCT configurations demonstrates that accuracy is vitally important, since it can increase speedup by a factor of 50% in the limit. The 4PerfCT and 8PerfCT cases show that there is marginal benefit to be gained from history-based predictors that track multiple values. Finally, the perfect configuration shows that dramatic speedups are possible for benchmarks that are limited by data flow.

10.4.6 Concluding Remarks

In summary, various schemes for speculative execution based on value prediction have been proposed. Researchers have described techniques for improving prediction accuracy and coverage and focusing predictor scope to where value predictions are perceived to be most useful. Many implementation issues, both in predictor design as well as effective microarchitectural support for value-speculative execution have been studied. At the same time, numerous unanswered questions and unexplored issues remain. No real designs that incorporate value prediction have yet emerged; only time will tell if the demonstrated performance potential of value prediction will compensate for the additional complexity required for its effective implementation.

10.5 Summary

This chapter has explored both speculative and nonspeculative techniques for improving register data flow beyond the classical data flow limit. These techniques are based on the program characteristic of value locality, which describes the likelihood that previously seen operand values will recur in later executions of static program instructions. This property is exploited to remove computations from a program's dynamic data flow graph, potentially reducing the height of the tree and allowing a compressed execution schedule that permits instructions to execute sooner than their position in the data flow graph might indicate. Whenever this scenario occurs, a program is said to be executing beyond the data flow limit, which is a rate computed by dividing the number of instructions in the data flow graph by the height of the graph. Since the height is reduced by these techniques, the rate of execution increases beyond the data flow limit.

The nonspeculative techniques range from memoization, which is a programming technique that stores and reuses the results of side-effect free computations; to instruction reuse, which implements memoization at the instruction level by reusing previously executed instructions whenever their operands match the current instance; to block, trace, and data flow region reuse, which extend instruction reuse to larger groups of instructions based on control or data flow relationships. Such techniques share the characteristic that they are only invoked when known to be safe for correctness; safety is determined by applying a *reuse test* that guarantees correctness. In contrast, the remaining value locality-based technique that we examined—value prediction—is speculative in nature, and removes computation from the data dependence graph whenever it can correctly predict the outcome of the computation. Value prediction introduces additional microarchitectural complexity, since speculative execution, misprediction detection, and recovery mechanisms must all be provided.

None of these techniques has yet been implemented in a real processor design. While published studies indicate that dramatic performance improvement is possible, it appears that industry practitioners have found that incremental implementations of these techniques that augment existing designs do not provide enough performance improvement to merit the additional cost and complexity. Only time will tell if future microarchitectures, perhaps more amenable to adaptation of these techniques, will actually do so and reap some of the benefits described in the literature.

REFERENCES

Burtscher, M., and B. Zorn: "Prediction outcome history-based confidence estimation for load value prediction," *Journal of Instruction Level Parallelism,* 1, 1999.

Calder, B., P. Feller, and A. Eustace: "Value profiling," *Proc. 30th Annual ACM/IEEE Int. Symposium on Microarchitecture,* 1997, pp. 259–269.

Calder, B., G. Reinman, and D. Tullsen: "Selective value prediction," *Proc. 26th Annual Int. Symposium on Computer Architecture (ISCA'99),* 27, 2 of *Computer Architecture News,* 1999, pp. 64–74, New York, ACM Press.

Connors, D. A., and W. mei W. Hwu: "Compiler-directed dynamic computation reuse: Rationale and initial results," *Int. Symposium on Microarchitecture,* 1999, pp. 158–169.

Fields, B., S. Rubin, and R. Bodik: "Focusing processor policies via critical-path prediction," *Proc. 28th Int. Symposium on Computer Architecture,* 2001, pp. 74–85.

Gabbay, F., and A. Mendelson: "Can program profiling support value prediction," *Proc. 30th Annual ACM/IEEE Int. Symposium on Microarchitecture,* 1997, pp. 270–280.

Gabbay, F., and A. Mendelson: "The effect of instruction fetch bandwidth on value prediction," *Proc. 25th Annual Int. Symposium on Computer Architecture,* Barcelona, Spain, 1998a, pp. 272–281.

Gabbay, F., and A. Mendelson: "Using value prediction to increase the power of speculative execution hardware," *ACM Trans. on Computer Systems,* 16, 3, 1998b, pp. 234–270.

Gonzalez, A., J. Tubella, and C. Molina: "Trace-level reuse," *Proc. Int. Conference on Parallel Processing,* 1999, pp. 30–37.

Huang, J., and D. J. Lilja: "Exploiting basic block value locality with block reuse," *HPCA,* 1999, pp. 106–114.

Lee, S.-J., and P.-C. Yew: "On table bandwidth and its update delay for value prediction on wide-issue ILP processors," *IEEE Trans. on Computers,* 50, 8, 2001, pp. 847–852.

Lipasti, M. H.: "Value Locality and Speculative Execution," PhD thesis, Carnegie Mellon University, 1997.

Lipasti, M. H., and J. P. Shen: "Exceeding the dataflow limit via value prediction," *Proc. 29th Annual ACM/IEEE Int. Symposium on Microarchitecture,* 1996, pp. 226–237.

Lipasti, M. H., and J. P. Shen: "Superspeculative microarchitecture for beyond AD 2000," *Computer,* 30, 9, 1997, pp. 59–66.

Lipasti, M. H., C. B. Wilkerson, and J. P. Shen: "Value locality and load value prediction," *Proc. Seventh Int. Conference on Architectural Support for Programming Languages and Operating Systems (ASPLOS-VII),* 1996, pp. 138–147.

Martin, M. M. K., D. J. Sorin, H. W. Cain, M. D. Hill, and M. H. Lipasti: "Correctly implementing value prediction in microprocessors that support multithreading or multiprocessing," *Proc. MICRO-34,* 2001, pp. 328–337.

Mendelson, A., and F. Gabbay: "Speculative execution based on value prediction," Technical report, Technion, 1997.

Sazeides, Y.: "An Analysis of Value Predictability and its Application to a Superscalar Processor," PhD Thesis, University of Wisconsin, Madison, WI, 1999.

Sazeides, Y., and J. E. Smith: "The predictability of data values," *Proc. 30th Annual ACM/IEEE Int. Symposium on Microarchitecture,* 1997, pp. 248–258.

Sodani, A.: "Dynamic Instruction Reuse," PhD thesis, University of Wisconsin, 2000.

Sodani, A., and G. S. Sohi: "Dynamic instruction reuse," *Proc. 24th Annual Int. Symposium on Computer Architecture,* 1997, pp. 194–205.

Sodani, A., and G. S. Sohi: "Understanding the differences between value prediction and instruction reuse," *Proc. 31st Annual ACM/IEEE Int. Symposium on Microarchitecture (MICRO-31),* 1998, pp. 205–215, Los Alamitos, IEEE Computer Society.

Tjaden, G. S., and M. J. Flynn: "Detection and parallel execution of independent instructions," *IEEE Trans. on Computers,* C19, 10, 1970, pp. 889–895.

Tomasulo, R.: "An efficient algorithm for exploiting multiple arithmetic units," *IBM Journal of Research and Development,* 11, 1967, pp. 25–33.

Tullsen, D., and J. Seng: "Storageless value prediction using prior register values," *Proc. 26th Annual Int. Symposium on Computer Architecture (ISCA'99),* vol. 27, 2 of *Computer Architecture News,* 1999, pp. 270–281, New York, ACM Press.

Wang, K., and M. Franklin: "Highly accurate data value prediction using hybrid predictors," *Proc. 30th Annual ACM/IEEE Int. Symposium on Microarchitecture,* 1999, pp. 281–290.

HOMEWORK PROBLEMS

P10.1 Figure 10.1 suggests it is possible to improve IPC from 1 to 4 by employing techniques such as instruction reuse or value prediction that collapse true data dependences. However, publications describing these techniques show speedups ranging from a few percent to a few tens of percent. Identify and describe one program characteristic that inhibits such speedups.

P10.2 As in Problem 10.1, identify and describe at least one implementation constraint that prevents best-case speedups from occurring.

P10.3 Assume you are implementing instruction reuse for integer instructions in the PowerPC 620. Assume you want to perform the reuse test based on value in the dispatch stage. Describe how many additional read and write ports you will need for the integer architected register file (ARF) and rename buffers.

P10.4 As in Problem 10.3, assume you are implementing instruction reuse in the PowerPC 620, and you wish to perform the reuse test by value in the dispatch stage. Show a design for the reuse buffer that integrates it into the 620 pipeline. How many read/write ports will this structure need?

P10.5 Assume you are building an instruction reuse mechanism that attempts to reuse load instructions by performing the reuse test by name in the PowerPC 620 dispatch stage. Since the addresses of all prior in-flight stores may not be known at this time, you have several design choices: (1) either disallow load reuse if stores with unknown addresses are still in flight, (2) delay dispatch of reused loads until such prior stores have computed their addresses, or (3) go ahead and allow such loads to be reused, relying on some other mechanism to guarantee correctness. Discuss these three alternatives from a performance perspective.

P10.6 Given the assumptions in Problem 10.5, describe what existing microarchitectural feature in the PowerPC 620 could be used to guarantee correctness for the third case. If you choose the third option, is your instruction reuse scheme still nonspeculative?

P10.7 Given the scenario described in Problem 10.5, comment on the likely effectiveness of load instruction reuse in a 5-stage pipeline like the PowerPC 620 versus a 20-stage pipeline like the Intel Pentium 4. Which of the three options outlined is likely to work best in a future deeply pipelined processor? Why?

P10.8 Construct a sequence of load value outcomes where a last-value predictor will perform better than a FCM predictor or a stride predictor. Compute the prediction rate for each type of predictor for your sequence.

P10.9 Construct a sequence of load value outcomes where an FCM predictor will perform better than a last-value predictor or a stride predictor. Compute the prediction rate for each type of predictor for your sequence.

P10.10 Construct a sequence of load value outcomes where a stride predictor will perform better than an FCM predictor or a last-value predictor. Compute the prediction rate for each type of predictor for your sequence.

P10.11 Consider the interaction between value predictors and branch predictors. Given a stride value predictor and a two-level GAg branch predictor with a 10-bit branch history register, write a C-code program snippet for which the stride value predictor can correct a branch that the branch predictor mispredicts.

P10.12 Consider further the interaction between value predictors and branch predictors. Given a last-value predictor and a two-level GAg branch predictor with a 10-bit branch history register, write a C-code program snippet for which the last-value predictor incorrectly resolves a branch that the branch predictor predicts correctly.

P10.13 Given that a value predictor can incorrectly redirect correctly predicted branches, suggest and discuss at least two microarchitectural alternatives for dealing with this problem.

P10.14 Assume you are implementing value prediction for integer instructions in the PowerPC 620. Describe how many additional read and write ports you will need for the integer architected register file (ARF) and rename buffers.

P10.15 As in Problem 10.14, assume you are implementing value prediction in the PowerPC 620. You have concluded that you need selective reissue via global broadcast as a recovery mechanism. In such a mechanism, each in-flight instruction must know precisely which earlier instructions it depends on, either directly or indirectly through multiple levels in the data flow graph. For the PowerPC 620, design a RAM/CAM hardware structure that tracks this information and enables direct selective reissue when a misprediction is detected. How many write ports does this structure need?

P10.16 For the hardware structure in Problem 10.15, determine the size of the hardware structure (number of bit cells it needs to store). Describe how this size would vary in a more aggressive microarchitecture like the Intel P6, which allows up to 40 instructions to be in flight at one time.

P10.17 Based on the data in Figure 10.10, provide and justify one possible explanation for why the *gawk* benchmark does not achieve higher speedups with more aggressive value prediction schemes.

P10.18 Based on the data in Figure 10.10, provide and justify one possible explanation for why the *swm256* benchmark achieves dramatically higher speedup with the perfect value prediction scheme.

P10.19 Based on the data in Figures 10.3 and 10.10, explain the apparent contradiction for the benchmark *sc*: even though roughly 60% of its register writes are predictable, no speedup is obtained from implementing value prediction. Discuss at least two reasons why this might be the case.

P10.20 Given your answer to Problem 10.19, propose a set of experiments that you could conduct to validate your hypotheses.

P10.21 Given the deadlock scenario described in Section 10.4.4.5, describe a possible solution that prevents deadlock without requiring all speculatively issued instructions to retain their reservation stations. Compare your proposed solution to the alternative solution that forces instructions to retain their reservation stations until they are deemed nonspeculative.

CHAPTER 11

Executing Multiple Threads

CHAPTER OUTLINE

11.1 Introduction
11.2 Synchronizing Shared-Memory Threads
11.3 Introduction to Multiprocessor Systems
11.4 Explicitly Multithreaded Processors
11.5 Implicitly Multithreaded Processors
11.6 Executing the Same Thread
11.7 Summary

References
Homework Problems

11.1 Introduction

Thus far in our exploration of high-performance processors, we have focused exclusively on techniques that accelerate the processing of a single thread of execution. That is to say, we have concentrated on compressing the latency of execution, from beginning to end, of a single serial program. As first discussed in Chapter 1, there are three fundamental interrelated terms that affect this latency: processor cycle time, available instruction-level parallelism, and the number of instructions per program. Reduced cycle time can be brought about by a combination of circuit design techniques, improvements in circuit technology, and architectural tradeoffs. Available instruction-level parallelism can be affected by advances in compilation technology, reductions in structural hazards, and aggressive microarchitectural techniques such as branch or value prediction that mitigate the negative effects of control and data dependences. Finally, the number of instructions per program is determined by algorithmic advances, improvements in compilation technology, and the fundamental characteristics of the instruction set being executed. All these factors assume a single thread of execution, where the processor traverses the static control flow graph of the program in a serial fashion from beginning to end, aggressively resolving control and data dependences but always maintaining the illusion of sequential execution.

In this chapter, we broaden our scope to consider an alternative source of performance that is widely exploited in real systems. This source, called *thread-level parallelism*, is primarily used to improve the throughput or instruction processing bandwidth of a processor or collection of processors. Exploitation of thread-level parallelism has its roots in the early time-sharing mainframe computer systems. These early systems coupled relatively fast CPUs with relatively slow input/output (I/O) devices (the slowest I/O device of all being the human programmer or operator sitting at a terminal). Since CPUs were very expensive, while slow I/O devices such as terminals were relatively inexpensive, operating system developers invented the concept of *time-sharing*, which allowed multiple I/O devices to connect to and share, in a time-sliced fashion, a single CPU resource. This allowed the expensive CPU to switch contexts to an alternative user thread whenever the current thread encountered a long-latency I/O event (e.g., reading from a disk or waiting for a terminal user to enter keystrokes). Hence, the most expensive resource in the system—the CPU—was kept busy as long as there were other users or threads waiting to execute instructions. The time-slicing policies—which also included time quanta that enforced fair access to the CPU—were implemented in the operating system using software, and hence introduced additional execution-time overhead for switching contexts. Hence, the latency of a single thread of execution (or the latency perceived by a single user) would actually increase, since it would now include context-switch and operating system policy management overhead. However, the overall instruction throughput of the processor would increase due to the fact that instructions were executed from alternative threads when an otherwise idle CPU would be waiting for a long-latency I/O event to complete.

From a microarchitectural standpoint, these types of time-sharing workloads provide an interesting challenge to a processor designer. Since they interleave the execution of multiple independent threads, they can wreak havoc on caches and other structures that rely on the spatial and temporal locality exhibited by the reference stream of a single thread. Furthermore, interthread conflicts in branch and value predictors can significantly increase the pressure on such structures and reduce their efficacy, particularly when these structures are not adequately sized. Finally, the large aggregate working set of large numbers of threads (there can be tens of thousands to hundreds of thousands of active threads in a modern, high-end time-shared system) can easily overwhelm the capacity and bandwidth provided by conventional memory subsystems, leading to designs with very large secondary and tertiary caches and extremely high memory bandwidth. These effects are illustrated in Figure 3.31.

Time-shared workloads that share data between concurrently active processes must serialize access to those shared data in a well-defined and repeatable manner. Otherwise, the workloads will generate nondeterministic or even erroneous results. We will consider some simple and widely used schemes for serialization or *synchronization* in Section 11.2; all these schemes rely on hardware support for *atomic* operations. An operation is considered *atomic* if all its suboperations are performed as an indivisible unit; that is to say, they are either all performed without interference by other operations or processes, or none of them are performed. Modern processors support primitives that can be used to implement various atomic operations that enable multiple processes or threads to synchronize correctly.

From the standpoint of system architecture, time-shared workloads create an additional opportunity for building systems that provide scalable throughput. Namely, the availability of large numbers of active and independent threads of execution motivates the construction of systems with multiple processors in them, since the operating system can distribute these ready threads to multiple processors quite easily. Building a *multiprocessor system* requires the designer to resolve a number of tradeoffs related primarily to the memory subsystem and how it provides each processor with a coherent and consistent view of memory. We will discuss some of these issues in Section 11.2 and briefly describe key attributes of the *coherence interface* that a modern processor must supply in order to support such a view of memory.

In addition to systems that simultaneously execute multiple threads of control on physically separate processors, processors that provide efficient, fine-grained support for interleaving multiple threads on a single physical processor have also been proposed and built. Such *multithreaded* processors come in various flavors, ranging from fine-grained multithreading, which switches between multiple thread contexts every cycle or every few cycles; to coarse-grained multithreading, which switches contexts only on long-latency events such as cache misses; to simultaneous multithreading, which does away with context switching by allowing individual instructions from multiple threads to be intermingled and processed simultaneously within an out-of-order processor's execution window. We discuss some of the tradeoffs and implementation challenges for proposed and real multithreaded processors in Section 11.4.

The availability of systems with multiple processors has also spawned a large body of research into parallel algorithms that use multiple collaborating threads to arrive at an answer more quickly than with a single serial thread. Many important problems, particularly ones that apply regular computations to massive data sets, are quite amenable to parallel implementations. However, the holy grail of such research—*automated parallelization of serial programs*—has yet to materialize. While automated parallelization of certain classes of algorithms has been demonstrated, such success has largely been limited to scientific and numeric applications with predictable control flow (e.g., nested loop structures with statically determined iteration counts) and statically analyzable memory access patterns (e.g., sequential walks over large multidimensional arrays of floating-point data). For such applications, a parallelizing compiler can decompose the total amount of computation into multiple independent threads by distributing partitions of the data set or the total set of loop iterations across multiple threads. Naturally, the partitioning algorithm must take care to avoid violating data dependences across parallel threads and may need to incorporate synchronization primitives across the threads to guarantee correctness in such cases. Successful automatic parallelization of scientific and numeric applications has been demonstrated over the years and is in fact in commercial use for many applications in this domain.

However, there are many difficulties in extracting thread-level parallelism from typical non-numeric serial applications by automatically parallelizing them at compile time. Namely, applications with irregular control flow, ones that tend to access data in unpredictable patterns, or ones that are replete with accesses to pointer-based data structures make it very difficult to statically determine memory data dependences between various portions of the original sequential program. Automatic parallelization of such codes is difficult because partitioning the serial algorithm into multiple parallel and independent threads becomes virtually impossible without exact compile-time knowledge of control flow and data dependence relationships.

Recently, several researchers have proposed shifting the process of automatic parallelization of serial algorithms from compile time to run time, or at least providing efficient hardware support for solving some of the thorny problems associated with the efficient extraction of multiple threads of execution. These *implicit multithreading* proposals range from approaches such as *dynamic multithreading* [Akkary and Driscoll, 1998], which advocates a pure hardware approach that automatically identifies and spawns speculative implicit threads of execution, to the *multiscalar* [Sohi et al., 1995] paradigm which uses a combination of hardware support and aggressive compilation to achieve the same purpose, to *thread-level speculation* [Steffan et al., 1997; 2000; Steffan and Mowry, 1998; Hammond et al., 1998; Krishnan and Torrellas, 2001], which relies on the compiler to create parallel threads but provides simple hardware support for detecting data dependence violations between threads. We will discuss some of these proposals for implicit multithreading in Section 11.5.

In another variation on this theme, researchers have proposed *preexecution*, which uses a second runahead thread to execute only critical portions of the main execution thread in order to prefetch data and instructions and to resolve difficult-to-predict conditional branches before the main thread encounters them. A similar approach has also been suggested for fault detection and fault-tolerant execution. We will discuss some of these proposals and their associated implementation challenges in Section 11.6.

11.2 Synchronizing Shared-Memory Threads

Time-shared workloads that share data between concurrently active processes must serialize access to those shared data in a well-defined and repeatable manner. Otherwise, the workloads will have nondeterministic or even erroneous results. Figure 11.1 illustrates four possible interleavings for the loads and stores performed against a shared variable A by two threads. Any of these four interleavings is possible on a time-shared system that is alternating execution of the two threads. Assuming an initial value of A = 0, depending on the interleaving, the final value of A can be either 3 [Figure 11.1(a)], 4 [Figure 11.1(b) and (c)], or 1 [Figure 11.1(d)]. Of course, a well-written program should have a predictable and repeatable outcome, instead of one determined only by the operating system's task dispatching policies.

EXECUTING MULTIPLE THREADS

```
Thread 0        Thread 1            Thread 0        Thread 1
                load r1, A          load r1, A
                addi r1, r1, 3      addi r1, r1, 1
                                    store r1, A
load r1, A                                          load r1, A
addi r1, r1, 1                                      addi r1, r1, 3
store r1, A                                         store r1, A
                store r1, A
        (a)                                 (b)

Thread 0        Thread 1            Thread 0        Thread 1
                load r1, A          load r1, A
                addi r1, r1, 3      addi r1, r1, 1
                store r1, A
load r1, A                                          load r1, A
addi r1, r1, 1                                      addi r1, rl, 3
store r1, A                                         store r1, A
                                    store r1, A
        (c)                                 (d)
```

This figure shows four possible interleavings of the references made by two threads to a shared variable A, resulting in 3 different final values for A.

Figure 11.1
The Need for Synchronization.

Table 11.1
Some common synchronization primitives

Primitive	Semantic	Comments
Fetch-and-add	Atomic load → add → store operation	Permits atomic increment; can be used to synthesize locks for mutual exclusion
Compare-and-swap	Atomic load → compare → conditional store	Stores only if load returns an expected value
Load-linked/store-conditional	Atomic load → conditional store	Stores only if load/store pair is atomic; that is, if there is no intervening store

This simple example motivates the need for well-defined synchronization between shared-memory threads.

Modern processors supply primitives that can be used to implement various atomic operations that enable multiple processes or threads to synchronize correctly. These primitives guarantee hardware support for *atomic* operations. An operation is considered *atomic* if all its suboperations are performed as an indivisible unit; that is to say, they are either all performed without interference by other operations or processes, or none of them are performed. Table 11.1 summarizes three commonly implemented primitives that can be used to synchronize shared-memory threads.

Thread 0	Thread 1	Thread 0	Thread 1	Thread 0	Thread 1
fetchadd A, 1	fetchadd A, 3	spin:	spin:	spin:	spin:
		cmpswp AL, 1	cmpswp AL, 1	ll r1, A	ll r1, A
		bfail spin	bfail spin	addi r1, r1, 1	addi r1, r1, 3
		load r1, A	load r1, A	stc r1, A	stc r1, A
		addi r1, r1, 1	addi r1, r1, 3	bfail spin	bfail spin
		store r1, A	store r1, A		
		store 0, AL	store 0, AL		
(a)		(b)		(c)	

Figure 11.2

Synchronization with (a) Fetch-and-Add, (b) Compare-and-Swap, and (c) Load-Linked/Store-Conditional.

The first primitive in Table 11.1, *fetch-and-add,* simply loads a value from a memory location, adds an operand to it, and stores the result back to the memory location. The hardware guarantees that this sequence occurs atomically; in effect, the processor must continue to retry the sequence until it succeeds in storing the sum before any other thread has overwritten the fetched value at the shared location. As shown in Figure 11.2(a), the code snippets in Figure 11.1 could be rewritten as "fetchadd A, 1" and "fetchadd A, 3" for the threads 0 and 1, respectively, resulting in a deterministic, repeatable shared-memory program. In this case, the only allowable outcome would be A = 4.

The second primitive, *compare-and-swap,* simply loads a value, compares it to a supplied operand, and stores the operand to the memory location if the loaded value matches the operand. This primitive allows the programmer to atomically swap a register value with the value at a memory location whenever the memory location contains the expected value. If the compare fails, a condition flag is set to reflect this failure. This primitive can be used to implement mutual exclusion for critical sections protected by locks. Critical sections are simply arbitrary sequences of instructions that are executed atomically by guaranteeing that no other thread can enter such a section until the thread currently executing a critical section has completed the entire section. For example, the updates in the snippets in Figure 11.1 could be made atomic by performing them within a critical section and protecting that critical section with an additional lock variable. This is illustrated in Figure 11.2(b), where the cmpswp instruction checks the AL lock variable. If it is set to 1, the cmpswp fails, and the thread repeats the cmpswp instruction until it succeeds, by branching back to it repeatedly (this is known as *spinning on a lock*). Once the cmpswp succeeds, the thread enters its critical section and performs its load, add, and store atomically (since mutual exclusion guarantees that no other processor is concurrently executing a critical section protected by the same lock). Finally, the thread stores a 0 to the lock variable AL to indicate that it is done with its critical section.

The third primitive, *load-linked/store-conditional* (ll/stc), simply loads a value, performs other arbitrary operations, and then attempts to store back to the same address it loaded from. Any intervening store by another thread will cause the store conditional to fail. However, if no other store to that address occurred, the load/store pair can execute atomically and the store succeeds. Figure 11.2(c) illustrates how the shared memory snippets can be rewritten to use `ll/stc` pairs. In this example, the `ll` instruction loads the current value from A, then adds to it, and then attempts to store the sum back with the `stc` instruction. If the stc fails, the thread spins back to the `ll` instruction until the pair eventually succeeds, guaranteeing an atomic update.

Any of the three examples in Figure 11.2 guarantee the same final result: memory location A will always be equal to 4, regardless of when the two threads execute or how their memory references are interleaved. This property is guaranteed by the atomicity property of the primitives being employed.

From an implementation standpoint, the `ll/stc` pair is the most attractive of these three. Since it closely matches the load and store instructions that are already supported, it fits nicely into the pipelined and superscalar implementations detailed in earlier chapters. The other two, fetch-and-add and compare-and-swap, do not, since they require two memory references that must be performed indivisibly. Hence, they require substantially specialized handling in the processor pipeline.

Modern instruction sets such as MIPS, PowerPC, Alpha, and IA-64 provide ll/stc primitives for synchronization. These are fairly easy to implement; the only additional semantic that has to be supported is that each `ll` instruction must, as a side effect, remember the address it loaded from. All subsequent stores (including stores performed by remote processors in a multiprocessor system) must check their addresses against this linked address and must clear it if there is a match. Finally, when the `stc` executes, it must check its address against the linked address. If it matches, the `stc` is allowed to proceed; if not, the `stc` must fail and set a condition code that reflects that failure. These changes are fairly incremental above and beyond the support that is already in place for standard loads and stores. Hence, `ll/stc` is easy to implement and is still powerful enough to synthesize both fetch-and-add and compare-and-swap as well as many other atomic primitives.

In summary, proper synchronization is necessary for correct, repeatable execution of shared-memory programs with multiple threads of execution. This is true not only for such programs running on a time-shared uniprocessor, but also for programs running on multiprocessor systems or multithreaded processors.

11.3 Introduction to Multiprocessor Systems

Building multiprocessor systems is an attractive proposition for system vendors for a number of reasons. First of all, they provide a natural, incremental upgrade path for customers with growing computational demands. As long as the key user applications provide thread-level parallelism, adding processors to a system or replacing a smaller system with a larger one that contains more processors provides the customer with a straightforward and efficient way to add computing capacity. Second, multiprocessor systems allow the system vendor to amortize the cost of a single microprocessor design across a wide variety of system design points that provide varying levels of performance and scalability. Finally, multiprocessors that provide coherent shared memory provide a programming model that is compatible with time-shared uniprocessors, making it easy for customers to deploy existing applications and develop new ones. In these systems, the hardware and operating system software collaborate to provide the user and programmer with the appearance of four *multiprocessor idealisms:*

- *Fully shared memory* means that all processors in the system have equivalent access to all the physical memory in the system.
- *Unit latency* means that all requests to memory are satisfied in a single cycle.
- *Lack of contention* means that the forward progress of one processor's memory references is never slowed down or affected by memory references from another processor.
- *Instantaneous propagation of writes* means that any changes to the memory image made by one processor's write are immediately visible to all other processors in the system.

Naturally, the system and processor designers must strive to approximate these idealisms as closely as possible so as to satisfy the performance and correctness expectations of the user. Obviously, factors such as cost and scalability can play a large role in how easy it is to reach these goals, but a well-designed system can in fact maintain the illusion of these idealisms quite successfully.

11.3.1 Fully Shared Memory, Unit Latency, and Lack of Contention

As shown in Figure 11.3, most conventional shared-memory multiprocessors that provide uniform memory access (UMA) are usually built using a *dancehall* organization, where a set of memory modules or banks is connected to the set of processors

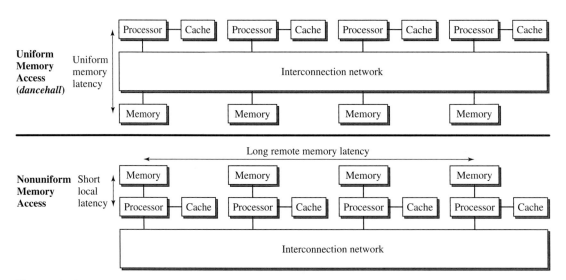

Figure 11.3
UMA versus NUMA Multiprocessor Architecture.

via a crossbar interconnect, and each processor incurs the same uniform latency in accessing a memory bank through this crossbar. The downsides of this approach are the cost of the crossbar, which increases as the square of the number of processors and memory banks, and the fact that every memory reference must traverse this crossbar. As an alternative, many system vendors now build systems with nonuniform memory access (NUMA), where the processors are still connected to each other via a crossbar interconnect, but each processor has a local bank of memory with much lower access latency. In a NUMA configuration, only references to remote memory must pay the latency penalty of traversing the crossbar.

In both UMA and NUMA systems, just as in uniprocessor systems, the idealism of unit latency is approximated with the use of caches that are able to satisfy references to both local and remote (NUMA) memories. Similarly, the traffic filtering effect of caches is used to mitigate contention in the memory banks, as is the use of intelligent memory controllers that combine and reorder requests to minimize latency. Hence, caches, which we have already learned are indispensable in uniprocessor systems, are similarly very effective in multiprocessor systems as well. However, the presence of caches in a multiprocessor system creates additional difficulties when dealing with memory writes, since these must now be somehow made visible to or propagated to other processors in the system.

11.3.2 Instantaneous Propagation of Writes

In a time-shared uniprocessor system, if one thread updates a memory location by writing a new value to it, that thread as well as any other thread that eventually executes will instantaneously see the new value, since it will be stored in the cache hierarchy of the uniprocessor. Unfortunately, in a multiprocessor system, this property does not hold, since subsequent references to the same address may now originate from different processors. Since these processors have their own caches that may contain private copies of the same cache line, they may not see the effects of the other processor's write. For example, in Figure 11.4(a), processor P1 writes a "1" to memory location A. With no coherence support, the copy of memory location A in P2's cache is not updated to reflect the new value, and a load at P2 would still observe the stale value of "0." This is known as the classic *cache coherence* problem, and to solve it, the system must provide a cache coherence protocol that ensures that all processors in the system gain visibility to all the other processors' writes, so that each processor has a coherent view of the contents of memory [Censier and Feautrier, 1978]. There are two fundamental approaches to cache coherence—update protocols and invalidate protocols—and these are discussed briefly in Section 11.3.3. These are illustrated in Figure 11.4(b) and (c).

11.3.3 Coherent Shared Memory

A coherent view of memory is a hard requirement for shared-memory multiprocessors. Without it, programs that share memory would behave in unpredictable ways, since the value returned by a read would vary depending on which processor performed the read. As already stated, the coherence problem is caused by the fact that writes are not automatically and instantaneously propagated to other processors'

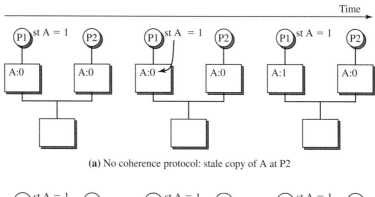

(a) No coherence protocol: stale copy of A at P2

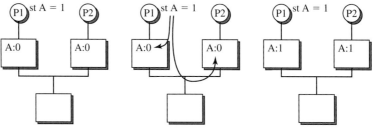

(b) Update protocol writes through to both copies of A

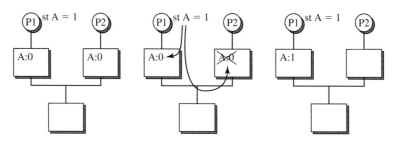

(c) Invalidate protocol eliminates stale remote copy

An update protocol updates all remote copies, while an invalidate protocol removes remote copies.

Figure 11.4
Update and Invalidate Protocols.

caches. To ensure that writes are made visible to other processors, two classes of coherence protocols exist.

11.3.3.1 Update Protocols. The earliest proposed multiprocessors employed a straightforward approach to maintaining cache coherence. In these systems, the processors' caches used a write-through policy, in which all writes were performed not just against the cache of the processor performing the write, but also against main memory. Such a protocol is illustrated in Figure 11.4(b). Since all processors were connected to the same electrically shared bus that also connected them to main memory, all other processors were able to observe the write-throughs as they occurred and were able to directly *update* their own copies of the data (if they had any such copies) by snooping the new values from the shared bus. In effect, these *update protocols* were based on a *broadcast write-through policy;* that is, every write by every processor was written through, not just to main memory, but also to any copy that existed in any other processor's cache. Obviously, such a protocol is not scalable beyond a small number of processors, since the write-through traffic from multiple processors will quickly overwhelm the bandwidth available on the memory bus.

A straightforward optimization allowed the use of writeback caching for private data, where writes are performed locally against the processor's cache and the changes are written back to main memory only when the cache line is evicted from the processor's cache. In such a protocol, however, any writes to shared cache lines (i.e., lines that were present in any other processor's cache) still had to be broadcast on the bus so the sharing processors could update their copies. Furthermore, a remote read to a line that was now dirty in the local cache required the dirty line to be flushed back to memory before the remote read could be satisfied.

Unfortunately, the excessive bandwidth demands of update protocols have led to their virtual extinction, as there are no modern multiprocessor systems that use an update protocol to maintain cache coherence.

11.3.3.2 Invalidate Protocols. Today's modern shared-memory multiprocessors all use invalidate protocols to maintain coherence. The fundamental premise of an invalidate protocol is simple: only a single processor is allowed to write a cache line at any point in time (such protocols are also often called *single-writer* protocols). This policy is enforced by ensuring that a processor that wishes to write to a cache line must first establish that its copy of the cache line is the only valid copy in the system. Any other copies must be invalidated from other processors' caches (hence the term *invalidate protocol*). This protocol is illustrated in Figure 11.4(c). In short, before a processor performs its write, it checks to see if there are any other copies of the line elsewhere in the system. If there are, it sends out messages to invalidate them; finally, it performs the write against its private and exclusive copy. Subsequent writes to the same line are streamlined, since no check for outstanding remote copies is required. Once again, as in uniprocessor writeback caches, the updated line is not written back to memory until it is evicted from the processor's cache. However, the coherence protocol must keep track of the fact that a modified copy of the line exists and must prevent other processors from attempting to read the stale version from memory. Furthermore, it must support flushing the modified data from the processor's cache so that a remote reference can be satisfied by the only up-to-date copy of the line.

Minimally, an invalidate protocol requires the cache directory to maintain at least two states for each cached line: modified (M) and invalid (I). In the invalid state, the requested address is not present and must be fetched from memory. In the modified state, the processor knows that there are no other copies in the system (i.e., the local copy is the exclusive one), and hence the processor is able to perform reads and writes against the line. Note that any line that is evicted in the modified state must be written back to main memory, since the processor may have performed a write against it. A simple optimization incorporates a dirty bit in the cache line's state, which allows the processor to differentiate between lines that are exclusive to that processor (usually called the E state) and ones that are exclusive and have been dirtied by a write (usually called the M state). The IBM/Motorola PowerPC G3 processors used in Apple's Macintosh desktop systems implement an MEI coherence protocol.

Note that with these three states (MEI), no cache line is allowed to exist in more than one processor's cache at the same time. To solve this problem, and to allow readable copies of the same line in multiple processors' caches, most invalidate protocols also include a shared state (S). This state indicates that one or more remote readable copies of a line may exist. If a processor wishes to perform a write against a line in the S state, it must first *upgrade* that line to the M state by invalidating the remote copies.

Figure 11.5 shows the state table and transition diagram for a straightforward MESI coherence protocol. Each row corresponds to one of the four states (M, E, S, or I), and each column summarizes the actions the coherence controller must perform in response to each type of bus event. Each transition in the state of a cache line is caused either by a local reference (read or write), a remote reference (bus read, bus write, or bus upgrade), or a local capacity-induced eviction. The cache directory or tag array maintains the MESI state of each line that is in that cache. Note that this allows each cache line to be in a different state at any point in time, enabling lines that contain strictly private data to stay in the E or M state, while lines that contain shared data can simultaneously exist in multiple caches in the S state. The MESI coherence protocol supports the single-writer principle to guarantee coherence but also allows efficient sharing of read-only data as well as silent upgrades from the exclusive (E) state to the modified (M) state on local writes (i.e., no bus upgrade message is required).

A common enhancement to the MESI protocol is achieved by adding an O, or owned state to the protocol, resulting in an MOESI protocol. The O state is entered following a remote read to a dirty block in the M state. The O state signifies that multiple valid copies of the block exist, since the remote requestor has received a valid copy to satisfy the read, while the local processor has also kept a copy. However, it differs from the conventional S state by avoiding the writeback to memory, hence leaving a stale copy in memory. This state is also known as *shared-dirty,* since the block is shared, but is still dirty with respect to memory. An owned block that is evicted from a cache must be written back, just like a dirty block in the M state, since the copy in main memory must be made up-to-date. A system that implements the O state can place either the requesting processor or the processor that supplies the dirty data in the O state, while placing the other copy in the S state, since only a single copy needs to be marked dirty.

Current State s	Event and Local Coherence Controller Responses and Actions (s' refers to next state)					
	Local Read (LR)	Local Write (LW)	Local Eviction (EV)	Bus Read (BR)	Bus Write (BW)	Bus Upgrade (BU)
Invalid (I)	Issue bus read if no sharers then s' = E else s' = S	Issue bus write s' = M	s' = I	Do nothing	Do nothing	Do nothing
Shared (S)	Do nothing	Issue bus upgrade s' = M	s' = I	Respond shared	s' = I	s' = I
Exclusive (E)	Do nothing	s' = M	s' = I	Respond shared s' = S	s' = I	Error
Modified (M)	Do nothing	Do nothing	Write data back; s' = I	Respond dirty; Write data back; s' = S	Respond dirty; Write data back; s' = I	Error

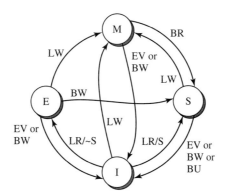

In response to local and bus events the coherence controller may need to change the local coherence state of a line, and may also need to fetch or supply the cache line data.

Figure 11.5
Sample MESI Cache Coherence Protocol.

11.3.4 Implementing Cache Coherence

Maintaining cache coherence requires a mechanism that tracks the state (e.g., MESI) of each active cache line in the system, so that references to those lines can be handled appropriately. The most convenient place to store the coherence state is in the cache tag array, since state information must be maintained for each line in the cache anyway. However, the local coherence state of a cache line needs to be available to other processors in the system so that their references to the line can be correctly satisfied as well. Hence, a cache coherence implementation must provide a means for distributed access to the coherence state of the lines in its cache. There are two overall approaches for doing so: *snooping* implementations and *directory* implementations.

11.3.4.1 Snooping Implementations.
The most straightforward approach for implementing coherence and consistency is via *snooping*. In a snooping implementation, all off-chip address events evoked by the coherence protocol (e.g., cache misses and invalidates in an invalidate protocol) are made visible to all other processors in the system via a shared address bus. In small-scale systems, the address bus is electrically shared and each processor sees all the other processors' commands as they are placed on the bus. More advanced point-to-point interconnect schemes that avoid slow multidrop busses can also support snooping by reflecting all commands to all processors via a hierarchical snoop interconnect. For notational convenience, we will simply refer to any such scheme as an address bus.

In a snooping implementation, the coherence protocol specifies if and how a processor must react to the commands that it observes on the address bus. For example, a remote processor's read to a cache line that is currently modified in the local cache must cause the cache controller to flush the line out of the local cache and transmit it either directly to the requester and/or back to main memory, so the requester will receive the latest copy. Similarly, a remote processor's invalidate request to a cache line that is currently shared in the local cache must cause the controller to update its directory entry to mark the line invalid. This will prevent all future local reads from consuming the stale data now in the cache.

The main shortcoming of snooping implementations of cache coherence is scalability to systems with many processors. If we assume that each processor in the system generates address bus transactions at some rate, we see that the frequency of inbound address bus transactions that must be snooped is directly proportional to the number of processors in the system.

$$\text{Outbound snoop rate} = s_o = (\text{cache miss rate}) + (\text{bus upgrade rate}) \quad (11.1)$$

$$\text{Inbound snoop rate} = s_i = n \times s_o \quad (11.2)$$

That is to say, if each processor generates s_o address transactions per second (consisting of read requests from cache misses and upgrade requests for stores to shared lines), and there are n processors in the system, then each processor must also snoop ns_o transactions per second. Since each snoop minimally requires a local cache directory lookup to check to see if the processor needs to react to the snoop (refer to Figure 11.5 for typical reactions), the aggregate lookup bandwidth required for large n can quickly become prohibitive. Similarly, the available link bandwidth connecting the processor to the rest of the system can be easily overwhelmed by this traffic; in fact, many snoop-based multiprocessors are performance-limited by address-bus bandwidth. Snoop-based implementations have been shown to scale to several dozen processors (up to 64 in the case of the Sun Enterprise 10000 [Charlesworth, 1997]), but scaling up to and beyond that number requires an expensive investment in increased address bus bandwidth.

Large-scale snoop-based systems can also suffer dramatic increases in memory latency when compared to systems designed for fewer processors, since the memory latency will be determined by the latency of the coherence response, rather than the DRAM and data interconnect latency. In other words, for large n, it often takes longer to snoop and collect snoop responses from all the processors in the system than it does to fetch the data from DRAM, even in a NUMA configuration that has long remote memory latencies. Even if the data from memory are transmitted speculatively to the requester, they are not known to be valid until all processors in the system have responded that they do not have a more up-to-date dirty copy of the line. Hence, the snoop response latency often determines how quickly a cache miss can be resolved, rather than the latency to retrieve the cache line itself from any local or remote storage location.

11.3.4.2 Directory Implementation.
The most common solution to the scalability and memory latency problems of snooping implementations is to use *directories*. In a directory implementation, coherence is maintained by keeping a copy of a cache line's coherence state collocated with main memory. The coherence state, which is stored in a directory that resides next to main memory, indicates if the line is currently cached anywhere in the system, and also includes pointers to all cached copies in a *sharing list* or *sharing vector*. Sharing lists can be either precise (meaning each sharer is individually indicated in the list) or coarse (meaning that multiple processors share an entry in the list, and the entry indicates that one or more of those processors has a shared copy of the line) and can be stored as linked lists or fixed-size presence vectors. Precise sharing vectors have the drawback of significant storage overhead, particularly for systems with large numbers of processors, since each cache line-size block of main memory requires directory storage proportional to the number of processors. For a large system with 64-byte cache lines and 512 processors, this overhead can be 100% just for the sharing vector.

Bandwidth Scaling. The main benefit of a directory approach is that directory bandwidth scales with memory bandwidth: Adding a memory bank to supply more memory data bandwidth also adds directory bandwidth. Another benefit is that demand for address bandwidth is reduced by filtering commands at the directory. In a directory implementation, address commands are sent to the directory first and are forwarded to remote processors only when necessary (e.g., when the line is dirty in a remote cache or when writing to a line that is shared in a remote cache). Hence, the frequency of inbound address commands to each processor is no longer proportional to the number of processors in the system, but rather it is proportional to the degree of data sharing, since a processor receives an address command only if it owns or has a shared copy of the line in question. Hence, systems with dozens to hundreds of processors can and have been built.

Memory Latency. Finally, latency for misses that are satisfied from memory can be significantly reduced, since the memory bank can respond with nonspeculative data as soon as it has checked the directory. This is particularly advantageous in a NUMA configuration where the operating and run-time systems have been optimized to place private or nonshared data in a processor's local memory. Since the latency to local memory is usually very low in such a configuration, misses can be resolved in dozens of nanoseconds instead of hundreds of nanoseconds.

Communication Miss Latency. The main drawback of directory-based systems is the additional latency incurred for cache misses that are found dirty in a remote processor's cache (called *communication misses* or *dirty misses*). In a snoop-based system, a dirty miss is satisfied directly, since the read request is transmitted directly to the responder that has the dirty data. In a directory implementation, the request is first sent to the directory and then forwarded to the current owner of the line; this results in an additional traversal of the processor/memory interconnect and increases latency. Applications such as database transaction processing that share data intensively are very sensitive to dirty miss latency and can perform poorly on directory-based systems.

Hybrid snoopy/directory systems have also been proposed and built. For example, the Sequent NUMA-Q system uses conventional bus-based snooping to maintain coherence within four-processor quads, but extends cache coherence across multiple quads with a directory protocol built on the scalable coherent interface (SCI) standard [Lovett and Clapp, 1996]. Hybrid schemes can obtain many of the scalability benefits of directory schemes while still maintaining a low average latency for communication misses that can be satisfied within a local snoop domain.

11.3.5 Multilevel Caches, Inclusion, and Virtual Memory

Most modern processors implement multiple levels of cache to trade off capacity and miss rate against access latency and bandwidth: the level-1 or primary cache is relatively small but allows one- or two-cycle access, frequently through multiple banks or ports, while the level-2 or secondary cache provides much greater capacity but with multicycle access and usually just a single port. The design of multilevel cache hierarchies is an exercise in balancing implementation cost and complexity to achieve the lowest average memory latency for references that both hit and miss the caches. As shown in Equation (11.3), the average memory reference latency lat_{avg} can be computed as the weighted sum of the latencies to each of n levels of the cache hierarchy, where each latency lat_i is weighted by the fraction of references ref_i satisfied by that level:

$$lat_{avg} = \sum_{i=1}^{n} ref_i \times lat_i \quad (11.3)$$

Of course, such an average latency measure is less meaningful in the context of out-of-order processors, where miss latencies to the secondary cache can often by overlapped with other useful work, reducing the importance of high hit rates in the primary cache. Besides reducing average latency, the other primary objective of primary caches is to reduce the bandwidth required to the secondary cache. Since the majority of references will be satisfied by a reasonably sized primary cache, only a small subset need to be serviced by the secondary cache, enabling a much narrower and usually single-ported access path to such a cache.

Guaranteeing cache coherence in a design with multiple levels of cache is only incrementally more complex than in the base case of only a single level of cache; some benefit can be obtained by maintaining *inclusion* between levels of the cache by forcing each line that resides in a higher level of cache to also reside in a lower level.

Noninclusive Caches. A straightforward approach to multilevel cache coherence which does not require inclusion treats each cache in the hierarchy as a peer in the coherence scheme, implying that coherence is maintained independently for each level. In a snooping implementation, this implies that all levels of the cache hierarchy must snoop all the address commands traversing the system's address bus. This can lead to excessive bandwidth demands on the level-1 tag array, since both the processor core and the inbound address bus can generate a high rate of references to the tag array. The IBM Northstar/Pulsar design [Storino et al., 1998], which is noninclusive and employs snoop-based coherence, maintains two copies of the level-1 tag array to provide what is effectively dual-ported access to this structure. In a noninclusive directory implementation, the sharing vector must maintain separate entries for each level of each processor (if the sharing vector is precise), or it can revert to a coarse sharing scheme which implies that messages must be forwarded to all levels of cache of the processor that has a copy of the line.

Inclusive Caches. A common alternative to maintaining coherence independently for each level of cache is to guarantee that the coherence state of each line in an upper level of cache is consistent with the lower private levels by maintaining *inclusion*. For example, in a system with two levels of cache, the cache hierarchy must ensure that each line that resides in the level-1 cache also resides in (or is *included* in) the level-2 cache in a consistent state. Maintaining inclusion is fairly straightforward: Whenever a line enters the level-1 cache, it must also be placed in the level-2 cache. Similarly, whenever a line leaves the level-2 cache (is evicted due to a replacement or is invalidated), it must also leave the level-1 cache. If inclusion is maintained, only the lower level of the cache hierarchy needs to participate directly in the cache coherence scheme. By definition, any coherence operation that pertains to lines in the level-1 cache also pertains to the corresponding line in the level-2 cache, and the cache hierarchy, upon finding such a line in the level-2 cache, must now apply that operation to the level-1 cache as well. In effect, snoop lookups in the tag array of the level-2 cache serve as a filter to prevent coherence operations that are not relevant from requiring a lookup in the level-1 tag array. In snoop-based implementations with lots of address traffic, this can be a significant advantage, since the tag array references are now mostly partitioned into two disjoint groups: 90% or more of processor core references are satisfied by the level-1 tag array as cache hits, while 90% or more of the address bus commands are satisfied by the level-2 tag array as misses. Only the level-1 misses require a level-2 tag lookup, and only coherence hits to shared lines require accesses to the level-1 tag array. This approach avoids having to maintain multiple copies of or dual-porting the level-1 tag array.

Cache Coherence and Virtual Memory. Additional complexity is introduced by the fact that nearly all modern processors implement virtual memory to provide access protection and demand paging. With virtual memory, the effective or virtual address generated by a user program is translated to a physical address using a mapping that is maintained by the operating system. Usually, this address translation is performed prior to accessing the cache hierarchy, but, for cycle time and capacity reasons, some processors implement primary caches that are virtually indexed or tagged. The access time for a virtually addressed cache can be lower since the cache can be indexed in parallel with address translation. However, since cache coherence is typically handled using physical addresses and not virtual addresses, performing coherence-induced tag lookups in such a cache poses a challenge. Some mechanism for performing reverse address translation must exist; this can be accomplished with a separate reverse address translation table that keeps track of all referenced real addresses and their corresponding virtual addresses, or—in a multilevel hierarchy—with pointers in the level-2 tag array that point to corresponding level-1 entries. Alternatively, the coherence controller can search all the level-1 entries in the congruence class corresponding to a particular real address. In the case of a large set-associative virtually addressed cache, this alternative can be prohibitively expensive, since the congruence class can be quite large. The interested reader is referred to a classic paper by Wang et al. [1989] on this topic.

11.3.6 Memory Consistency

In addition to providing a coherent view of memory, a multiprocessor system must also provide support for a predefined *memory consistency model*. A consistency model specifies an agreed-upon convention for ordering the memory references of one processor with respect to the references of another processor and is an integral part of the instruction set architecture specification of any multiprocessor-capable system [Lamport, 1979]. Consistent ordering of memory references across processors is important for the correct operation of any multithreaded applications that share memory, since without an architected set of rules for ordering such references, such programs could not correctly and reliably synchronize between threads and behave in a repeatable, predictable manner. For example, Figure 11.6 shows a simple

```
Reorder      Proc0                    Proc1
load         st A=1                   st B=1
before       if (load B==0) {         if (load A==0) {
store            ...critical section      ...critical section
             }                        }
```

If either processor reorders the load and executes it before the store, both processors can enter the mutually exclusive critical section simultaneously.

Figure 11.6

Dekker's Algorithm for Mutual Exclusion.

serialization scheme that guarantees mutually exclusive access to a critical section, which may be updating a shared datum. Dekker's mutual exclusion scheme for two processors consists of processor 0 setting a variable A, testing another variable B, and then performing the mutually exclusive access (the variable names are reversed for processor 1). As long as each processor sets its variable before it tests the other processor's variable, mutual exclusion is guaranteed. However, without a consistent ordering between the memory references performed here, two processors could easily get confused about whether the other has entered the critical section. Imagine a scenario in which both tested each other's variables at the same time, but neither had yet observed the other's write, so both entered the critical section, continuing with conflicting updates to some shared object. Such a scenario is possible if the processors are allowed to reorder memory references so that loads execute before independent stores (termed *load bypassing* in Chapter 5).

11.3.6.1 Sequential Consistency. The simplest consistency model is called sequential consistency, and it requires imposing a total order among all references being performed by all processors [Lamport, 1979]. Conceptually, a sequentially consistent (SC) system behaves as if all processors take turns accessing the shared memory, creating an interleaved, totally ordered stream of references that also obeys program order for each individual processor. This approach is illustrated in Figure 11.7 and is in principle similar to the interleaving of references from multiple threads executing on a single time-shared processor. Because of this similarity, it is easier for programmers to reason about the behavior of shared-memory programs on SC systems, since multithreaded programs that operate correctly on time-shared uniprocessors will also usually operate correctly on a sequentially consistent multiprocessor.

However, sequential consistency is challenging to implement efficiently. Consider that imposing a total order requires not only that each load and store must issue in program order, effectively crippling a modern out-of-order processor, but that each reference must also be ordered with respect to all other processors in the system, naively requiring a very-high-bandwidth interconnect for establishing

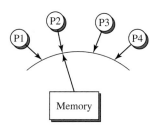

Each processor accesses memory in program order, and accesses from all processors are interleaved as if memory serviced requests from only one processor at a time.

Figure 11.7
Sequentially Consistent Memory Reference Ordering.
Source: Lamport, 1979.

this global order. Fortunately, the same principle that allows us to relax instruction ordering within an out-of-order processor also allows us to relax the requirement for creating a total memory reference order. Namely, just as sequential execution of the instruction stream is an overspecification and is not strictly required for correctness, SC total order is also overly rigorous and not strictly necessary. In an out-of-order processor, register renaming and the reorder buffer enable relaxed execution order, gated only by true data dependences, while maintaining the illusion of sequential execution. Similarly, SC total order can be relaxed so that only those references that must be ordered to enforce data dependences are in fact ordered, while others can proceed out of order. This allows programs that expect SC total order to still run correctly, since the failures can only occur when the reference order of one processor is exposed to another via data dependences expressed as accesses to shared locations.

11.3.6.2 High-Performance Implementation of Sequential Consistency.
There are a number of factors that enable efficient implementation of this relaxation of SC total order. The first is the presence of caches and a cache coherence mechanism. Since cache coherence guarantees each processor visibility to other processors' writes, it also reveals to us any interprocessor data dependences; namely, if a read on one processor is data-dependent on a write from another processor [i.e., there is a read-after-write (RAW) dependence], the coherence mechanism must intervene to satisfy that dependence by first invalidating the address in question from the reader's cache (to guarantee single-writer coherence) and then, upon the subsequent read that misses the invalidated line, supplying the updated line by flushing it from the writer's cache and transmitting it to the reader's cache.

Conveniently, in the absence of such coherence activity (invalidates and/or cache misses), we know that no dependence exists. Since the vast majority of memory references are satisfied with cache hits which require no such intervention, we can safely relax the reference order between cache hits. This decomposes the problem of reference ordering into a local problem (ordering local references with respect to boundaries formed by cache misses and remote invalidate requests) and a global problem (ordering cache misses and invalidate requests). The former is accomplished by augmenting a processor's load/store queue to monitor global address events in addition to processor-local addresses, which it monitors anyway to enforce local store-to-load dependences. Cache misses and upgrades are ordered by providing a global ordering point somewhere in the system. For small-scale systems with a shared address bus, arbitration for the single shared bus establishes a global order for misses and invalidates, which must traverse this bus. In a directory implementation, commands can be ordered upon arrival at the directory or at some other shared point in the system's interconnect.

However, we must still solve the local problem by ordering all references with respect to coherence events. Naively, this requires that we must ensure that all prior memory references result in a cache hit before we can perform the current reference. Clearly, this degenerates into in-order execution of all memory references and precludes high-performance out-of-order execution. Here, we can apply speculation to solve this problem and enable relaxation of this ordering requirement.

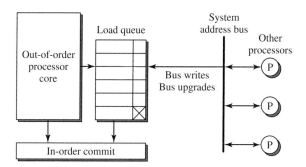

Loads issue out of order, but loaded addresses are tracked in the load queue. Any remote stores that occur before the loads retire are snooped against the load queue. Address matches indicate a potential ordering violation and trigger refetch-based recovery when the load attempts to commit.

Figure 11.8
Read Set Tracking to Detect Consistency Violations.

Namely, we can speculate that a particular reference in fact need not be ordered, execute it speculatively, and recover from that speculation only in those cases where we determine that it needed to be ordered. Since a canonical out-of-order processor already supports speculation and recovery, we need only to add a mechanism that detects ordering violations and initiates recovery in those cases.

The most straightforward approach for detecting ordering violations is to monitor global address events and check to see if they conflict with local speculatively executed memory references. Since speculatively executed memory references are already tracked in the processor's load/store queue, a simple mechanism that checks global address events (invalidate messages that correspond to remote writes) against all unretired loads is sufficient. As shown in Figure 11.8, a matching address causes the load to be marked for a potential ordering violation. As instructions are retired in program order at completion time, they are checked for ordering violations. If the processor attempts to retire such a load, the processor treats the load as if it were a branch misprediction and refetches the load and all subsequent instructions. Upon re-execution, the load is now ordered after the conflicting remote write. A mechanism similar to this one for guaranteeing adherence to the memory consistency model is implemented in the MIPS R10000 [Yeager, 1996], HP PA-8000, and Intel Pentium Pro processors and their later derivatives.

11.3.6.3 Relaxed Consistency Models. An architectural alternative to sequential consistency is to specify a more relaxed consistency model to the programmer. A broad variety of *relaxed consistency* (RC) models have been proposed and implemented, with various subtle differences. The interested reader is referred to Adve and Gharachorloo's [1996] consistency model tutorial for a detailed discussion of several relaxed models. The underlying motivation for RC models is to simplify implementation of the hardware by requiring the programmer to identify and label those references that need to be ordered, while allowing the hardware to proceed with unordered execution of all unlabeled references.

Memory Barriers. The most common and practical way of labeling ordered references is to require the programmer to insert memory barrier instructions or fences in the code to impose ordering requirements. Typical memory barrier semantics (e.g., the `sync` instruction in the PowerPC instruction set) require all memory references that precede the barrier to complete before any subsequent memory references are allowed to begin. A simple and practical implementation of a memory barrier stalls instruction issue until all earlier memory instructions have completed. Care must be taken to ascertain that all memory instructions have in fact completed; for example, many processors retire store instructions into a store queue, which may arbitrarily delay performing the stores. Hence, checking that the reorder buffer does not contain stores is not sufficient; checking must be extended to the queue of retired stores. Furthermore, invalidate messages corresponding to a store may still be in flight in the coherence interconnect, or may even be delayed in an invalidate queue on a remote processor chip, even though the store has already been performed against the local cache and removed from the store queue. For correctness, the system has to guarantee that all invalidates originating from stores preceding a memory barrier have actually been applied, hence preventing any remote accesses to stale copies of the line, before references following the memory barrier are allowed to issue. Needless to say, this can take a very long time, even into hundreds of processor cycles for systems with large numbers of processors.

The main drawback of relaxed models is the additional burden placed on the programmer to identify and label references that need to be ordered. Reasoning about the correctness of multithreaded programs is a difficult challenge to begin with; imposing subtle and sometimes counterintuitive correctness rules on the programmer can only hurt programmer productivity and increase the likelihood of subtle errors and problematic race conditions.

Benefits of Relaxed Consistency. The main advantage of relaxed models is better performance with simpler hardware. This advantage can disappear if memory barriers are frequent enough to require implementations that are more efficient than simply stalling issue and waiting for all pending memory references to complete. A more efficient implementation of memory barriers can look very much like the invalidation tracking scheme illustrated in Figure 11.8; all load addresses are snooped against invalidate messages, but a violation is triggered only if a memory barrier is retired before the violating load is retired. This can be accomplished by marking a load in the load/store queue twice: first, when a conflicting invalidate occurs, and second, when a local memory barrier is retired and the first mark is already present. When the load attempts to retire, a refetch is triggered only if both marks are present, indicating that the load may have retrieved a stale value from the data cache.

The fundamental advantage of relaxed models is that in the absence of memory barriers, the hardware has greater freedom to overlap the latency of store misses with the execution of subsequent instructions. In the SC execution scheme outlined in Section 11.3.6.2, such overlap is limited by the size of the out-of-order instruction window; once the window is full, no more instructions can be executed until the pending store has completed. In an RC system, the store can be retired into a store queue, and subsequent instructions can be retired from the instruction window to make room for new ones. The relative benefit of this distinction depends on the frequency of memory barriers. In the limiting case, when each store is followed by a memory barrier, RC will provide no performance benefit at all, since the instruction window will be full whenever it would be full in an equivalent SC system. However, even in applications such as relational databases with a significant degree of data sharing, memory barriers are much less frequent than stores.

Assuming relatively infrequent memory barriers, the performance advantage of relaxed models varies with the size of the instruction window and the ability of the instruction fetch unit to keep it filled with useful instructions, as well as the relative latency of retiring a store instruction. Recent trends indicate that the former is growing with better branch predictors and larger reorder buffers, but the latter is also increasing due to increased clock frequency and systems with many processors interconnected with multistage networks. Given what we know, it is not clear if the fundamental advantage of RC systems will translate into a significant performance advantage in the future. In fact, researchers have recently argued against relaxed models, due to the difficulty of reasoning about their correctness [Hill, 1998]. Nevertheless, all recently introduced instruction sets specify relaxed consistency (Alpha, PowerPC, IA-64) and serve as existence proofs that the relative difficulty of reasoning about program correctness with relaxed consistency is by no means an insurmountable problem for the programming community.

11.3.7 The Coherent Memory Interface

A simple uniprocessor interfaces to memory via a bus that allows the processor to issue read and write commands as single atomic bus transactions. With a simple bus, once a processor has successfully arbitrated for the bus, it places the appropriate command on the bus, and then holds the bus until it receives all data and address responses, signaling completion of the transaction. More advanced uniprocessors add support for *split transactions,* where requests and responses are separated to expose greater concurrency and allow better utilization of the bus.

On a split-transaction bus, the processor issues a request and then releases the bus before it receives a data response from the memory controller, so that subsequent requests can be issued and overlapped with the response latency. Furthermore, requests can be split from coherence responses as well, by releasing the address bus before the coherence responses have returned. Figure 11.9 illustrates the benefits of a split-transaction bus. In Figure 11.9(a), a simple bus serializes the request, snoop response, DRAM fetch, and data transmission latencies for two requests, one to address A and one to address B. Figure 11.9(b) shows how a split-transaction bus that releases the bus after every request, and receives snoop responses and data responses on separate busses, can satisfy four requests in a pipelined fashion in less time than the simple bus can satisfy two requests. Of course, the design is significantly more complex, since multiple concurrent split transactions are in flight and have to be tracked by the coherence controller.

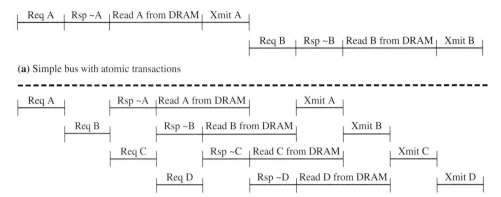

(a) Simple bus with atomic transactions

(b) Split-transaction bus with separate requests and responses

A split-transaction bus enables higher throughput by pipelining requests, responses, and data transmission.

Figure 11.9
Simple Versus Split-Transaction Busses.

Usually, a tag that is unique systemwide is associated with each outstanding transaction; this tag, which is significantly shorter than the physical address, is used to identify subsequent coherence and data messages by providing additional signal lines or message headers on the data and response busses. Each outstanding transaction is tracked with a miss-status handling register (MSHR), which keeps track of the miss address, critical word information, and rename register information that are needed to restart execution once the memory controller returns the data needed by the missing reference. MSHRs are also used to merge multiple requests to the same line to prevent transmitting the same request multiple times. In addition, writeback buffers are used to delay writing back evicted dirty lines from the cache until after the corresponding demand miss has been satisfied; and fill buffers are used to collect a packetized data response into a whole cache line, which is then written into the cache. An example of an advanced split-transaction bus interface is shown in Figure 11.10.

This relatively simple uniprocessor interface must be augmented in several ways to handle coherence in a multiprocessor system. First of all, the bus arbitration mechanism will have to be enhanced to support multiple requesters or bus masters. Second, there must be support for handling inbound address commands that originate at other processors in the system. In a snooping implementation, these are all the commands placed on the bus by other processors, while in a directory implementation these are commands forwarded from the directory. Minimally, this command set must provide functionality for probing the processor's tag array to check the current state of a line, for flushing modified data from the cache, and for invalidating a line. While earlier microprocessor designs required external board-level coherence controllers that issued such low-level commands to the processor's cache, virtually all modern processors support *glueless multiprocessing* by integrating the coherence controller directly on the processor chip. This on-chip

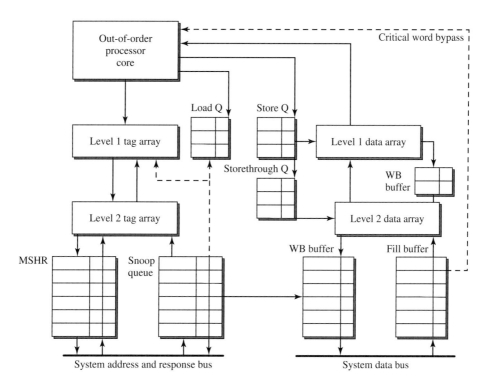

A processor may communicate with memory through two levels of cache, a load queue, store queue, storethrough queue (needed if L1 is write-through), MSHR (miss-status handling registers), snoop queue, fill buffers, and write-back buffers. Not shown is the complex control logic that coordinates all this activity.

Figure 11.10
Processor-Memory Interface.

coherence controller reacts to higher-level commands observed on the address bus (e.g., remote read, read exclusive, or invalidate), and then issues the appropriate low-level commands to the local cache. To expose as much concurrency as possible, modern processors implement snoop queues (see Figure 11.10) that accept snoop commands from the bus and then process their semantics in a pipelined fashion.

11.3.8 Concluding Remarks

Systems that integrate multiple processors and provide a coherent and consistent view of memory have enjoyed tremendous success in the marketplace. They provide obvious advantages to system vendors and customers by enabling scalable, high-performance systems that are straightforward to use and write programs for and provide a growth path from entry-level to enterprise-class systems. The abundance of thread-level parallelism in many important applications is a key enabler for such systems. As the demand for performance and scalability continues to grow, designers of such systems are faced with a myriad of tradeoffs for implementing cache coherence and shared memory while minimizing the latency of communication misses and misses to memory.

11.4 Explicitly Multithreaded Processors

Given the prevalence of applications with plentiful thread-level parallelism, an obvious next step in the evolution of microprocessors is to make each processor chip capable of executing more than a single thread. The primary motivation for doing so is to further increase the utilization of the expensive execution resources on the processor chip. Just as time-sharing operating systems enable better utilization of a CPU by swapping in another thread while the current thread waits on a long-latency I/O event (illustrated in Figure 3.31), chips that execute multiple threads are able to keep processor resources busy even while one thread is stalled on a cache miss or branch misprediction. The most straightforward approach for achieving this capability is by integrating multiple processor cores on a single processor chip [Olukotun et al., 1996]; at least two general-purpose microprocessor designs that do so have been announced (the IBM POWER4 [Tendler et al., 2001] and the Hewlett-Packard PA-8900). While relatively straightforward, some interesting design questions arise for chip multiprocessors. Also, as we will discuss in Section 11.5, several researchers have proposed extending chip multiprocessors to support speculative parallelization of single-threaded programs.

While chip multiprocessors (CMPs) provide one extreme of supporting execution of more than one thread per processor chip by replicating an entire processor core for each thread, other less costly alternatives exist as well. Various approaches to multithreading a single processor core have been proposed and even realized in commercial products. These range from fine-grained multithreading (FGMT), which interleaves the execution of multiple threads on a single execution core on a cycle-by-cycle basis; coarse-grained multithreading (CGMT), which also interleaves multiple threads, but on coarser boundaries delimited by long-latency events like cache misses; and simultaneous multithreading (SMT), which eliminates context switching between multiple threads by allowing instructions from multiple simultaneously active threads to occupy a processor's execution window. Table 11.2 summarizes the context switch mechanism and degree of resource sharing for several approaches to on-chip multithreading. The assignment of execution resources for each of these schemes is illustrated in Figure 11.11.

11.4.1 Chip Multiprocessors

Historically, improvements in transistor density have made it possible to incorporate increasingly complex and area-intensive architectural features such as out-of-order execution, highly accurate branch predictors, and even sizable secondary caches directly onto a processor chip. Recent designs have also integrated coherence controllers to enable glueless multiprocessing, tag arrays for large off-chip cache memories, as well as memory controllers for direct connection of DRAM. System-on-a-chip designs further integrate graphics controllers, other I/O devices, and I/O bus interfaces directly on chip. An obvious next step, as transistor dimensions

Table 11.2
Various approaches to resource sharing and context switching

MT Approach	Resources Shared between Threads	Context Switch Mechanism
None	Everything	Explicit operating system context switch
Fine-grained	Everything but register file and control logic/state	Switch every cycle
Coarse-grained	Everything but I-fetch buffers, register file, and control logic/state	Switch on pipeline stall
SMT	Everything but instruction fetch buffers, return address stack, architected register file, control logic/state, reorder buffer, store queue, etc.	All contexts concurrently active; no switching
CMP	Secondary cache, system interconnect	All contexts concurrently active; no switching

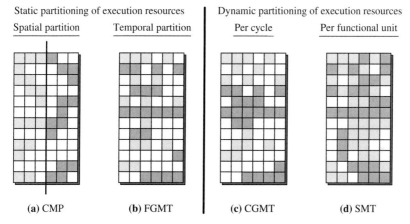

Four possible alternatives are: chip multiprocessing (a), which statically partitions execution bandwidth; fine-grained multiprocessing (b), which executes a different thread in alternate cycles; coarse-grained multithreading (c), which switches threads to tolerate long-latency events; and simultaneous multithreading (d), which intermingles instructions from multiple threads. The CMP and FGMT approaches partition execution resources statically, either with a spatial partition by assigning a fixed number of resources to each processor, or with a temporal partition that time-multiplexes multiple threads onto the same set of resources. The CGMT and SMT approaches allow dynamic partitioning, with either a per-cycle temporal partition in the CGMT approach, or a per-functional unit partition in the SMT approach. The greatest flexibility and highest resource utilization and instruction throughput are achieved by the SMT approach.

Figure 11.11
Running Multiple Threads on One Chip.
Source: Tullsen et al., 1996.

continue to shrink, is to incorporate multiple processor cores onto the same piece of silicon. Chip multiprocessors provide several obvious advantages to system designers: Integrating multiple processor cores on a single chip eases the physical challenges of packaging and interconnecting multiple processors; tight integration reduces off-chip signaling and results in reduced latencies for processor-to-processor communication and synchronization; and finally, chip-scale integration provides interesting opportunities for rethinking and perhaps sharing elements of the cache hierarchy and coherence interface [Olukotun et al., 1996].

Shared Caches. One obvious design choice for CMPs is to share the on- or off-chip cache memory between multiple cores (both the IBM POWER4 and HP PA-8800 do so). This approach reduces the latency of communication misses between the on-chip processors, since no off-chip signaling is needed to resolve such misses. Of course, sharing misses to remote processors are still a problem, although their frequency should be reduced. Unfortunately, it is also true that if the processors are executing unrelated threads that do not share data, a shared cache can be overwhelmed by conflict misses. The operating system's task scheduler can mitigate conflicts and reduce off-chip sharing misses by scheduling for *processor affinity;* that is, scheduling the same and related tasks on processors sharing a cache.

Shared Coherence Interface. Another obvious choice is to share the coherence interface to the rest of the system. The cost of the interface is amortized over two processors, and it is more likely to be efficiently utilized, since multiple independent threads will be driving it and creating additional memory-level parallelism. Of course, an underengineered coherence interface is likely to be even more overwhelmed by the traffic from two processors than it is from a single processor. Hence, designers must pay careful attention to make sure the bandwidth demands of multiple processors can be satisfied by the coherence interface. On a different tack, assuming an on-chip shared cache and plenty of available signaling bandwidth, designers ought to reevaluate write-through and update-based protocols for maintaining coherence on chip. In short, there is no reason to assume that on-chip coherence should be maintained using the same approach with which chip-to-chip coherence is maintained. Similarly, advanced schemes for synchronization between on-chip processors should be investigated.

CMP Drawbacks. However, CMP designs have some drawbacks as well. First of all, one can always argue that given equivalent silicon technology, one can always build a uniprocessor that executes a single thread faster than a CMP of the same cost, since the available die area can be dedicated to better branch prediction, larger caches, or more execution resources. Furthermore, the area cost of multiple cores can easily lead to a very large die that may cause yield or manufacturability issues, particularly when it comes to speed-binning parts for high frequency; empirical evidence suggests that the CMP part, even though designed for the same nominal target frequency, may suffer from a yield-induced frequency disadvantage. Finally, many argue that operating system and software scalability constraints place a ceiling on the total number of processors in a system that is well below the one imposed by packaging and other physical constraints. Hence, one might conclude that CMP is left as a niche approach that may make sense from a cost/performance perspective for a subset of a system vendor's product range, but offers no fundamental advantage at the high end or low end. Nevertheless, several system vendors have announced CMP designs, and they do offer some compelling advantages, particularly in the commercial server market where applications contain plenty of thread-level parallelism.

IBM POWER4. Figure 11.12 illustrates the IBM POWER4 chip multiprocessor. Each processor chip contains two deeply pipelined out-of-order processor cores, each with a private 64K-byte level-1 instruction cache and a private 32K-byte data cache. The level-1 data caches are write-through; writes from both processors are collected and combined in store queues within each bank of the shared level-2 cache (shown as P0 STQ and P1 STQ). The store queues have four 64-byte entries that allow arbitrary write combining. Each of the three level-2 banks is approximately 512K bytes in size and contains multiple MSHRs for tracking outstanding transactions, multiple

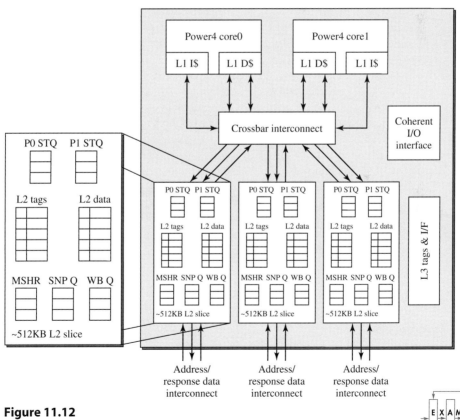

Figure 11.12
IBM POWER4: Example Chip Multiprocessor.
Source: Tendler et al., 2001.

writeback buffers, and multiple snoop queue entries for handling incoming coherence requests. The processors also share the coherence interface to the other processors in the system, a separate interface to the coherent I/O subsystem, as well as the interface to the off-chip level-3 cache and its on-chip tag array. Because of the store-through policy for the level-1 data caches, all coherence requests from remote processors as well as reads from the other on-chip core can be satisfied from the level-2 cache. The level-2 tag array maintains a sharing vector for the two on-chip processors that records which of the two cores contains a shared copy of any cache line in the inclusive level-2 cache. This sharing vector is referenced whenever one of the local cores or a remote processor issues a write to a shared line; an invalidate message is forwarded to one or both of the local cores to guarantee single-writer cache coherence. The POWER4 design supplies tremendous bandwidth (in excess of 100 Gbytes/s) from the level-2 to the processor cores, and also provides multiple high-bandwidth interfaces (each in excess of 10 Gbytes/s) to the level-3 cache and to surrounding processor chips in a multiprocessor configuration.

11.4.2 Fine-Grained Multithreading

A fine-grained multithreaded processor provides two or more thread contexts on chip and switches from one thread to the next on a fixed, fine-grained schedule, usually processing instructions from a different thread on every cycle. The origins of fine-grained multithreading can be traced all the way back to the mid-1960s, when Seymour Cray designed the CDC-6600 supercomputer [Thornton, 1964]. In the CDC-6600, 10 I/O processors shared a single central processor in a round-robin fashion, interleaving work from each of the I/O processors on the central processing unit. In the 1970s, Burton Smith proposed and built the Denelcor HEP, the first true multithreaded processor, which interleaved instructions from a handful of thread contexts in a single pipeline to mask memory latency and avoid the need to detect and resolve interinstruction dependences [Smith, 1991].

A more recent yet similar machine by Burton Smith, the Tera MTA, focused on maximizing the utilization of the memory access path by interleaving references from multiple threads on that path [Tera Computer Company, 1998]. The recent MTA design was targeted for high-end scientific computing and invested heavily in a high-bandwidth, low-latency path to access memory. In fact, the memory bandwidth provided by the MTA machine is the most expensive resource in the system; hence, it is reasonable to design the processor to maximize its utilization. The MTA machine is a fine-grained multithreaded processor; that is, it switches threads on a fixed schedule, on every processor clock cycle. It has enough register contexts (128) to fully mask the main memory latency, making a data cache unnecessary. The path to memory is fully pipelined, allowing each of the 128 threads to have an outstanding access to main memory at all times. The main advertised benefit of the machine is its very lack of data cache; since there is no cache, and all threads access memory with uniform latency, there is no need for algorithmic or compiler transformations that restructure access patterns to maximize utilization of a data cache hierarchy. Instead, the compiler concentrates on identifying independent threads of computation (e.g., do-across loops in scientific programs) to schedule into each of the 128 contexts. While some early performance success has been reported for the Tera MTA machine, its future is currently uncertain due to delays in its second-generation CMOS implementation (the first generation used an exotic gallium arsenide technology).

Single-Thread Performance. The main drawback of fine-grained multithreaded processors like the Tera MTA is that they sacrifice single-thread performance for overall throughput. Since each memory reference takes 128 cycles to complete, the latency to complete the execution of a single thread on the MTA can be longer by a factor of more than 100 when compared to a conventional cache-based design, where the majority of references are satisfied from cache in a few cycles. Of course, for programs with poor cache locality, the MTA will perform no worse than a cache-based system with similar memory latency but will achieve much higher throughput for the entire set of threads. Unfortunately, there are many applications where single-thread performance is very important. For example, most commercial workloads restrict access to shared data by limiting shared references to critical sections protected by locks. To maintain high throughput for software systems with frequent sharing (e.g., relational database systems), it is very important to execute those critical sections as quickly as possible to reduce the occurrence of lock contention. In a fine-grained multithreaded processor like the MTA, one would expect contention for locks to increase to the point where system throughput would be dramatically and adversely affected. Hence, it is unlikely that fine-grained multithreading will be successfully applied in the general-purpose computing domain unless it is somehow combined with more conventional means of masking memory latency (e.g., caches). However, fine-grained multithreading of specific pipe stages can play an important role in hybrid multithreaded designs, as we will see in Section 11.4.4.

11.4.3 Coarse-Grained Multithreading

Coarse-grained multithreading (CGMT) is an intermediate approach to multithreading that enjoys many of the benefits of the fine-grained approach without imposing severe limits on single-thread performance. CGMT, first proposed at the Massachusetts Institute of Technology and incorporated in several research machines there [Agarwal et al., 1990; Fillo et al., 1995], was successfully commercialized in the Northstar and Pulsar PowerPC processors from IBM [Eickemeyer et al., 1996; Storino et al., 1998]. A CGMT processor also provides multiple thread contexts within the processor core, but differs from fine-grained multithreading by switching contexts only when the currently active thread stalls on a long-latency event, such as a cache miss. This approach makes the most sense on an in-order processor that would normally stall the pipeline on a cache miss. Rather than stall, the pipeline is filled with ready instructions from an alternate thread, until, in turn, one of those threads also misses the cache. In this manner, the execution of two or more thread contexts is interleaved in the processor, resulting in better utilization of the processor's execution resources and effectively masking a large fraction of cache miss latency.

Thread-Switch Penalty. One key design issue in a CGMT processor is the cost of performing a context switch between threads. Since context switches occur in response to dynamic events such as cache misses, which may not be detected until late in the pipeline, a naive context-switch implementation will incur several penalty cycles. Since instructions following the missing instruction may already be in the pipeline, they need to be drained from the pipeline. Similarly, instructions from the new thread will not reach the execution stage until they have traversed the earlier pipeline stages. Depending on the length of the pipeline, this results in one or more pipeline bubbles. A straightforward approach for avoiding a thread-switch penalty is to replicate the processor's pipeline registers for each thread and to save the current state of the pipeline at each context switch. Hence, an alternate thread context can be switched back in the very next cycle, avoiding any pipeline bubbles (a similar approach was employed in the Motorola 88000 processor to reduce interrupt latency). Of course, the area and complexity cost of shadowing all the pipeline state is considerable. With a fairly short pipeline and a context-switch penalty of only three cycles, the IBM Northstar/Pulsar designers found that such complexity was not merited; eliminating the three-cycle switch penalty provided only marginal performance benefit. This is reasonable, since the switches are triggered to cover the latency of cache misses that can take a hundred or more processor cycles to resolve; saving a few cycles out of hundreds does not translate into a worthwhile performance gain. Of course, a design with a longer pipeline and a larger switch penalty could face a very different tradeoff and may need to shadow pipeline registers or mitigate switch penalty in some other fashion.

Guaranteeing Fairness. One of the challenges of building a CGMT processor is to provide some guarantee of fairness in the allocation of execution resources to prevent starvation from occurring. As long as each thread has comparable cache miss rates, the processor pipeline will be shared fairly among the thread contexts, since each thread will surrender the CPU to an alternate thread at a comparable rate. However, the cache miss rate of a thread is not a property that is easily controlled by the programmer or operating system; hence, additional features are needed to provide fairness and avoid starvation. Standard techniques from operating system scheduling policies can be adopted: Threads with low miss rates can be preempted after a time slice expires, forcing a thread switch; and the hardware can enforce a minimum quantum to avoid starvation caused by premature preemption.

Beyond guaranteeing fairness, a CGMT processor should provide a scheme for minimizing useless execution bandwidth and also for maximizing execution bandwidth for situations where single-thread throughput is critical for performance. The former can occur whenever a thread is in a busy-wait state (e.g., spinning on a lock held by some other thread or processor) or when a thread enters the operating system idle loop. Clearly, in both these cases, all available execution resources should be dedicated to an alternate thread that has useful work, instead of expending them on a busy-wait or idle loop. The latter can occur whenever a thread is holding a critical resource (e.g., a highly contested lock) and there are other threads in the system waiting for that resource to be released. In such a scenario, the execution of the high-priority thread should not be preempted, even if it is stalled on a cache miss, since the alternate threads may slow down the primary thread either directly (due to thread-switch penalty overhead) or indirectly (by causing additional conflict misses or contention in the memory hierarchy).

Thread Priorities. A CGMT processor can avoid these pitfalls of performance by architecting a priority scheme that assigns at least three levels of priority—high, medium, and low—to the active threads. Note that these are not priorities in the operating system sense, where a thread or process has a fixed priority set by the operating system or system administrator. Rather, these thread priorities vary dynamically and reflect the relative importance of execution of the current execution phase of the thread. Hence, programmer intervention is required to notify the hardware whenever a thread undergoes a priority transition. For example, when a thread enters a critical section after acquiring a lock, it should transition to high priority; conversely, when it exits, it should reduce its priority level. Similarly, when a thread enters the idle loop or begins to spin on a lock that is currently held by another thread, it should lower its priority. Of course, such communication requires that an interface be specified, usually through special instructions in the ISA that identify these phase transitions, and also requires programmers to place these instructions in the appropriate locations in their programs. Alternatively, implicit pattern-matching mechanisms that recognize execution sequences that usually accompany these transitions can also be devised. The former approach was employed by the IBM Northstar/Pulsar processors, where specially encoded NOP instructions are used to indicate thread priority level. Fortunately, the required software changes are concentrated in a relatively few locations in the operating system and middleware (e.g., database) and have been realized with minimal effort.

Thread-Switch State Machine. Figure 11.13 illustrates a simple thread-switch state machine for a CGMT processor. As shown, there are four possible states for each processor thread: running, ready, stalled, and swapped. Threads transition between states whenever a cache miss is initiated or completed, and when the thread switch logic decides to switch to an alternate thread. In a well-designed CGMT processor, the following conditions can cause a thread switch to occur:

- A cache miss has occurred in the primary thread, and there is an alternate thread in the ready state.

- The primary thread has entered the idle loop, and there is an alternate nonidle thread in the ready state.

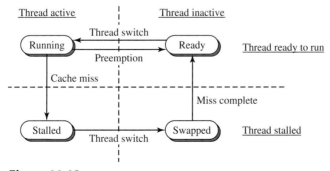

Figure 11.13
CGMT Thread Switch State Machine.

- The primary thread has entered a synchronization spin loop (busy wait), and there is an alternate nonidle thread in the ready state.

- A swapped thread has transitioned to the ready state, and the swapped thread has a higher priority than the primary thread.

- An alternate ready, nonidle thread has not retired an instruction in the last n cycles (avoiding starvation).

Finally, forward progress can be guaranteed by preventing a preemptive thread switch from occurring if the running thread has been active for less than some fixed number of cycles.

Performance and Cost. CGMT has been shown to be a very cost-effective technique for improving instruction throughput. IBM reports that the Northstar/Pulsar line of processors gains about 30% additional instruction throughput at the expense of less than 10% die area and negligible effect on cycle time. The only complexity introduced by CGMT in this incarnation is control complexity for managing thread switches and thread priorities, as well as a doubling of the architected register file to hold two thread contexts instead of one. Finally, the minor software changes required to implement thread priorities must also be figured into the cost equation.

11.4.4 Simultaneous Multithreading

The final and most sophisticated approach for on-chip multithreading is to allow fine-grained and dynamically varying interleaving of instructions from multiple threads across shared execution resources. This technology has recently been commercialized in the Intel Pentium 4 processor but was first proposed in 1995 by researchers at the University of Washington [Tullsen, 1996; Tullsen et al., 1996]. They argued that prior approaches to multithreading shared hardware resources across threads inefficiently, since the thread-switch paradigm restricted either the entire pipeline or minimally each pipeline stage to contain instructions from only a single thread. Since instruction-level parallelism is unevenly distributed, this led to unused instruction slots in each stage of the pipeline and reduced the efficiency of multithreading. Instead, they proposed simultaneous multithreading (SMT), which allows instructions to be interleaved within and across pipeline stages to maximize utilization of the processor's execution resources.

Several attributes of a modern out-of-order processor enable efficient implementation of simultaneous multithreading. First of all, instructions traverse the intermediate pipeline stages out of order, decoupled from program or fetch order; this enables instructions from different threads to mingle within these pipe stages, allowing the resources within these pipe stages to be more fully utilized. For example, when data dependences within one thread restrict a wide superscalar processor from issuing more than one or two instructions per cycle, instructions from an alternate independent thread can be used to fill in empty issue slots. Second, architected registers are renamed to share a common pool of physical registers; this renaming removes the need for tracking threads when resolving data dependences dynamically. The rename table simply maps the same architected register from each thread to a different physical register, and the standard out-of-order execution hardware takes care of the rest, since dependences are resolved using renamed physical register names. Finally, the extensive buffers (i.e., reorder buffer, issue queues, load/store queue, retired store queue) present in an out-of-order processor to extract and smooth out uneven and irregular instruction-level parallelism can be utilized more effectively by multiple threads, since serializing data and control dependences that can starve the processor now only affect the portion of instructions that belong to the thread that is encountering such a dependence; instructions from other threads are still available to fill the processor pipeline.

11.4.4.1 SMT Resource Sharing. The primary goal of an SMT design is to improve processor resource utilization by sharing those resources across multiple active threads; in fact, the increased parallelism created by multiple simultaneously active threads can be used to justify deeper and wider pipelines, since the additional resources are more likely to be useful in an SMT configuration. However, it is less clear which resources should be shared and which should not or perhaps cannot be shared. Figure 11.14 illustrates a few alternatives, ranging from the design on the left that shares everything but the fetch and retire stages, to the design on the right that shares only the execute and memory stages. Regardless of which design point is chosen, instructions from multiple threads have to be joined before the pipeline stage where resources are shared and must be separated out at the end of the pipeline to preserve precise exceptions for each thread.

Interstage Buffer Implementation. One of the key issues in SMT design, just as in superscalar processor design, is the implementation of the interstage buffers that track instructions as they traverse the pipeline. If the fetch or decode stages

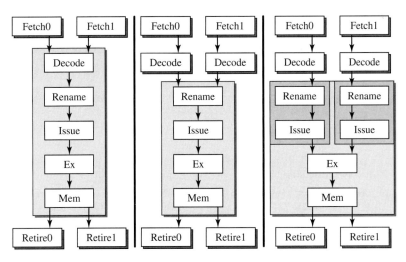

Figure 11.14
SMT Resource Sharing Alternatives.

are replicated, as shown in the left and middle options of Figure 11.14, the stage where the replicated pipelines meet must support multiple simultaneous writers into its buffer. This will complicate the design over a baseline non-SMT processor, since there is only a single writer in that case. Furthermore, the load/store queue and reorder buffer (ROB), which are used to track instructions in program order, must also be redesigned or partitioned to accommodate multiple threads. If they are partitioned per thread, their design will be very similar to the analogous conventional structures. Of course, a partitioned design will preclude best-case single-thread performance, since a single thread will no longer be able to occupy all available slots. Sharing a reorder buffer among multiple threads introduces additional complexity, since program order must be tracked separately for each thread, and the ROB must support selective flushing of nonconsecutive entries to support per-thread branch misprediction recovery. This in turn requires complex free-list management, since the ROB can no longer be managed as a circular queue. Similar issues apply to the load/store queue, but these are further complicated by memory consistency model implications on how the load/store queue resolves memory data dependences; these are discussed briefly here.

SMT Sharing of Pipeline Stages. There are a number of issues that affect how sensible or feasible it is to attempt to share the resources in each pipeline stage; we will discuss some of these issues for each stage, based on the pipeline structure outline in Figure 11.14.

- *Fetch.* The most expensive resource in the instruction fetch stage is the instruction cache port. Since a cache port is limited to accessing a contiguous range of addresses, it would be very difficult to share a single port between multiple threads, as it is very unlikely that more than one thread would be fetching instructions from contiguous or even spatially local addresses. Hence, an SMT design would most likely either provide a dedicated fetch stage per thread or would time-share a single port in a fine-grained or coarse-grained manner. The cost of dual-porting the instruction cache is quite high and difficult to justify, so it is likely that real SMT designs will employ a time-sharing approach. The other expensive resource is the branch predictor. Likewise, multiporting the branch predictor is equivalent to halving its effective size, so a time-shared approach probably makes most sense. However, certain elements of modern branch predictors rely on serial thread semantics and do not perform well if the semantics of multiple threads are interleaved in an arbitrary fashion. For example, the return address stack relies on FIFO (first-in, first-out) behavior for program calls and returns and will not work reliably if calls and returns from multiple threads are interleaved. Similarly, any branch predictor that relies on a global branch history register (BHR) has been shown to perform poorly if branch outcomes from interleaved threads are shifted arbitrarily into the BHR. Hence, it is likely that in a time-shared branch predictor design, at least these elements will need to be replicated for each thread.

- *Decode.* For simple RISC instruction sets, the primary task of the decode stage is to identify source and destination operands and resolve dependences between instructions in a decode group. This involves logic with $O(n^2)$ complexity with respect to decode group width to implement operand specifier comparators and priority decoders. Since there are, by definition, no such inter-instruction dependences between instructions from different threads, it may make sense to partition this resource across threads in order to reduce its complexity. For example, two four-wide decoders could operate in parallel on two threads with much less logic complexity than a single, shared eight-wide decoder. Of course, this design tradeoff could compromise single-thread performance in those cases where a single thread is actually able to supply eight instructions for decoding in a single cycle. For a CISC instruction set, the decode stage is much more complex since it requires determining the semantics of the complex instructions and (usually) decomposing it into a sequence of simpler, RISC-like primitives. Since this can be a very complex task, it may make sense to share the decode stage between threads. However, as with the fetch stage, it may be sensible to time-share it in a fine-grained or coarse-grained manner, rather than attempting to decode instructions from multiple threads simultaneously.

- *Rename.* The rename stage is responsible for allocating physical registers and for mapping architected register names to physical register names. Since physical registers are most likely allocated from a common pool, it makes perfect sense to share the logic that manages the free list between SMT threads. However, mapping architected register names to physical register names is done by indexing into a rename or mapping table with the architected register number and either updating the mapping (for destination operands) or reading it (for source operands). Since architected register numbers are disjoint across threads, the rename table could be partitioned across threads, thus providing high bandwidth into the table at a much lower cost than true multiporting. However, this would imply partitioning the rename stage across threads and, just as with the decode stage, potentially limiting single-thread throughput for programs with abundant instruction-level parallelism.

- *Issue.* The issue stage implements Tomasulo's algorithm for dynamic scheduling of instructions via a two-phase *wakeup-and-select* process: waking up instructions that are data-ready, and then selecting issue candidates from the data-ready pool to satisfy structural dependences. Clearly, if multiple threads are to simultaneously share functional units, the selection process must involve instructions from more than one thread. However, instruction wakeup is clearly limited to intrathread interaction; that is, an instruction wakes up only in response to the execution of an earlier instruction from that same thread. Hence, it may make sense to partition the issue window across threads, since wakeup events will never cross such partitions anyway. Of course, as with the earlier pipe stages, partitioning can have a negative impact on single-thread performance. However, some researchers have argued that issue window logic will be one of the critical cycle-time-limiting paths in future process technologies. Partitioning this logic to exploit the presence of multiple data-flow-disjoint threads may enable a much larger overall issue window for a fixed cycle-time budget, resulting in better SMT throughput.

- *Execute.* The execute stage realizes the semantics of the instructions by executing each instruction on a functional unit. Sharing the functional units themselves is fairly straightforward, although even here there is an opportunity for multithread optimization: The bypass network that connects functional units to allow back-to-back execution of dependent instructions can be simplified, given that instructions from different threads need never bypass results. For example, in a clustered microarchitecture along the lines of the Alpha 21264, issue logic could be modified to direct instructions from the same thread to the same cluster, hence reducing the likelihood of cross-cluster result bypassing. Alternatively, issue logic could prevent back-to-back issue of dependent instructions, filling the gaps with independent instructions from alternate threads, and hence avoiding the need for the cycle-time critical ALU-output-to-ALU-input bypass path. Again, such optimizations may compromise single-thread performance, except to the extent that they enable higher operating frequency.

- *Memory.* The memory stage performs cache accesses to satisfy load instructions but is also responsible for resolving memory dependences between loads and stores and for performing other memory-related bookkeeping tasks. Sharing cache access ports between threads to maximize their utilization is one of the prime objectives of an SMT design and can be accomplished in a fairly straightforward manner. However, sharing the hardware that detects and resolves memory dependences is more complex. This hardware consists of the processor's load/store queue, which keeps track of loads and stores in program order and detects if later loads alias to earlier stores. Extending the load/store queue to handle multiple threads requires an understanding of the architected memory consistency model, since certain models (e.g., sequential consistency, see Section 11.3.6) prohibit forwarding a store value from one thread to a load from another. To handle such cases, the load/store queue must be enhanced to be thread-aware, so that it will forward values when it can and will stall the dependent load when it cannot. It may be simpler to provide separate load/store queues for each thread; of course, this will reduce the degree to which the SMT processor is sharing resources across threads and will restrict the effective window size for a single thread to the capacity of its partition of the load/store queue.

- *Retire.* In the retire pipeline stage, instruction results are committed in program order. This involves checking for exceptions or other anomalous conditions and then committing instruction results by updating rename mappings (in a physical register file-based design) or copying rename register values to architected registers (in a rename register-based design). In either case, superscalar retirement requires checking and prioritizing write-after-write (WAW) dependences (since the last committed write of a register must win) and multiple ports into the rename table or the architected register file. Once again, partitioning this hardware across threads can ease implementation, since WAW dependences can only occur within a thread, and commit updates do not conflict across threads. A viable alternative, provided that retirement latency and bandwidth are not critical, is to time-share the retirement stage in a fine-grained or coarse-grained manner.

In summary, the research to date does not make a clear case for any of the resource-sharing alternatives discussed here. Based on the limited disclosure to date, the Pentium 4 SMT design appears to simultaneously share most of the issue, execute, and memory stages, but performs coarse-grained sharing of the processor front end and fine-grained sharing of the retire pipe-stages. Hence, it is clearly a compromise between the SMT ideal of sharing as many resources as possible and the reality of cycle-time and complexity challenges presented by attempting to maximize sharing.

SMT Support for Serializing Instructions. All instruction sets contain instructions with serializing semantics; typically, such instructions affect the global state (e.g., by changing the processor privilege level or invalidating an address translation) or impose ordering constraints on memory operations (e.g., the memory barriers discussed in Section 11.3.6.3). These instructions are often implemented in a brute-force manner, by draining the processor pipeline of active instructions, applying the semantics of the instruction, and then resuming issue following the instruction. Such a brute-force approach is used because these instructions are relatively rare, and hence even an inefficient implementation does not affect performance very much. Furthermore, the semantics required by the instructions can be quite subtle and difficult to implement correctly in a more aggressive manner, making it difficult to justify a more aggressive implementation.

However, in an SMT design, the frequency of serializing instructions can increase dramatically, since it is proportional to the number of threads. For example, in a single-threaded processor, let's assume that a serializing instruction occurs once every 600 cycles, while in a four-threaded SMT processor that achieves three times the instruction throughput of the single-threaded processor, they will now occur once every 200 cycles. Obviously, a more efficient and aggressive implementation for such instructions may now be required to sustain high performance, since draining the pipeline every 200 cycles will severely degrade performance. The execution of serializing instructions that update the global state can be streamlined by renaming the global state, just as register renaming streamlines execution by removing false dependences between instructions. Once the global state is renamed, only those subsequent instructions that read that state will be delayed, while earlier instructions can continue to read the earlier instance. Hence, instructions from before and after the serializing instruction can be intermingled in the processor's instruction window. However, renaming the global state may not be as easy as it sounds. For example, serializing updates to the translation-lookaside buffer (TLB) or other address-translation and protection structures may require wholesale or targeted renaming of large array structures. Unfortunately, this will increase the latency of accessing these structures, and such access paths may already be cycle-time-critical. Finally, streamlining the execution of memory barrier instructions, which are used to serialize memory references, requires resolving numerous subtle issues related to the system's memory consistency model; some of these issues are discussed in Section 11.3.6.3. One possible approach for memory barriers is to drain the pipeline selectively for each thread, while still allowing concurrent execution of other threads. This has obvious implications for the reorder buffer design, as well as the issue logic, which must now selectively block issue of instructions from a particular thread while allowing issue to continue from alternate threads. In any case, the complexity implications are nontrivial and largely unexplored in the research literature.

Managing Multiple Threads. Many of the same issues discussed in Section 11.4.3 on coarse-grained multithreading also apply, at least to some extent, to SMT designs. Namely, the processor's issuing policies must provide some guarantee of fairness and forward progress for all active threads. Similarly, priority policies that prevent useless instructions (spin loops, idle loop) from consuming execution resources should be present; similarly, an elevated priority level that provides maximum throughput to thread phases that are performance-critical may also be needed. However, since a pure SMT design has no notion of thread-switching, the mechanism for implementing such policies will be different: rather than switching out a low-priority thread or switching in a high-priority thread, an SMT design can govern execution resource allocation at a much finer granularity, by prioritizing a particular thread in the issue logic's instruction selection phase. Alternatively, threads at various priority levels can be prevented from occupying more than some fixed number of entries in the processor's execution window by gating instruction fetch from those threads. Similar restrictions can be placed on any dynamically allocated resource within the processor. Examples of such resource limits are load/store queue occupancy, to restrict a thread's ability to stress the memory subsystem; or MSHR occupancy, to restrict the number of outstanding cache misses per thread; or entries in a branch or value prediction structure, in order to dedicate more of those resources to high-priority threads.

SMT Performance and Cost. Clearly, there are many subtle issues that can affect the performance of an SMT design. One example is interference between threads in caches, predictors, and other structures. Some published evidence indicates such interference is not excessive, particularly for larger structures such as secondary caches, but the effect on primary caches and other smaller structures is less clear. To date, the only definitive evidence on the performance potential of SMT designs is the preliminary announcement from Intel that claims 16% to 28% throughput improvement for the Pentium 4 design when running server workloads with abundant thread-level parallelism. The following paragraph summarizes some of the details of the Pentium 4 SMT design that have been released. Since the Pentium 4 design has limited machine parallelism, supports only two threads, and only implements true SMT for parts of the issue, execute, and memory stages, it is perhaps not surprising that this gain is much less than the factor of 2 or 3 improvement reported in the research literature. However, it is not clear that the proposals described in the literature are feasible, or that SMT designs that deal with all the real implementation issues discussed before are scalable beyond two or perhaps three simultaneously active threads. Certainly the cost of implementing SMT, both in terms of implementation complexity as well as resource duplication, has been understated in the research literature to date.

The Pentium 4 Hybrid Multithreading Implementation. The Intel Pentium 4 processor incorporates a hybrid form of multithreading that enables two logical processors to share some of the execution resources of the processor. Intel's implementation—named *hyperthreading*—is conceptually similar to the SMT proposals that have appeared in academic literature, but differs in substantial ways. The limited disclosure to date indicates that the in-order portions of the Pentium 4 pipeline (i.e., the front-end fetch and decode engine and the commit stages) are multithreaded in a fine-grained fashion. That is, the two logical threads fetch, decode, and retire instructions in alternating cycles, unless one of the threads is stalled for some reason. In the latter case a single thread is able to consume all the fetch, decode, or commit resources of the processor until the other thread resolves its stall. Such a scheme could also be described as coarse-grained with a single-cycle time quantum. The Pentium 4 also implements two-stage scheduling logic, where instructions are placed into five issue queues in the first stage and are issued to functional units from these five issue queues in the second stage. Here again, the first stage of scheduling is fine-grained multithreaded: Only one thread can place instructions into the issue queues in any given cycle. Once again, if one thread is stalled, the other can continue to place instructions into the issue queues until the stall is resolved. Similarly, stores are retired from each thread in alternating cycles, unless one thread is stalled. In essence, the Pentium 4 implements a combination of fine-grained and coarse-grained multithreading of all these pipe stages. However, the Pentium 4 does implement true simultaneous multithreading for the second issue stage as well as the execute and memory stages of the pipeline, allowing instructions from both threads to be interleaved in an arbitrary fashion.

Resource sharing in the Pentium 4 is also somewhat complicated. Most of the buffers in the out-of-order portion of the pipeline (i.e., reorder buffer, load queue, store queue) are partitioned in half rather than arbitrarily shared. The scheduler queues are partitioned in a less rigid manner, with high-water marks that prevent either thread from consuming all available entries. As discussed earlier, such partitioning of resources sacrifices maximum achievable single-thread performance in order to achieve high throughput when two threads are available. At a high level, such partitioning can work well if the two threads are largely symmetric in behavior, but can result in poor performance if they are asymmetric and have differing resource utilization needs. However, this effect is mitigated by the fact that the Pentium 4 supports a single-threaded mode in which all resource partitioning is disabled, enabling the single active thread to consume all available resources.

11.5 Implicitly Multithreaded Processors

So far we have restricted our discussion of multithreaded processors and multiprocessor systems to designs that exploit explicit, programmer-created threads to improve instruction throughput. However, there are many important applications where single-thread performance is still of paramount importance. One approach for improving the performance of a single-threaded application is to break that thread down into multiple threads of execution that can be executed concurrently. Rather than relying on the programmer to explicitly create multiple threads by manually parallelizing the application, proposals for *implicit multithreading* (IMT) describe techniques for automatically spawning such threads by exploiting attributes in the program's control flow.

In contrast to automatic compiler-based or manual parallelization of scientific and numeric workloads, which typically attempt to extract thread-level parallelism to occupy dozens to hundreds of CPUs and achieve orders of magnitude speedup, implicit multithreading attempts to sustain up to only a half-dozen or dozen threads simultaneously. This difference in scale is driven primarily by the tightly coupled nature of implicit multithreading, which is caused by threads of execution that tend to be relatively short (tens of instructions) and that often need to communicate large amounts of state with other active threads to resolve data and control dependences. Furthermore, heavy use of speculation in these proposed systems requires efficient recovery from misspeculation, which also requires a tight coupling between the processing elements. All these factors conspire to make it very difficult to scale implicit multithreading beyond a handful of concurrently active threads. Nevertheless, implicit multithreading proposals have claimed nontrivial speedups for applications that are not amenable to conventional approaches for extracting instruction-level parallelism.

Some IMT proposals are motivated by a desire to extract as much instruction-level parallelism as possible, and achieve this goal by filling a large shared execution window with instructions sequenced from multiple disjoint locations in the program's control flow graph. Other IMT proposals advocate IMT as a means for building more scalable instruction windows: Implicit threads that are independently sequenced can be assigned to and executed in separate processing elements, eliminating the need for a centralized, shared execution window that poses many implementation challenges. Of course, such decentralized designs must still provide a means for satisfying data dependences between the processing elements; much of the research has focused on efficient solutions to this problem.

Fundamentally, there are three main challenges that must be faced when designing an IMT processor. Not surprisingly, these are the same challenges faced by a superscalar design: resolving control dependences, resolving register data dependences, and resolving memory data dependences. However, due to some unique characteristics of IMT designs, resolving them can be substantially more difficult. Some of the proposals rely purely on hardware mechanisms for resolving these problems, while others rely heavily on compilation technology supported by critical hardware assists. We will discuss each of these challenges and describe some of the solutions that have been proposed in the literature.

11.5.1 Resolving Control Dependences

One of the main arguments for IMT designs is the difficulty of effectively constructing and traversing a single thread of execution that is large enough to expose significant amounts of instruction-level parallelism. The conventional approach for constructing a single thread—using a branch predictor to speculatively traverse a program's control flow graph—is severely limited in effectiveness by cumulative branch prediction accuracy. For example, even a 95% accurate branch predictor deteriorates to a cumulative prediction accuracy of only 60% after 10 consecutive branch predictions. Since many important programs have only five or six instructions between conditional branches, this allows the branch predictor to construct a window of only 50 to 60 instructions before the likelihood of a branch misprediction becomes unacceptably high. The obvious solution of improving branch prediction accuracy continues to be an active field of research; however, the effort and hardware required to incrementally improve the accuracy of predictors that are already 95% accurate can be prohibitive. Furthermore, it is not clear if significant improvements in branch prediction accuracy are possible.

Control Independence. All proposed IMT designs exploit the program attribute of control independence to increase the size of the instruction window beyond joins in the control flow graph. A node in a program's control flow graph is said to be control-independent if it post-dominates the current node, that is, if execution will eventually reach that node regardless of how intervening conditional branches are resolved. Figure 11.15 illustrates several sources of control independence in programs. In the proposed IMT designs, implicit threads can be spawned at joins in the control flow, at subroutine return addresses, across loop iterations, or at the loop fall-through point. These threads can often be spawned nonspeculatively, since control independence guarantees that the program will eventually reach these

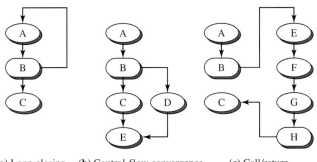

(a) Loop-closing   (b) Control-flow convergence   (c) Call/return

There are multiple sources of control independence: in (a), block C eventually follows block B since the loop has a finite number of iterations; in (b) block E always follows B independent of which way the branch resolves; and in (c), block C eventually follows block B after the subroutine call to E completes.

Figure 11.15
Sources of Control Independence.

initiation points. However, they can also be spawned speculatively, to encompass cases where the intervening control flow cannot be fully determined at the time the thread is spawned. For example, a loop that traverses a linked list may have a data-dependent number of iterations: Spawning speculative threads for multiple iterations into the future will often result in better performance, even when some of those speculative threads need to eventually be squashed as incorrect.

Spawning an implicit *future thread* at a subsequent control-independent point in the program's control flow has several advantages. First of all, any intermediate branch instructions that may be mispredicted will not directly affect the control independent thread, since it will be executed no matter what control flow path is used to reach it. Hence, exploiting control independence allows the processor to skip ahead past hard-to-predict branches to find useful instructions. Second, skipping ahead can have a positive prefetching effect. That is to say, the act of fetching instructions from a future point in the control flow can effectively overlap useful work from the current thread with instruction cache misses caused by the future thread. Conversely, the current thread may also encounter instruction cache misses which can now be overlapped with the execution of the future thread. Note that such prefetching effects are impossible with conventional single-threaded execution, since the current and future thread's instruction fetches are by definition serialized. This prefetching effect can be substantial; Akkary reports that a DMT processor fetches up to 40% of its committed instructions from beyond an intervening instruction cache miss [Akkary and Driscoll, 1998].

Disjoint Eager Execution. An interesting alternative for creating implicit threads is proposed in the disjoint eager execution (DEE) architecture [Uht and Sindagi, 1995]. Conventional eager execution attempts to overcome conditional branches by executing both paths following a branch. Of course, this results in a combinatorial explosion of paths as multiple branches are traversed. In the DEE proposal, the eager execution decision tree is pruned by comparing cumulative branch prediction rates along each branch in the tree and choosing the branch path with the highest cumulative prediction rate as the next path to follow; this process is illustrated in Figure 11.16. The branch prediction rates for each static branch can be estimated using profiling, and the cumulative rates can be computed by multiplying the rates for each branch used to reach that branch in the tree. However, for practical implementation reasons, Uht has found that assuming a uniform static prediction rate for each branch works quite well, resulting in a straightforward fetch policy that always backtracks a fixed number of levels in the branch tree and interleaves execution of these alternate paths with the main path provided by a conventional branch predictor. These alternate paths are introduced into the DEE core as implicit threads.

Table 11.3 summarizes four IMT proposals in terms of the control flow attributes they exploit; what the sources of implicit threads are, how they are created, sequenced, and executed; and how dependences are resolved. In cases where threads are created by the compiler, program control flow is statically analyzed to determine opportune thread creation points. Most simply, the thread-level speculation (TLS) proposals create a thread for each iteration of a loop at compile time to harness parallelism [Steffan et al., 1997]. The multiscalar proposal allows much

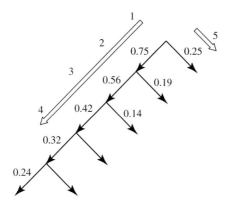

Assuming each branch is predicted with 75% accuracy, the cumulative branch prediction rate is shown; after fetching branch paths 1, 2, 3, and 4, the next-highest cumulative rate is along branch path 5, so it is fetched next.

Figure 11.16
Disjoint Eager Execution.
Source: Uht and Sindagi, 1995.

Table 11.3
Attributes of several implicit multithreading proposals

	Multiscalar	Disjoint Eager Execution (DEE)	Dynamic Multi-threading (DMT)	Thread-Level Speculation (TLS)
Control flow attribute exploited	Control independence	Control independence, cumulative branch misprediction	Control independence	Control independence
Source of implicit threads	Loop bodies, control flow joins	Loop bodies, control flow joins, cumulative branch mispredictions	Loop exits, subroutine returns	Loop bodies
Thread creation mechanism	Software/compiler	Implicit hardware	Implicit hardware	Software/compiler
Thread creation and sequencing	Program order	Out of program order	Out of program order	Program order
Thread execution	Distributed processing elements	Shared processing elements	Shared multithreaded processing elements	Separate CPUs
Register data dependences	Software with hardware speculation support	Hardware; no speculation	Hardware; data dependence prediction and speculation	Disallowed; compiler must avoid
Memory data dependences	Hardware-supported speculation	Hardware	Hardware; prediction and speculation	Dependence speculation; checked with simple extension to MESI coherence

greater flexibility to the compiler by providing architected primitives for spawning threads (called *tasks* in the multiscalar literature) at arbitrary points in the program's control flow [Sohi et al., 1995; Franklin, 1993]. The DEE proposal dynamically detects control independence and exploits that within a single instruction window, but also creates implicit threads by backtracking through the branch prediction tree, as illustrated in Figure 11.16 [Uht and Sindagi, 1995]. Finally, the dynamic multithreading (DMT) proposal uses hardware detection heuristics to spawn threads at procedure calls as well as backward loop branches [Akkary and Driscoll, 1998]. In these cases execution continues simultaneously within the procedure call as well as following it, at the return site, and similarly, within the next loop iteration as well as at the code following the loop exit.

Out-of-Order Thread Creation. One challenge that is unique to the DMT approach is that threads are spawned out of program order. For example, in the case of nested procedure calls, the first call will spawn a thread for executing the call, as well as executing the code at the subroutine return site, resulting in two active threads. The code in the called procedure now encounters the nested procedure call and spawns an additional thread to execute that call, resulting in three active threads. However, this thread, though created third, actually occurs before the second thread in program order. As a result, the logical reorder buffer used in this design now has to support out-of-order insertion of an arbitrary number of instructions into the middle of a set of already active instructions. As we will see, the process of resolving register and memory data dependences is also substantially complicated by out-of-order thread creation. Whether such an approach is feasible remains to be seen.

Physical Organization. Of course, constructing a large window of instructions is only half the battle; any design that attempts to detect and exploit parallelism from such a window must demonstrate that it is feasible to build hardware that accomplishes such a feat. Many IMT proposals partition the execution resources of a processor so that each thread executes independently on a partition, enabling distributed and scalable extraction of instruction-level parallelism. Since each partition need only contain the instruction window of a single thread, it need not be more aggressive than a current-generation design. In fact, it may even be less aggressive. Additional parallelism is extracted by overlapping the execution of multiple such windows. For TLS proposals, each partition is actually an independent microprocessor core in a system that is very similar to a multiprocessor, or chip multiprocessor (CMP, as discussed in Section 11.4.1). In contrast, the DMT proposal relies on an SMT-like multithreaded execution core that tracks and interleaves implicit threads instead of explicit threads. DMT also proposes a hierarchical two-level reorder buffer that enables a very large instruction window; threads that have finished execution but cannot be committed migrate to the second level of the reorder buffer and are only fetched out of the second level in case they need to re-execute due to data mispredictions. Finally, the DEE processor has a centralized execution window that tracks multiple implicit threads simultaneously by organizing the window basic on the static program structure rather than a dynamic single path. That is to say, the instruction window of the DEE prototype design, Levo, captures a static view of the program and includes hardware for simultaneously tracking multiple dynamic instances of the same static control flow constructs (e.g., loop bodies).

Finally, the multiscalar proposal is structured as a circular queue of processing elements. The tail of the queue is considered nonspeculative and executes the current thread or task; other nodes are executing future tasks that can be speculative with respect to both control and data dependences. As the tail thread completes execution, its results are retired, and the next node becomes the nonspeculative tail node. Simultaneously, a new future thread is spawned to occupy the processing element that was freed up as the tail thread completed execution. In this way, by overlapping execution across multiple processing elements, additional parallelism is exposed beyond what can be extracted by a single processing element.

Thread Sequencing and Retirement. One of the most challenging aspects of IMT designs is the control and/or prediction hardware that must sequence threads and retire them in program order. Relying on compiler assistance for creating threads can ease this task. Similarly, a queue-based machine organization such as multiscalar can at least conceptually simplify the task of sequencing and retiring tasks. However, all proposals share the need for control logic that determines that no correctness violations have occurred before a task is allowed to retire and update the architected state. Control dependence violations are fairly straightforward; as long as nonspeculative control flow eventually reaches the thread in question, and as long as control flow leaves that thread and proceeds to the next speculative thread, the thread can safely be retired. However, resolving data dependences can be quite a bit more complex and is discussed in the following.

11.5.2 Resolving Register Data Dependences

Register data dependences consist of name or false (WAR and WAW) dependences and true data dependences (RAW). In IMT designs, just as in conventional superscalar processors, the former are solved via register renaming and in-order commit. The only complication is that in-order commit has to be coordinated across multiple threads, but this is easily resolved by committing threads in program order.

True register data dependences can be broken down into two types: dependences within a thread or *intrathread dependences,* and dependences across threads or *interthread dependences*. Intrathread dependences can be resolved with standard techniques studied in earlier chapters, since instructions within a thread are sequenced in program order, and can be renamed, bypassed, and eventually committed using conventional means. Interthread dependences, however, are complicated by the fact that instructions are now sequenced out of program order. For this reason, it can be difficult to identify the correct producer-consumer relationships, since the producer or register-writing instruction may not have been decoded yet at the time the consumer or register-reading instruction becomes a candidate for execution. For example, this can happen when a register value is read near the beginning of a new thread, while the last write to that register value does not occur until near the end of the prior thread. Since the prior thread is still busy executing older instructions, the instruction that performs the last write has not even been fetched yet. In such a scenario, conventional renaming hardware fails to correctly capture the true dependence, since the producing instruction has not updated the renaming information to reflect its pending write. Hence, either simplifications to the programming model or more sophisticated renaming solutions are necessary to maintain correct execution.

The easiest solution for resolving interthread register data dependences is to simplify the programming model by disallowing them at compile-time. Thread-level speculation proposals take this approach. As the compiler creates implicit threads for parallel execution, it is simply required to communicate all shared operands through memory with loads and stores. Register dependences are tracked within threads only, using well-understood techniques like register renaming and Tomasulo's algorithm, just as in a single-threaded uniprocessor.

In contrast, the multiscalar proposal allows register communication between implicit threads, but also enlists the compiler's help by requiring it to identify interthread register dependences explicitly. This is done by communicating to the future thread, as it is created, which registers in the register file have pending writes to them, and also marking the last instruction to write to any such register so that the prior thread's processing element knows to forward it to future tasks once the write occurs. Transitively, pending writes from older threads must also be forwarded to future threads as they arrive at a processing element. The compiler embeds this information in a write mask that is provided to the future thread when it is spawned. Thus, with helpful assistance from the compiler, it is possible to effectively implement a distributed, scalable dependence resolution scheme with relatively straightforward hardware implementation.

The DEE and DMT proposals assume no compiler assistance, however, and are responsible for dynamically resolving data dependences. The DEE proposal constructs a single, most likely thread of execution, and fetches and decodes all the instructions along that path in program order. Hence, identifying data dependences along that path is relatively straightforward. The alternate *eager execution* paths, which we treat as implicit threads in our discussion, have similar sequential semantics, so forward dependence resolution is possible. However, the DEE proposal also detects control independence by implementing *minimal control dependences* (MCD). The hardware for MCD is capable of identifying and resolving data dependences across divergent control flow paths that eventually join, as these paths are introduced into the execution window by the DEE fetch policy. The interested reader is referred to Uht and Sindagi [1995] for a description of this novel hardware scheme.

The DMT proposal, on the other hand, does not have a sequential instruction stream to work with. Hence, the most challenging task is identifying the last write to a register that is read by a future thread, since the instruction performing that write may not have been fetched or decoded yet. The simplistic solution is to assume that all registers will be written by the current thread and to delay register reads in future threads until all instructions in the current thread have been fetched and decoded. Of course, this will result in miserable performance. Hence, the DMT proposal relies on data dependence speculation, where future threads assume that their register operands are already stored in the register file and proceed to execute speculatively with those operands. Of course, the future threads must recover by re-executing such instructions if an older thread performs a write to any such register. The DMT proposal describes complex dependence resolution mechanisms that enable such re-execution whenever a dependence violation is detected. In addition, researchers have explored adaptive prediction mechanisms that attempt to identify pending register writes based on historical information. Whenever such a predictor identifies a pending write, dependent instructions in future threads are stalled, and hence prevented from misspeculating with stale data. Furthermore, the register dependence problem can also be eased by employing value prediction; in cases of pending or unknown but likely pending writes, the operand's value can be predicted, forwarded to dependent operands, and later verified. Many of the issues discussed in Chapter 10 regarding value prediction, verification, and recovery will apply to any such design.

11.5.3 Resolving Memory Data Dependences

Finally, an implicit multithreading design must also correctly resolve memory data dependences. Here again, it is useful to decompose the problem into intrathread and interthread memory dependences. Intrathread memory dependences, just as intrathread register dependences, can be resolved with conventional and well-understood techniques from prior chapters: WAW and WAR dependences are resolved by buffering stores until retirement, and RAW dependences are resolved by stalling dependent loads or forwarding from the load/store queue.

Interthread false dependences (WAR and WAW) are also solved in a straightforward manner, by buffering writes from future threads and committing them when those threads retire. There are some subtle differences among the proposed alternatives. The DEE and DMT proposals use structures similar to conventional load/store queues to buffer writes until commit. The multiscalar design uses a complex mechanism called the *address resolution buffer* (ARB) to buffer in-flight writes. Finally, the TLS proposal extends conventional MESI cache coherence to allow multiple instances of cache lines that are being written by future threads. These future instances are tagged with an epoch number that is incremented for each new thread. The epoch number is appended to the cache line address, allowing conventional MESI coherence to support multiple modified instances of the same line. The retirement logic is then responsible for committing these modified lines by writing them back to memory whenever a thread becomes nonspeculative.

True (RAW) interthread memory dependences are significantly more complex than true register dependences, although conceptually similar. The fundamental difficulty is the same: since instructions are fetched and decoded out of program order, later loads are unable to obtain dependence information with respect to earlier stores, since those stores may not have computed their target addresses yet or may not have even been fetched yet.

TLS Memory RAW Resolution. Again, the simplest solution is employed by the TLS design: Future threads simply assume that no dependence violations will occur and speculatively consume the latest available value for a particular memory address. This is accomplished by a simple extension to conventional snoop-based cache coherence: When a speculative thread executes a load that causes a cache miss, the caches of the other processors are searched in reverse program order for a matching address. By searching in reverse program order (i.e., reverse thread creation order), the latest write, if any, is identified and used to satisfy the load. If no match is found, the load is simply satisfied from memory, which holds the committed state for that cache line. In effect the TLS scheme is predicting that any actual store to load dependences occur far enough apart that the older thread will already have performed the relevant store, resulting in a snoop hit when the newer thread issues its load miss. Only those cases where the store and load are actually executed out of order across the speculative threads will result in erroneous speculation.

Of course, since TLS is employing a simple form of data dependence speculation, a mechanism is needed to detect and recover from violations that may occur. Again, a simple extension to the existing cache coherence protocol is employed. There are two cases that need to be handled: first, if the future load is satisfied from memory, and second, if the future load is satisfied by a modified cache line written to by an earlier thread. In the former case, the cache line is placed in the future thread's cache in the exclusive state, since it is the only copy in the system. Subsequently, an older thread performs a store to the same cache line, hence causing a potential dependence violation. In order to perform the store, the older thread must snoop the other caches in the system to obtain exclusive access to the line. At this point, the future thread's copy of the line is discovered, and that thread is squashed due to the violation. The latter case, where the future thread's load was satisfied from a modified line written by an older thread, is very similar. The line is placed in the future thread's cache in the shared state and is also downgraded to the shared state in the older thread's cache. This is exactly what would happen when satisfying a remote read to a modified line, as shown earlier in Figure 11.5. When the older thread writes to the line again, it has to upgrade the line by snooping the other processor's caches to invalidate their copies. At this point, again, the future thread's shared copy is discovered and a violation is triggered. The recovery mechanism is simple: the thread is squashed and restarted.

DMT Memory RAW Resolution. The DMT proposal handles true memory dependences by tracking the loads and stores from each thread in separate per-thread load and store queues. These queues are used to handle intrathread memory dependences in a conventional manner, but are also used to resolve interthread dependences by conducting cross-thread associative searches of earlier threads' store queues whenever a load issues and later threads' load queues whenever a store issues. A match in the former case will forward the store data to the dependent load; a match in the latter case will signal a violation, since the load has already executed with stale data, and will cause the later thread to reissue the load and its dependent instructions. Effectively, the DMT mechanism achieves memory renaming, since multiple instances of the same memory location can be in flight at any one time, and dependent loads will be satisfied from the correct instance as long as all the writes in the sequence have issued and are present in the store queues. Of course, if an older store is still pending, the mechanism will fail to capture dependence information correctly and the load will proceed with potentially incorrect data and will have to be restarted once the missing store does issue.

DEE Memory RAW Resolution. The DEE proposal describes a mechanism that is conceptually similar to the DMT approach but is described in greater detail. DEE employs an address-interleaved, high-throughput structure that is capable of tracking program order and detecting dependence violations whenever a later load reads a value written by an earlier store. Again, since these loads and stores can be performed out of order, the mechanism must logically sort them in program order and flag violations only when they actually occurred. This is complicated by the fact that implicit threads spawned by the DEE fetch policy can also contain stores and must be tracked separately for each thread.

Multiscalar ARB. The multiscalar address resolution buffer (ARB) is a centralized, multiported, address-interleaved structure that allows multiple in-flight stores to the same address to be correctly resolved against loads from future threads. This structure allocates a tracking entry for each speculative load as it is performed by a future thread and checks subsequent stores from older threads against such entries. Any hit will flag a violation and cause the violating thread and all future threads to be squashed and restarted. Similarly, each load is checked against all prior unretired stores, which are also tracked in the ARB, and any resulting data dependence is satisfied with data from the prior store, rather than from the data cache. It should be noted that such prior stores also form visibility barriers to older unexecuted stores, due to WAW ordering. For example, let's say a future thread n + 1 stores to address A. This store is placed in the ARB. Later on, future thread n + 2 reads from address A; this read is satisfied by the ARB from thread n + 1's store entry. Eventually, current thread n performs a store against A. A naive implementation would find the future load from thread n + 2, and squash and refetch thread n + 2 and all newer future threads. However, since thread n + 1 performed an intervening store to address A, no violation has actually occurred and thread n + 2 need not be squashed.

Implementation Challenges. The main drawback of the ARB and similar, centralized designs that track all reads and writes is scalability. Since each processing element needs high bandwidth into this structure, scaling to a significant number of processing elements becomes very difficult. The TLS proposal avoids this scalability problem by using standard caching protocols to filter the amount of traffic that needs to be tracked. Since only cache misses and cache upgrades need to be made visible outside the cache, only a small portion of references are ordered and checked against the other processing elements. Ordering within threads is provided by conventional load and store queues within the processor. An analogous cache-based enhancement of the ARB, the *speculative versioning cache,* has also been proposed for multiscalar. Of course, the corresponding drawback of cache-based filtering is that false dependences arise due to address granularity. That is to say, since cache coherence protocols operate on blocks that are larger than a single word (usually 32 to 128 bytes), a write to one word in the block can falsely trigger a violation against a read from a different word in the same block, causing additional recovery overhead that would not occur with a more fine-grained dependence mechanism.

Other problems involved with memory dependence checking are more mundane. For example, limited buffer space can stall effective speculation, just as a full load or store queue can stall instruction fetch in a superscalar processor. Similarly, commit bandwidth can cause limitations, particularly for TLS systems, since commit typically involves writing modified lines back to memory. If a speculative thread modifies a large number of lines, writeback bandwidth can limit performance, since a future thread cannot be spawned until all commits have been performed. Finally, TLS proposals as well as more fine-grained proposals all suffer from the inherently serial process of searching for the newest previous write when resolving dependences. In the TLS proposal, this is accomplished by serially snooping the other processors in reverse thread creation order. The other IMT proposals suggest parallel associative lookups, which are faster, but more expensive and difficult to scale to large numbers of processing elements.

11.5.4 Concluding Remarks

To date, implicit multithreading exists only in research proposals. While it shows dramatic potential for improving performance beyond what is achievable with single-threaded execution, it is not clear if all the implementation issues discussed here, as well as others that may not be discovered until someone attempts a real implementation, will ultimately prevent the adoption of IMT. Certainly, as chip multiprocessor designs become widespread, it is quite likely that the simple enhancements required for thread-level speculation in such systems will in fact become available. However, these changes will only benefit applications that have execution characteristics that match TLS hardware and that can be recompiled to exploit such hardware. The more complex schemes—DEE, DMT, and multiscalar—require much more dramatic changes to existing processor implementations, and hence must meet a higher standard to be adopted in real designs.

11.6 Executing the Same Thread

So far, we have discussed both explicitly and implicitly multithreaded processor designs that attempt to sequence instructions from multiple threads of execution to maximize processor throughput. An interesting alternative that several researchers have proposed is to execute the same instructions in multiple contexts. Although it may seem counterintuitive, there are several potential benefits to such an approach. The first proposal to suggest doing so [Rotenberg, 1999], active-stream/redundant-stream simultaneous multithreading (AR-SMT), focused on fault detection. By executing an instruction stream twice in separate thread contexts and comparing

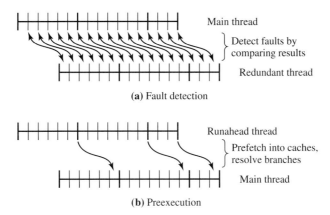

Figure 11.17
Executing the Same Thread.

execution results across the threads, transient errors in the processing pipeline can be detected. That is to say, if the pipeline hardware flips a bit due to a soft error in a storage cell, the likelihood of the same bit being flipped in the redundant stream is very low. Comparing results across threads will likely detect many such transient errors. An interesting observation grew out of this work on fault detection: namely, that the active and redundant streams end up helping each other execute more effectively. That is to say, they can prefetch memory references for each other and can potentially resolve branch mispredictions for each other as well. This cooperative effect has been exploited in several research proposals. We will discuss some of these proposals in the context of these benefits—fault detection, prefetching, and branch resolution—in this section. Figure 11.17 illustrates these uses for executing the same thread; Figure 11.17(a) shows how a redundant thread can be used to check the main thread for transient faults, while Figure 11.17(b) shows how a runahead thread can prefetch cache misses and resolve mispredicted branches for the main thread.

11.6.1 Fault Detection

As described, the original work in redundant execution of the same thread was based on the premise that inconsistencies in execution between the two thread instances could be used to detect transient faults. The AR-SMT proposal assumes a baseline SMT processor and enhances the front end of the SMT pipeline to replicate the fetched instruction stream into two separate thread contexts. Both contexts then execute independently and store their results in a reorder buffer. The commit stage of the pipeline is further enhanced to compare instruction outcomes, as they are committed, to check for inconsistencies. Any such inconsistencies are used to identify transient errors in the execution pipeline. A similar approach is used in real processor designs that place emphasis on fault detection and fault tolerance. For example, the IBM S/390 G5 processor also performs redundant execution of all instructions, but achieves this by replicating the pipeline hardware on chip and running both pipelines in lock step. Similar system-level designs are available from Hewlett Packard's Tandem division; in these designs, two physical processor chips are coupled to run the same threads in a lockstep manner, and faults are detected by comparing the results of the processors to each other. In fact, there is a long history of such designs, both real and proposed, in the fault-tolerant computing domain.

The DIVA proposal [Austin, 1999] builds on the AR-SMT concept, but instead of using two threads running on an SMT processor, it employs a simple processor that dynamically checks the computations of a complex processor by re-executing the instruction stream. At first glance, it appears that the throughput of the pair of processors would be limited by the simpler one, resulting in poor performance. In fact, however, the simple processor can easily keep up with the complex processor if it exploits the fact that the complex processor has speculatively resolved all control and data flow dependences. Since this is the case, it is trivial to parallelize the code running on the simple processor, since all dependences are removed: All conditional branches are resolved, and all data dependences disappear since input and output operand values are already known. The simple processor need only verify each instruction in isolation, by executing with the provided inputs and comparing the output to the provided output. Once each instruction is verified in this manner, then, by induction, the entire instruction stream is also verified. Since the simple processor is by definition easy to verify for correctness, it can be trusted to check the operation of the much more complex and design-error-prone runahead processor. Hence, this approach is able to cover design errors in addition to transient faults.

Dynamic verification with a simple, slower processor does have one shortcoming that has not been adequately addressed in the literature. As long as the checker processor is only used to verify computation (i.e., ALU operations, memory references, branches), it is possible to trivially parallelize the checking, since each computation that is being checked is independent of all others. However, this relies on the complex processor's ability to provide correct operands to all these computations. In other words, the operand communication that occurs within the complex processor is not being checked, since the checker relies on the complex processor to perform it correctly. Since operand communication is one of the worst sources of complexity in a modern out-of-order processor, one could argue that the checker is focusing on the wrong problem. In other words, in terms of fault coverage, one could argue that checking communication is much more important than checking computation, since it is relatively straightforward to verify the correctness of ALUs and other computational paths that can be viewed as combinational delay paths. On the other hand, verifying the correctness of complex renaming schemes and associative operand bypassing is extremely difficult. Furthermore, soft errors in the complex processor's register file would also not be detected by a DIVA checker that does not check operand communication.

To resolve this shortcoming, the DIVA proposal also advocates checking operand communication separately in the checker processor. The checker decodes each instruction, reads its source operands from a register file, and writes its result operands to the same checker register file. However, the process of reading and writing register operands that may have read-after-write (RAW), write-after-read (WAR), and write-after-write (WAW) dependences with instructions immediately preceding or following the instruction being checked is not trivial to parallelize. As explained in detail in Chapter 5, such dependences have to be detected and resolved with complex dependence-checking logic that is $O(n^2)$ in complexity with respect to pipeline width n. Hence, parallelizing this checking process will require hardware equivalent in complexity to the hardware in the complex processor. Furthermore, if, as the DIVA proposal advocates, the checker processor runs slower than the baseline processor, it will have to support a wider pipeline to avoid becoming the execution bottleneck. In this case, the checker must actually implement more complex logic than the processor it is checking. Further investigation is needed to determine how much of a problem this will be and whether it will prevent the adoption of DIVA as a design technique for enhancing fault tolerance and processor performance.

11.6.2 Prefetching

One positive side effect of redundant execution can be prefetching, since both threads are generating the same stream of instruction and data memory references. Whenever one thread runs ahead of the other, it prefetches useful instructions and data into the processor's caches. This can result in a net speedup, since additional memory-level parallelism is exposed. The key to extracting significant performance benefit is to maximize the degree of runahead, or slip, between the two threads. The *slipstream* processor proposal [Sundaramoorthy et al., 2000] does exactly that, by specializing the runahead thread; instead of redundantly executing all instructions in the program, the runahead thread is stripped down so that instructions that are considered nonessential are removed from execution. Nonessential instructions are ones that have no effect on program outcome or only contribute to resolving predictable branches. Since the runahead thread no longer needs to execute these instructions, it is able to get further ahead in the control flow of the program, increasing the slip between the two threads and improving the timeliness of the prefetches that it creates.

The principle of maximizing slip to ensure timeliness has been further refined in proposals for *preexecution* [Roth, 2001; Zilles, 2002; Collins et al., 2001]. In these proposals, profiling information is used to identify problematic instructions like branch instructions that are frequently mispredicted or load instructions that frequently cause cache misses. The backward dynamic data flow slice for such instructions is then constructed at compile time. The instructions composing that backward slice then form a speculative preexecution thread that is spawned at run time in an available thread context on an SMT-like processor. The preexecuted slice will then precompute the outcome for the problematic instruction and issue a prefetch to memory if it misses. Subsequently, the worker thread catches up to the preexecuted instruction and avoids the cache miss.

The main benefit of slipstreaming and preexecution over the implicit multithreading proposals discussed in Section 11.5 is that the streamlined runahead thread has no correctness requirement. That is, since it is only serving to generate prefetches and "assist" the main thread's execution, and it has no effect on the architected program state, generating and executing the thread is much easier. None of the issues regarding control and data dependence resolution have to be solved exactly. Of course, precision in dependence resolution is likely to result in a more useful runahead thread, since it is less likely to issue useless prefetches from paths that the real thread never reaches; but this is a performance issue, rather than a correctness issue, and can be solved much more easily.

Intel has described a fully functional software implementation of preexecution for the Pentium 4 SMT processor. In this implementation, a runahead thread is spawned and assigned to the same physical processor as the main thread; the runahead thread then prefetches instructions and data for the main thread, resulting in a measurable speedup for some programs.

An alternative and historically interesting approach that uses redundant execution for data prefetching is the datascalar architecture [Burger et al., 1997]. In this architecture, memory is partitioned across several processors that all execute the same program. The processors are connected by a fast broadcast network that allows them to communicate memory operands to each other very quickly. Each processor is responsible for broadcasting all references to its local partition of memory to all the other processors. In this manner, each reference is broadcast once, and each processor is able to satisfy all its references either from its local memory or from a broadcast initiated by the owner of that remote memory. With this policy, all remote memory references are satisfied in a request-free manner. That is to say, no processor ever needs to request a copy of a memory location; if it is not available locally, the processor need only wait for it to show up on the broadcast interconnect, since the remote processor that owns the memory will eventually execute the same reference and broadcast the result. The net result is that average memory latency no longer includes the request latency, but consists simply of the transfer latency over the broadcast interconnect. In many respects, this is conceptually similar to the redundant-stream prefetching used in the slipstream and preexecution proposals.

11.6.3 Branch Resolution

The other main benefit of both slipstreaming and preexecution is early resolution of branch instructions that are hard to predict with conventional approaches to branch prediction. In the case of slipstreaming, instructions that are data flow antecedents of the problematic branch instructions are considered essential and are therefore executed in the runahead thread. The branch outcome is forwarded to the real thread so that when it reaches the branch, it can use the precomputed outcome to avoid the misprediction. Similarly, preexecution constructs a backward program slice for the branch instruction and spawns a speculative thread to preexecute that slice. The main implementation challenge for early resolution of branch outcomes stems from synchronizing the two threads. For instruction and data prefetching, no synchronization is necessary, since the real thread's instruction fetch or memory reference will benefit by finding its target in the instruction or data cache, instead of experiencing a cache miss. In effect, the threads are synchronized through the instruction cache or data cache, which tolerates some degree of inaccuracy in both the fetch address (due to spatial locality) as well as the timing (due to temporal locality). As long as the prefetches are timely, that is to say they occur neither too late (failing to cover the entire miss latency) or too early (where the prefetched line is evicted from the cache before the real thread catches up and references it), they are beneficial.

However, for branch resolution, the preexecuted branch outcome must be exactly synchronized with the same branch instance in the real thread; otherwise, if it is applied to the wrong branch, the early resolution-based prediction may fail. The threads cannot simply synchronize based on the static branch (i.e., branch PC), since multiple dynamic instances of the same static branch can exist in the slip-induced window of instructions between the two threads. Hence, a reference-counting scheme must be employed to make sure that a branch is resolved with the correct preexecuted branch outcome. Such a reference-counting scheme must keep track of exactly how many instances of each static branch separate the runahead thread from the main thread. The outcome for each instance is stored in an in-order queue that separates the two threads; the runahead thread inserts new branch outcomes into one end of this queue, while the main thread removes outcomes from the other end. If the queue length is incorrect, and the two threads become unsynchronized, the predicted outcomes are not likely to be very useful. Building this queue and the associated control logic, as well as mechanisms for flushing it whenever mispredictions are detected, is a nontrivial problem that has not been satisfactorily resolved in the literature to date.

Alternatively, branch outcomes can be communicated indirectly through the existing branch predictor by allowing the runahead thread to update the predictor's state. Hence, the worker thread can benefit from the updated branch predictor state when it performs its own branch predictions, since the two threads synchronize implicitly through the branch predictor. However, the likelihood that the runahead thread's predictor update is both timely and accurate are low, particularly in modern branch predictors with multiple levels of history.

11.6.4 Concluding Remarks

Redundant execution of the same instructions has been proposed and implemented for fault detection. It is quite likely that future fault-tolerant implementations will employ redundant execution in the context of SMT processors, since the overhead for doing so is quite reasonable and the fault coverage can be quite helpful, particularly as smaller transistor dimensions lead to increasing vulnerability to soft errors. Exploiting redundant-stream execution to enhance performance by generating prefetches or resolving branches early has not yet reached real designs. It is likely that purely software-based redundant-stream prefetching will materialize in the near future, since it is at least theoretically possible to achieve without any hardware changes; however, the performance benefits of a software-only scheme are less clear. The reported performance benefits for the more advanced preexecution and slipstream proposals are certainly attractive; assuming that baseline SMT and

CMP designs become commonplace in the future, the extensions required for supporting these schemes are incremental enough that it is likely they will be at least partially adopted.

11.7 Summary

This chapter discusses a wide range of both real and proposed designs that execute multiple threads. Many important applications, particularly in the server domain, contain abundant thread-level parallelism and can be efficiently executed on such systems. We discussed explicit multithreaded execution in the context of both multiprocessor systems and multithreaded processors. Many of the challenges in building multiprocessor systems revolve around providing a coherent and consistent view of memory to all threads of execution while minimizing average memory latency. Multithreaded processors enable more efficient designs by sharing execution resources either at the chip level in chip multiprocessors (CMP), in a fine-grained or coarse-grained time-sharing manner in multithreaded processors that alternate execution of multiple threads, or seamlessly in simultaneous multithreaded (SMT) processors. Multiple thread contexts can also be used to speed up the execution of serial programs. Proposals for doing so range from complex hardware schemes for implicit multithreading to hybrid hardware/software schemes that employ compiler transformations and critical hardware assists to parallelize sequential programs. All these approaches have to deal correctly with control and data dependences, and numerous implementation challenges remain. Finally, multiple thread contexts can also be used for redundant execution, both to detect transient faults and to improve performance by preexecuting problematic instruction sequences to resolve branches and issue prefetches to memory.

Many of these techniques have already been adopted in real systems; many others exist only as research proposals. Future designs are likely to adopt at least some of the proposed techniques to overcome many of the implementation challenges associated with building high-throughput, high-frequency, and power-efficient computer systems.

REFERENCES

Adve, S. V., and K. Gharachorloo: "Shared memory consistency models: A tutorial," *IEEE Computer,* 29, 12, 1996, pp. 66–76.

Agarwal, A., B. Lim, D. Kranz, and J. Kubiatowicz: "APRIL: a processor architecture for multiprocessing," *Proc. ISCA-17,* 1990, pp. 104–114.

Akkary, H., and M. A. Driscoll: "A dynamic multithreading processor," *Proc. 31st Annual Int. Symposium on Microarchitecture,* 1998, pp. 226–236.

Austin, T.: "DIVA: A reliable substrate for deep-submicron processor design," *Proc. 32nd Annual ACM/IEEE Int. Symposium on Microarchitecture (MICRO-32),* Los Alamitos, IEEE Computer Society, 1999.

Burger, D., S. Kaxiras, and J. Goodman: "Datascalar architectures," *Proc. 24th Int. Symposium on Computer Architecture,* 1997, pp. 338–349.

Censier, L., and P. Feautrier: "A new solution to coherence problems in multicache systems," *IEEE Trans. on Computers,* C-27, 12, 1978, pp. 1112–1118.

Charlesworth, A.: "Starfire: extending the SMP envelope," *IEEE MICRO,* vol. 18 no. 1, 1998, pp. 39–49.

Collins, J., H. Wang, D. Tullsen, C. Hughes, Y. Lee, D. Lavery, and J. Shen: "Speculative precomputation: Long-range prefetching of delinquent loads," *Proc. 28th Annual Int. Symposium on Computer Architecture,* 2001, pp. 14–25.

Eickemeyer, R. J., R. E. Johnson, S. R. Kunkel, M. S. Squillante, and S. Liu: "Evaluation of multithreaded uniprocessors for commercial application environments," *Proc. 23rd Annual Int. Symposium on Computer Architecture,* Philadelphia, ACM SIGARCH and IEEE Computer Society TCCA, 1996, pp. 203–212.

Fillo, M., S. Keckler, W. Dally, and N. Carter: "The M-Machine multicomputer," *Proc. 28th Annual Int. Symposium on Microarchitecture (MICRO-28),* 1995, pp. 146–156.

Franklin, M.: "The multiscalar architecture," Ph.D. thesis, University of Wisconsin-Madison, 1993.

Hammond, L., M. Willey, and K. Olukotun: "Data speculation support for a chip-multiprocessor," *Proc. 8th Symposium on Architectural Support for Programming Languages and Operating Systems,* 1998, pp. 58–69.

Hill, M.: "Multiprocessors should support simple memory consistency models," *IEEE Computer,* 31, 8, 1998, pp. 28–34.

Krishnan, V., and J. Torrellas: "The need for fast communication in hardware-based speculative chip multiprocessors," *Int. Journal of Parallel Programming,* 29, 1, 2001, pp. 3–33.

Lamport, L.: "How to make a multiprocessor computer that correctly executes multiprocess programs," *IEEE Trans. on Computers,* C-28, 9, 1979, pp. 690–691.

Lovett, T., and R. Clapp: "STiNG: A CC-NUMA Computer System for the Commercial Marketplace," *Proc. 23rd Annual Int. Symposium on Computer Architecture,* 1996, pp. 308–317.

Olukotun, K., B. A. Nayfeh, L. Hammond, K. Wilson, and K. Chang: "The case for a single-chip multiprocessor," *Proc. 7th Int. Conf. on Architectural Support for Programming Languages and Operating Systems (ASPLOS-VII),* 1996, pp. 2–11.

Rotenberg, E.: "AR-SMT: A microarchitectural approach to fault tolerance in microprocessors," *Proc. 29th Fault-Tolerant Computing Symposium,* 1999, pp. 84–91.

Roth, A.: "Pre-execution via speculative data-driven multithreading," Ph.D. Thesis, University of Wisconsin, Madison, WI, 2001.

Smith, B.: "Architecture and applications of the HEP multiprocessor computer system," *Proc. Int. Society for Optical Engineering,* 1991, pp. 241–248.

Sohi, G., S. Breach, and T. Vijaykumar: "Multiscalar processors," *Proc. 22nd Annual Int. Symposium on Computer Architecture,* 1995, pp. 414–425.

Steffan, J., C. Colohan, and T. Mowry: "Architectural support for thread-level data speculation," Technical report, School of Computer Science, Carnegie Mellon University, 1997.

Steffan, J. G., C. Colohan, A. Zhai, and T. Mowry: "A scalable approach to thread-level speculation," *Proc. 27th Int. Symposium on Computer Architecture,* 2000.

Steffan, J. G., and T. C. Mowry: "The potential for using thread-level data speculation to facilitate automatic parallelization," *Proc. of HPCA,* 1998, pp. 2–13.

Storino, S., A. Aipperspach, J. Borkenhagen, R. Eickemeyer, S. Kunkel, S. Levenstein, and G. Uhlmann: "A commercial multi-threaded RISC processor," *Int. Solid-State Circuits Conference*, 1998.

Sundaramoorthy, K., Z. Purser., and E., Rotenberg: "Slipstream processors: Improving both performance and fault tolerance," *Proc. 9th Int. Conf. on Architectural Support for Programming Languages and Operating Systems*, 2000, pp. 257–268.

Tendler, J. M., S. Dodson, S. Fields, and B. Sinharoy: "IBM eserver POWER4 system microarchitecture," IBM Whitepaper, 2001.

Tera Computer Company: "Hardware characteristics of the Tera MTA," 1998.

Thornton, J. E.: "Parallel operation in the Control Data 6600," *AFIPS Proc. FJCC*, part 2, 26, 1964, pp. 33–40.

Tullsen, D., S. Eggers, J. Emer, H. Levy, J. Lo, and R. Stamm: "Exploiting choice: instruction fetch and issue on an implementable simultaneous multithreading processor," *Proc. 23rd Annual Symposium on Computer Architecture*, 1996, pp. 191–202.

Tullsen, D. M.: "Simultaneous multithreading," Ph.D. Thesis, University of Washington, Seattle, WA, 1996.

Uht, A. K., and V. Sindagi: "Disjoint eager execution: An optimal form of speculative execution," *Proc. 28th Annual ACM/IEEE Int. Symposium on Microarchitecture*, 1995, pp. 313–325.

Wang, W.-H., J.-L. Baer, and H. Levy: "Organization and performance of a two-level virtual-real cache hierarchy," *Proc. 16th Annual Int. Symposium on Computer Architecture*, 1989, pp. 140–148.

Yeager, K.: "The MIPS R10000 superscalar microprocessor," *IEEE Micro*, 16, 2, 1996, pp. 28–40.

Zilles, C.: "Master/slave speculative parallelization and approximate code," Ph.D. Thesis, University of Wisconsin, Madison, WI, 2002.

HOMEWORK PROBLEMS

P11.1 Using the syntax in Figure 11.2, show how to use the load-linked/store conditional primitives to synthesize a compare-and-swap operation.

P11.2 Using the syntax in Figure 11.2, show how to use the load-linked/store conditional primitives to acquire a lock variable before entering a critical section.

P11.3 A processor such as the PowerPC G3, widely deployed in Apple Macintosh systems, is primarily intended for use in uniprocessor systems, and hence has a very simple MEI cache coherence protocol. Identify and discuss one reason why even a uniprocessor design should support cache coherence. Is the MEI protocol of the G3 adequate for this purpose? Why or why not?

P11.4 Apple marketed a G3-based dual-processor system that was mostly used for running asymmetric workloads. In other words, the second processor was only used to execute parts of specific applications, such as Adobe Photoshop, rather than being used in a symmetric manner by the operating system to execute any ready thread or process. Assuming a multiprocessor-capable operating system (which the MacOS, at the time, was not), explain why symmetric use of a G3-based dual-processor system might result in very poor performance. Propose a software solution implemented by the operating system that would mitigate this problem, and explain why it would help.

P11.5 Given the MESI protocol described in Figure 11.5, create a similar specification (state table and diagram) for the much simpler MEI protocol. Comment on how much easier it would be to implement this protocol.

P11.6 Many modern systems use a MOESI cache coherence protocol, where the semantics of the additional O state are that the line is shared-dirty: i.e., multiple copies may exist, but the other copies are in the S state, and the cache that has the line in the O state is responsible for writing the line back if it is evicted. Modify the table and state diagram shown in Figure 11.5 to include the O state.

P11.7 Explain what benefit accrues from the addition of the O state to the MESI protocol.

P11.8 Real coherence controllers include numerous transient states in addition to the ones shown in Figure 11.5 to support split-transaction busses. For example, when a processor issues a bus read for an invalid line (I), the line is placed in an IS transient state until the processor has received a valid data response that then causes the line to transition into the shared state (S). Given a split-transaction bus that separates each bus command (bus read, bus write, and bus upgrade) into a request and response, augment the state table and state transition diagram of Figure 11.5 to incorporate all necessary transient states and bus responses. For simplicity, assume that any bus command for a line in a transient state gets a negative acknowledge (NAK) response that forces it to be retried after some delay.

P11.9 Given Problem 11.8, further augment Figure 11.5 to eliminate at least three NAK responses by adding necessary additional transient states. Comment on the complexity of the resulting coherence protocol.

P11.10 Assuming a processor frequency of 1 GHz, a target CPI of 2, a per-instruction level-2 cache miss rate of 1% per instruction, a snoop-based cache coherent system with 32 processors, and 8-byte address messages (including command and snoop addresses), compute the inbound and outbound snoop bandwidth required at each processor node.

P11.11 Given the assumptions of Problem 11.10, assume you are planning an enhanced system with 64 processors. The current level-2 cache design has a single-ported tag array with a lookup latency of 3 ns. Will the 64-processor system have adequate snoop bandwidth? If not, describe an alternative design that will.

P11.12 Using the equation in Section 11.3.5, compute the average memory latency for a three-level hierarchy where hits in the level-1 cache take one cycle, hits in the level-2 cache take 12 cycles, hits in the level-3 cache take 50 cycles, and misses to memory take 250 cycles. Assume a level-1 miss rate of 5% misses per program reference, a level-2 miss rate of 2% per program reference, and a level-3 miss rate of 0.5% per program reference.

P11.13 Given the assumptions of Problem 11.12, compute the average memory latency for a system with no level-3 cache and only 200 cycle latency to memory (since the level-3 lookup is no longer performed before initiating the fetch from memory). Which system performs better? What is the breakeven miss rate per program reference for the two systems (i.e., the level-3 miss rate at which both systems provide the same performance)?

P11.14 Assume a processor similar to the Hewlett-Packard PA-8500, with only a single level of data cache. Assume the cache is virtually indexed but physically tagged, is four-way associative with 128-byte lines, and is 512 KB in size. In order to snoop coherence messages from the bus, a reverse-address translation table is used to store physical-to-virtual address mappings stored in the cache. Assuming a fully associative reverse-address translation table and 4K-byte pages, how many entries must it contain so that it can map the entire data cache?

P11.15 Given the assumptions of Problem 11.14, describe a reasonable set-associative organization for the RAT that is still able to map the entire data cache.

P11.16 Given the assumptions of Problem 11.14, explain the implications of a reverse-address translation table that is not able to map all possible entries in the data cache. Describe the sequence of events that must occur whenever a reverse-address translation table entry is displaced due to replacement.

Problems 17 through 19

In a two-level cache hierarchy, it is often convenient to maintain inclusion between the primary cache and the secondary cache. A common mechanism for tracking inclusion is for the level-2 cache to maintain presence bits for each level-2 directory entry that indicate the line is also present in the level-1 cache. Given the following assumptions, answer the following questions:

- Presence bit mechanism for maintaining inclusion
- 4K virtual memory page size
- Physically indexed, physically tagged 2-Mbyte eight-way set-associative cache with 64-byte lines

P11.17 Given a 32K-byte eight-way set-associative level-1 data cache with 32-byte lines, outline the steps that the level-2 controller must follow whenever it removes a cache line from the level-2 cache. Be specific, explain each step, and make sure the level-2 controller has the information it needs to complete each step.

P11.18 Given a virtually indexed, physically tagged 16K-byte direct-mapped level-1 data cache with 32-byte lines, how does the level-2 controller's job change?

P11.19 Given a virtually indexed, virtually tagged 16K-byte direct-mapped level-1 data cache with 32-byte lines, are presence bits still a reasonable solution or is there a better one? Why or why not?

P11.20 Figure 11.8 explains read-set tracking as used in high-performance implementations of sequentially consistent multiprocessors. As shown, a potential ordering violation is detected by snooping the load queue and refetching a marked load when it attempts to commit. Explain why the processor should not refetch right away, as soon as the violation is detected, instead of waiting for the load to commit.

P11.21 Given the mechanism referenced in Problem 11.20, false sharing (where a remote processor writes the lower half of a cache line, but the local processor reads the upper half) can cause additional pipeline refetches. Propose a hardware scheme that would eliminate such refetches. Quantify the hardware cost of such a scheme.

P11.22 Given Problem 11.21, describe a software approach that would derive the same benefit.

P11.23 A chip multiprocessor (CMP) implementation enables interesting combinations of on-chip and off-chip coherence protocols. Discuss all combinations of the following coherence protocols and implementation approaches and their relative advantages and disadvantages. On-chip, consider update and invalidate protocols, implemented with snooping and directories. Off-chip, consider invalidate protocols, implemented with snooping and directories. Which combinations make sense? What are the tradeoffs?

P11.24 Assume that you are building a fine-grained multithreaded processor similar to the Tera MTA that masks memory latency with a large number of concurrently active threads. Assuming your processor supports 100 concurrently active threads to mask a memory latency of one hundred 1-ns processor cycles. Further assume that you are using conventional DRAM chips to implement your memory subsystem. Assume the DRAM chips you are using have a 30-ns command occupancy, i.e., each command (read or write) occupies the DRAM chip interface for 30 ns. Compute the minimum number of independent DRAM chip interfaces your memory controller must provide to prevent your processor from stalling by turning around a DRAM request for every processor cycle.

P11.25 Assume what is described in Problem 11.24. Further, assume your DRAM chips support page mode, where sequential accesses of 8 bytes each can be made in only 10 ns. That is, the first access requires 30 ns, but subsequent accesses to the same 512-byte page can be satisfied in 10 ns. The scientific workloads your processor executes tend to perform unit stride accesses to large arrays. Given this memory reference behavior, how many independent DRAM chip interfaces do you need now to prevent your processor from stalling?

P11.26 Existing coarse-grained multithreaded processors such as the IBM Northstar and Pulsar processors only provide in-order execution in the core. Explain why or why not coarse-grained multithreading would be effective with a processor that supports out-of-order execution.

P11.27 Existing simultaneous multithreaded processors such as the Intel Pentium 4 also support out-of-order execution of instructions. Explain why or why not simultaneous multithreading would be effective with an in-order processor.

P11.28 An IMT design with distributed processing elements (e.g., multiscalar or TLS) must perform some type of load balancing to ensure that each processing element is doing roughly the same amount of work. Discuss hardware- and software-based load balancing schemes and comment on which might be most appropriate for both multiscalar and TLS.

P11.29 An implicit multithreaded processor such as the proposed DMT design must insert instructions into the reorder buffer out of program order. This implies a complex free-list management scheme for tracking the available entries in the reorder buffer. The physical register file that is used in existing out-of-order processors also requires a similar free-list management scheme. Comment on how DMT ROB management differs, if at all, from free-list management for the physical register file. Describe such a scheme in detail, using a diagram and pseudocode that implements the management algorithm.

P11.30 The DEE proposal appears to rely on fairly uniform branch prediction rates for its limited eager execution to be effective. Describe what happens if branch mispredictions are clustered in a nonuniform distribution (i.e., a mispredicted branch is likely to be followed by one or more other mispredictions). What happens to the effectiveness of this approach? Use an example to show whether or not DEE will still be effective.

P11.31 A recent study shows that the TLS architecture benefits significantly from silent stores. Silent stores are store instructions that write a value to memory that is already stored at that location. Create a detailed sample execution that shows how detecting and eliminating a silent store can substantially improve performance in a TLS system.

P11.32 Preexecution of conditional branches in a redundant runahead thread allows speedy resolution of mispredicted branches, as long as branch instances from both threads are properly synchronized. Propose a detailed design that will keep the runahead thread and the main thread synchronized for this purpose. Identify the design challenges and quantify the cost of such a hardware unit.

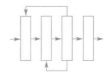

Index

A

Aargon, J. L., 481
Abraham, Santosh G., 472
accelerated graphics port (AGP), 155, 162
accumulator, 7
Acosta, R., 208, 382
active-/redundant-stream SMT, 610–612
addressability, 524
address bellow, 418
address resolution buffer (ARB), 607, 609
Adve, S. V., 267, 579
affector branches, 489
affector register file (ARF), 489–490
Agarwal, A., 589
Agerwala, Tilak, 23, 87, 91, 378, 380, 397, 440
aggregate degree of parallelism, 24
agree predictor, 477–478
Akkary, H., 562, 602, 604
aliasing, 266, 472
    from load bypassing, 269–270
    reducing with predictors, 472–481
    virtual address, 142
Allen, Fran, 373, 376, 440
Allen, M., 186, 187, 224, 423
allocator, 353–355
alloyed history predictors, 482–483
Alpert, Don, 381, 405, 407, 440
Alpha (AXP) architecture, 62, 387–392
    PowerPC vs., 321–322
    synchronization, 565

translation misses, 145
value locality, 525
Alpha 21064, 383, 386, 388–389
Alpha 21064A, 387
Alpha 21164, 322, 389–390
Alpha 21264, 236, 382, 385, 390–391, 471, 512
Alpha 21264A, 387
Alpha 21364, 391
Alpha 21464, 383, 391–392, 477
Alsup, Mitch, 440
AltiVec extensions, 428, 431
ALU forwarding paths, 79–84
ALU instructions, 62–65
    leading, in pipeline hazards, 79–80, 82–84
    processing. See register data flow techniques
    RAW hazard worst-case penalties, 77–78, 80
    specifications, 63, 65
    unifying procedure and, 66–67
ALU penalties
    in deep pipelines, 94
    reducing via forwarding paths, 79–84
Amdahl, Gene, 6, 18, 373–374, 376–377, 440
Amdahl's law, 17–18, 21, 220
AMD 29000, 410
AMDAHL 470V/7, 60–61
AMD K5, 196, 381, 383, 385, 410–411
    predecoding, 198, 199

AMD K6 (NexGen Nx686), 233, 411–412
AMD K7 (Athlon), 382, 383, 385, 412–413, 454
AMD K8 (Opteron), 134–135, 383, 417
analysis, digital systems design, 4
Anderson, Willie, 422
Ando, Hisashige, 435
anomalous decisions, 461–462
anti-dependence. See WAR data dependence
Apollo DN10000, 382
Apple Computer
    Macintosh, 156, 570
    PowerPC alliance, 302, 398, 424–425
architected register file (ARF), 239
architecture, 5–6. See also instruction set architecture (ISA)
A ring, 436
arithmetic operation tasks, 61–62
arithmetic pipelines, 40, 44–48
A/R-SMT, 610–612
Asprey, T., 393
assembly instructions, 7
assembly language, 7
assign tag, 241
associative memory, 119–120, 146
Astronautics ZS-1, 378–379
Athlon (AMD K7), 382, 383, 385, 412–413, 454
atomic operations, 561, 563
AT&T Bell Labs, 381
Auslander, Marc, 380

623

Austin, T., 277, 612
average memory reference
  latency, 574
average value locality, 526
Avnon, D., 407

B

backwards branch, 456
Baer, J., 153, 276
Bailey, D., 386
Ball, Thomas, 456–457
Ball/Larus heuristics, 456–457
bandwidth, 40, 107–110
  defined, 108
  improving, 273–274
    with caching, 117–118
    cost constraints on, 109
  infinite, 110
  instruction fetch, 192–195, 223
  measurement of, 108–109
  peak *vs.* sustainable, 108–109
Bannon, Pete, 389, 390, 440
Barreh, J., 402
baseline scalar pipelined machine,
  28–29
Bashe, C., 370
Bashteen, A., 30
Beard, Doug, 409
Becker, M., 425
Benschneider, B., 390
Bentley, B., 440
Bertram, Jack, 373
biasing bit, 477
bimodal branch predictor, 323–324,
  459–462, 473–474
binding, 354–355
Bishop, J., 432
bit-line precharging, 130
Blaauw, Gerrit, 370
Black, B., 15
Blanchard, T., 394
Blanck, Greg, 203, 433
Blasgen, M., 401
Bloch, Erich, 39, 370, 440
blocks. *See* caching/cache memory
Bluhm, Mark, 407, 440
Bodik, R., 537, 541
bogus branch, 499

Boney, Joel, 435, 440
Borkenhagen, J., 432
Bose, Pradip, 380, 440
Bowhill, W., 390
brainiacs, 321–322, 384, 386–387
branch address calculator (BAC),
  339, 346
branch classification, 494–495
branch condition resolution
  penalties, 222–223
branch condition speculation, 223,
  224–226, 228–229
branch confidence prediction,
  501–504
branch delay slot, 455–456
branch filtering, 480
branch folding, 227–228
branch hints, 456
branch history register (BHR),
  232–233, 341–343, 462–463
branch history shift register
  (BHSR), 232–233
branch history table (BHT),
  230–231, 465
  in PowerPC 620, 307–309
branch instruction address (BIA), 223
branch instructions, 62–63
  anticipating, 375
  conditional. *See* conditional
    branch instructions
  control dependences and, 72
  IBM ACS-1 techniques, 374–375
  leading, in pipeline hazards, 81,
    86–87
  PC-relative addressing mode,
    92–93
  prepare-to-branch, 375
  processing. *See* instruction flow
    techniques
  unifying procedure and, 66–67
branch penalties. *See also*
  instruction flow techniques
  in deep pipelines, 94–95, 453–454
  reducing, 91–93
  worst-case, RAW hazard, 77, 80
branch prediction, 223–228, 453
  advanced techniques, 231–236
  backwards taken/forwards
    not-taken, 456

bimodal, 323–324, 459–462,
  473–474
branch condition speculation,
  223, 224–226, 228–229
branch folding, 227–228
branch target speculation,
  223–224
branch validation, 229–230
correlated (two-level adaptive),
  232–236, 462–469
counter-based predictors, 228
decoder, 346
dynamic. *See* dynamic branch
  prediction
experimental studies on, 226–228
gshare scheme, 235–236, 323–324
history-based predictors, 225–228
hybrid branch predictors, 491–497
  branch classification, 494–495
  multihybrid predictor,
    495–496
  prediction fusion, 496–497
  static predictor selection, 493
  tournament predictor, 491–493
in Intel P6, 341–343
misprediction recovery, 228–231
multiple branches and, 236
in PowerPC 620, 307–311
preexecution and, 614–615
slipstreaming and, 614
static. *See* static branch prediction
taken branches and, 236–237
training period, 471
Yeh's algorithm, 341–343
branch-resume cache, 421
branch target address (BTA),
  220–221, 223
branch target address cache
  (BTAC), 230–231,
  307–309, 497
branch target buffer (BTB),
  223–224, 497–500
  block diagram, 343
  history-based branch
    prediction, 226
  in Intel P6, 339–343
branch target prediction, 497–501
branch target speculation, 223–224
Brennan, John, 418

INDEX **625**

broadcast write-through policy, 569
Brooks, Fred, 370, 440
BTFNT branch prediction, 456
Bucholtz, W., 39, 370
buffers
    branch target. *See* branch target buffer (BTB)
    cache, for disk drives, 156–157
    collapsing, 504–506
    completion, 190, 305, 312, 313
    dispatch, 190, 201–203
    elastic history (EHB), 485
    instruction buffer network, 195
    in PowerPC 620, 303–304, 311, 312
    instruction reuse, 530–533
    interpipeline-stage, 186–190
    multientry, 187–190
    for out-of-order designs, 385
    reorder. *See* reorder buffer (ROB)
    reservation stations. *See* reservation stations
    single-entry, 186–187, 188
    translation lookaside. *See* translation lookaside buffer (TLB)
Burger, D., 614
Burgess, Brad, 425, 426, 428, 440
Burkhardt, B., 409
Burtscher, M., 527, 539, 553
busses, 161–165
    bus turnaround, 131
    common data (CDB), 247–248
    design parameters, 163–165
    design trends, 165
    I/O, 162
    processor busses, 161–162
    simple, 163
    split-transaction, 163, 581–582
    storage, 162–163
busy vector, 374

C

cache coherence, 567–576
    average memory reference latency, 574
    hardware cache coherence, 168
    implementation, 571–574
    bandwidth scaling and, 573
    communication (dirty) misses and, 574
    directories, 573–574
    hybrid snoopy/directory systems, 574
    memory latency and, 573–574
    snooping, 572–573
    inclusive caches, 575–576
    invalidating protocols, 568, 569–571
    multilevel caches, 574–576
    noninclusive caches, 575
    relaxation of SC total order, 578–579
    software, 168
    updating protocols, 568–569
    virtual memory and, 576
caching/cache memory, 112, 115–127
    attributes of, 111
    average latency, 115, 574
    bandwidth benefits, 117–118
    blocks
        block (line) size, 118
        block offset, 118–119
        block organization, 119–120, 123
        evicting, 120–121
        FIFO replacement, 120
        locating, 118–120
        LRU (least recently used) replacement, 120–121
        multiword per block, 146–147
        NMRU (not most recently used) replacement, 121
        random replacement, 121
        replacement policies, 120–121
        single word per block, 146–147
        updating, 121–122
    branch-resume cache, 421
    branch target address cache, 230–231, 307–309, 497
    cache buffer, 156–157
    cache coherence. *See* cache coherence
    cache misses, 115, 265
        cache organization and, 125
        miss classification, 122–125
        miss rates, 115–117
    conflict aliasing solution, 473
    CPI estimates, 116–117
    data cache. *See* D-cache
    design parameters, 123
    direct-mapped, 119, 146–147, 509
    dual-ported data cache, 273–274
    fully-associative, 119–120, 147
    global *vs.* local hit rates, 115–116
    in IBM ACS-1, 375–376
    implementation, 146–147
    instruction cache. *See* I-cache
    multilevel caches, 574–576
    in multiprocessor systems, 567
    nonblocking cache, 274–275, 319–320
    noninclusive cache, 575
    organization and design of, 118–122
    in PowerPC 620, 318–320
    prefetching cache, 274, 275–277
    to reduce memory latency, 274–277
    row buffer cache, 131
    set-associative, 119, 120, 146–148
    shared caches, 586
    speculative versioning cache, 609–610
    trace cache, 236–237, 415, 506–508
    translation lookaside buffer, 149–150
    two-level example, 125–127
    virtually indexed, 152–153
    write-allocate policy, 121
    writeback cache, 122
    write-no-allocate policy, 121–122
    write-through cache, 121–122
caching-inhibited (Ca) bit, 142–143
Calder, Brad, 261, 262, 458, 509, 527, 541
call rule, 457
call-subgraph identities, 524
capacity, of memory, 110
capacity aliasing, 473
capacity misses, 123–125

Capek, Peter, 440
Carmean, D., 96
CDC 6400, 226
CDC 6600, 29, 39, 185–186, 372, 373, 376, 588
Censier, L., 567
centralized reservation stations, 201–203
Chan, K., 395
Chang, Al, 380
Chang, Po-Yung, 480, 493, 494
change (Ch) bit, 142–143
Charlesworth, A., 572
Chen, C., 436
Chen, T., 276
Chessin, Steve, 435
Chin Ching Kau, 429
chip multiprocessors (CMPs), 324, 584–588
    cache sharing in, 586
    coherence interface sharing, 586
    drawbacks of, 586–587
chip select (CS) control line, 132
chip set, 162
choice PHT, 478–479
choice predictor, 473
Christie, Dave, 411, 417
Chrysos, G., 273
Circello, Joe, 423, 424, 440
CISC architecture, 9, 16
    instruction decoding in, 196–198
    Intel 486 example, 89–91
    pipelined, 39
    superscalar retrofits, 384–385
Clapp, R., 574
clock algorithm, 140
clock frequency
    deeper pipelining and, 94
    increasing with pipelining, 13, 17
    microprocessor evolution, 2
    speed demon approach, 321–322, 384, 386
clustered reservation stations, 202
coarse-grained disk arrays, 157–158
coarse-grained multithreading (CGMT), 584, 585, 589–592
    cost of, 592
    fairness, guaranteeing, 590
    performance of, 592

thread priorities, 591
thread-switch penalty, 589–590
thread switch state machine, 591–592
Cocke, John, 23, 87, 91, 370, 372, 373–374, 376, 378, 380, 397, 440
code generation, 237–238
coherence interface, 561, 586
Cohler, E., 379
cold misses, 123–125
collapsing buffer, 504–506
Collins, J., 613
column address strobe (CAS), 130
Colwell, R., 9, 329, 413, 440
common data bus (CDB), 247–248
communication misses, 574
Compaq Computer, 383, 387, 493
compare-and-swap primitive, 563–564
completed store buffer, 268–269
completion buffer, 426
    in PowerPC 620, 305, 312, 313
completion stage. *See* instruction completion stage
complex instructions, 346
complex instruction set computer (CISC). *See* CISC architecture
compulsory aliasing, 472–473
compulsory misses, 123–125
computed branches, 524
computer system overview, 106–107
conditional branch instructions
    control dependences and, 72
    resolution penalties, 222–223
    specifications, 63–65
    target address generation penalties, 220–221
    unifying procedure and, 66–67
conflict aliasing, 473
conflict misses, 123–125
Connors, D. A., 534
Conte, Thomas M., 236, 504
contender stack, 374
control dependences, 72
    examining for pipeline hazards, 76–77

resolving control hazards, 78, 86–87
resolving in IMT designs, 601–605
control flow graph (CFG), 218, 535
control independence, 601–602
convergent algorithms, 524
Conway, Lynn, 374, 440
correct/incorrect registers (CIRs), 502
corrector predictor, 490
correlated (two-level adaptive) branch predictors, 232–236, 462–469
Corrigan, Mike, 431
cost/performance tradeoff model
    deeper pipelines, 95–96
    pipelined design, 43–44
Cox, Stan, 440
Crawford, John, 39, 87, 89, 90, 181, 405
Cray, Seymour, 588
CRAY-1, 29
CRISP microprocessor, 381
CRT (cathode-ray tube) monitor, 107, 154
CSPI MAP 200, 379
Culler, Glen, 379
Culler-7, 379
cumulative multiply, 370
cycles per instruction (CPI), 11
    deeper pipelining and, 94, 96
    perfect cache, 117
    reducing, 12, 91–93
cycle time, 11, 12–13
Cydrome Cydra-5, 418
Cyrix 6×86 (M1), 407–409

D

dancehall organization, 566–567
data cache. *See* D-cache
data-captured scheduler, 260
data dependence graph (DDG), 244
data dependences, 71–72
    false. *See* false data dependences
    memory. *See* memory data dependences
    register. *See* register data dependences
data flow branch predictor, 489–491

INDEX

data flow eager execution, 548, 549
data flow execution model, 245
data flow graph (DFG), 535
data flow limit, 244–245, 520–521
    exceeding, 261–262
data flow region reuse, 534–535
data movement tasks, 61–62
data redundancy, 523
datascalar architecture, 614
data translation lookaside buffer
    (DTLB), 362, 364
Davies, Peter, 417
DAXPY example, 266–267
D-cache, 62
    multibanked, 205
    Pentium processor, 184
    PowerPC 620, 318–319
    TYP instruction pipeline, 68–69
DEC Alpha. *See* Alpha (AXP)
    architecture
DEC PDP-11, 226
DEC VAX 11/780, 6
DEC VAX architecture, 6, 39,
    380, 387
decoder shortstop, 414
decoupled access-execute
    architectures, 378–379
DEE. *See* disjoint eager execution
    (DEE)
deeply pipelined processors, 94–97
    branch penalties in, 94–95,
        453–454
    Intel P6 microarchitecture, 331
Dehnert, Jim, 418
Dekker's algorithm, 576–577
DeLano, E., 393
delayed branches, 78
demand paging, 137, 138–141
Denelcor HEP, 588
Denman, Marvin, 427, 440
dependence prediction, 273
destination allocate, 240, 241
destructive interference, 477
Diefendorff, Keith, 186, 187, 224,
    410, 413, 422, 423, 424–425,
    428, 437, 440
Diep, T. A., 301, 302
digital systems design, 4–5
direction agreement bit, 477

direct-mapped memory, 119,
    146–147
direct-mapped TLB, 149–150
direct memory access (DMA), 168
directory approach, cache
    coherence, 573–574
DirectPath decoder, 412
dirty bit, 122
dirty misses, 574
disjoint eager execution (DEE), 236,
    503, 602–604
    attributes of, 603
    control flow techniques, 603, 604
    data dependences, resolving, 606
    interthread false dependences,
        resolving, 607
    physical organization, 604–605
disk arrays, 157–161
disk drives, 111, 155–157
disk mirroring, 159
disks, 111, 153
dispatch buffers, 190, 201–203
dispatching. *See* instruction
    dispatching
dispatch stack, 382
distributed reservation stations,
    201–203
Ditzel, Dave, 381, 435
DIVA proposal, 612–613
diversified pipelines, 179, 184–186
DLX processor, 71
Dohm, N., 440
Domenico, Bob, 373
done flag, 420
DRAM, 108, 112
    access latency, 130–131
    addressing, 130, 133–134
    bandwidth measurement, 108–109
    capacity per chip, 128–129
    chip organization, 127–132
    memory controller organization,
        132–136
    Rambus (RDRAM), 131–132
    synchronous (SDRAM), 129, 131
    in 2-level cache hierarchy, 126
Dreyer, Bob, 405
Driscoll, M. A., 562, 602, 604
dual inline memory module
    (DIMM), 127

dual operation mode, 382
dual-ported data cache, 273–274
Dunwell, Stephen, 370
dynamic branch prediction, 228–231,
    458–491
    with alternative contexts, 482–491
        alloyed history predictors,
            482–483
        data flow predictors, 489–491
        DHLF predictors, 485–486
        loop counting predictors,
            486–487
        path history predictors,
            483–485
        perceptron predictors, 487–489
        variable path length
            predictors, 485
    basic algorithms, 459–472
        global-history predictor,
            462–465
        gshare, 469–471
        index-sharing, 469–471
        local-history predictor,
            465–468
        mispredictions, reasons for,
            471–472
        per-set branch history
            predictor, 468
        pshare, 471
        Smith's algorithm, 459–462
        two-level prediction tables,
            462–469
    branch speculation, 228–229
    branch validation, 229–230
    interference-reducing predictors,
        472–481
        agree predictor, 477–478
        bi-mode predictor, 473–474
        branch filtering, 480
        gskewed predictor, 474–477
        selective branch inversion,
            480–482
        YAGS predictor, 478–480
    PowerPC 604 implementation,
        230–231
dynamic execution core, 254–261
    completion phase, 254, 256
    dispatching phase, 255
    execution phase, 255

dynamic execution core—*Cont.*
  instruction scheduler, 260–261
  instruction window, 259
  micro-dataflow engine, 254
  reorder buffer, 256, 258–259
  reservation stations, 255–258
dynamic history length fitting
    (DHLF), 485–486
dynamic instruction reuse, 262
dynamic instruction scheduler,
    260–261
dynamic multithreading (DMT), 562
  attributes of, 603
  control flow techniques, 603, 604
  interthread false dependences, 607
  memory RAW resolution,
    608–609
  out-of-order thread creation, 604
  physical organization, 604
  register data dependences,
    606–607
dynamic pipelines, 180, 186–190
dynamic random-access memory.
    See DRAM
dynamic-static interface (DSI),
    8–10, 32

E
Earle, John, 377
Earle latch, 41
Eden, M., 407
Eden, N. A., 478
Edmondson, John, 389, 390, 440
Eickemeyer, R. J., 589
elastic history buffer (EHB), 485
11-stage instruction pipeline,
    57–58, 59
Emer, Joel, 273, 392
Emma, P. B., 500
engineering design components, 4
enhanced gskewed predictor,
    476–477
EPIC architecture, 383
error checking, 524
error-correction codes (ECCs), 159
Ethernet, 155
Evers, Marius, 471, 495
exceptions, 207–208, 265, 266

exclusive-OR hashing function,
    460*n*, 469
execute permission (Ep) bit, 142
execute stage. *See* instruction
    execution (EX) stage
execution-driven simulation, 14–15
explicitly parallel instruction
    computing, 383
external fragmentation, 50, 53, 61–71

F
Faggin, Federico, 405
Fagin, B., 500
false data dependences
  in IMT designs, 605
  register reuse and, 237–239
  resolved by Tomasulo's
    algorithm, 253–254
  write-after-read. *See* WAR data
    dependence
  write-after-write. *See* WAW data
    dependence
fast Fourier transform (FFT),
    244–245
FastPath decoder, 417
fast schedulers, 415
fault tolerance, of disk arrays,
    157–160
Favor, Greg, 410, 411
Feautrier, P., 567
fetch-and-add primitive, 563–564
fetch group, 192–195
fetch stage. *See* instruction
    fetch (IF) stage
Fetterman, Michael, 413
Fields, B., 537, 541
Fillo, M., 589
fill time, 50
final reduction, floating-point
    multiplier, 46
fine-grained disk arrays, 157–158
fine-grained multithreading, 584,
    585, 588–589
fine-grained parallelism, 27
finished load buffer, 272–273
finished store buffer, 268–269
finite-context-method (FCM)
    predictor, 539

finite state machine (FSM), 225–226
first in, first out (FIFO), 120, 140
first opcode markers, 345
Fisher, Josh, 26, 31, 378, 440, 458
Fisher's optimism, 26
five-stage instruction pipeline, 55–56
  Intel 486 example, 89–91
  MIPS R2000/R3000 example,
    87–89
floating-point buffers (FLBs), 246
floating-point instruction
    specifications, 63, 179
floating-point multiplier, 45–48
floating-point operation stack
    (FLOS), 247
floating-point registers (FLRs), 246
  tag fields used in, 248–250
  value locality of, 526
floating-point unit (FPU), 203–205
  IBM 360 design, 246–247
  IBM RS/6000 implementation,
    242–243
Flynn, Mike, 9, 25, 45–48, 373, 374,
    377, 440, 519
Flynn's bottleneck, 25
forwarding paths, 79–81
  ALU, 79–84
  critical, 81
  load, 84–86
  pipeline interlock and, 82–87
forward page tables, 143–144
Foster, C., 25, 377
four-stage instruction pipeline, 28,
    56–58
frame buffer, 154–155
Franklin, M., 262, 273, 527,
    540, 604
free list (FL) queue, 243
Freudenberger, Stephan M., 458
Friendly, Daniel H., 506
fully-associative memory,
    119–120, 146
fully-associative TLB, 150–151
functional simulators, 13, 14–15
function calls, 500–501, 524
fused multiply-add (FMA)
    instructions, 243, 370
fusion table, 496–497
future thread, 602

G

Gabbay, F., 261, 527, 540
Gaddis, N., 396, 397
Garibay, Ty, 407, 409
Garner, Robert, 432, 440
Gelsinger, Patrick, 405
generic computations, 61–62
GENERIC (GNR) pipeline, 55–56
Gharachorloo, K., 267, 579
Gibson, G., 158
Gieseke, B., 391
Glew, Andy, 413, 440
global completion table, 430
global (G) BHSR, 233, 235–236
global (g) PHT, 233, 469
global-history two-level branch predictor, 462–465
global hit rates, 115–116
Gloy, N. C., 234
glue code, 524
glueless multiprocessing, 582–583
Gochman, Simcha, 416, 417, 486
Goldman, G., 438
Golla, Robert, 426
Gonzalez, A., 533
Goodman, J., 118
Goodrich, F., 424
Gowan, M., 391
graduation, 420
graphical display, 153, 154–155
Gray, R., 440
Green, Dan, 409
Greenley, D., 438
Grochowski, Ed, 405
Grohoski, Greg, 193, 224, 242, 380, 397, 399, 401, 409, 440
Gronowski, P., 390
Groves, R. D., 193, 224, 242
Grundmann, W., 386, 440
Grunwald, Dirk, 493, 509
gskewed branch predictor, 474–477
guard rule, 457
Gwennap, L., 324, 409, 411, 415

H

Haertel, Mike, 440
Halfhill, T., 411, 412, 415, 421, 431
Hall, C., 400
HaL SPARC64, 382, 385, 435–437
Hammond, L., 562
Hansen, Craig, 417
Harbison, S. P., 262
hardware cache coherence, 168
hardware description language (HL), 5
hardware instrumentation, 14
hardware RAID controllers, 160–161
hardware TLB miss handler, 145
Hartstein, A., 96, 454
hashed page tables, 143, 144–145
Hauck, Jerry, 395
hazard register, 77
Heinrich, J., 455
Hennessy, John, 71, 417, 456
Henry, Dana S., 496
Henry, Glenn, 410
Hester, P., 401
high-performance substrate (HPS), 196, 380–381, 409
Hill, G., 435
Hill, M., 120, 123, 472, 581
Hinton, Glenn, 382, 403, 404, 413, 415, 416, 454, 501, 508
history-based branch prediction, 225–228
hits, 115–116, 135
Hochsprung, Ron, 424
Hoevel, L., 9
Hollenbeck, D., 394
Hopkins, Marty, 380, 440
Horel, T., 122, 439
Horowitz, M., 380
Horst, R., 382
HP Precision Architecture (PA), 62, 392–397
HP PA-RISC Version 1.0, 392–395
    PA 7100, 384, 392, 393, 394
    PA 7100LC, 384, 393–394
    PA 7200, 394–395
    PA 7300LC, 394
HP PA-RISC Version 2.0, 395–397
    PA 8000, 199, 324, 382, 385, 395–397, 579
    PA 8200, 397
    PA 8500, 397
    PA 8600, 397
    PA 8700, 397, 478
    PA 8800, 397, 586
    PA 8900, 397, 584
HP Tandem division, 612
Hrishikesh, M. S., 454
Hsu, Peter, 417, 418, 419
Huang, J., 533
Huck, J., 395
Huffman, Bill, 417
Hunt, D., 397
Hunt, J., 397
Hwu, W., 196, 380, 534
HyperSPARC, 384, 434
hyperthreading, 599

I

IBM
    pipelined RISC machines, 91–93
    PowerPC alliance, 302, 383, 398, 424–425
IBM 360, 6
IBM 360/85, 115
IBM 360/91, 6, 41, 201, 247–254
IBM 360/370, 7
IBM 370, 7, 226
IBM 7030. *See* IBM Stretch computer
IBM 7090, 372, 376
IBM 7094, 377
IBM ACS-1, 369, 372–377
IBM ACS-360, 377
IBM America, 380
IBM Cheetah, 380
IBM ES/9000/520, 377, 382
IBM/Motorola PowerPC.
    See PowerPC
IBM Northstar, 302, 432, 575, 589, 592
IBM OS/400, 302
IBM Panther, 380
IBM POWER architecture, 62, 383, 397–402
    brainiac approach, 386–387
    PowerPC. *See* Power PC
    PowerPC-AS extension, 398, 431–432
    RIOS pipelines, 399–401
    RS/6000. *See* IBM RS/6000
    RSC implementation, 401
IBM POWER2, 322, 387, 401–402

630 INDEX

IBM POWER3, 301–302, 322–323, 385, 429–430
IBM POWER4, 122, 136, 301–302, 381, 382, 430–431, 584
   branch prediction, 471
   buffering choices, 385
   chip multiprocessor, 587–588
   key attributes of, 323–324
   shared caches, 586
IBM pSeries 690, 109
IBM Pulsar, 575, 589, 592
IBM Pulsar/RS64-III, 302, 432
IBM 801 RISC, 380
IBM RS/6000, 375, 380
   branch prediction, 224, 227
   first superscalar workstation, 382
   FPU register renaming, 242–243
   I-cache, 193–195
   MAF floating-point unit, 203
IBM S/360, 6, 372
IBM S/360 G5, 611–612
IBM S/360/85, 375, 377
IBM S/360/91, 372–373, 375, 376, 380, 520
IBM S/390, 167
IBM s-Star/RS64-IV, 302, 432
IBM Star series, 302, 432
IBM Stretch computer, 39, 369–372
IBM Unix (AIX), 302
IBM xSeries 445, 125–126
I-cache, 62
   in Intel P6, 338–341
   in TEM superscalar pipeline, 191–195
   in TYP instruction pipeline, 68–69
identical computations, 48, 50, 53, 54
idling pipeline stages. *See* external fragmentation
IEEE Micro, 439
implementation, 2, 5–6
implicit multithreading (IMT), 562, 600–610
   control dependences, 601–605
      control independence, 601–602
      disjoint eager execution (DEE), 602–604

      out-of-order thread creation, 604
      physical organization, 604–605
      thread sequencing/ retirement, 605
   dynamic. *See* dynamic multithreading (DMT)
   memory data dependences, 607–610
      implementation challenges, 609–610
      multiscalar ARB, 607, 609
      true (RAW) interthread dependences, 607–609
   multiscalar. *See* multiscalar multithreading
   register data dependences, 605–607
   thread-level speculation. *See* thread-level speculation (TLS)
IMT. *See* implicit multithreading (IMT)
inclusion, 575–576
independent computations, 48, 50–51, 53, 54
independent disk arrays, 157–158
indexed memory, 146
index-sharing branch predictors, 469–471
indirect branch target, 497
individual (P) BHSR, 233–235
individual (p) PHT, 233
inertia, 461
in-order retirement (graduation), 420
input/output (I/O) systems, 106–107, 153–170
   attributes of, 153
   busses, 160–165
   cache coherence, 168
   communication with I/O devices, 165–168
   control flow granularity, 167
   data flow, 167–168
   direct memory access, 168
   disk arrays, 157–161
   disk drives, 155–157
   graphical display, 153, 154–155

   inbound control flow, 166–167
   interrupts, 167
   I/O busses, 162
   keyboard, 153–154
   LAN, 153, 155
   long latency I/O events, 169–170
   magnetic disks, 153
   memory hierarchy and, 168–170
   memory-mapped I/O, 166
   modem, 153, 155
   mouse, 153–154
   outbound control flow, 165–166
   polling system, 166–167
   processor busses, 161–162
   RAID levels, 158–161
   snooped commands, 168
   storage busses, 162–163
   time sharing and, 169
instruction buffer network, 195
   in PowerPC 620, 303–304, 311, 312
instruction cache. *See* I-cache
instruction completion stage
   completion buffer, 426
   defined, 207
   in dynamic execution core, 254, 256
   PowerPC 620, 305, 312, 313, 318–320
   in superscalar pipeline, 206–209
instruction control unit (ICU), 412–413
instruction count, 11–12, 17
instruction cycle, 51–52, 55
instruction decode (ID) stage, 28, 55. *See also* instruction flow techniques
   Intel P6, 343–346
   in SMT design, 595
   in superscalar pipelines, 195–199
instruction dispatching, 199–203
   dispatching, defined, 202–203
   in dynamic execution core, 254–255, 256–257
   in PowerPC 620, 304, 311–315
instruction execution (EX) stage, 28, 55
   in dynamic execution core, 254, 255

INDEX **631**

in Intel P6, 355–357
for multimedia applications, 204–205
in PowerPC 620, 305, 316–318
in SMT design, 596
in superscalar pipelines, 203–206
instruction fetch (IF) stage, 28, 55
   instruction flow techniques. *See* instruction flow techniques
   in Intel P6, 334–336, 338–343
   in PowerPC 620, 303, 307–311
   in SMT design, 594
   in superscalar pipelines, 191–195
instruction fetch unit (IFU), 338–343
instruction flow techniques, 218–237, 453–518
   branch confidence prediction, 501–504
   branch prediction. *See* branch prediction
   high-bandwidth fetch mechanisms, 504–508
      collapsing buffer, 504–506
      trace cache, 506–508
   high-frequency fetch mechanisms, 509–512
      line prediction, 509–510
      overriding predictors, 510–512
      way prediction, 510
   performance penalties, 219–223, 453–454
      condition resolution, 222–223
      target address generation, 220–221
   program control flow, 218–219
   target prediction, 497–501
      branch target buffers, 497–500
      return address stack, 500–501
instruction grouper, 381–382
instruction groups, 430
instruction length decoder (ILD), 345
instruction-level parallelism (ILP), 3, 16–32
   data flow limit and, 520–521
   defined, 24
   Fisher's optimism, 26

Flynn's bottleneck, 25
limits of, 24–27
machines for, 27–32
   baseline scalar pipelined machine, 28–29
   Jouppi's classifications, 27–28
   superpipelined machine, 29–31
   superscalar machine, 31
   VLIW, 31–32
scalar to superscalar evolution, 16–24
   Amdahl's law, 17–18
   parallel processors, 17–19
   pipelined processors, 19–22
   superscalar proposal, 22–24
studies of, 377–378
instruction loading, 381
instruction packet, 425
instruction pipelining, 40, 51–54
instruction retirement stage
   defined, 207
   in IMT designs, 605
   in SMT design, 596–597
   in superscalar pipelines, 206–209
instruction reuse buffer, 530–533
   coherence mechanism, 532–533
   indexing/updating, 530–531
   organization of, 531
   specifying live inputs, 531–532
instruction select, 258
instruction sequencing tasks, 61–62
instruction set architecture (ISA), 1–2, 4, 6–8
   as design specifications, 7
   DSI placement and, 8–10
   innovations in, 7
   instruction pipelining and, 53–54
   instruction types and, 61–62
   of modern RISC processors, 62
   processor design and, 4
   as software/hardware contract, 6–7
   software portability and, 6–7
instruction set processor (ISP), 1–2
instruction set processor (ISP) design, 4–10
   architecture, 5–6. *See also* instruction set architecture (ISA)
   digital systems design, 4–5

   dynamic-static interface, 8–10
   implementation, 5–6
   realization, 5–6
instructions per cycle (IPC), 17
   brainiac approach, 321–322, 384, 386–387
   microprocessor evolution, 2
instruction splitter, 378
instruction steering block (ISB), 343–344
instruction translation lookaside buffer (ITLB), 339–341
instruction type classification, 61–65
instruction wake up, 257
instruction window, 259
integer functional units, 203–205
integer future file and register file (IFFRF), 413
integer instruction specifications, 63
Intel 386, 89, 90, 405
Intel 486, 39, 87, 89–91, 181–183, 405, 455, 456
Intel 860, 382
Intel 960, 402–405
Intel 960 CA, 382, 384, 403–404
Intel 960 CF, 405
Intel 960 Hx, 405
Intel 960 MM, 384, 405
Intel 4004, 2
Intel 8086, 89
Intel Celeron, 332
Intel IA32 architecture, 6, 7, 89, 145, 165, 329, 381. *See also* Intel P6 microarchitecture
   64-bit extension, 417, 565, 581
   decoding instructions, 196–198
   decoupled approaches, 409–417
      AMD K5, 410–411
      AMD K7 (Athlon), 412–413
      AMD K6 (NexGen Nx686), 411–412
      Intel P6 core, 413–415
      Intel Pentium 4, 415–416
      Intel Pentium M, 416–417
      NexGen Nx586, 410
      WinChip series, 410
   native approaches, 405–409
      Cyrix 6x86 (M1), 407–409
      Intel Pentium, 405–407

Intel Itanium, 383
Intel Itanium 2, 125–126
Intel P6 microarchitecture, 6,
    196–197, 329–367, 382
  basic organization, 332–334
  block diagram, 330
  decoupled IA32 approach,
      413–415
  front-end pipeline, 334–336,
      338–355
    address translation, 340–341
    allocator, 353–355
    branch misspeculation
        recovery, 339
    branch prediction,
        341–343, 467
    complex instructions, 346
    decoder branch prediction, 346
    flow, 345
    I-cache and ITLB, 338–341
    instruction decoder (ID),
        343–346
    MOB allocation, 354
    register alias table. *See* register
        alias table (RAT)
    reservation station allocation,
        354–355
    ROB allocation, 353–354
    Yeh's algorithm, 341–343
  memory operations, 337,
      361–364
    deferring, 363–364
    load operations, 363
    memory access ordering, 362
    memory ordering buffer,
        361–362
    page faults, 364
    store operations, 363
  novel aspects of, 331
  out-of-order core pipeline,
      336–337, 355–357
    cancellation, 356–357
    data writeback, 356
    dispatch, 355–356
    execution unit data paths, 356
    reservation station, 355–357
    scheduling, 355
  Pentium Pro block diagram, 331
  pipeline stages, 414

  pipelining, 334–338
  product packaging formats, 332
  reorder buffer (ROB), 357–361
    event detection, 360–361
    implementation, 359–360
    placement of, 357–358
    retirement logic, 358–360
    stages in pipeline, 358–360
  reservation station, 336, 355–357
  retirement pipeline, 337–338,
      357–361
    atomicity rule, 337
    external event handling,
        337–338
Intel Pentium, 136, 181–184, 196,
    382, 384, 405–407
  D-cache, 205
  pipeline stages, 406
Intel Pentium II, 329, 413–415
Intel Pentium III, 329, 413–415
Intel Pentium 4, 381, 382, 415–416
  branch misprediction penalty, 454
  buffering choices, 385
  data speculation support, 550
  hyperthreading, 599
  preexecution, 614
  resource sharing, 599
  SMT attributes of, 592, 597
  trace caching, 236–237, 508
Intel Pentium M, 416–417, 486–487
Intel Pentium MMX, 407
Intel Pentium Pro, 233, 329,
    381, 385
  block diagram, 331
  centralized reservation
      station, 201
  instruction decoding, 196, 197
  memory consistency
      adherence, 579
  P6 core, 413–415
Intel Xeon, 125
interference correction, 480–481
internal fragmentation, 49, 53,
    55–58
interrupts, 207
interthread memory dependences,
    607–609
interthread register dependences,
    605–607

intrablock branch, 505–506
Intrater, Gideon, 440
intrathread memory
    dependences, 607
intrathread register
    dependences, 605
invalidate protocols, 568, 569–571
inverted page tables, 143, 144–145
I/O systems. *See* input/output (I/O)
    systems
iron law of processor performance,
    10–11, 17, 96
issue latency (IL), 27–28
issue parallelism (IP), 28
issue ports, 414
issuing
  defined, 202–203
  in SMT design, 595–596

J

Jacobson, Erik, 502
Jimenez, Daniel A., 487, 488, 510
Johnson, L., 394
Johnson, Mike, 217, 271, 380,
    410, 426
Jouppi, Norm, 27–28, 276, 380, 440
Jourdan, S., 206
Joy, Bill, 435
Juan, Toni, 486
jump instructions, 63–65

K

Kaeli, David R., 500
Kagan, M., 407
Kahle, Jim, 425, 430
Kaminski, T., 378
Kane, G., 56, 87, 455
Kantrowitz, M., 440
Katz, R., 158
Keller, Jim, 236, 389, 390, 417
Keltcher, C., 134, 417
Kennedy, A., 428
Kessler, R., 391, 471, 493, 512
keyboard, 153–154
Kilburn, T., 136
Killian, Earl, 417, 421, 440
Kissell, Kevin, 440

INDEX 633

Klauser, A., 503
Kleiman, Steve, 435
Knebel, P., 394
Kogge, Peter, 40, 43
Kohn, Les, 437
Kolsky, Harwood, 370
Krewell, K., 437
Krishnan, V., 562
Kroft, D., 274
Krueger, Steve, 203, 433, 435, 440
Kuehler, Jack, 431
Kurpanek, G., 395

L

Laird, Michael, 440
Lamport, L., 267, 577
LAN (local area network), 107, 153, 155
Larus, James R., 456–457
last committed serial number (CSN), 436
last issued serial number (ISN), 436
last-n value predictor, 539
last value predictor, 538–539
latency, 107–110
 average memory reference, 574
 cache hierarchy, 115
 defined, 108
 disk drives, 156
 DRAM access, 130–131
 improving, 109
 input/output systems, 169–170
 issue (IL), 27–28
 load instruction processing, 277
 memory, 130–131, 274–279, 573–574
 operation (OL), 27
 override, 512
 queueing, 156
 rotational, 156
 scheduling, 548–549
 seek, 156
 time-shared systems, 169–170
 transfer, 156
 zero, 110
Lauterbach, Gary, 122, 438, 439
lazy allocation, 139
LCD monitor, 107, 154

least recently used (LRU) policy, 120–121, 140
Lee, Chih-Chieh, 473
Lee, J., 226, 342, 497
Lee, R., 395, 440
Lee, S.-J., 541
Leibholz, D., 391
Lempel, O., 405
Lesartre, G., 397
Lev, L., 438
Levitan, Dave, 301, 302, 429
Levy, H., 153
Lichtenstein, Woody, 379, 440
Lightner, Bruce, 434, 435
Lilja, D. J., 533
Lin, Calvin, 487, 488
linear address, 340
linearly separable boolean functions, 488
line prediction, 509–510
LINPAC routines, 267
Lipasti, Mikko H., 261, 278, 523, 527, 535
Liptay, J., 115, 377, 382
live inputs, 529, 531–532
live range, register value, 238
Livermore Automatic Research Computer (LARC), 370
load address prediction, 277–278
load buffer (LB), 361
load forwarding paths, 84–86
load instructions
 bypassing/forwarding, 267–273, 577
 leading, in pipeline hazards, 80–81, 84–86
 processing. See memory data flow techniques
 RAW hazard worst-case penalties, 77–78, 80
 specifications, 63–65
 unifying procedure and, 66–67
 value locality of, 525–527
 weak ordering of, 319, 321
load-linked/store-conditional (ll/stc) primitive, 563–565
load penalties
 in deep pipelines, 94

 reducing via forwarding paths, 79–81, 84–86
load prediction table, 277, 278
load/store queue (LSQ), 533, 596
load value prediction, 278
local-history two-level branch predictor, 465–468
local hit rate, 115–116
locality of reference, 113
local miss rate, 116
Loh, Gabriel H., 496
long decoder, 411
lookahead unit, 371
loop branch, 486
loop branch rule, 457
loop counting branch predictors, 486–487
loop exit rule, 457
loop header rule, 457
Lotz, J., 396, 397
Lovett, T., 574
Ludden, J., 440

M

machine cycle, 51–52
machine parallelism (MP), 22, 27
MacroOp, 412–413
Mahon, Michael, 392, 395
main memory, 111–112, 127–136
 computer system overview, 106–107
 DRAM. See DRAM
 memory controller, 132–136
 memory module organization, 132–134
  interleaved (banked), 133–134, 136
  parallel, 132–134
 organization of, 128
 reference scheduling, 135–136
 weak-ordering accesses, 319, 321
Mangelsdorf, S., 440
Manne, Srilatha, 481
map table, 239, 242–243
Markstein, Peter, 380
Markstein, Vicky, 380
Martin, M. M. K., 542n
Matson, M., 391

Maturana, Guillermo, 437
May, C., 302
May, D., 382
Mazor, Stanley, 405
McFarland, Mack, 410
McFarling, Scott, 234–235, 456, 458, 469, 491
McGeady, S., 404, 405
McGrath, Kevin, 417
McKee, S. A., 129
McLellan, H., 381, 389
McMahan, S., 409
Mehrotra, Sharad, 440
MEI coherence protocol, 570
Meier, Stephan, 412
Melvin, Steve, 380
memoization, 522, 527–528
memory alias resolution, 524
memory barriers, 580
memory consistency models, 576–581
memory cycles per instruction (MCPI), 117
memory data dependences, 72
    enforcing, 266–267
    examining for pipeline hazards, 75
    predicting, 278–279
    resolving in IMT designs, 607–610
memory data flow techniques, 262–279
    caching to reduce latency, 274–277
    high-bandwidth systems, 273–274
    load address prediction, 277–278
    load bypassing/forwarding, 267–273, 577
    load value prediction, 278
    memory accessing instructions, 263–266
    memory dependence prediction, 278–279
    ordering of memory accesses, 266–267
    store instruction processing, 265–266
memory hierarchy, 110–136
    cache memory. *See* caching/cache memory

    components of, 111–113
    computer system overview, 106–107
    implementation, 145–153
        accessing mechanisms, 146
        cache memory, 146–147
        TLB/cache interaction, 151–153
        translation lookaside buffer (TLB), 149–153
    locality, 113–114
    magnetic disks, 111
    main memory. *See* main memory
    memory idealisms, 110, 126–127
    register file. *See* register file
    SMT sharing of resources, 596
    virtual memory. *See* virtual memory
memory interface unit (MIU), 361
memory-level parallelism (MLP), 3
memory order buffer (MOB), 354, 361–362
memory reference prediction table, 275
memory-time-per-instruction (MTPI), 116–117
memory wall, 129
MEM pipeline stage, 67, 69
Mendelson, A., 261, 527, 540
Mergen, Mark, 380
MESI coherence protocol, 570–571, 607
Metaflow Lightning and Thunder, 434–435
meta-predictor M, 491–492
Meyer, Dirk, 412, 454
Michaud, Pierre, 472, 473, 474
microarchitecture, 6
microcode read-only memory (UROM), 345
microcode sequence (MS), 345
micro-dataflow engine, 254
micro-operations (μops), 196–197, 413–416
    in Intel P6, 331, 333–334
microprocessor evolution, 2–4
Microprocessor Reports, 387
Microsoft X Box, 131
millicoding, 431

Mills, Jack, 405
minimal control dependences (MCD), 606
minor cycle time, 29
MIPS architecture, 417–422
    synchronization, 565
    translation miss handling, 145
MIPS R2000/R3000, 56, 59–60, 71, 87–89
MIPS R4000, 30–31
MIPS R5000, 384, 421–422
MIPS R8000, 418–419
MIPS R10000, 199, 202, 324, 382, 419–421
    buffering choices, 385
    memory consistency adherence, 579
    pipeline stages, 419
Mirapuri, S., 30, 418
mismatch RAT stalls, 353
missed load queue, 275
miss-status handling register (MSHR), 582
MMX instructions, 329, 405, 409
modem, 153, 155
MOESI coherence protocol, 570
Monaco, Jim, 434, 440
monitors, 154–155
Montanaro, J., 389
Montoye, Bob, 380
Moore, Charles, 401, 425, 430, 439
Moore, Gordon, 2
Moore's Law, 2, 3
Moshovos, A., 273, 278
Motorola, 302, 383, 398, 422–425
Motorola 68K/M68K, 6, 39
Motorola 68040, 6
Motorola 68060, 381, 385, 423–424
Motorola 88110, 186, 187, 204, 224, 382, 385, 422–423
mouse, 153–154
Moussouris, J., 56, 87
Mowry, T. C., 562
Moyer, Bill, 422
MTPI metric, 116–117
Mudge, Trevor N., 478
Muhich, John, 425
Multiflow TRACE computer, 26

INDEX **635**

multihybrid branch predictor, 495–496
multimedia applications, 204–205
multiple threads, executing, 559–622
    explicit multithreading, 561, 584–599
        chip multiprocessors, 324, 584–588
        coarse-grained (CGMT), 584, 585, 589–592
        fine-grained (FGMT), 584, 585, 588–589
        SMT. *See* simultaneous multithreading (SMT)
    implicit multithreading. *See* implicit multithreading (IMT)
    multiprocessor systems. *See* multiprocessor systems
    multiscalar proposal. *See* multiscalar multithreading
    same thread execution. *See* redundant execution
    serial program parallelization, 561–562
    synchronization, 561, 562–565
multiply-add-fused (MAF) unit, 203–204
multiprocessor systems, 561, 565–584
    cache coherence. *See* cache coherence
    coherent memory interface, 581–583
    glueless multiprocessing, 582–583
    idealisms of, 566
    instantaneous write propagation, 567
    memory consistency, 576–581
        memory barriers, 580
        relaxed consistency, 579–581
        sequential consistency, 577–579
    uniform *vs.* nonuniform memory access, 566–567
multiscalar multithreading, 562
    address resolution buffer, 607, 609
    attributes of, 603
    control flow techniques, 602–604
    physical organization, 605
    register data dependences, 606
Myers, Glen, 402

N

Nair, Ravi, 227–228, 484
National Semiconductor Swordfish, 381, 395
negative interference, 477
neutral interference, 477
NexGen Nx586, 381, 410
NexGen Nx686, 233, 411–412
Nicolau, A., 26
Noack, L., 440
nonblocking cache, 274–275, 319–320
noncommitted memory serial number pointer, 436
non-data-captured scheduler, 260–261
noninclusive caches, 575
nonspeculative exploitation, value locality, 527–535
    basic block/trace reuse, 533–534
    data flow region reuse, 534–535
    indexing/updating reuse buffer, 530–531
    instruction reuse, 527, 529–533
    live inputs, specifying, 531–532
    memoization, 522, 527–528
    reuse buffer coherence mechanism, 532–533
    reuse buffer organization, 531
    reuse history mechanism, 529–533
    reuse mechanism, 533
nonuniform memory access (NUMA), 566–567
nonvolatility of memory, 110
Normoyle, Kevin, 440
not most recently used (NMRU), 121

O

O'Brien, K., 400
O'Connell, F., 302, 322, 430
O'Connor, J. M., 438
Oehler, Rich, 193, 224, 242, 397, 401, 424
Olson, Tim, 440
Olukotun, K., 584, 586
op code rule, 457
operand fetch (OF), 55
operand store (OS), 55
operation latency (OL), 27
Opteron (AMD K8), 134–135, 383, 417
out-of-order execution, 180. *See also* dynamic execution core
output data dependence. *See* WAW data dependence
override latency, 512
overriding predictors, 510–512

P

page faults, 138, 140, 141, 265
    Intel P6, 364
    TLB miss, 151
page miss handler (PMH), 362
page-mode accesses, 131
page table base register (PTBR), 143
page tables, 142–145, 147–153, 151
page walk, 364
Paley, Max, 373
Pan, S. T., 462
Papworth, Dave, 413, 415
parallel pipelines, 179, 181–184. *See also* superscalar machines
partial product generation, 45
partial product reduction, 45–46
partial resolution, 500
partial update policy, PHT, 474, 476
partial write RAT stalls, 352–353
path history branch predictors, 483–485
Patkar, N., 436
Patt, Yale, 8, 196, 232–233, 341, 380, 409, 415, 440, 458, 462, 468–469
pattern history table (PHT), 232, 463
    choice, 478–479
    global (g), 233, 469
    individual (p), 233
    organization alternatives, 468
    partial update policy, 474, 476
    per-address (p), 469, 471
    per-set (s), 469
    shared (s), 233, 234–235

pattern table (PT), 342
Patterson, David, 71, 158, 160, 432
PC mod 2^m hashing function, 460n
PC-relative addressing mode, 91–93
Peleg, A., 405
pending target return queue (PTRQ), 243
Peng, C. R., 429
per-address pattern history table (pPHT), 469, 471
per-branch (P) BHSR, 233–235
perceptron branch predictor, 487–489
performance simulators, 13–16
   trace-driven, 13–14, 306
   VMW-generated, 301, 305–307
per-instruction miss rate, 116–117
peripheral component interface (PCI), 108
permission bits, 142–143
per-set branch history table (SBHT), 468
per-set pattern history table (sPHT), 469
persistence of memory, 110
personal computer (PC), 3
phantom branch, 499
PHT. See pattern history table (PHT)
physical address, 136–137
physical address buffer (PAB), 361
physical destinations (PDst's), 351–352
pipelined processor design, 54–93
   balancing pipeline stages, 53, 55–61
      example instruction pipelines, 59–61
      hardware requirements, 58–59
      stage quantization, 53, 55–58
   commercial pipelined processors, 87–93
      CISC example, 89–91
      RISC example, 87–89
      scalar processor performance, 91–93
   deeply pipelined processors, 94–97
      optimum pipeline depth, 96

   pipeline stall minimization, 71–87
      forwarding paths, 79–81
      hazard identification, 73–77
      hazard resolution, 77–78
      pipeline interlock hardware, 82–87
      program dependences, 71–73
   pipelining fundamentals. See pipelining fundamentals
   pipelining idealism, 54
   trends in, 61
   unifying instruction types, 61–71
      classification, 61–65
      instruction pipeline implementation, 68–71
      optimization objectives, 67–68
      procedure for, 65–68
      resource requirements, 65–68
      specifications, 63–65
pipelined processors, 39–104. See also pipelined processor design; pipelining fundamentals
   Amdahl's law, 21
   commercial, 87–93
   deep pipelines, 94–97
   effective degree of pipelining, 22
   execution profiles, 19–20
   performance of, 19–22
   stall cycles. See pipeline stalls
   superpipelined machines, 29–31
   superscalar. See superscalar machines
   TYP pipeline, 21–22
pipeline hazards
   data dependences, 71–72
   hazard register, 77
   identifying, 73–77
   resolving, 77–78, 82–87
   TYP pipeline example, 75–77
pipeline interlock, 82–87
pipeline stalls, 20–21, 51
   dispatch stalls, 311–314
   issue stalls, PowerPC 620, 316–317
   minimizing, 53, 71–87

   RAT stalls, 352–353
   rigid pipelines and, 179–180
pipelining fundamentals, 40–54
   arithmetic pipelines, 40, 44–48
      nonpipelined floating-point multiplier, 45–46, 47
      pipelined floating-point multiplier, 46–48
   instruction pipelining, 51–54
      instruction pipeline design, 51–53
      ISA impacts, 53–54
      pipelining idealism and, 52–54
   pipelined design, 40–44
      cost/performance tradeoff, 43–44
      limitations, 42–43
      motivations for, 40–42
   pipelining defined, 12–13
   pipelining idealism. See pipelining idealism
pipelining idealism, 40, 48–51
   identical computations, 48, 50, 53, 54
   independent computations, 48, 50–51, 53, 54
   instruction pipeline design and, 52–54
   pipelined processor design and, 54
   uniform subcomputations, 48–49, 53
Pleszkun, A., 208
pointer rule, 457
polling algorithms, 524
pooled register file, 242–243
Popescu, V., 208, 435
Potter, T., 425
Poursepanj, A., 426
power consumption, 3
   branch mispredictions and, 503
   optimum pipeline depth and, 96–97
PowerPC, 6, 62, 145, 302–305
   32-bit architecture, 424–429
   64-bit architecture, 429–431
   relaxed memory consistency, 581
   RISC attributes, 62

INDEX **637**

synchronization, 565
value locality, 525
PowerPC e500 Core, 428–429
PowerPC 601, 302, 382, 425
PowerPC 603, 302, 425–426
PowerPC 603e, 426
PowerPC 604, 6, 230–231, 302, 426–427
   buffering choices, 385
   pipeline stages, 427
PowerPC 604e, 427
PowerPC 620, 199, 301–327, 429
   Alpha AXP *vs.*, 321–322
   architecture, 302–305
   block diagram, 303
   bottlenecks, 320–321
   branch prediction, 307–311
   buffering choices, 385
   cache effects, 318–320
   complete stage, 305, 318–320
   completion buffer, 305, 312, 313
   conclusions/observations, 320–322
   dispatch stage, 304, 311–315
   execute stage, 305, 316–318
   experimental framework, 305–307
   fetch stage, 303, 307–311
   IBM POWER3 *vs.*, 322–323
   IBM POWER4 *vs.*, 323–324
   instruction buffer, 303–304
   instruction pipeline diagram, 304
   latency, 317–318
   parallelism, 315, 317, 318
   reservation stations, 201, 304–305
   SPEC 92 benchmarks, 305–307
   weak-ordering memory access, 319, 321
   writeback stage, 305
PowerPC 750 (G3), 302, 385, 428, 570
PowerPC 970 (G5), 112, 431
PowerPC 7400 (G4), 302, 428
PowerPC 7450 (G4+), 428
PowerPC-AS, 398, 431–432
PowerPC-AS A10 (Cobra), 432
PowerPC-AS A30 (Muskie), 432
PowerPC-AS A35 (Apache, RS64), 432

PowerPC-AS A50 (Star series), 432
precise exceptions, 208, 385
predecoding, 198–199
prediction fusion, 496–497
preexecution, 562, 613–615
prefetching, 90, 109
   IBM POWER3, 323
   prefetching cache, 274, 275–277
   prefetch queue, 275
   in redundant execution, 613–614
Prener, Dan, 380
Preston, R., 392
primary (L1) cache, 111, 112, 274
primitives, synchronization, 563–565
Probert, Dave, 440
processor affinity, 586
processor performance, 17
   Amdahl's law, 17–18, 21, 220
   baseline scalar pipelined machine, 28–29
   cost *vs.*, 43–44, 95–96, 598–599
   equation for, 10–11
   evaluation methods, 13–16
   iron law of, 10–11, 17, 96
   optimizing, 11–13
   parallel processors, 17–19
   pipelined processors, 19–22
   principles of, 10–16
   scalar pipelined RISC machines, 91–93
   sequential bottleneck and, 19
   simulators. *See* performance simulators
   vectorizability and, 18–19
program constants, 524
program counter (PC), 76–77, 192
program parallelism, 22
Project X, 372–373
Project Y, 373
pseudo-operand, 250
pshare algorithm, 471
Pugh, E., 373, 377
Puzak, Thomas R., 96, 454

Q

QED RM7000, 421–422
quadavg instruction, 204–205

queuing latency, 156
queuing time, 108

R

RAID levels, 158–161
Rambus DRAM (RDRAM), 131–132
RAM digital-to-analog converter (RAMDAC), 154
Randell, Brian, 373
Rau, Bob, 378
RAW data dependence, 71–72
   interthread dependences
      memory, 607–609
      register, 605–607
   intrathread dependences
      memory, 607
      register, 605
   between load/store instructions, 266–267
   in memory controller, 135
   register data flow and, 244–245
RAW hazard, 73
   detecting, 83–84
   necessary conditions for, 74–75
   penalty reduction, 79–81
   resolving, 77–78
   in TYP pipeline, 76–77
   worst-case penalties, 77, 80
Razdan, R., 391
read-after-write. *See* RAW data dependence
read permission (Rp) bit, 142
ReadQ command, 134–135
realization, 5–6
Reches, S., 484
reduced instruction set computer. *See* RISC architecture
redundant arrays of inexpensive disks. *See* RAID levels
redundant execution, 610–616
   A/R-SMT, 610–612
   branch resolution, 614–615
   datascalar architecture, 614
   DIVA proposal, 612–613
   fault detection, 611–613
   preexecution, 562, 613–615
   prefetching, 613–614
   slipstreaming, 613–615

reference (Ref) bit, 142
refetching, 546
register alias table (RAT), 333, 336, 346–353
    basic operation, 349–351
    block diagram, 347
    floating-point overrides, 352
    implementation details, 348–349
    integer retirement overrides, 351
    new PDst overrides, 351–352
    stalls, 352–353
register data dependences, 72
    in IMT designs, 605–607
    pipeline hazards of, 75–76
register data flow techniques, 237–262, 519–558
    data flow limits, 244–245
    dynamic execution core. *See* dynamic execution core
    dynamic instruction reuse, 262
    false data dependences, 237–239
    register allocation, 237–238
    register renaming. *See* register renaming
    register reuse problems, 237–239
    Tomasulo's algorithm, 246–254
    true data dependences, 244–245
    value locality. *See* value locality
    value prediction, 261–262, 521–522
register file, 112–113, 119. *See also* register data flow techniques
    attributes of, 111
    definition (writing) of, 238
    pooled, 242–243
    read port saturation, 312
    TYP instruction pipeline interface, 69–70
    use (reading) of, 238
register recycling, 237–239
register renaming, 239–244
    destination allocate, 240, 241
    in dynamic execution core, 255
    instruction scheduling and, 261
    map table approach, 242–243
    pooled register file, 242–243
    register update, 240, 241–242

rename register file (RRF), 239–240
    registers in, 360
    saturation of, 313
    source read, 240–241
register spill code, 524
register transfer language (RTL), 5, 15
register update, 240
Reilly, M., 440
Reininger, Russ, 425
relaxed consistency (RC) models, 579–581
reorder buffer (ROB), 208, 209
    in dynamic execution core, 256, 258–259
    in Intel P6, 353–354, 357–361
    and reservation station, combined, 259
    with RRF attached, 239–240
    in SMT design, 594
reservation stations, 201–203, 209
    dispatch step, 256–257
    in dynamic execution core, 255–258
    entries, 255
    IBM 360/91, 246–248
    instruction wake up, 257
    Intel P6, 336, 355–357
    issuing hazards, 316–317
    issuing step, 258
    PowerPC 620, 304–305
    and reorder buffer, combined, 259
    saturation of, 313
    tag fields used in, 248–250
    waiting step, 257
resource recovery pointer (RRP), 436
response time, 106, 108. *See also* latency
RespQ command, 134
restricted data flow, 380
retirement stage. *See* instruction retirement stage
return address stack (RAS), 500–501
return rule, 457
return stack buffer (RSB), 501
reuse test, 554
Richardson, S. E., 262
Riordan, Tom, 417, 421

RISC architecture, 9
    IBM study on, 91–93
    instruction decoding in, 195–196
    MIPS R2000/3000 example, 87–89
    modern architecture, 62–65
    predecoding, 198–199
    RISC86 operation group, 411–412
    RISC operations (ROPs), 196
    superscalar retrofits, 384–385
Riseman, E., 25, 377
Robelen, Russ, 373
Rodman, Paul, 418
Rosenblatt, F., 487
rotational latency, 156
Rotenberg, Eric, 236, 506, 610
Roth, A., 613
row address strobe (RAS), 130
row buffer cache, 131
Rowen, Chris, 419
row hits, 135
Rubin, S., 537, 541
Ruemmler, C., 157
Russell, K., 500
Russell, R. M., 29
Ruttenberg, John, 418, 440
Ryan, B., 409, 410, 429

S

safe instruction recognition, 407
Sandon, Peter, 431
saturating k-bit counter, 461–462
Sazeides, Y., 261, 527, 539
scalar computation, 18
scalar pipelined processors, 16
    limitations, 178–180
    performance, 91–93, 179–180
    pipeline rigidity, 179–180
    scalar instruction pipeline, defined, 73
    single-entry buffer, 186–187
    unifying instruction types, 179
    upper bound throughput, 178–179
scheduling latency, 548–549
scheduling matrices, 374
Schorr, Herb, 373, 374, 376

SECDED codes, 159
secondary (L2) cache, 111, 112, 274
seek latency, 156
selective branch inversion (SBI), 480–482, 502
selective eager execution, 503
selective reissue, 546–551
select logic, 258
Sell, John, 424
Seng, J., 541
sense amp, 130
sequential bottleneck, 19, 22–23, 220
sequential consistency model, 577–578
Sequent NUMA-Q system, 574
serialization constraints, 311, 316
serializing instructions, 597–598
serial program parallelization, 561–562
service time, 108
set-associative memory, 119, 120, 146–148
set-associative TLB, 150
set busy bit, 241
Seznec, Andre, 392, 477
shared-dirty state, 570
shared (s) PHT, 233, 234–235
sharing list (vector), 573
Shebanow, Mike, 196, 380, 409, 435, 437
Shen, John Paul, 15, 261, 384, 523, 527, 535
Shima, Masatoshi, 405
Shippy, D., 402
short decoders, 411
Shriver, B., 412
Silha, E., 425
Simone, M., 436
simulators. *See* performance simulators
simultaneous multithreading (SMT), 584, 585, 592–599
    active-/redundant-stream (A/R-SMT), 610–612
    branch confidence, 504
    cost of, 598–599
    instruction serialization support, 597–598

    interstage buffer implementation, 593–594
    multiple threads, managing, 598
    Pentium 4 implementation, 599
    performance of, 598–599
    pipeline stage sharing, 594–597
    resource sharing, 593–599
Sindagi, V., 236, 602, 603, 604, 606
single-assignment code, 238
single-direction branch prediction, 455–456
single-instruction serialization, 311
single-thread performance, 589
single-writer protocols, 569–571
sink, 249
Sites, Richard, 387, 389
six-stage instruction pipeline. *See* TYPICAL (TYP) instruction pipeline
six-stage template (TEM) superscalar pipeline, 190–191
Skadron, Kevin, 482
Slavenburg, G., 204
slipstreaming, 613–615
slotting stage, 389
small computer system interface (SCSI), 108
Smith, Alan Jay, 120, 226, 342, 497
Smith, Burton, 412, 588
Smith, Frank, 403
Smith, Jim E., 208, 225, 228, 261, 378–379, 387, 401, 402, 425, 440, 460, 527, 539
Smith, M., 234, 380
Smith's algorithm, 459–462
SMT. *See* simultaneous multithreading (SMT)
snooping, 168, 572–573
Snyder, Mike, 428
Sodani, A., 262, 527, 529, 534, 541, 545
soft interrupts, 375
software cache coherence, 168
software instrumentation, 13–14
software portability, 6–7
software RAID, 160
software TLB miss handler, 145
Sohi, G. S., 208, 262, 273, 277, 527, 529, 541, 545, 562, 604

Soltis, Frank, 431, 432
Song, Peter, 427, 430, 437, 440
Sony Playstation 2, 131
source, 249
source read, 240–241
SPARC Version 8, 432–435
SPARC Version 9, 435–439
spatial locality, 113–114
spatial parallelism, 181–182, 205–206
SPEC benchmarks, 26, 227, 305–307
special-purpose register (mtspr) instruction, 311
specification, 1–2, 4–5
speculative exploitation, value locality, 535–554
    computational predictors, 540
    confidence estimation, 538
    data flow region verification, 545–546
    history-based predictors, 539–540
    hybrid predictors, 540
    implementation issues, 541
    prediction accuracy, 538
    prediction coverage, 538–541
    prediction scope, 541–542
    speculative execution using predicted values, 542–551
    data flow eager execution, 548
    data speculation support, 550–551
    memory data dependences, 549
    misprediction penalty, selective reissue, 547–548
    prediction verification, 543–545
    propagating verification results, 544–545
    refetch-based recovery, 546
    scheduling latency effect, 548–549
    scheduling logic, changes to, 549–550
    selective reissue recovery, 546–551
    speculative verification, 544

speculative execution using predicted values—*Cont.*
  straightforward value speculation, 542
  value prediction. *See* value prediction
  weak dependence model, 535–536
speculative versioning cache, 609–610
speed demons, 321–322, 384, 386
spill code, 262–263
spinning on a lock, 564
split-transaction bus, 581–582
Sporer, M., 382
Sprangle, Eric, 96, 477
SRAM, 112, 130
SSE instructions, 329
stage quantization, 53, 55–61
stall cycles. *See* pipeline stalls
Standard Performance Evaluation Corp. benchmarks. *See* SPEC benchmarks
Stark, Jared, 485
static binding with load balancing, 355
static branch prediction, 346, 454–458
  backwards taken/forwards not-taken, 456
  Ball/Larus heuristics, 456–457
  profile-based, 455, 457–458
  program-based, 456–457
  rule-based, 455–457
  single-direction prediction, 455–456
static predictor selection, 493
static random-access memory (SRAM), 112, 130
Steck, Randy, 329
Steffan, J. G., 562, 602
Stellar GS-1000, 382
Stiles, Dave, 410
Stone, Harold, 19
store address buffer (SAB), 361
store buffer (SB), 265–266, 268–272, 361
store coloring, 362
store data buffer (SDB), 246–248, 361

store instructions. *See also* memory data flow techniques
  processing, 265–266
  senior, 354
  specifications, 63–65
  unifying procedure, 66–67
  weak ordering of, 319, 321
Storer, J., 379
store rule, 457
Storino, S., 302, 575, 589
streaming SIMD extension (SSE2), 416
stride predictor, 540
strong dependence model, 535–536
subcomputations
  for ALU instructions, 63, 65
  for branch instructions, 63–65
  generic, 55
  for load/store instructions, 63–64
  merging, 55–58
  subdividing, 56–57
  uniform, 48–49, 53, 54
Sugumar, Rabin A., 472
Sundaramoorthy, K., 613
Sun Enterprise 10000, 572
Sun UltraSPARC. *See* UltraSPARC
superpipelined machines, 29–31
superscalar machines, 16, 31
  brainiacs, 321–322, 384, 386–387
  development of, 369–384
    Astronautics ZS-1, 378–379
    decoupled architectures access-execute, 378–379
    microarchitectures, 380–382
    IBM ACS-1, 372–377
    IBM Cheetah/Panther/America, 380
    IBM Stretch, 369–372
    ILP studies, 377–378
    instruction fission, 380–381
    instruction fusion, 381–382
    multiple-decoding and, 378–379
    1980s multiple-issue efforts, 382
    superscalar design, 372–377
    timeline, 383

uniprocessor parallelism, 369–372
wide acceptance, 382–384
goal of, 24
instruction flow. *See* instruction flow techniques
memory data flow. *See* memory data flow techniques
pipeline organization. *See* superscalar pipeline organization
recent design classifications, 384–387
register data flow. *See* register data flow techniques
RISC/CISC retrofits, 384–385
  dependent integer issue, 385
  extensive out-of-order issue, 385
  floating-point coprocessor style, 384
  integer with branch, 384
  multiple function, precise exceptions, 385
  multiple integer issue, 384
  speed demons, 321–322, 384, 386
verification of, 439–440
VLIW processors *vs.*, 31–32
superscalar pipeline organization, 177–215
  diversified pipelines, 184–186
  dynamic pipelines, 186–190
  fetch group misalignment, 191–195
  instruction completion/retirement, 206–209
    exceptions, 207–208
    interrupts, 207
  instruction decoding, 195–199
  instruction dispatching, 199–203
  instruction execution, 203–206
    hardware complexity, 206
    memory configurations, 205
    optimal mix of functional units, 205
    parallelism and, 205–206
  instruction fetching, 190–195
  overview, 190–209

parallelism, 181–184
predecoding, 198–199
reservation stations, 201–203
scalar pipeline limitations, 178–180
six-stage template, 190–191
SuperSPARC, 203, 381–382, 385, 433
Sussenguth, Ed, 373, 374, 375, 376, 440
synchronization, 561, 562–565
synchronous DRAM (SDRAM), 129, 131
synthesis, 4

T

tag, 119
tag fields, 248–250
Talmudi, Ran, 440
Tandom Cyclone, 382
Tarlescu, Maria-Dana, 485
Taylor, S., 440
temporal locality, 113–114
temporal parallelism, 181–182, 205–206
Tendler, Joel M., 122, 136, 302, 323–324, 431, 471, 584, 587
Tera MTA, 588–589
think time, 169
third level (L3) cache, 274
Thomas, Renju, 489
Thompson, T., 429
Thornton, J. E., 29, 185, 588
thread-level parallelism (TLP), 3, 560. *See also* multiple threads, executing
thread-level speculation (TLS), 562
attributes of, 603
control flow techniques, 602, 603
memory RAW resolution, 608
physical organization, 604
register data dependences, 606
thread switch state machine, 591–592
3 C's model, 123–125
throughput. *See* bandwidth
time-sharing, 560–561
Tirumalai, P., 438
TI SuperSPARC. *See* SuperSPARC

Tjaden, Gary, 25, 377, 519
TLB miss, 265
Tobin, P., 394
Tomasulo, Robert, 201, 373, 520
Tomasulo's algorithm, 246–254, 535
common data bus, 246–248
IBM 360 FPU original design, 246–247
instruction sequence example, 250–254
reservation stations, 246–248
Torng, H. C., 382
Torrellas, J., 562
total sequential execution, 73
total update, 476
tournament branch predictor, 390, 491–493
Towle, Ross, 418, 440
trace cache, 236–237, 415, 506–508
trace-driven simulation, 13–14, 306
trace prediction, 508
trace scheduling, 26
training threshold, 487
transfer latency, 156
transistor count, 2
translation lookaside buffer (TLB), 145, 149–153, 265
data cache interaction, 151–153
data (DTLB), 362, 364
fully-associative, 150–151
instruction (ITLB), Intel P6, 339–341
set-associative, 150
translation memory, 142–145, 147–153
Transputer T9000, 381–382
trap barrier instruction (TRAPB), 387
Tremblay, Marc, 437, 438, 440
TriMedia-1 processor, 204
TriMedia VLIW processor, 204
true dependence. *See* RAW data dependence
Tsien, B., 440
Tullsen, D. M., 541, 585, 592
Tumlin, T. J., 440
Turumella, B., 440
two-level adaptive (correlated) branch prediction, 232–236, 462–469

TYPICAL (TYP) instruction pipeline, 67–71
logical representation, 66
memory subsystem interface, 69
MIPS R2000/R3000 *vs.*, 89
physical organization, 68–69
register file interface, 69–70
from unified instruction types, 65–68

U

Uhilg, Richard, 473
Uht, A. K., 236, 455, 503, 602, 603, 604, 606
UltraSPARC, 199
UltraSPARC-I, 382, 437–438
UltraSPARC-III, 122, 382, 438–439
UltraSPARC-IV, 439
Undy, S., 394
uniform memory access (UMA), 566–567
Univac A19, 382
universal serial bus (USB), 154
μops. *See* micro-operations (μops)
update map table, 241
update protocols, 568–569
up-down counter, 461–462

V

Vajapeyam, S., 208
value locality, 261, 521, 523–527
average, 526
causes of, 523–525
nonspeculative exploitation. *See* nonspeculative exploitation, value locality
quantifying, 525–527
speculative exploitation. *See* speculative exploitation, value locality
value prediction, 261–262, 521–522, 536–537
idealized machine model, 552–553
performance of, 551–553
value prediction table (VPT), 538–539

value prediction unit (VPU), 537–542
Van Dyke, Korbin, 410
variable path length predictors, 485
Vasseghi, N., 421
vector computation, 18–19
vector decoder, 411
VectorPath decoder, 412, 417
Vegesna, Raju, 434
very large-scale integration (VLSI) processor, 455
very long instruction word (VLIW) processor, 26, 31–32
virtual address, 136–137
virtual function calls, 524
virtually indexed data cache, 152–153
virtual memory, 127, 136–145
  accessing backing store, 140–141
  address translation, 136–137, 147–153, 263–264
  in Intel P6, 340–341
  cache coherence and, 576
  demand paging, 137, 138–141
  evicting pages, 140
  lazy allocation, 139
  memory protection, 141–142
  page allocation, 140
  page faults, 138, 140, 141, 265
  page table architectures, 142–145, 147–153
  permission bits, 142–143
  translation memory, 142–145, 147–153
  virtual address aliasing, 142
visual instruction set (VIS), 437
VMW-generated performance simulators, 301, 305–307

W

wakeup-and-select process, 595
wake-up logic, 258
Waldecker, Don, 429
Wall, D. W., 27, 380
Wang, K., 262, 527, 540
Wang, W., 153, 576
WAN (wide area network), 107
WAR data dependence, 71–72
  enforcing, 238–239
  in IMT designs, 605, 607
  between load/store instructions, 266–267
  in memory controller, 135
  pipeline hazard caused by, 73–74
  resolved by Tomasulo's algorithm, 252–254
  in TYP pipeline, 76
Waser, Shlomo, 45–48
Watson, Tom, 372
WAW data dependence, 72
  enforcing, 238–239
  in IMT designs, 605, 607
  between load/store instructions, 266–267
  in memory controller, 135
  pipeline hazard caused by, 73–74
  resolved by Tomasulo's algorithm, 253–254
  in TYP pipeline, 75–76
Wayner, P., 438
way prediction, 510
weak dependence model, 535–536
Weaver, Dave, 435
Weber, Fred, 412, 417
Weiser, Uri, 405, 440
Weiss, S., 387, 401, 402, 425, 484
White, S., 302, 322, 402, 430
Wilcke, Winfried, 435, 436
Wilkerson, C. B., 261
Wilkes, J., 157
Wilkes, M., 115
Williams, T., 436
Wilson, J., 380
WinChip microarchitecture, 410
Witek, Rich, 387, 389
Wolfe, A., 384
word line, 129
Worley, Bill, 392, 440
Wottreng, Andy, 431
write-after-read. *See* WAR data dependence
write-after-write. *See* WAW data dependence
writeback cache, 122
write back (WB) stage, 28, 305
write permission (Wp) bit, 142
WriteQ command, 134–135
write-through cache, 121–122
wrong-history mispredictions, 482
Wulf, W. A., 129

Y

YAGS predictor, 478–480
Yates, John, 440
Yeager, K., 324, 419, 421, 579
Yeh, T. Y., 232–233, 341, 458, 462, 468–469
Yeh's algorithm, 341–343
Yew, P., 541
Young, C., 234
Yung, Robert, 435, 437

Z

Zilles, C., 613
Zorn, B., 527, 539, 553